Lecture Notes in Computer Science   1970

Edited by G. Goos, J. Hartmanis and J. van Leeuwen

Springer
Berlin
Heidelberg
New York
Barcelona
Hong Kong
London
Milan
Paris
Singapore
Tokyo

Mateo Valero   Viktor K. Prasanna
Sriram Vajapeyam (Eds.)

High Performance Computing – HiPC 2000

7th International Conference
Bangalore, India, December 17-20, 2000
Proceedings

Springer

Series Editors

Gerhard Goos, Karlsruhe University, Germany
Juris Hartmanis, Cornell University, NY, USA
Jan van Leeuwen, Utrecht University, The Netherlands

Volume Editors

Mateo Valero
Technical University of Catalonia, DAC-UPC, Computer Architecture Department
Campus Nord, D6, Jordi Girona, 1-3 Barcelona, Spain
E-mail: mateo@ac.upc.es

Viktor K. Prasanna
University of Southern California, Department of EE-Systems
Computer Engineering Division, 3740 McClintok Ave, EEB 200C
Los Angeles, CA 90089-2562, USA
E-mail: prasana@ganges.usc.edu

Sriram Vajapeyam
Indian Institute of Science
Sir C.V. Raman Avenue, Bangalore 560012, India
E-mail: sriram@csa.iisc.ernet.in

Cataloging-in-Publication Data applied for

Die Deutsche Bibliothek - CIP-Einheitsaufnahme

High performance computing : 7th international conference, Bangalore,
India, December 17 - 20, 2000 ; proceedings / HiPC 2000. Mateo Valero
... (ed.). - Berlin ; Heidelberg ; New York ; Barcelona ; Hong Kong ;
London ; Milan ; Paris ; Singapore ; Tokyo : Springer, 2000
  (Lecture notes in computer science ; Vol. 1970)
  ISBN 3-540-41429-0

CR Subject Classification (1998): C.1-4, D.1-4, F.1-2, G.1-2

ISSN 0302-9743
ISBN 3-540-41429-0 Springer-Verlag Berlin Heidelberg New York

Springer-Verlag Berlin Heidelberg New York
a member of BertelsmannSpringer Science+Business Media GmbH
© Springer-Verlag Berlin Heidelberg 2000
Printed in Germany

Typesetting: Camera-ready by author, data conversion by Steingräber Satztechnik GmbH, Heidelberg
Printed on acid-free paper   SPIN: 10781111   06/3142   5 4 3 2 1 0

Message from the Program Chair

I wish to welcome all the delegates to the 7th International Conference on High-Performance Computing to be held in Bangalore, India, 17–20 December 2000. This edition of the conference consists of ten sessions with contributed papers, organized as two parallel tracks. We also have six keynotes and an invited papers session by leading researchers, industry keynotes, two banquet speeches, a poster session, and eight tutorials.

The technical program was put together by a distinguished program committee consisting of five program vice-chairs, nine special session organizers, and 36 program committee members. We received 127 submissions for the contributed sessions. Each paper was reviewed by three members of the committee and supervised by either a program vice-chair or a special session organizer. After rigorous evaluation, 46 papers were accepted for presentation at the conference, of which 32 were regular papers (acceptance rate of 25%) and 14 were short papers (acceptance rate of 11%). The papers that will be presented at the conference are authored by researchers from 11 countries, which is an indication of the true international flavor of the conference.

Half of the technical sessions are special sessions. All papers submitted to these special sessions were subject to the same review process as described above. The sessions are: Applied Parallel Processing organized by Partha Dasgupta and Sethuraman Panchanathan (Arizona State University, USA), Cluster Computing and Its Applications organized by Hee Yong Youn (Information and Communication University, South Korea), High-Performance Middleware organized by Shikharesh Majumdar (Carleton University, Canada) and Gabriel Kotsis (University of Vienna, Austria), Large-Scale Data Mining organized by Gautam Das (Microsoft Research, USA) and Mohammad Zaki (Rennselaer Polytechnic Institute, USA) and Wireless and Mobile Communication Systems organized by Azzedine Boukerche (University of North Texas, Denton, USA). The program includes an invited papers session by leading computer architects, organized by Sriram Vajapeyam (Indian Institute of Science) and myself. In a plenary session titled "Future General-Purpose and Embedded Processors," the speakers will share their visions for future processors. The speakers are: Trevor Mudge (University of Michigan, Ann Arbor, USA), Bob Rau (Hewlett-Packard HP Labs, USA), Jim Smith (University of Wisconsin, Madison, USA) and Guri Sohi (University of Wisconsin, Madison, USA).

I wish to thank the program vice-chairs and special session organizers for their time and effort in the process of selecting the papers and preparing an excellent technical program. They are Nader Bagherzadeh (University of California at

Irvine), Jack Dongarra (University of Tennessee at Knoxville and Oak Ridge National Lab), David Padua (University of Illinois at Urbana-Champaign), Assaf Schuster (Israel Institute of Technology, Technion), and Satish Tripathi (University of California at Riverside). Viktor Prasanna, Sriram Vajapeyam and Sajal Das provided excellent feedback about the technical program in their roles as general co-chairs and vice general chair, respectively. I would also like to thank Sartaj Sahni (University of Florida), Manavendra Misra (KBkids.com), and Vipin Kumar (University of Minnesota) for performing their roles as poster session, tutorials session, and keynote address chairs, respectively. I would like to express my gratitude to Nalini Venkatasubramanian (University of California at Irvine) for compiling the proceedings of the conference.

I also wish to acknowledge the tremendous support provided by Eduard Ayguade (Technical University of Catalonia). He managed all the work relating to receiving papers through mail and web, arranging the electronic reviews, collecting the reviews, organizing the reviews in summary tables for the program committee, and informing all authors of the decisions.

Finally, I would like to thank Viktor Prasanna and Sriram Vajapeyam for inviting me to be part of HiPC 2000 as program chair.

Mateo Valero

Message from the General Co-Chairs

It is our pleasure to welcome you to the Seventh International Conference on High Performance Computing. We hope you enjoy the meeting as well as the rich cultural heritage of Karnataka State and India.

The meeting has grown from a small workshop held six years ago that addressed parallel processing. Over the years, the quality of submissions has improved and the topics of interest have been expanded in the general area of high performance computing. The growth in the participation and the continued improvement in quality are primarily due to the excellent response from researchers from around the world, enthusiastic volunteer effort, and support from IT industries worldwide.

Mateo Valero agreed to be the Program Committee Chair despite his tight schedule. We are thankful to him for taking this responsibility and organizing an excellent technical program. His leadership helped attract excellent Program Committee members. It encouraged high quality submissions to the conference and invited papers by leading researchers for the session on future processors.

As Vice General Chair, Sajal Das interfaced with the volunteers and offered his thoughtful inputs in resolving meeting-related issues. In addition, he was also in charge of the special sessions. It was a pleasure to work with him.

Eduard Ayguade has redefined the term "volunteer" through his extraordinary efforts for the conference. Eduard was instrumental in setting up the Web-based paper submission and review process and the conference website, and also helped with publicity. We are indeed very grateful to Eduard.

Vipin Kumar invited the keynote speakers and coordinated the keynotes. Sartaj Sahni handled the poster/presentation session. Nalini Venkatasubramanian interfaced with the authors and Springer-Verlag in bringing out these proceedings. Manav Misra put together the tutorials. Venugopal handled publicity within India and local arrangements. As in the past, Ajay Gupta did a fine job in handling international financial matters. C. P. Ravikumar administered scholarships for students from Indian academia. M. Amamiya and J. Torelles handled publicity in Asia and Europe, respectively.

R. Govindarajan coordinated the industrial track and exhibits and also interfaced with sponsors. Dinakar Sitaram, Novell India, provided invaluable inputs regarding conference planning.

We would like to thank all of them for their time and efforts. Our special thanks go to A. K. P. Nambiar for his continued efforts in handling financial matters as well as coordinating the activities within India.

Major financial support for the meeting was provided by several leading IT companies. We would like to thank the following individuals for their support:

N. R. Narayana Murthy, Chairman, Infosys; Avinash Agrawal, SUN Microsystems (India); Konrad Lai, Intel Microprocessor Research Labs; Karthik Ramarao, HP India; Amitabh Shrivastava, Microsoft Research; and Uday Shukla, IBM (India).

Continued sponsorship of the meeting by the IEEE Computer Society and ACM are much appreciated. Finally, we would like to thank Henryk Chrostek and Bhaskar Srinivasan for their assistance over the past year.

October 2000   Sriram Vajapeyam
                                                                Viktor K. Prasanna

Conference Organization

GENERAL CO-CHAIRS
Viktor K. Prasanna, University of Southern California
Sriram Vajapeyam, Indian Institute of Science

VICE GENERAL CHAIR
Sajal K. Das, The University of Texas at Arlington

PROGRAM CHAIR
Mateo Valero, Technical University of Catalonia

PROGRAM VICE CHAIRS
Nader Bagherzadeh, University of California at Irvine
Jack Dongrarra, University of Tennessee, Knoxville and Oak Ridge National Lab
David Padua, University of Illinois at Urbana-Champaign
Assaf Schuster, Israel Insitute of Technology, Technion
Satish Tripathi, University of California at Riverside

KEYNOTE CHAIR
Vipin Kumar, University of Minnesota

POSTER/PRESENTATION CHAIR
Sartaj Sahni, University of Florida

TUTORIALS CHAIR
Manavendra Misra, KBkids.com

EXHIBITS CHAIR
R. Govindarajan, Indian Institute of Science

SCHOLARSHIPS CHAIR
C.P. Ravikumar, Indian Institute of Technology, Delhi

AWARDS CHAIR
Arvind, MIT

FINANCE CO-CHAIRS
A.K.P. Nambiar, Software Technology Park, Bangalore
Ajay Gupta, Western Michigan University

PUBLICITY CHAIRS
Europe: Eduard Ayguade, Technical University of Catalonia
USA: Josep Torrelles, University of Illinois
Asia: Makoto Amamiya, Kyushu University

LOCAL ARRANGEMENTS CHAIR
K.R. Venugopal, UVCE

PUBLICATIONS CHAIR
Nalini Venkatasubramanian, University of California at Irvine

STEERING CHAIR
Viktor K. Prasanna, University of Southern California

Steering Committee

Jose Duato, Universidad Politecnica de Valencia
Viktor K. Prasanna, USC Chair
N. Radhakrishnan, US Army Resarch Lab
Sartaj Sahni, University of Florida
Assaf Schuster, Israel Institute of Technology, Technion

National Advisory Committee

Alok Aggarwal, IBM Solutions Research Centre
R.K. Bagga, DRDL, Hyderabad
N. Balakrishnan, Supercomputer Education and Research Centre, Indian
Institute of Science
Ashok Desai, Silican Graphics Systems (India) Private Ltd.
Kiran Deshpande, Mahindra British Telecom Ltd.
H.K. Kaura, Bhabha Atomic Research Centre
Hans H. Krafka, Siemens Communication Software Ltd.
Ashish Mahadwar, PlanetAsia Ltd.
Susanta Misra, Motorola India Electronics Ltd.
Som Mittal, Digital Equipment (India) Ltd.
B.V. Naidu, Software Technology Park, Bangalore
N.R. Narayana Murthy, Infosys Technologies Ltd.
S.V. Raghavan, Indian Institute of Technology, Madras
V. Rajaraman, Jawaharlal Nehru Centre for Advanced Scientific Research
S. Ramadorai, Tata Consultancy Services, Mumbai
K. Ramani, Future Software Private Ltd.
S. Ramani, National Centre for Software Technology
Karthik Ramarao, Hewlett-Packard (India) Private Ltd.
Kalyan Rao, Satyam Computers Ltd.
S.B. Rao, Indian Statistical Institute
Uday Shukla, IBM (India) Ltd.
U.N. Sinha, National Aerospace Laboratories

Program Committee

Eduard Ayguade, UPC Technical University of Catalonia
R. Badrinath, Indian Istitute of Technology, Kharagpur
David Bailey, NERSC, Lawrence Berkeley National Lab
John K. Benett, Rice University
Luc Bouge, Ecole Normale Superieure de Lyon
Jose Duato, University of Valencia
Iain Duff, Rutherford Appleton Laboratory
Charbel Farhat, University of Colorado
Eliseu Chaves Filho, Federal University of Rio de Janeiro
Sharad Gavali, NASA Ames Research Center
Dipak Ghosal, University of California at Davis
James R. Goodman, University of Wisconsin
Mary Hall, University of Southern California
Omar Hammami, University of Aizu
Ulrich Herzog, University Erlangen
Jay Hoeflinger, University of Illinois at Urbana-Champaign
Laxmikant Kale, University of Illinois at Urbana-Champaign
Vijay Karamcheti, New York University
Stefanos Kaxiras, Bell Labs/Lucent Technologies
David Keyes, ICASE, NASA Langley Research Center
Klara Nahrstedt, University of Illinois at Urbana-Champaign
Fatourou Panagiota, Max Planck Institute at Saarbrucken
Keshav Pingali, Cornell University
Krithi Ramamritham, Indian Institute of Technology, Bombay
Bhaskar Ramamurthi, Indian Institute of Technology, Madras
Abhiram Ranade, Indian Institute of Technology, Bombay
Lawrence Rauchwerger, Texas A&M University
Arny Rosenberg, University of Massachusetts at Amherst
Catherine Rosenberg, Purdue University
Dheeraj Sanghi, Indian Institute of Technology, Kanpur
Giuseppe Serazzi, Politechnic di Milano
Gabby Silberman, IBM Toronto
Horst Simon, NERSC, Lawrence Berkely National Lab
Kumar Sivarajan, Indian Institute of Science, Bangalore
Theo Ungerer, University of Karlsruhe
Uzi Vishkin, University of Maryland
Hans P. Zima, University of Vienna

HiPC 2000 Reviewers

Istabrak Abdul-Fatah
Georg Acher
George Almasi
Ghita Amor
Emmanuelle Anceaume
Rob Andrews
Amrinder Arora
Eduard Ayguade
R. Badrinath
Nader Bagherzadeh
David H. Bailey
Valmir C. Barbosa
Anindya Basu
Riccardo Bettati
Milind Bhandarkar
Bobby Bhattacharjee
Luc Bouge
Azzedine Boukerche
Peter Brezany
Robert Brunner
Guohong Cao
Kai Chen
Yogesh L. Chobe
Toni Cortes
Vitor Santos Costa
Paolo Cremonesi
T. A. Dahlberg
S. Dandamudi
Alain Darte
Gautam Das
Samir Das
Partha Dasgupta
Ewa Deelman
Giorgio Delzanno
Jack Dongarra
Jose Duato
Iain Duff
Winfried Dulz
Ines de Castro Dutra
Guy Edjlali
Rudolf Eigenmann
Thomas Fahringer
Panagiota Fatourou
Eliseu M. C. Filho
Edil S. T. Fernandes
Venkatesh Ganti
Serazzi Giuseppe
Jonathan Goldstein

Antonio Gonzalez
Rama K. Govindaraju
Guerin
Dimitrios Gunopulos
Arobinda Gupta
Mary Hall
O. Hammami
Ulrich Herzog
Jay Hoeflinger
Haiming Huang
Yvon Jegou
Won J. Jeon
Alin Jula
Vijay Karamcheti
Abhay Karandikar
Stefanos Kaxiras
D. Keyes
Stephan Kindermann
S. C. Kothari
Gabriele Kotsis
Thomas Kunz
Kwan
John L. Larson
Ben Lee
Dongman Lee
Baochun Li
Nikolai Likhanov
Pedro Lopez
S. Majumdar
Manil Makhija
Xavier Martorell
Ravi R. Mazumdar
Dominique Mery
Sangman Moh
David Mount
Klara Nahrstedt
B. Nandy
Girija Narlikar
Mirela
S. M. A. Notare
Yunheung Paek
S. Panchanathan
Eric Parsons
Himadri Sekhar Paul
Fabrizio Petrini
Dorina Petriu
Keshav Pingali
Paulo F. Pires

Steve Plimpton
Viktor K. Prasanna
Antonio Puliafito
Guenther Rackl
Ramesh Radhakrishnan
Krithi Ramamritham
A. Ranade
S. S. Rappaport
Nouhad J. Rizk
Wonwoo Ro
Jerry Rolia
Arnold Rosenberg
Azriel Rosenfeld
Emilia Rosti
Silvius Rus
Huzur Saran
Assaf Schuster
Giuseppe Serazzi
Samarth H. Shah
Elizabeth Shriver
Federico Silla
Horst Simon
Kumar N. Sivarajan
Paul Stodghill
Hong Sungbum
Denis Talay
S. Tripathi
F. Tronel
Theo Ungerer
U. Vishkin
Yul Williams
Terry Wilmarth
Bogdan Wiszniewski
Kastner Wolfgang
Yang Xiao
Yang Xiaod
Ramesh Yerraballi
Kee-Young Yoo
Hee Yong Youn
Cliff Young
Chansu Yu
Hao Yu
Vladimir Zadorozhny
Mohammed Zaki
A. Zeidler
Jiajing Zhu

Table of Contents

FUTURE PROCESSORS: INVITED SESSION
Co-Chairs: S. Vajapeyam (Indian Institute of Science),
*   M. Valero (Thechnical University of Catalonia)*

SESSION III-A: Cluster Computing and Its Applications
Chair: H.Y. Yuon (Information and Communications University, Korea)

SESSION III-B: Architecture
Chair: E. Ayguade (Technical University of Catalonia)

SESSION IV-A: Applied Parallel Processing
Chair: P. Dasgupta (Arizona State University)

SESSION IV-B: Networks
Chair: C.S. Raghavendra (University of Southern California)

SESSION V-A:
Wireless and Mobile Communciation Systems
Chair: Azzedine Boukerche (University of North Texas, Denton)

SESSION V-B: Large-Scale Data Mining
Chair: G. Das (Microsoft Research)

Session I-A

Systems Software
Chair: Wei Hsu
University of Minnesota

Charon Message-Passing Toolkit
for Scientific Computations

Rob F. Van der Wijngaart

Computer Sciences Corporation
NASA Ames Research Center, Moffett Field, CA 94035, USA
wijngaar@nas.nasa.gov

Abstract. Charon is a library, callable from C and Fortran, that aids the conversion of structured-grid legacy codes—such as those used in the numerical computation of fluid flows—into parallel, high-performance codes. Key are functions that define distributed arrays, that map between distributed and non-distributed arrays, and that allow easy specification of common communications on structured grids. The library is based on the widely accepted MPI message passing standard. We present an overview of the functionality of Charon, and some representative results.

1 Introduction

A sign of the maturing of the field of parallel computing is the emergence of facilities that shield the programmer from low-level constructs such as message passing (MPI, PVM) and shared memory parallelization directives (P-Threads, OpenMP, etc.), and from parallel programming languages (High Performance Fortran, Split C, Linda, etc.). Such facilities include: 1) (semi-)automatic tools for parallelization of legacy codes (e.g. CAPTools [8], ADAPT [5], CAPO [9], SUIF compiler [6], SMS preprocessor [7], etc.), and 2) application libraries for the construction of parallel programs from scratch (KeLP [4], OVERTURE [3], PETSc [2], Global Arrays [10], etc.).

The Charon library described here offers an alternative to the above two approaches, namely a mechanism for *incremental* conversion of legacy codes into high-performance, scalable message-passing programs. It does so without the need to resort up front to explicit parallel programming constructs. Charon is aimed at applications that involve structured discretization grids used for the solution of scientific computing problems. Specifically, it is designed to help parallelize algorithms that are not naturally data parallel—i.e., that contain complex data dependencies—which include almost all advanced flow solver methods in use at NASA Ames Research Center. While Charon provides strictly a set of user-callable functions (C and Fortran), it can nonetheless be used to convert serial legacy codes into highly-tuned parallel applications. The crux of the library is that it enables the programmer to codify information about existing multi-dimensional arrays in legacy codes and map between these non-distributed arrays and newly defined, truly distributed arrays *at runtime*. This allows the

M. Valero, V.K. Prasanna, and S. Vajapeyam (Eds.): HiPC 2000, LNCS 1970, pp. 3–14, 2000.

programmer to keep most of the serial code unchanged and only use distributed arrays in that part of the code of prime interest. This code section, which is parallelized using more functions from the Charon library, is gradually expanded, until the entire code is converted. The major benefit of incremental parallelization is that it is easy to ascertain consistency with the serial code. In addition, the user keeps careful control over data transfer between processes, which is important on high-latency distributed-memory machines[1].

The usual steps that a programmer takes when parallelizing a code using Charon are as follows. First, define a distribution of the arrays in the program, based on a division of the grid(s) among all processors. Second, select a section of the code to be parallelized, and construct a so-called parallel bypass: map from the non-distributed (legacy code) array to the distributed array upon entry of the section, and back to the non-distributed array upon leaving it. Third, do the actual parallelization work for the section, using more Charon functions.

The remainder of this paper is structured as follows. In Section 2 we explain the library functions used to define and manipulate distributed arrays (*distributions*), including those that allow the mapping between non-distributed and distributed arrays. In Section 3 we describe the functions that can be used actually to parallelize an existing piece of code. Some examples of the use and performance of Charon are presented in Section 4.

2   Distributed Arrays

Parallelizing scientific computations using Charon is based on domain decomposition. One or more multi-dimensional grids are defined, and arrays—representing computational work—are associated with these grids. The grids are divided into nonoverlapping pieces, which are assigned to the processors in the computation. The associated arrays are thus distributed as well. This process takes place in several steps, illustrated in Fig. 1, and described below.

First (Fig. 1a), the logically rectangular discretization *grid* of a certain dimensionality and extent is defined, using `CHN_Create_grid`. This step establishes a geometric framework for all arrays associated with the grid. It also attaches to the grid an MPI [11] communicator, which serves as the context and processor subspace within which all subsequent Charon-orchestrated communications take place. Multiple coincident or non-coincident communicators may be used within one program, allowing the programmer to assign the same or different (sets of) processors to different grids in a multiple-grid computation.

Second (Fig. 1b), tessellations of the domain (*sections*) are defined, based on the grid variable. The associated library call is `CHN_Create_section`. Sections contain a number of cutting planes (*cuts*) along each coordinate direction. The grid is thus carved into a number of *cells*, each of which contains a logically rectangular block of grid points. Whereas the programmer can specify any number of cuts and cut locations, a single call to a high-level routine

---

[1] We will henceforth speak of *processors*, even if *processes* is the more accurate term.

often suffices to define all the cuts belonging to a particular domain decomposition. For example, defining a section with just a single cell (i.e. a non-divided grid with zero cuts) is accomplished with CHN_Set_solopartition_cuts. Using CHN_Set_unipartition_cuts divides the grid evenly into as many cells as there are processors in the communicator—nine in this case.

Third (Fig. 1c), cells are assigned to processors, resulting in a *decomposition*. The associated function is CHN_Create_decomposition. The reason why the creation of section and decomposition are separated is to provide flexibility. For example, we may divide a grid into ten slices for execution on a parallel computer, but assign all slices to the same processor for the purpose of debugging on a serial machine. As with the creation of sections, we can choose to assign each cell to a processor individually, or make a single call to a high-level routine. For example, CHN_Set_unipartition_owners assigns each cell in the unipartition section to a different processor. But regardless of the number of processors in the communicator, CHN_Set_solopartition_ _owners assigns all cells to the same processor.

Finally (Fig. 1d), arrays with one or more spatial dimensions (same as the grid) are associated

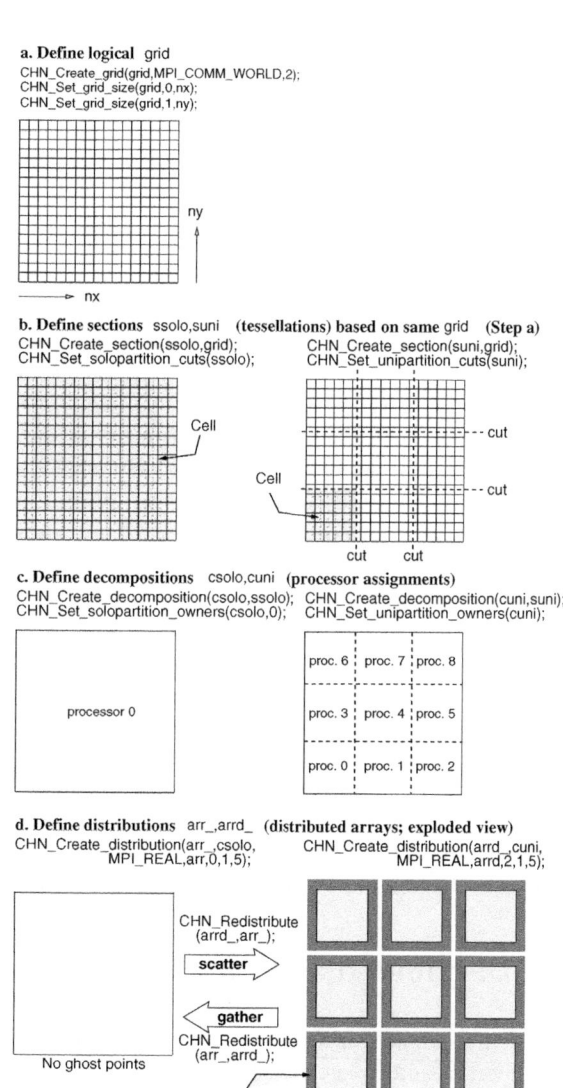

Fig. 1. Defining and mapping distributed arrays

with a decomposition, resulting in *distributions*. The associated function is CHN_Create_distribution. The arrays may represent scalar quantities at each grid point, or higher-order tensors. In the example in Fig. 1d the tensor rank is 1, and the thusly defined vector has 5 components at each grid point. A distribution has one of a subset of the regular MPI data types (MPI_REAL in this case). Since Charon expressly supports stencil operations on multi-dimensional grids, we also specify a number of *ghost points*. These form a border of points (shaded area) around each cell, which acts as a cache for data copied from adjacent cells. In this case the undivided distribution has zero ghost points, whereas the unipartition distribution has two, which can support higher-order difference stencils (e.g. a 13-point 3D star).

The aspect of distributions that sets Charon apart most from other parallelization libraries is the fact that the programmer also supplies the memory occupied by the distributed array; Charon provides a *structuring interpretation* of user space. In the example in Fig. 1d it is assumed that arr is the starting address of the non-distributed array used in the legacy code, whereas arrd is the starting address of a newly declared array that will hold data related to the unipartition distribution. By mapping between arr and arrd—using CHN_Redistribute (see also Fig. 1d)—we can dynamically switch from the serial legacy code to truly distributed code, and back. All that is required is that the programmer define distribution arr_ such that the memory layout of the (legacy code) array arr coincide exactly with the Charon specified layout. This act of reverse engineering is supported by functions that allow the programmer to specify array padding and offsets, by a complete set of query functions, and by Charon's unambiguously defined memory layout model (see Section 3.3).

CHN_Redistribute can be used not only to construct parallel 'bypasses' of serial code of the kind demonstrated above, but also to map between *any* two compatible distributions (same grid, data type, and tensor rank). This is shown in Fig. 2, where two stripwise distributions, one aligned with the first coordinate axis, and the other with the second, are mapped into each other, thereby establishing a dynamic transposition. This is useful when there are very strong but mutually incompatible data dependencies in different parts of the code (e.g. 2D FFT). By default, the unipartition decomposition divides all coordinate directions evenly, but by excluding certain directions from partitioning (CHN_Exclude_partition_direction) we can force a stripwise distribution.

3   Distributed and Parallel Execution Support

While it is an advantage to be able to keep most of a legacy code unchanged and focus on a small part at a time for parallelization, it is often still nontrivial to arrive at good parallel code for complicated numerical algorithms. Charon offers support for this process at two levels.

The first concerns a set of wrapping functions that allows us to keep the serial logic and structure of the legacy code unchanged, although the data is truly distributed. These functions incur a significant overhead, and are meant to be removed in the final version of the code. They provide a stepping stone in the parallelization, and may be skipped by the more intrepid programmer.

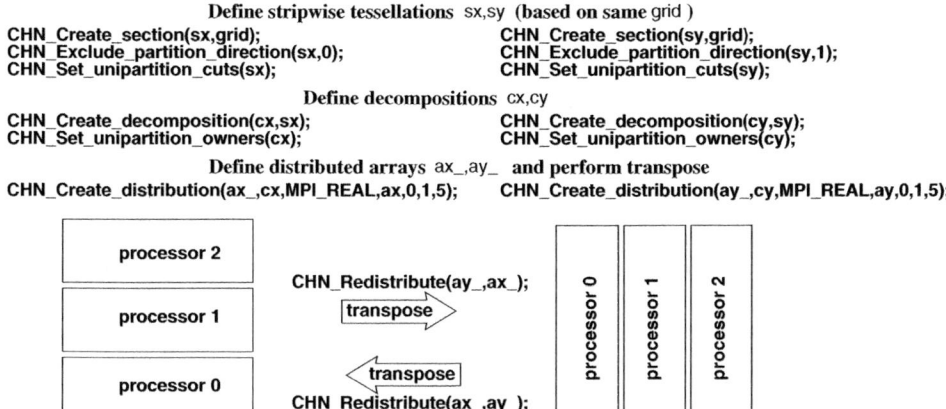

Fig. 2. Transposing distributed arrays

The second is a set of versatile, highly optimized bulk communication functions that support the implementation of sophisticated data-parallel and—more importantly—non-data-parallel numerical methods, such as pipelined algorithms.

3.1 Wrapping Functions

In a serial program it is obvious what the statement a(i,j,k) = b(i+2,j-1,k) means, provided a and b have been properly dimensioned, but when these arrays are distributed across several processors the result is probably not what is expected, and most likely wrong. This is due to one of the fundamental complexities of message-passing, namely that the programmer is responsible for defining explicitly and managing the data distribution. Charon can relieve this burden by allowing us to write the above assignment as:

call CHN_Assign(CHN_Address(a_,i,j,k),CHN_Value(b_,i+2,j-1,k))

with no regard for how the data is distributed (assuming that a_ and b_ are distributions related to arrays a and b, respectively). The benefit of this wrapping is that the user need not worry (yet) about communications, which are implicitly invoked by Charon, as needed.

The three functions introduced here have the following properties. CHN_Value inspects the distribution b_, determines which (unique) processor owns the grid point that holds the value, and broadcasts that value to all processors in the communicator ('*owner serves*' rule). CHN_Address inspects the distribution a_ and determines which processor owns the grid point that holds the value. If the calling processor is the owner, the actual address—an *lvalue*—is returned, and NULL otherwise[2]. CHN_Assign stores the value of its second argument at the address in the first argument **if** the address is not NULL. Consequently, only the

---

[2] In Fortran return values cannot be used as lvalues, but this problem is easily circumvented, since the address is immediately passed to a C function.

point owner of the left hand side of the assignment stores the value (*'owner assigns'* rule). No distinction is made between values obtained through CHN_Value and local values, or expressions containing combinations of each; all are *rvalues*. Similarly, no distinction is made between local addresses and those obtained through CHN_Address. Hence, the following assignments are all legitimate:

```
    call CHN_Assign(CHN_Address(a_,i,j,k),5.0)                     !1
    call CHN_Assign(aux,CHN_Value(b_,i,j-1,k)+1.0)                 !2
    aux = CHN_Value(b_,i,j-1,k)+1.0                                !3
```

It should be observed that assignments 2 and 3 are equivalent. An important feature of wrapped code is that it is completely serialized. All processors execute the same statements, and whenever an element of a distributed array occurs on the right hand side of an assignment, it is broadcast. As a result, it is guaranteed to have the correct serial logic of the legacy code.

3.2   Bulk Communications

The performance of wrapped code can be improved by removing the need for the very fine-grained, implicitly invoked communications, and replacing them with explicitly invoked bulk communications. Within structured-grid applications the need for non-local data is often limited to (spatially) nearest-neighbor communication; stencil operations can usually be carried out without any communication, provided a border of ghost points (see Fig. 1d) is filled with array values from neighboring cells. This fill operation is provided by CHN_Copyfaces, which lets the programmer specify exactly which ghost points to update. The function takes the following arguments:

- the thickness of layer of ghost points to be copied; this can be at most the number of ghost points specified in the definition of the *distribution*,
- the components of the tensor to be copied. For example, the user may wish only to transfer the diagonal elements of a matrix,
- the coordinate direction in which the copying takes place,
- the sequence number of the cut (defined in the *section*) across which copying takes place,
- the rectangular subset of points within the cut to be copied.

In general, all processors within the grid's MPI communicator execute CHN_Copyfaces, but those that do not own points involved in the operation may safely skip the call. A useful variation is CHN_Copyfaces_all, which fills all ghost points of the distribution in all coordinate directions. It is the variation most commonly encountered in other parallelization packages for structured-grid applications, since it conveniently supports data parallel computations. But it is not sufficient to implement, for example, the pipeline algorithm of Section 4.2.

The remaining two bulk communications provided by Charon are the previously described CHN_Redistribute, and CHN_Gettile. The latter copies a subset of a distributed array—which may be owned by several processors—into the local memory of a specified processor (cf. Global Arrays' ga_get [10]). This is useful for applications that have non-nearest-neighbor remote data dependencies, such as non-local boundary conditions for flow problems.

3.3 Parallelizing Distributed Code Segments

Once the remote-data demand has been satisfied through the bulk copying of ghost point values, the programmer can instruct Charon to suppress broadcasts by declaring a section of the code *local* (see below). Within a local section not all code should be executed anymore by all processors, since assignments to points not owned by the calling processor will usually require remote data not present on that calling processor; the code must be restructured to restrict the index sets of loops over (parts of) the grid. This is the actual process of parallelization, and it is left to the programmer. It is often conceptually simple for structured-grid codes, but the bookkeeping matters of changing all data structures at once have traditionally hampered such parallelization. The advantage of using Charon is that the restructuring focuses on small segments of the code at any one time, and that the starting point is code that already executes correctly on distributed data sets. The parallelization of a loop nest typically involves the following steps.

1. Determine the order in which grid cells should be visited to resolve all data dependencies in the target parallel code. For example, during the x-sweep in the SP code described in Section 4.1 (Fig. 3), cells are visited layer by layer, marching in the positive x-direction. In this step all processors still visit all cells, and no explicit communications are required, thanks to the wrapper functions. This step is supported by query functions that return the number of cells in a particular coordinate direction, and also the starting and ending grid indices of the cells (useful for computing loop bounds).
2. Fill ghost point data in advance. If the loop is completely data parallel, a single call to `CHN_Copyfaces` or `CHN_Copyfaces_all` before entering the loop is usually sufficient to fill ghost point values. If a non-trivial data dependence exists, then multiple calls to `CHN_Copyfaces` are usually required. For example, in the x-sweep in the SP code `CHN_Copyfaces` is called between each layer of cells. At this stage all processors still execute all statements in the loop nest, so that they can participate in broadcasts of data not resident on the calling processor. However, whenever it is a ghost point value that is required, it is served by the processor that owns it, rather than the processor that owns the cell that has that point as an interior point. This seeming ambiguity is resolved by placing calls to `CHN_Begin_ghost_access` and `CHN_End_ghost_access` around the code that accesses ghost point data, which specify the index of the cell whose ghost points should be used.
3. Suppress broadcasts. This is accomplished by using the bracketing construct `CHN_Begin_local`/`CHN_End_local` to enclose the code that accesses elements of distributed arrays. For example:
```
call CHN_Begin_local(MPI_COMM_WORLD)
call CHN_Assign(CHN_Address(a_,i),CHN_Value(b_,i+1)-CHN_Value(b_,i-1))
call CHN_End_local(MPI_COMM_WORLD)
```
At the same time, the programmer restricts accessing lvalues to points actually owned by the calling processor. This is supported by the query functions `CHN_Point_owner` and `CHN_Cell_owner`, which return the MPI rank of the processor that owns the point and the grid cell, respectively.

Once the code segment is fully parallelized, the programmer can strips the wrappers to obtain the final, high-performance code. Stripping effectively consists of translating global grid coordinates into local array indices, a chore that is again easily accomplished, due to Charon's transparent memory layout model. By default, all subarrays of the distribution associated with individual cells of the grid are dimensioned identically, and these dimensions can be computed in advance, or obtained through query functions. For example, assume that the number of cells owned by each processor is nmax, the dimensions of the largest cell in the grid are nx×ny, and the number of ghost points is gp. Then the array w related to the scalar distribution w_ can be dimensioned as follows:

```
dimension w(1-gp:nx+gp,1-gp:ny+gp,nmax)
```

Assume further that the programmer has filled the arrays beg(2,nmax) and end(2,nmax) with the beginning and ending point indices, respectively (using Charon query functions), of the cells in the grid owned by the calling processor. Then the following two loop nests are equivalent, provided n ≤ nmax.

```
do j=beg(2,n),end(2,n)                 do j=1,end(2,n)-beg(2,n)+1
  do i=beg(1,n),end(1,n)                 do i=1,end(1,n)-beg(1,n)+1
    call CHN_Assign(CHN_Address(w_,i,j),5.0)   w(i,j,n) = 5.0
  end do                                 end do
end do                                 end do
```

The above example illustrates the fact that Charon minimizes encapsulation; it is always possible to access data related to distributed arrays directly, without having to copy data or call access functions. This is a programming convenience, as well as a performance gain. Programs parallelized using Charon usually ultimately only contain library calls that create and query distributions, and that perform high-level communications.

Finally, it should be noted that it is not necessary first to wrap legacy code to take advantage of Charon's bulk communication facilities for the construction of parallel bypasses. Wrappers and bulk communications are completely independent.

4   Examples

We present two examples of numerical problems, SP and LU, with complicated data dependencies. Both are taken from the NAS Parallel Benchmarks (NPB) [1], of which hand-coded MPI versions (NPB-MPI) and serial versions are freely available. They have the form: $Au^{n+1} = b(u^n)$, where u is the time-dependent solution, n is the number of the time step, and b is a nonlinear 13-point-star stencil operator. The difference is in the shape of A, the discretization matrix that defines the 'implicitness' of the numerical scheme. For SP it is effectively: $A_{SP} = L_z L_y L_x$, and for LU: $A_{LU} = L_+ L_-$.

4.1   SP Code

L_z, L_y and L_x are fourth-order difference operators that determine data dependencies in the z, y, and x directions, respectively. A_{SP} is numerically inverted in three corresponding phases, each involving the solution of a large number of independent banded (penta-diagonal) matrix equations, three for each grid line. Each equation is solved using Gaussian elimination, implemented as two sweeps along the grid line, one in the positive (forward elimination), and one in the negative direction (backsubstitution). The method chosen in NPB-MPI is the *multipartition* (MP) decomposition strategy. It assigns to each processor multiple cells such that, regardless of sweep direction, each processor has work to do during each stage of the solution process. An example of a 9-processor 3D MP is shown in Fig. 3. Details can be found in [1]. While most packages we have studied do not allow the definition of MP, it is easily specified in Charon:

```
call CHN_Create_section(multi_sec,grid)
call CHN_Set_multipartition_cuts(multi_sec)
call CHN_Create_decomposition(multi_cmp,multi_sec)
call CHN_Set_multipartition_owners(multi_cmp)
```

The solution process in the x direction is as follows (Fig. 3). All processors start the forward elimination on the left side of the grid. When the boundary of the first layer of cells is reached, the elements of the penta-diagonal matrix that need to be passed to the next layer of cells are copied in bulk using CHN_Copyfaces. Then the next layer of cells is traversed, followed by another copy operation, etc. The number of floating point operations and words communicated in the Charon version of the code is exactly the same as in NPB-MPI. The only difference is that Charon copies the values into ghost points, whereas in NPB-MPI they are directly used to update the matrix system without going to main memory. The latter is more efficient, but comes at the cost of a much greater program complexity, since the communication must be fully integrated with the computation.

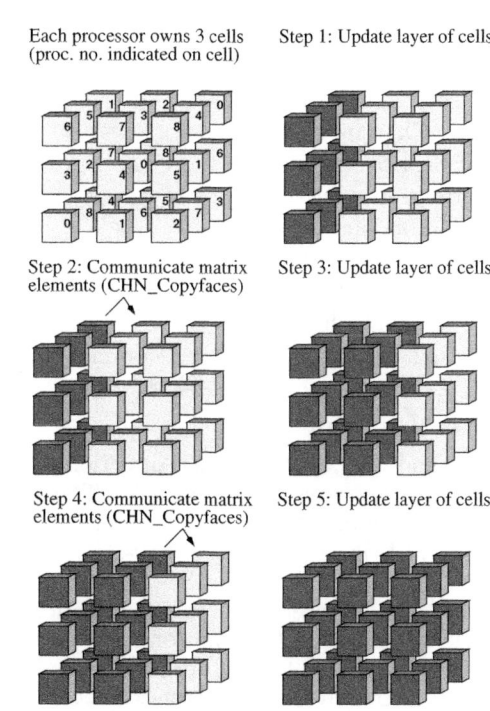

Fig. 3. Nine-processor multipartition decomposition, and solution process for SP code (L_x: forward elimination)

The results of running both Charon and NPB-MPI versions of SP for three different grid sizes on an SGI Origin2000 (250 MHz MIPS R10000) are shown in Fig. 4. Save for a deterioration at 81 processors for class B (102^3 grid) due to a bad stride, the results indicate that the Charon version achieves approximately 70% of the performance of NPB-MPI, with roughly the same scalability characteristics. While 30% difference is significant, it should be noted that the Charon version was derived from the serial code in three days, whereas NPB-MPI took more than one month (both by the same author). Moreover, since Charon and MPI calls can be freely mixed, it is always possible for the programmer who is not satisfied with the performance of Charon communications to do (some of) the message passing by hand.

If SP is run on a 10-CPU Sun Ultra Enterprise 4000 Server (250 MHz Ultrasparc II processor), whose cache structure differs significantly from the R10000, the difference between results for the Charon and NPB-MPI versions shrinks to a mere 3.5%; on this machine there is no gain in performance through hand coding.

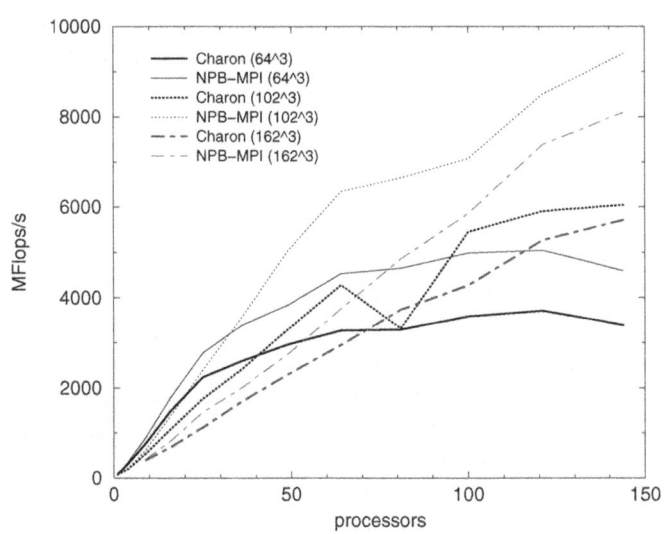

Fig. 4. Performance of SP benchmark on 250 MHz SGI Origin2000

4.2   LU Code

L_- and L_+ are first-order direction-biased difference operators. They define two sweeps over the entire grid. The structure of L_- dictates that no point (i, j, k) can be updated before updating all points (i_p, j_p, k_p) with smaller indices: $\{(i_p, j_p, k_p) | i_p \leq i, j_p \leq j, k_p \leq k, (i_p, j_p, k_p) \neq (i, j, k)\}$. This data dependency is the same as for the Gauss-Seidel method with lexicographical point ordering. L_+ sweeps in the other direction. Unlike for SP, there is no concept of independent grid lines for LU. The solution method chosen for NPB-MPI is to divide the grid into pencils, one for each processor, and pipeline the solution process, Fig. 5;

```
call CHN_Create_section(pencil_sec,grid)
call CHN_Exclude_partition_direction(pencil_sec,2)
call CHN_Set_unipartition_cuts(pencil_sec)
call CHN_Create_decomposition(uni_cmp,uni_sec)
call CHN_Set_unipartition_owners(uni_cmp)
```

Each unit of computation is a single plane of points (*tile*) of the pencil. Once a tile is updated, the values on its boundary are communicated to the pencil's Eastern and Northern (for L_-) neighbors. Subsequently, the next tile in the pencil is updated. Of course, not all boundary points of the whole pencil should be transferred after completion of each tile update, but only those of the 'active' tile. This is easily specified in CHN_Copyfaces.

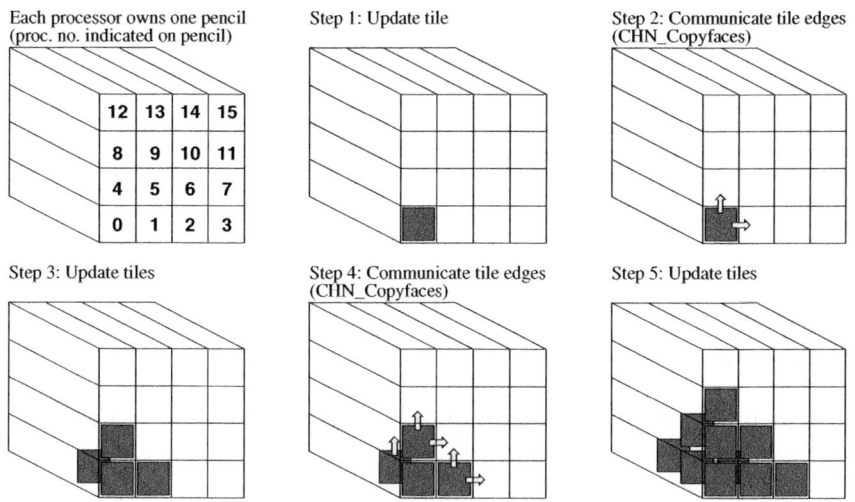

Fig. 5. Unipartition pencil decomposition (16 processors); Start of pipelined solution process for LU code (L_-)

The results of both Charon and NPB-MPI versions of LU for three different grid sizes on the SGI Origin are shown in Fig. 6. Now the performance of the Charon code is almost the same as that of NPB-MPI. This is because both use ghost points for transferring information between neighboring pencils. Again, on the Sun E4000, the performances of the hand coded and Charon parallelized programs are nearly indistinguishable.

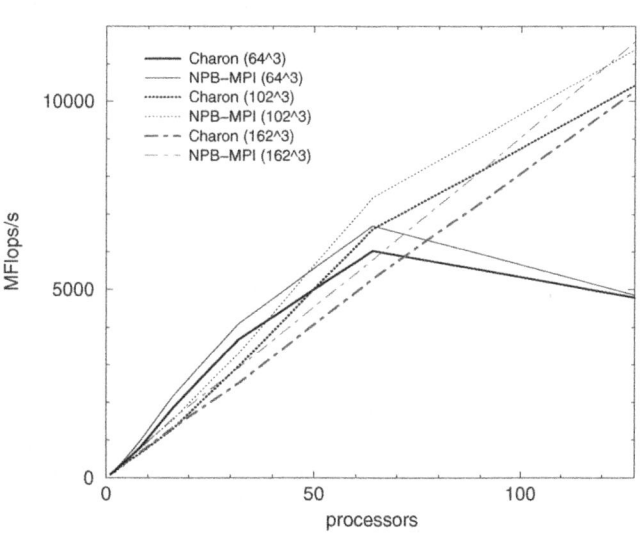

Fig. 6. Performance of LU benchmark on 250 MHz SGI Origin2000

5 Discussion and Conclusions

We have given a brief presentation of some of the major capabilities of Charon, a parallelization library for scientific computing problems. It is useful for those applications that need high scalability, or that have complicated data dependencies that are hard to resolve by analysis engines. Since some hand coding is required, it is more labor intensive than using a parallelizing compiler or code transformation tool. Moreover, the programmer must have some knowledge about the application program structure in order to make effective use of Charon. When the programmer does decide to make the investment to use the library, the results are close to the performance of hand-coded, highly tuned message passing implementations, at a fraction of the development cost. More information on the toolkit and a user guide are being made available by the author [12].

References

1. D.H. Bailey, T. Harris, W.C. Saphir, R.F. Van der Wijngaart, A.C. Woo, M. Yarrow, "The NAS parallel benchmarks 2.0," Report NAS-95-020, NASA Ames Research Center, Moffett Field, CA, December 1995.
2. S. Balay, W.D. Gropp, L. Curfman McInnes, B.F. Smith, "PETSc 2.0 Users manual," Report ANL–95/11 - Revision 2.0.24, Argonne National Laboratory, Argonne, IL, 1999.
3. D.L. Brown, G.S. Chesshire, W.D. Henshaw, D.J. Quinlan, "Overture: An object-oriented software system for solving partial differential equations in serial and parallel environments," 8th SIAM Conf. Parallel Proc. for Scientific Computing, Minneapolis, MN, March 1997.
4. S.B. Baden, D. Shalit, R.B. Frost, "KeLP User Guid Version 1.3," Dept. Comp. Sci. and Engin., UC San Diego, La Jolla, CA, January 2000.
5. M. Frumkin, J. Yan, "Automatic Data Distribution for CFD Applications on Structured Grids," NAS Technical Report NAS-99-012, NASA Ames Research Center, CA, 1999.
6. M.W. Hall, J.M. Anderson, S.P. Amarasinghe, B.R. Murphy, S.-W. Liao, E. Bugnion, M.S. Lam, "Maximizing Multiprocessor Performance with the SUIF Compiler," IEEE Computer, Vol. 29, pp. 84–89, December 1996.
7. T. Henderson, D. Schaffer, M. Govett, L. Hart, "SMS Users Guide," NOAA/Forecast Systems Laboratory, Boulder, CO, January 2000.
8. C.S. Ierotheou, S.P. Johnson, M. Cross, P.F. Leggett, "Computer aided parallelisation tools (CAPTools)—conceptual overview and performance on the parallelisation of structured mesh codes," Parallel Computing, Vol. 22, pp. 163–195, 1996.
9. H. Jin, M. Frumkin, J. Yan, "Use Computer-aided tools to parallelize large CFD applications," NASA High Performance Computing and Communications Computational Aerosciences (CAS) Workshop 2000, NASA Ames Research Center, Moffett Field, CA, February 2000.
10. J. Nieplocha, R.J. Harrison, R.J. Littlefield, "The global array programming model for high performance scientific computing," SIAM News, Vol. 28, August-September 1995.
11. M. Snir, S.W. Otto, S. Huss-Lederman, D.W. Walker, J. Dongarra, "MPI: The Complete Reference," MIT Press, 1995.
12. R.F. Van der Wijngaart, Charon home page, http://www.nas.nasa.gov/~wijngaar/charon.

Dynamic Slicing of Concurrent Programs

D. Goswami and R. Mall

Department of Computer Science and Engineering
IIT Kharagpur, Kharagpur - 721 302, INDIA
{diganta, rajib}@cse.iitkgp.ernet.in

Abstract. We present a framework for computing dynamic slices of concurrent programs using a form of dependence graph as intermediate representations. We introduce the notion of a Dynamic Program Dependence Graph ($DPDG$) to represent various *intra-* and *inter*process dependences of concurrent programs. We construct this graph through three hierarchical stages. Besides being intuitive, this approach also enables us to display slices at different levels of abstraction. We have considered interprocess communication using both shared memory and message passing mechanisms.

1 Introduction

Program slicing is a technique for extracting only those statements from a program which may affect the value of a chosen set of variables at some point of interest in the program. Development of an efficient program slicing technique is a very important problem, since slicing finds applications in numerous areas such as program debugging, testing, maintenance, re-engineering, comprehension, program integration and differencing. Excellent surveys on the applications of program slicing and existing slicing methods are available in [1,2].

The slice of a program P is computed with respect to a slicing criterion $< s, v >$, where s is a statement in P and v is a variable in the statement. The backward slice of P with respect to the slicing criterion $< s, v >$ includes only those statements of P which affect the value of v at the statement s. Several methods for computing slices of programs have been reported [3,4]. The seminal work of Weiser proposed computation of slices of a program using its control flow graph (CFG) representation [3]. Ottenstein and Ottenstein were the first to define slicing as a graph reachability problem [4]. They used a Program Dependence Graph (PDG) for static slicing of single-procedure programs. The PDG of a program P is a directed graph whose vertices represent either assignment statements or control predicates, and edges represent either control dependence or flow dependence [4]. As originally introduced by Weiser, slicing (static slicing) considered all possible program executions. That is, static slices do not depend on the input data to a program. While debugging however, we typically deal with a particular incorrect execution and are interested in locating the cause of incorrectness in that execution. Therefore, we are interested in a slice that preserves the program behavior for a specific program input, rather than that

M. Valero, V.K. Prasanna, and S. Vajapeyam (Eds.): HiPC 2000, LNCS 1970, pp. 15–26, 2000.

for all possible inputs. This type of slicing is referred to as dynamic slicing. Korel and Laski were the first to introduce the idea of dynamic slicing [5]. A dynamic slice contains all statements that actually affect the value of a variable at a program point for a particular execution of the program. Korel and Laski extended Weiser's static slicing algorithm based on data-flow equations for the dynamic case [5]. Agrawal and Horgan were the first to propose a method to compute dynamic slices using PDG [6].

Present day software systems are becoming larger and complex and usually consist of concurrent processes. It is much more difficult to debug and understand the behavior of such concurrent programs than the sequential ones. Program slicing techniques promise to come in handy at this point. In this paper, we present a framework to compute slices of concurrent programs for specific program inputs by introducing the notion of Dynamic Program Dependence Graph ($DPDG$). To construct a $DPDG$, we proceed by first constructing a process graph and a Static Program Dependence Graph ($SPDG$) at compile time. Trace files are generated at run-time to record the information regarding the relevant events that occur during the execution of concurrent programs. Using this information stored in the trace files, the process graph is refined to realize a *concurrency graph*. The $SPDG$, the information stored in trace files, and the concurrency graph are then used to construct the $DPDG$. Once the $DPDG$ is constructed, it is easy to compute slices of the concurrent program using a simple graph reachability algorithm.

2 Static Graph Representation

Before we can compute slices of a concurrent program, we need to construct a general model of the program containing all information necessary to construct a slice. We do this through three hierarchical levels. In the first stage we graphically represent static aspects of concurrent programs which can be extracted from the program code. We will enhance this representation later to construct a $DPDG$. In our subsequent discussions, we will use primitive constructs for process creation, interprocess communication and synchronization which are similar to those available in the Unix environment [7]. The main motivation behind our choice of Unix-like primitives is that the syntax and semantics of these primitive constructs are intuitive, well-understood, easily extensible to other parallel programming models and also can be easily tested. The language constructs that we consider for message passing are *msgsend* and *msgrecv*. The syntax and semantics of these two constructs are as follows:

- *msgsend(msgqueue, msg):* When a msgsend statement is executed, the message *msg* is stored in the message queue *msgqueue*. The msgsend statement is nonblocking, i.e. the sending process continues its execution after depositing the message in the message queue.

- *msgrecv(msgqueue, msg):* When a msgrecv statement is executed, the variable *msg* is assigned the value of the corresponding message from the message queue *msgqueue*. The msgrecv statement is blocking, i.e. if the *msgqueue* is found to be empty, it waits for the corresponding sending process for depositing the message.

We have considered nonblocking send and blocking receive semantics of inter-process communication because these have traditionally been used for concurrent programming applications. In this model, no assumptions are made regarding the order in which messages arrive in a message queue from the msgsend statements belonging to different processes except that messages sent by one process to a message queue are stored in the same order in which they were sent by the process.

A fork() call creates a new process called *child* which is an exact copy of the parent. It returns a nonzero value (process ID of the child process) to the parent process and zero to the child process [7]. Both the child and the parent have separate copies of all variables. However, shared data segments acquired by using the shmget() and shmat() function calls are shared by the concerned processes. Parent and child processes execute concurrently. A wait() call can be used by the parent process to wait for the termination of the child process. In this case, the parent process would not proceed until the child terminates. Semaphores are synchronization primitives which can be used to control access to shared variables. In the Unix environment, semaphores are realized through the semget() call. The value of a semaphore can be set by semctl() call. The increment and decrement operations on semaphores are carried out by the semop() call [7]. However, for simplicity of notation, in the rest of the paper we shall use P(sem) and V(sem) as the semaphore decrement and increment operations respectively.

2.1   Process Graph

A *process graph* captures the basic process structure of a concurrent program. In this graph, we represent the process creation, termination, and joining of processes. More formally, a process graph is a 5-tuple (En, T, N, E, C) where En is special entry node denoting start of the program. T is the set of terminal nodes. N is the set of non-terminal, non-entry nodes, E is the set of edges, and C is a function that assigns statements to edges in the process graph. A node from N represents any one of the following types of statements: a fork() statement (denoted as F), a wait() statement (denoted as W) or a loop predicate whose loop body contains a fork() statement (denoted as L). Each individual terminal node from T will be denoted as t. Edges may be of three types: *process edge, loop edge*, and *join edge*. A process edge represents the sequence of statements starting from the statement represented by the source node of the edge till the statement represented by the sink node of the edge. Loop and join edges are dummy edges and do not represent any statement of the program but are included to represent the control flow. The source nodes of both these two types of edges are terminal nodes. The sink node of a loop edge is of L-type and that of a join edge is of W-type. Direction of an edge represents direction of control flow. We will use solid edges to denote process edges, dashed edges to denote loop edges, and

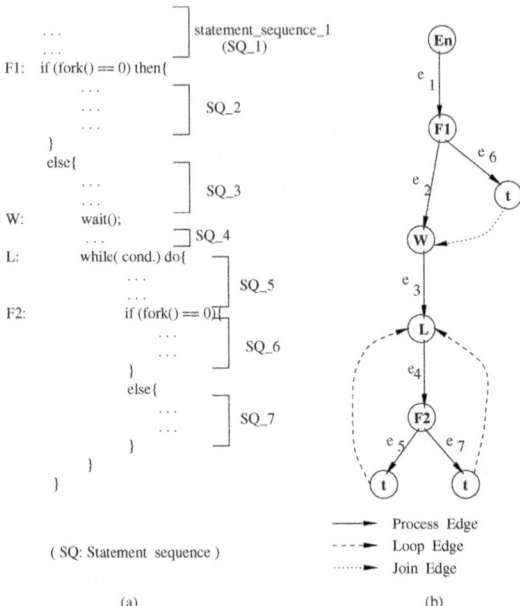

Fig. 1. (a) An Example Concurrent Program (b) Its Process Graph

dotted edges to represent join edges. Figure 1(b) shows the process graph of the example concurrent program given in figure 1(a). In the example of figure 1, the labels of the statements indicate the type of nodes represented by the concerned statements.

2.2   Static Program Dependence Graph

A static program dependence graph ($SPDG$) represents the program dependences which can be determined statically. The $SPDG$ is constructed by extending the process graph discussed in the previous section. It can easily be observed that control dependences among statements are always fixed and do not vary with the choice of input values and hence can be determined statically at compile time. A major part of data dependence edges can also be determined statically, excepting the statements appearing under the scope of selection and loop constructs which have to be handled at run-time. To construct the $SPDG$ of a program, we first construct a dummy node S for each edge whose source node is a fork node F to represent the beginning of the statement sequence represented by that edge. Control dependence edges from node F to each node S is then constructed and the statements which were earlier assigned to the edges beginning with the fork node F are now assigned to the corresponding edges beginning with the dummy nodes S.

Let W be a node in the process graph representing a wait call. Let m be the first node of type F or L or t found by traversing the process graph from node

W along the directed edges. All the statements which are represented by the edge from W to m are said to be *controlled* by the node W.

We now construct a control dependence subgraph for each statement sequence representing an edge (or a sequence of edges) beginning with node En or S and ending with a node F or a t in the process graph. Informally, in each of these control dependence subgraphs, a control dependence edge from a node x to a node y exists, iff any one of the following holds:

- x is the entry node (En) or a process start node (S) and y is a statement which is not nested within any loops or conditional statements and is also not controlled by a wait call.

- x is a predicate and y is a statement which is immediately nested within this predicate.

- x is a wait call and y is a statement which is controlled by this wait call and is not immediately nested within any loop or conditional statement.

After construction of control dependence subgraph for each edge of the process graph, all data dependence edges between the nodes are then constructed. Data dependence edges can be classified into two type: *deterministic* and *potential*. A data dependence edge which can be determined statically is said to be deterministic and the one which might or might not exist depending on the exact execution path taken at run-time is said to be potential. We construct all potential data dependence edges at compile time but whether a potential data dependence edge is actually taken can be determined only based on run-time information. To mark an edge as a potential edge in the graph, we store this information at the sink node of the edge.

3   Dynamic Program Dependence Graph

In this section, we discuss how dynamic information can be represented in an enhanced graph when a concurrent program is executed. We call this enhanced graph with dynamic information, a Dynamic Program Dependence Graph ($DPDG$). A $DPDG$ of a concurrent program is a directed graph. The nodes of $DPDG$ represent the individual statements of the program. There is a special entry node and one or more terminal nodes. Further for every process there is a special start node as already discussed in the context of $SPDG$ definition. The edges among the nodes may be of data/control dependence type or synchronization/communication dependence type. The dynamic data dependence edges are constructed by analyzing the run-time information available. Data dependences due to shared data accesses, synchronization due to semaphore operations, and communication due to message passing are considered and corresponding edge types are added in the graph for those instructions which are actually executed. After construction of the $DPDG$, we apply a reachability criterion to compute dynamic slices of the program. To construct $DPDG$, it is necessary to first construct a concurrency graph to determine the concurrent components in the graph.

3.1   Concurrency Graph

A concurrency graph is a refinement of a process graph and is built using the program's run-time information. A concurrency graph is used to perform concurrency analysis which is necessary to resolve shared dependences existing across process boundaries. A concurrency graph retains all the nodes and edges of a process graph. Besides these, it contains two new types of edges: *synchronization edges* and *communication edges*. It also contains a new type of node called a *synchronization node*. The purpose and construction procedure for these new edges and nodes are explained in the following subsections. The concurrency graph also represents the processes which get dynamically created during run-time.

3.1.1 Communication Edge

The execution of a msgrecv statement depends on the execution of a corresponding msgsend statement. This dependence is referred as *communication dependence*. A msgrecv statement s_r is communication dependent on a msgsend statement s_s, if communication occurs between s_r and s_s during execution. Communication dependence cannot be determined statically because messages sent from different processes to a message queue may arrive in any order. So, communication dependences can only be determined using run-time information. The information needed to construct a communication edge from a send node x in a process P to a receive node y in a process Q is the pair (P, x) stored at node y. To be able to construct the communication edges, the sending process P needs to append the pair (*sending process ID, sending statement no.*) to the sent message. After receiving the message, the receiving process Q needs to store this information in a trace file. From this information, the communication edges can immediately be established. Each message passing statement (both send and recieve) which gets executed is represented in the concurrency graph as a node and this node is referred as a *synchronization node, Sn*.

3.1.2 Synchronization Edge

Semaphore operations (P and V) are used for controlling accesses to shared variables by either acquiring resources (through a P operation) or releasing resources (through a V operation). We construct a synchronization edge from the node representing each V operation to the node representing the corresponding P operation on the same semaphore. A synchronization edge depicts which P operation depends on which V operation. We define a source node for the V operation and a sink node for the corresponding P operation. Identification of a pair of related semaphore operations can be done by matching the nth V operation to the $(n + i)$th P operation on the same semaphore variable, i being the initial value of the semaphore variable. To record the information required to determine this semaphore pairing, each semaphore operation records (for the process it belongs to) the number of operations on the given semaphore which have already occurred. From these recorded information, the semaphore operations can easily be paired and synchronization edges can then be constructed from these pairings. Each semaphore operation which gets executed is represented as a node in the concurrency graph, and will be referred as a *synchronization node, Sn*.

3.1.3 Handling Dynamic Creation of Processes

Parallel and distributed programs may dynamically create processes during run-time. If a fork call is nested within a loop or conditional, it is difficult to determine during compile time, the exact number of fork calls that would be made. This information can only be obtained at run-time. For example, if the loop predicate L of the process graph shown in figure 2(a) gets executed for zero, one or two times, the number of processes created varies and the corresponding concurrency graphs for these new situations are depicted in figure 2(b), 2(c), and 2(d). The concurrency graph incorporates all these dynamically created processes from the information stored in the trace files.

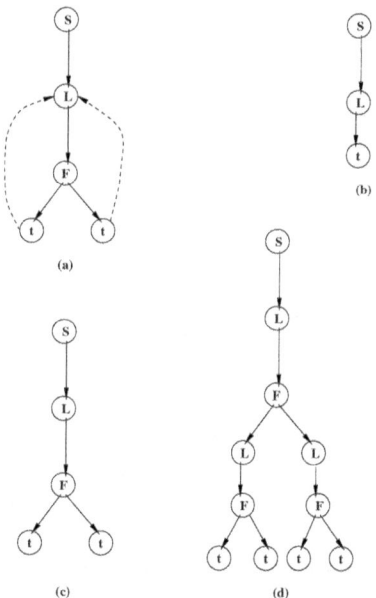

Fig. 2. (a) A Sample Process Graph. The Corresponding Concurrency Graphs which Result After Execution of the Loop for (b) Zero, (c) One, and (d) Two Times (assuming that there exists no interprocess communication among processes)

3.2   Code Instrumentation for Recording Events

For construction of the $DPDG$, we have to record the necessary information related to process creation, interprocess communication, and synchronization aspects of concurrent programs at run-time. When a program is executed, a trace file for each process is generated to record the event history for that process. We have identified the following types of events to be recorded in the trace file along with all relevant information . The source code would have to be instrumented to record these events.

• **Process creation:** Whenever a process gets created it has to be recorded. The information recorded are the process ID and parent process's ID. A process may create many child processes by executing fork calls. To determine the parent process for any child process the parent process would record (after execution of the fork call) the node (statement) number of the first statement to be executed by the child process after its creation.

• **Shared data access:** In case of access to shared data by a process, the information required to be recorded are the variable accessed and whether it is being defined or used.

• **Semaphore operation:** For each semaphore
operation (P or V), semaphore ID, and the value of a counter associated with the semaphore indicating the number of synchronization operations performed on that semaphore so far.

• **Message passing:** In case of interprocess communication by message passing, the recorded information should include the type of the message (e.g. send/ receive) and the pair (sending_process_ID, sending_node_no.) in case of a receive type. For this, the process executing a *msgsend* statement would have to append the above mentioned pair with the message. Whenever a process executes a *msgrecv* statement, it would extract the pair and store it in the trace file.

The concurrency graph of a concurrent program is constructed from the execution trace of each process recorded at run-time. An execution trace, TX of a program is a sequence of nodes (statements) that has actually been executed in that order for some input. Node Y at position p is written as Y^p or $TX(p)$ and is referred to as an *event*. By v^q we denote variable v at position q. A use of variable v is an event denoting that this variable is referenced. A definition of variable v is an event denoting an assignment of a value to that variable. $D(Y^p)$ is a set of variables whose values are defined in event Y^p. By the term *most recent definition* $MR(v^k)$ of variable v^k in TX we mean Y^p such that $v \in D(Y^p)$ and v is not defined in any of the events after Y^p upto position k.

We now discuss how to handle dynamic process creation in the process graph and the corresponding control and data dependence subgraphs in the $SPDG$ for later use. Let (x, y) and (y, z) be two directed edges. We define *transformation by doing fusion on node y* as the replacement of these two edges by the single edge (x, z). Then, by doing a fusion on synchronization nodes we get the *dynamic process graph* i.e., the process graph which contains all the dynamically created processes. The $SPDG$ can then be modified to incorporate the control and data dependence subgraphs for these dynamically created processes. From this, we get a mapping of every event in the execution trace to the nodes of $SPDG$. More than one event may map to a single node in the $SPDG$, if the event corresponds to a loop predicate or a statement within a loop which gets executed more than once.

In a concurrent or distributed environment, different processes might write the significant events to their individual trace files without using any globally synchronized timestamp. Therefore, it is necessary to order the recorded events properly for use in constructing the $DPDG$. In particular, ordering of events is necessary to construct shared dependence edges among nodes accessing shared variables.

3.3   Event Ordering

The structure of the concurrency graph and the information stored in the event history (i.e. execution trace) help us to order the events. Please note that we will be dealing with partial ordering of events only [8]. This type of ordering with respect to a logical clock is sufficient to determine the causal relationships between the events. We can partially order the nodes and edges of the concurrency graph by using the *happened-before* relation $\rightarrow$ [8] as follows:

- Let x and y be any two nodes of the concurrency graph. $x \rightarrow y$ is true, if y is reachable from x following any sequence of edges in the concurrency graph.

- Let e_i be the edge from node x_1 to node x_2 and e_j be the edge from node y_1 to node y_2; $e_i \rightarrow e_j$ is true, if $x_2 \rightarrow y_1$ is true.

If two edges e_i and e_j cannot be ordered using the happened-before relationship, then these two edges are said to be *incomparable.* This is expressed as $e_i \| e_j$. Let SV denotes a shared variable. Then, SV_i indicates an access of the shared variable SV in the edge i; SV_{ir} and SV_{iw} denote respectively a read and a write operation respectively on the shared variable SV in the edge i. Data race leads to ambiguous and erroneous conditions in programming. A concurrent program is said to be data race free, if every pair of incomparable edges in the concurrency graph is data race free. The condition given below leads to data race in a concurrent program. The condition states that if a shared variable SV is accessed in two edges which are incomparable and if at least one of these accesses is a write, then this leads to a data race.

$$((SV_{ir} \wedge SV_{jw}) \vee (SV_{iw} \wedge SV_{jw})) \wedge (e_i \| e_j) \text{ for any } i, j.$$

It is reasonable to assume that the program to be sliced is data race free. If the program under consideration is not data race free, then this can be identified during the process of event ordering and can be reported to the user.

3.4   Construction of $DPDG$

Once the concurrency graph is constructed, the $DPDG$ for the concurrent program can be constructed from the $SPDG$ and the concurrency graph. For every event in the execution trace, a corresponding node in the $DPDG$ is created. All the data and control dependence edges to the corresponding nodes from this node are then constructed in the $DPDG$. If there exists any potential data dependence edge, only the edge which has actually been taken during execution is created by finding out the *most recent definition* of the variable from the trace file. Data dependence edges may cross the process boundaries when a statement in a process accesses a shared data item defined by a statement in some other process. Shared variable accesses can either be synchronized or unsynchronized.

● Synchronized access: In case of synchronized access to shared resources using semaphores, shared dependence edges are constructed by referring to the concurrency graph. Consider an event SV_{ir} within a (P, V)-block, (P_i, V_i). The concurrency graph is then traversed backwards from synchronization node P_i along synchronization edges to reach the corresponding source synchronization node V_j. A shared dependency edge from the event SV_{kw} within the (P, V)-block, (P_j, V_j) to the event SV_{ir} is then constructed.

● Unsynchronized access: When a statement in a process uses a shared variable, we need to identify the statement which modified this shared variable most recently to establish the shared dependence. The task is to trace the event history searching for the event(s) modifying this variable. If more than one modification to the same variable occurred, then these are ordered by referring to the concurrency graph. A data dependence edge is then constructed from the statement which performed the most recent modification to this variable to the statement reading it. Consider the unsynchronized event SV_{ir}. Let the events SV_{kw}, SV_{mw}, and SV_{tw} represent the modifications to the same variable. Let the concurrency graph reveal that $e_k \rightarrow e_m$, $e_m \rightarrow e_t$, and $e_t \rightarrow e_i$. In this case, we need to construct a shared dependence edge from SV_{tw} to SV_{ir}.

Communication edges are established from the pairs (*sending process ID, sending node no.*) which are stored in the trace file when msgrecv statements get executed.

4   Related Work

Cheng proposed a representation for concurrent programs where he generalized the notions of CFG and PDG [9]. Cheng's algorithm for computing dynamic slices is basically a generalization of the initial approach taken by Agrawal and Horgan [6], which computes a dynamic slice using a static graph. Therefore, their slicing algorithm may compute inaccurate slices in the presence of loops. Miller and Choi use a dynamic dependence graph, similar to ours to perform flow-back analysis in their parallel program debugger [10]. Our method, however, differs from theirs in the way the intermediate graphs are constructed. Our graph representation is substantially different from theirs to take care of dynamically created processes and message passing using message queues. In their approach, a branch dependence graph is constructed statically. In the branch dependence graph data dependence edges for individual basic blocks are included. Control dependence edges are included during execution and the dynamic dependence graph is built by combining in order the data dependence graphs of all basic blocks reached during execution. We find that it is possible to resolve the control dependences and a major part of inter-block data dependences statically and we take advantage of this. The other differences are that we construct the $DPDG$ in a hierarchical manner and consider the Unix primitives for process creation and interprocess communication. Another significant difference arises because of our choice of Unix primitives for message passing. In this model messages get stored in message queues and are later retrieved from the queue by the receiving process. This is a more elaborate message passing mechanism. Duesterwald et al. represent distributed programs using a Distributed Dependence Graph (DDG). Run-time behavior of a program is analyzed to add data and communication dependence features to DDG. They have not considered interprocess communication using shared variables. DDG constructs a single vertex for each statement and control predicate in the program [11]. Since it uses a single vertex for all occurrences of a statement, the slice computed would be inaccurate if the program contains loops.

5 Conclusion

We have proposed a hierarchical graph representation of concurrent programs that lets us efficiently compute dynamic slices. Our method can handle both shared memory and message passing constructs. Unix semantics of message passing where messages get stored in a message queue in a partially ordered manner introduces complications. We have proposed a solution to this by attaching the source process ID and the statement number of the msgsnd statement along-with the message. Since we create a node in $DPDG$ for each occurrence of a statement in the execution trace, the resulting slices are more precise for programs containing loops compared to traditional dynamic slicing algorithms such as those by Duesterwald et al [11] and Cheng's [9] which use a single vertex for all occurrences of a statement.

References

1. F. Tip, "A survey of program slicing techniques," *Journal of Programming Languages*, vol. 3, no. 3, pp. 121-189, September 1995.
2. D. Binkley and K. B. Gallagher, "Program slicing," *Advances in Computers, Ed. M. Zelkowitz*, Academic Press, San Diego, CA, vol. 43, 1996.
3. M. Weiser, "Program slicing," *IEEE Trans. on Software Engg.*, vol. 10, no. 4, pp. 352-357, July 1984.
4. K. Ottenstein and L. Ottenstein, "The program dependence graph in software development environment," In *Proceedings of the ACM SIGSOFT/SIGPLAN Software Engineering Symposium on Practical Software Development Environments, SIG- PLAN Notices*, vol. 19, no. 5, pp. 177-184, 1984.
5. B. Korel and J. Laski, "Dynamic program slicing," *Information Processing Letters*, vol. 29, no. 3, pp. 155-163, October 1988.
6. H. Agrawal and J. Horgan, "Dynamic program slicing," In *Proceedings of the ACM SIGPLAN '90 Conf. on Programming Language Design and Implementation, SIG-*

PLAN Notices, Analysis and Verification, White Plains, New York, vol. 25, no. 6, pp. 246-256, June 1990.

7. M. J. Bach, *The Design Of The Unix Operating System*. Prentice Hall India Ltd., New Delhi, 1986.

8. L. Lamport, "Time, clocks, and the ordering of events in a distributed system," *Communications of the ACM*, vol. 21, no. 7, pp. 558-565, July 1978.

9. J. Cheng, "Slicing concurrent programs-a graph theoretical approach," In *Proceedings of the First International Workshop on Automated and Algorithmic Debugging, P. Fritzson Ed., Lecture Notes in Comp. Sc., Springer-Verlag*, vol. 749, pp. 223-240, 1993.

10. J. D. Choi, B. Miller, and R. Netzer, "Techniques for debugging parallel programs with owback analysis," *ACM Trans. on Programming Languages and Systems*, vol. 13, no. 4, pp. 491-530, 1991.

11. E. Duesterwald, R. Gupta, and M. L. Soffa, "Distributed slicing and partial reexecution for distributed programs," In *Proceedings of the Fifth Workshop on Languages and Compilers for Parallel Computing, New Haven Connecticut, LNCS Springer-Verlag*, vol. 757, pp. 329-337, August 1992.

An Efficient Run-Time Scheme for Exploiting Parallelism on Multiprocessor Systems

Tsung-Chuan Huang[1], Po-Hsueh Hsu[2], and Chi-Fan Wu[1]

[1] Department of Electrical Engineering, National Sun Yat-sen University, Taiwan
tch@mail.nsysu.edu.tw
[2] Department of Electronic Engineering, Cheng Shiu Institute of Technology, Taiwan
phhsu@atm.ee.nsysu.edu.tw

Abstract. High performance computing capability is crucial for the advanced calculations of scientific applications. A parallelizing compiler can take a sequential program as input and automatically translate it into a parallel form. But for loops with arrays of irregular (i.e., indirectly indexed), nonlinear or dynamic access patterns, no state-of-the-art compilers can determine their parallelism at compile-time. In this paper, we propose an efficient run-time scheme to compute a high parallelism execution schedule for those loops. This new scheme first constructs a predecessor iteration table in inspector phase, and then schedules the whole loop iterations into wavefronts for parallel execution. For non-uniform access patterns, the performance of the inspector/executor methods usually degrades dramatically, but it is not valid for our scheme. Furthermore, this scheme is especially suitable for multiprocessor systems because of the features of high scalability and low overhead.

1 Introduction

Recently, automatic parallelization is a key enabling technique for parallel computing. How to exploit the parallelism in a loop, or the *loop parallelization*, is an important issue in this area. Current parallelizing compilers demonstrate their effectiveness for loops that have no cross-iteration dependences or have only uniform dependences. But there are some limitations in the parallelization of loops with complex or statically insufficiently defined access patterns.

In order to convert sequential programs into their parallel equivalents, parallelizing compilers must perform data dependence analysis first to determine whether a loop, or part of it, can be executed in parallel without violating the original semantics. This analysis is mainly focused on array subscript expressions, i.e., array access patterns. Basically, the application programs can be classified into two types:

1. Regular programs, in which memory accesses are described by linear equations of variables (usually loop index variables).
2. Irregular programs, in which memory accesses are described by indirection mapping (e.g., index arrays) or computation dependent.

Regular programs are much easier to deal with because they can be statically analyzed at compile-time. Unfortunately, many scientific programs performing complex modeling or simulations, such as DYNA-3D and SPICE, are usually irregular programs. The form of irregular accesses looks like A(w(i)) or A(idx), where

M. Valero, V.K. Prasanna, and S. Vajapeyam (Eds.): HiPC 2000, LNCS 1970, pp. 27–36, 2000
© Springer-Verlag Berlin Heidelberg 2000

w is an index array and idx is computed inside the loop but not an induction variable. In the circumstances, we can only resort to run-time parallelization techniques as complementary solutions because information for analysis is not available until program execution.

2 Related Work

Two different approaches have been developed for run-time loop parallelization: the speculative doall execution and the inspector/executor parallelization method. The former approach assumes the loop to be fully parallelizable and executes it speculatively, then examines the correctness of parallel execution after loop termination. In the latter approach, the inspector examines cross-iteration dependences and produces a parallel execution schedule first, then the executor performs actual loop operations based on the schedule arranged by the inspector.

Speculative Doall Parallelization. The speculative doall parallelization speculatively executes the loop operations in parallel, accompanied with a marking mechanism to track the accesses to the target arrays. After loop termination, an analysis mechanism is applied to examine whether this speculative doall parallelization passes or not, i.e., to check whether no cross-iteration dependence occurs in the loop. If it passes, a significant speedup will be obtained. Otherwise, the altered variables should be restored and the loop is re-executed serially. The reader who is interested in this topic may refer to Huang and Hsu [2], and Rauchwerger and Padua [9].

Run-Time Doacross Parallelization (The Inspector/Executor Method). According to the scheduling unit, we classify the inspector/executor methods into two types: the reference-level and the iteration-level. The reference-level type assumes a memory reference of the loop body as the basic unit of scheduling and synchronization in the inspector. Busy-waits are used to ensure values are produced before used during the executor phase. This type of method has the advantage of increasing the overlap of dependent iterations, but at the expense of more synchronization overhead. The reader who is interested in this topic may see Chen et al. [1], and Xu and Chaudhary [11].

   The iteration-level type assumes loop iteration as the basic scheduling unit in the inspector. The inspector schedules the source loop iterations into appropriate wavefronts at run-time; wavefronts will be executed serially but the iterations in a wavefront are executed in parallel. The executor then performs actual execution according to the wavefront sequence. The reader who is interested in this topic may see Zhu and Yew [12], Midkiff and Padua [6], Polychronopoulos [7], Saltz *et al.* [10], Leung and Zahorjan [4], Leung and Zahorjan [5], Rauchwerger *et al.* [8], and Huang *et al.* [3].

   In general, speculative doall execution gains significant speedup if the target loop is intrinsically fully parallel; otherwise a hazard arises when cross-iteration dependences occur [2, 9]. In contrast, inspector/executor methods are profitable in extracting doacross loop parallelism, but may suffer a relative amount of processing overhead and synchronization burden [1, 3-8, 10-12].

3 Our Efficient Run-Time Scheme

In retrospect of the development of run-time doacross parallelization, we find some inefficient factors as follows:

- sequential inspector,
- synchronization overhead on updating shared variables,
- overhead in constructing dependence chains for all array elements,
- large memory space in operation,
- inefficient scheduler,
- possible load migration from inspector to executor.

In this paper, we propose an efficient run-time scheme to overcome these problems. Our new scheme constructs an immediate predecessor table first, and then schedules the whole loop iterations efficiently into wavefronts for parallel execution. Owing to the characteristics of high scalability and low overhead, our scheme is especially suitable for multiprocessor systems.

3.1 Design Considerations

As described in the previous section, the benefits coming from run-time parallelization may be offset by a relative amount of processing overhead. Consequently, how to reduce the processing overhead is a major concern of our method, and parallel inspectors, which can exploit the capability of multiprocessing, becomes the objective that we pursue.

Rauchwerger *et al.*[8] designed a run-time parallelization method, which is fully parallel, with no synchronization, and can be applied on any kind of loop. This scheme is based on the operation of predecessor references; we call it PRS (Predecessor_Reference_Scheme) for short. Their inspector encodes the predecessor/successor information, for the references to A(x), in a reference array R_x and a hierarchy vector H_x so that the scheduler can arrange loop iterations according to this dependence information. The scheduler is easy to implement but not efficient enough. In order to identify the iterations belonging to the ith wavefront, *all the references* must be examined to determine the ready states of corresponding unscheduled iterations in the ith step. Thus, the scheduler takes $O((numref/numproc)*cpl)$ time for this processing, where *numref* is the number of references, *numproc* is the number of processors, and *cpl* is the length of the critical path in the directed acyclic graph that describes the cross-iteration dependency in the loop. In fact, the above processing overhead can be completely eliminated if we use the data representation of *predecessor iteration* instead of *predecessor reference*.

Hereby, we develop a new run-time scheme to automatically extract the parallelism of doacross loops under the following considerations:

1. Devising a high efficient parallel inspector to construct a predecessor iteration table for recording the information of iteration dependences.
2. Devising a high efficient parallel scheduler to quickly produce the wavefront schedule with the help of predecessor iteration table.
3. The scheme should be no synchronization to ensure good scalability.

3.2 An Example

We demonstrate the operation of our run-time scheme by an example. Fig. 1(a) is an irregular loop to be parallelized. The access pattern of the loop is shown in Fig. 1(b). In our scheme, a predecessor iteration table, with the help of an auxiliary array *la*(which will be explained in the next subsection), is constructed first as shown in Fig. 1(c) at run-time. Then, the wavefronts of loop iterations can be quickly scheduled out there in Fig. 1(d). The predecessor iteration table is implemented by two arrays, *pw(1:numiter)* and *pr(1:numiter)*, where *numiter* is the number of iterations of the target loop. Element pw(i) records the predecessor iteration of iteration i that accesses(either *read* or *write*) the same array element x(w(i)). A similar explanation applies to pr(i). After the predecessor iteration table is built, the scheduler uses this information to arrange iterations into wavefronts: firstly, the iterations with no predecessor iteration (i.e. whose associated elements in *pw* and *pr* are both zero) are scheduled into the first wavefront, then the iterations whose predecessor iterations have been scheduled are assigned into the next wavefront. The procedure is repeated until all the iterations are scheduled.

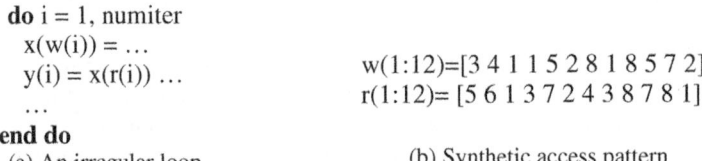

    do i = 1, numiter
      x(w(i)) = ...
      y(i) = x(r(i)) ...   w(1:12)=[3 4 1 1 5 2 8 1 8 5 7 2]
      ...   r(1:12)= [5 6 1 3 7 2 4 3 8 7 8 1]
    end do
     (a) An irregular loop   (b) Synthetic access pattern

iter. / Pred. Iter.	1	2	3	4	5	6	7	8	9	10	11	12
pw(i)	0	0	0	3	1	0	0	4	7	5	10	6
pr(i)	0	0	0	1	0	0	2	4	7	5	9	8

(c) Predecessor iteration table

Iter. / Wavefront	1	2	3	4	5	6	7	8	9	10	11	12
wf(i)	1	1	1	2	2	1	2	3	3	3	4	4

(d) Wavefront schedule

Fig. 1. An example to demonstrate the operations of our scheme

3.3 The Inspector

The goal of our inspector is to construct a predecessor iteration table *in parallel*. The iterations of target loop are distributed in block faction into processors; each processor takes charge of *blksize* contiguous iterations. We use an auxiliary array *la(1:numproc,1:arysize)*, which is initially set to zero, to keep the latest iteration, all the way, that access the array element for each processor. Namely, la(p,j) records the latest iteration getting access to the array element j for processor p (*la* is the abbreviation of *latest access*). The algorithm of our parallel inspector is shown in Fig. 2, which consists of two phases:

1. Parallel recording phase (lines 3 to 10).

    Step 1. If $la(p,w(i))$ or $la(p,r(i))$ is not zero, let them be c and d respectively, then record $pw(i)$ as c and $pr(i)$ as d, meaning that c is the predecessor iteration of iteration i for accessing the array element $w(i)$ and d is the predecessor iteration of iteration i for accessing the array element $r(i)$. The arrays *pw* and *pr* are initially set to zero.

    Step 2. Set $la(p,w(i))$ and $la(p,r(i))$ to current iteration i. This means that, for processor p, iteration i is the latest iteration that access the array elements $w(i)$ and $r(i)$.

2. Parallel patching phase (lines 11 to 24). For each iteration i, if $pw(i)=0$ then find the largest-numbered processor q, where q < the current processor p, such that $la(q,w(i))$ is not zero. Assume now that $la(q,w(i))=j$, then set $pw(i)=j$. This means that j is the real predecessor iteration of iteration i in accessing the array element $w(i)$. In the same way, if $pr(i)=0$ we find the largest-numbered processor t < the current processor p such that $la(t,r(i))=k! \, 0$, and set $pr(i)=k$ to mean that, in reality, k is the predecessor iteration of iteration i in accessing the array element $r(i)$.

For the example in previous subsection, we assume that there are two processors. By block distribution, iterations 1 to 6 are charged by processor 1 and iterations 7 to 12 by processor 2. The contents of the auxiliary array *la* and the predecessor iteration table *pw* and *pr* for processor 1 and processor 2 are shown in Fig. 3(a) and Fig. 3(b), respectively, when the parallel recording phase is in progress. Remember that processor 1 is in charge of iterations 1 to 6 and processor 2 in charge of iterations 7 to 12. For easier reference, the access pattern in Section 3.2 is also included in Fig. 3. We can see that, when processor1 is dealing with iteration 4 it will write array element 1 (since $w(4)=1$) and read array element 3 (since $r(4)=3$). But array element 1 has been accessed (both read and written in this example) by iteration 3 (this can be seen from $la(1,1)=3$), and array element 3 has been accessed (written in this example) by iteration 1 (since $la(1,3)=1$). Therefore, the predecessor iterations of iteration 4 are iteration 3 for writing array element 1, and iteration 1 for reading array element 3. The predecessor iteration table is hence recorded as $pw(4)=3$ and $pr(4)=1$. This behavior is encoded in lines 5 and 6 of our inspector algorithm in Fig. 2. After this, $la(1,1)$ will be updated from 3 to 4 to mean that iteration 4 is now the latest iteration of accessing the array element 1, and $la(1,3)$ will be updated from 1 to 4 to mean that iteration 4 is now the latest iteration of accessing the array element 3. This behavior is coded in lines 7 and 8 in Fig. 2. Since the values in array *la* after iteration i is processed might be changed when iteration i+1 is dealt with, for clear demonstration, the new updated values are **bold-faced** in Fig. 3(a).

In the parallel patching phase, let us see, in Fig. 3(b), why $pw(7)$ remains 0 (unchanged) and how $pr(7)$ is changed to 2 by processor 2. Since $pw(7)$ in Fig. 3(a) is zero, this means no prior iteration in processor 2 accesses the same array element with iteration 7. But from $w(7)=8$, we know that iteration 7 *writes* array element 8. Hence, we have to trace back to processor 1 to check whether it has accessed this element. By checking $la(1,8)=0$, we find that no iteration in processor 1 has ever

```
      /* The construction of predecessor iteration table */
1     pr(1:numiter)=0
2     pw(1:numiter)=0
      /* Parallel recording phase */
3     doall p=1,numproc
4        do i=(p-1)*(numiter/numproc)+1,p*(numiter/numproc)
5           if (la(p,w(i)).ne.0) then pw(i)=la(p,w(i))
6           if (la(p,r(i)).ne.0) then pr(i)=la(p,r(i))
7           la(p,w(i))=i
8           la(p,r(i))=i
9        enddo
10    enddoall
      /* Parallel patching phase */
11    doall p=2,numproc
12       do i=(p-1)*(numiter/numproc)+1,p*(numiter/numproc)
13          if(pw(i).eq.0) then
14             do j=p-1,1,-1
15                if(la(j,w(i)).ne.0) then
16                   pw(i)=la(j,w(i))
17                   goto S1
18                endif
19             enddo
20 s1:      endif
21          if(pr(i).eq.0) then
22             do j=p-1,1,-1
23                if(la(j,r(i)).ne.0) then
24                   pr(i)=la(j,r(i))
25                   goto S2
26                endif
27             enddo
28 S2:      endif
29       enddo
30    enddoall
```

Fig. 2. The algorithm of inspector

accessed array element 8. Therefore, pw(7) remains unchanged. In the same argument, pr(7) in Fig. 3(a) is zero, which means no prior iteration in processor 2 accesses the same array element with iteration 7. But r(7)=4 indicates that iteration 7 *reads* array element 4. By tracing back to processor 1, we find la(1,4)=2. This means that, in processor 1, iteration 2 also accesses array element 4. Hence, the predecessor iteration of iteration 7 should be 2 for *read* access and pr(7) is changed to 2. The behavior is coded in lines 13 to 20 and lines 21 to 28 for *write* access and *read* access respectively. The predecessor iterations that have been changed are also **bold-faced** in Fig. 3(b) for illustration. As for the time complexity, it is easy to see that our parallel inspector takes O(numiter) time.

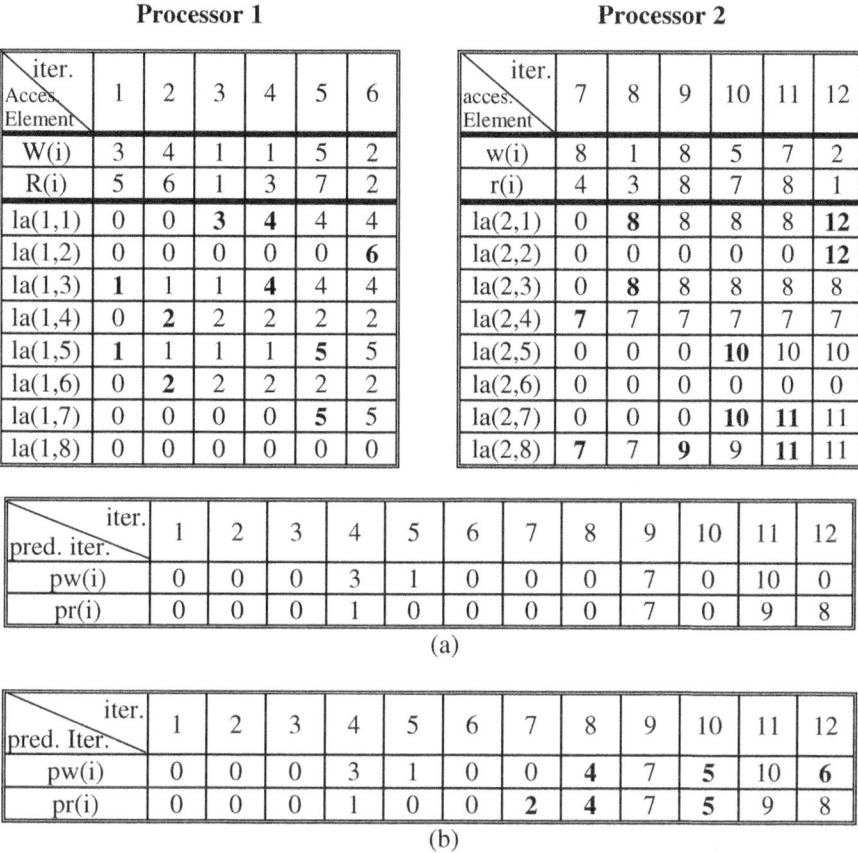

Fig. 3. (a) The contents of array *la* and the predecessor iteration table for processor 1 and processor 2 when the parallel recording phase is in progress. (b) The predecessor iteration table after parallel patching phase is finished

3.4 The Scheduler

The algorithm of our parallel scheduler is presented in Fig. 4. Scheduling the loop iterations into wavefronts becomes very easy once the predecessor iteration table is available. In the beginning, the wavefront table wf(1:numiter) is set to zero to indicate that all loop iterations have not been scheduled.

Like our parallel inspector, the loop iterations are distributed into processors in block faction. For each wavefront number, all processors simultaneously examine the iterations they are charged in series (lines 8 and 9) to see if they can be assigned into the current wavefront. Only the iterations that have not been scheduled yet (line 10) and whose predecessor iterations, for both *write* and *read* access, have been scheduled can be arranged into the current wavefront (lines 11 and 12). This procedure repeats until all loop iterations have been scheduled (line 5). The maximum wavefront number will be referred as *cpl* (critical path length) because it is the length of critical path in the directed acyclic graph representing the cross-iteration dependency in the target loop.

```
    /* Schedule iterations into wavefronts in parallel */
1   wf(1:numiter)=0
2   wf(0)=1
3   done=.false.
4   wfnum=0
    /* Repeated until all iterations are scheduled */
5   do while (done.eq..false.)
6      done=.true.
7      wfnum=wfnum+1
8      doall p=1,numproc
9        do i=(p-
1)*(numiter/numproc)+1,p*(numiter/numproc)
10         if (wf(i).eq.0) then
11           if (wf(pw(i)).ne.0.and.wf(pr(i)).ne.0) then
12             wf(i)=wfnum
13           else
14             done=.false.
15           endif
16         endif
17       enddo
18     enddoall
19 enddo
```

Fig. 4. The algorithm of the scheduler

4 Experimental Results

We performed our experiments on ALR Quad6, a shared-memory multiprocessor machine with four Pentium Pro 200MHz processors and 128MB global memory. The synthetic loop in Fig. 1(a) was added with run-time procedures and OpenMP parallelization directives, and then compiled using the pgf77 compiler.

We intended to evaluate the impact of parallelism degree on the performance of two run-time schemes: PRS (Predecessor_Reference_Scheme) by Rauchwerger *et al.*[8], and PIS (Predecessor_Iteration_Scheme) by us. Let the grain size (workload) be 200us, the array size be 2048, and the iteration number vary from 2048 to 65536. The access patterns are generated using a probabilistic method. Since the accessed area (array size) is fixed to 2048, a loop with 65536 iterations will be more serial in comparison to a loop with 2048 iterations. Table 1 shows the execution time measured in each run-time phase.

Fig. 5 is a speedup comparison for two run-time schemes. We can see that PIS always obtains a higher speedup and still has a satisfactory speedup of 2.8 in the worst case. The overhead comparison in Fig. 6 shows that the processing overhead ((inspector time + scheduler time) / sequential loop time) of PRS is dramatically larger than that of PIS. This is because the scheduler of PRS examines all the references repeatedly as mentioned in Section 3.1.

Table 1. Execution time measured on two run-time schemes

Number of Iterations		2048	4096	8192	16384	32768	65536
Critical Path Length		9	16	30	52	105	199
Sequential Loop Time		413	828	1657	3322	6651	13322
Inspector Time (msec)	PRS	164	274	326	332	390	484
	PIS	3	6	10	31	48	90
Scheduler Time (msec)	PRS	4	14	59	314	2283	8470
	PIS	2	5	11	47	117	460
Executor Time (msec)	PRS	117	229	469	945	1914	4010
	PIS	114	231	477	980	1975	4210
Total Execution Time	PRS	285	517	854	1591	4587	12964
	PIS	119	242	498	1058	2140	4760
Speedup	PRS	1.45	1.60	1.94	2.09	1.45	1.03
	PIS	3.47	3.42	3.33	3.14	3.11	2.80

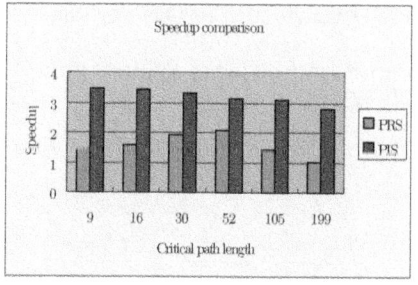

Fig. 5. Speedup comparison   **Fig. 6.** Overhead comparison

5 Conclusion

In the parallelization of partially parallel loops, the biggest challenge is to find a parallel execution schedule that can fully extract the potential parallelism but incurs run-time overhead as little as possible. In this paper, we present a new run-time parallelization scheme PIS, which use the information of predecessor/successor iterations instead of predecessor references to eliminate the processing overhead of repeatedly examining all the references in PRS. The predecessor iteration table can be constructed in parallel with no synchronization and our parallel scheduler can quickly produce the wavefront schedule with its help. From either theoretical time analysis or experimental results, our run-time scheme reveals better speedup and less processing overhead than PRS.

References

1. Chen, D. K., Yew, P. C., Torrellas, J.: An Efficient Algorithm for the Run-Time Parallelization of Doacross Loops. Proc. 1994 Supercomputing (1994) 518-527
2. Huang, T. C., Hsu, P. H.: The SPNT Test: A New Technology for Run-Time Speculative Parallelization of Loops. Lecture Notes in Computer Science Vol. 1366. Springer-Verlag, Berlin Heidelberg New York (1998) 177-191
3. Huang, T. C., Hsu, P. H., Sheng, T. N.: Efficient Run-Time Scheduling for Parallelizing Partially Parallel Loops," J. Information Science and Engineering 14(1), (1998) 255-264
4. Leung, S. T., Zahorjan, J.: Improving the Performance of Run-Time Parallelization. Proc. 4th ACM SIGPLAN Symp. Principles and Practice of Parallel Programming, (1993) 83-91
5. Leung, S. T., Zahorjan, J.: Extending the Applicability and Improving the Performance of Run-Time Parallelization. Tech. Rep. 95-01-08, Dept. CSE, Univ. of Washington (1995)
6. Midkiff, S., Padua, D.: Compiler Algorithms for Synchronization. IEEE Trans. Comput. C-36, 12, (1987) 1485-1495
7. Polychronopoulos, C.: Compiler Optimizations for Enhancing Parallelism and Their Impact on Architecture Design. IEEE Trans. Comput. C-37, 8, (1988) 991-1004
8. Rauchwerger, L., Amato, N., Padua, D.: A Scalable Method for Run-Time Loop Parallelization. Int. J. Parallel Processing, 26(6), (1995) 537-576
9. Rauchwerger, L., Padua, D.: The LRPD Test: Speculative Run-Time Parallelization of Loops with Privatization and Reduction Parallelization. IEEE Trans. Parallel and Distributed Systems, 10(2), (1999) 160-180
10. Saltz, J., Mirchandaney, R., Crowley, K.: Run-time Parallelization and Scheduling of Loops. IEEE Trans. Comput. 40(5), (1991) 603-612
11. Xu, C., Chaudhary, V.: Time-Stamping Algorithms for Parallelization of Loops at Run-Time. Proc. 11th Int. Parallel Processing Symp. (1997)
12. Zhu, C. Q., Yew, P. C.: A Scheme to Enforce Data Dependence on Large Multiprocessor Systems. IEEE Trans. Software Eng. 13(6), (1987) 726-739

Characterization and Enhancement of Static Mapping Heuristics for Heterogeneous Systems*

Praveen Holenarsipur, Vladimir Yarmolenko, Jose Duato, D.K. Panda, and P. Sadayappan

Dept. of Computer and Information Science
The Ohio State University, Columbus, OH 43210-1277

{holenars, yarmolen, duato, panda, saday}@cis.ohio-state.edu

Abstract. Heterogeneous computing environments have become attractive platforms to schedule computationally intensive jobs. We consider the problem of mapping independent tasks onto machines in a heterogeneous environment where expected execution time of each task on each machine is known. Although this problem has been much studied in the past, we derive new insights into the effectiveness of different mapping heuristics by use of two metrics – efficacy (E) and utilization (U). Whereas there is no consistent rank ordering of the various previously proposed mapping heuristics on the basis of total task completion time, we find a very consistent rank ordering of the mapping schemes with respect to the new metrics. Minimization of total completion time requires maximization of the product $E{\times}U$. Using the insights provided by the metrics, we develop a new matching heuristic that produces high-quality mappings using much less time than the most effective previously proposed schemes.

Keywords: Heterogeneous Computing, Cluster Computing, Scheduling for Heterogeneous Systems, Task Assignment, Mapping Heuristics, Performance Evaluation

1   Introduction

The steady decrease in cost and increase in performance of commodity workstations and personal computers have made it increasingly attractive to use clusters of such systems as compute servers instead of high-end parallel supercomputers [3,4,7]. For example, the Ohio Supercomputer Center has recently deployed a cluster comprising of 128 Pentium processors to serve the high-end computing needs of its customers. Due to the rapid advance in performance of commodity computers, when such clusters are upgraded by addition of nodes, they become heterogeneous. Thus, many organizations today have large and heterogeneous collections of networked computers. The issue of effective scheduling of tasks onto such heterogeneous

---

* This work was supported in part by a grant from the Ohio Board of Regents.

M. Valero, V.K. Prasanna, and S. Vajapeyam (Eds.): HiPC 2000, LNCS 1970, pp. 37–48, 2000

clustered systems is therefore of great interest [1,18,22], and several recent research studies have addressed this problem [2,6,7,12,13,14,15,16,19,20,21].

Although a number of studies have focused on the problem of mapping tasks onto heterogeneous systems, there is still an incomplete understanding of the relative effectiveness of different mapping heuristics. In Table 1, we show results from a simulation study comparing four mapping heuristics that have previously been studied by other researchers [2,6,7]. Among these four heuristics, the Min-Min heuristic has been reported to be superior to the others [6] on the basis of simulation studies, using randomly generated matrices to characterize the expected completion times of tasks on the machines. However, when we carried out simulations, varying the total number of tasks to be mapped, we discovered that the Min-Min scheme was not consistently superior. When the average number of tasks-per-processor was small, one of the schemes (Max-Min) that performed poorly for a high task-per-processor ratio, was generally the best scheme. For intermediate values of the task-per-processor ratio, yet another of the previously studied schemes (Fast Greedy) was sometimes the best. The following table shows the completion time (makespan, as defined later in the paper) for the mapping produced by the four heuristics (they are all explained later in Sec. 2 of this paper).

Table 1. The best mapping scheme depends on the average number of tasks per processor

Heuristic	Makespan			
	16 tasks	32 tasks	64 tasks	256 tasks
Max-Min	*381.8*	719.8	1519.6	7013.6
Fast Greedy	414.7	*680.8*	1213.4	4427.1
Min-Min	431.8	700.4	*1146.3*	*3780.6*
UDA	496.3	832.1	1424.8	4570.3

Our observations prompted us to probe further into the characteristics of the mappings produced by the different schemes. In a heterogeneous environment, the total completion time for a set of tasks depends on two fundamental factors:
1) The utilization of the system, which is maximized by balancing the, and
2) The extent to which tasks are mapped to machines that are the most effective in executing them.

These two factors are generally difficult to simultaneously optimize. For instance, if there is a particular fast processor on which all tasks have the lowest completion time, it would be impossible for a mapping to both maximize "effectiveness" and achieve high utilization of the entire system. This is because maximizing execution effectiveness would require mapping all tasks to the fast processor, leaving all others idle.

In this study, we use two performance metrics (defined later in Sec. 3) – efficacy (E) and utilization (U), to develop a better understanding of the characteristics of different mapping schemes. We first characterize several previously proposed static mapping heuristics using these metrics and show that some of them score highly on one metric while others score highly on the other metric. But it is the product $E{\times}U$ that determines the makespan. Using the insights provided by use of these metrics, we develop a new heuristic for task mapping that is both effective and computationally very efficient. Over a wide range of machine/task parameters, it provides consistently superior mappings (with improvement in makespan of up to 20%), and takes considerably less time to generate the mappings than the Min-Min heuristic.

The rest of the paper is organized as follows. Section 2 provides necessary definitions. The performance metrics used are defined in Section 3. Enhancements to an existing matching heuristics are proposed in Sections 4. Experimental results are discussed in Section 5. Section 6 provides conclusions.

2   Background and Notations

In this section, we provide some background on previously proposed task matching heuristics and explain the notation used. We use the same notation as in [6]. A set of t independent tasks is to be mapped onto a system of m machines, with the objective of minimizing the total completion time of the tasks. It is assumed that the execution time for each task on each machine is known prior to execution [3,6,8,9,17] and contained within an ETC (Expected Time to Compute) matrix. Each row of the ETC matrix contains the estimated execution times for a given task on each machine. Similarly, each column of the ETC matrix consists of the estimated execution times for each task on a given machine. Thus, $ETC[i,j]$ is the estimated execution time for task i on machine j. The *machine availability time* for machine j, $MAT[j]$, is the earliest time a machine j can complete the execution of all tasks that have previously been assigned to it. The *completion time*, $CT[i,j]$, is the sum of machine j's availability time prior to assignment of task i and the execution time of task i on machine j, i.e. $CT[i,j] = MAT[j] + ETC[i,j]$. The performance criterion usually used for comparison of the heuristics is the maximum value of $CT[i,j]$, for $0 \leq i < t$ and $0 \leq j < m$, also called the *makespan*[6]. The goal of the matching heuristics is to minimize the makespan, i.e. completion time of the entire set of tasks.

A number of heuristics have been proposed in previous studies. However, in this paper we only consider a subset of the heuristics reported, that run within a few seconds on a Pentium PC for mapping several hundred tasks. Below, we briefly describe the heuristics considered.

UDA: User-Direct Assignment assigns each task to a machine with the lowest execution time for that task [3,6]. Its computational complexity is $O(N)$, where N is the number of tasks being mapped.

Fast Greedy: The Fast Greedy heuristic assigns tasks in arrival order, with each task being assigned to the machine which would result in the minimum completion time for that task [3,6], based on the partial assignment of previous tasks. Its computational complexity is also $O(N)$.

Min-Min: The Min-Min heuristic begins with the set U of all unmapped tasks. The set of minimum completion times, $M = \{m_i : m_i = min_{0 \leq j < m}(CT[i, j]),$ for each $i \in U\}$ is found. The task i with the overall minimum completion time from M is selected and assigned to the corresponding machine which would result in the minimum completion time, based on the partial assignment of tasks so far. The newly mapped task is removed from U and the process repeated until all tasks are mapped (i.e. $U = \emptyset$) [6]. The computational complexity of Min-Min is $O(N^2)$.

Max-Min: The Max-Min heuristic is similar to Min-Min and at each step finds the set of minimum completion times, $M = \{m_i : m_i = min_{0 \leq j < m}(CT[i, j])$, for each $i \in U\}$, but the task i with the *highest* completion time from M is selected and assigned to the corresponding machine. The newly mapped task is removed from U and the process repeated until all tasks are mapped (i.e. $U = \emptyset$) [6]. The computational complexity of Max-Min is $O(N^2)$.

3   Performance Metrics

In this section, we define two metrics that we use to quantitatively capture the fundamental factors that affect the quality of mapping for a heterogeneous environment. Using these metrics, we then characterize the four static mapping heuristics described in Sec. 2.

Efficacy Metric: In a heterogeneous system, the execution time of a task depends on the machine to which it gets mapped. For any given task, there exist one or more machines that are the most effective for that task, i.e. the execution time for that task is minimum on that(those) machine(s). The metric proposed below is a collective measure of *efficacy* for a mapping. We call this metric efficacy in order to distinguish it from the term efficiency, which generally denotes the fraction of peak machine performance achieved.

We denote the *Total Execution Time* (TET), as the sum of execution times of all the tasks on their assigned machines, i.e., $TET = \Sigma ETC[i,Map[i]]$, where $Map[i]$ denotes the machine onto which task i is mapped. It represents the total number of computing cycles spent on the collection of tasks by all the processors combined. This is also equal to the sum of the machine availability times $\Sigma MAT[j]$ for all machines, for the mapping used.

We define the *Best Execution Time* (BET), as the sum of machine availability times for all machines when all tasks are mapped to their optimal machines. $BET = \Sigma ETC[i,Best[i]]$, where $Best[i]$ represents a machine on which task i has the lowest execution time.

The system efficacy metric is defined as:

$$E = \frac{BET}{TET} \tag{1}$$

Note that E always lies between 0 and 1. This is because for any task i, $ETC[i,Best[i]] \leq ETC[i,Map[i]]$. A value of 1 for E implies that all tasks have been assigned to maximally effective machines. A low value for E suggests that a large fraction of the work has been mapped to machines that are considerably slower for those tasks than the best suited machines for them.

Utilization Metric: The utilization metric is a measure of overall load-balance achieved by the mapping. It is defined as:

$$U = \frac{TET}{\text{makespan} \times m} \qquad (2)$$

This characteristic reflects the percentage of useful computing cycles as a fraction of the total available cycles. Note that U always lies between 0 and 1. This is because $MAT[j] \leq makespan$, for every machine j. Since $TET = \Sigma MAT[j]$, and there are m machines, the result follows. A value of 1 for U implies perfect load balancing among all the machines. A low value for U implies poor load balancing.

Although high values of efficacy and utilization are inherently desirable, it is generally impossible to achieve mappings that simultaneously maximize both. Consider, for example, a situation where a particular processor happens to be the fastest for every task. The only mapping that would maximize E is the one that maps all tasks to that fast processor. This would clearly result in very low utilization ($1/m$ for a system with m machines). In terms of the metrics E and U, minimization of makespan requires the maximization of their product, since:

$$makespan = \frac{BET}{m} \times \frac{1}{E \times U} \qquad (3)$$

In the following subsection, we characterize the previously mentioned mapping heuristics in terms of these two metrics.

Characterization of Previously Proposed Mapping Heuristics

A number of simulations were performed using the four mapping heuristics (UDA, Fast Greedy, Min-Min and Max-Min). Various combinations of heterogeneity parameters were experimented with. The observed trends did not change significantly with the values for the heterogeneity parameters for task and machine heterogeneity. Hence we present results for just one combination of parameters – low task heterogeneity, low machine heterogeneity and an inconsistent ETC matrix. Table 2 shows the efficacy and utilization values for the same mappings whose makespans were summarized earlier in Table 1.

Whereas there is no consistently superior scheme as far as the makespan is concerned, the picture is very consistent when viewed in terms of the metrics of efficacy and utilization. For all four cases, over a range of variation of parameters, there is a single consistent rank ordering of the four schemes in relation to each of the metrics. The ordering by increasing efficacy is: Max-Min, Fast Greedy, Min-Min, UDA. The ordering by increasing utilization is exactly the reverse: UDA, Min-Min, Fast Greedy, Max-Min.

The analysis of the efficacy and utilization of the four methods helps explain the fact that no single method is consistently superior in relation to the overall makespan. When the average number of tasks per processor is low, the utilization is lower than when the average number of tasks per processor is high, but the difference in utilization of the different schemes is higher. However, the difference in efficacy across schemes is lower when the average number of tasks per processor is low. So for a low task-per-processor ratio, the difference in the $E \times U$ product for different schemes is dominated by the differences in utilization. Max-Min achieves better utilization than Min-Min. It tends to assign the longest tasks first and therefore is better able to balance load at the end of the schedule by assigning the shorter jobs. In contrast, Min-Min tends to assign the shorter jobs first, and is generally not as effective in creating a load-balanced schedule because the longest jobs are assigned last and can cause greater load imbalance. Max-Min achieves an overall higher product of $E{\times}U$ and therefore lower makespan when the average number of tasks per processor is low. The reason is that Max-Min's utilization superiority over the other schemes more than makes up for its lower efficacy. As the average number of tasks per processor increases, the utilization of all schemes improves, tending asymptotically to 1.0. Thus the difference between the utilization of the various schemes decreases and the efficacy dominates the $E{\times}U$ product.

Table 2. Efficacy, utilization and makespan values for mapping heuristics

Metrics	Heuristics	16 tasks	32 tasks	64 tasks	256 tasks
E	Max-Min	0.69	0.62	0.56	0.48
	Greedy	0.85	0.83	0.82	0.80
	Min-Min	0.93	0.93	0.95	0.98
	UDA	1.00	1.00	1.00	1.00
U	Max-Min	0.79	0.94	0.98	1.00
	Greedy	0.60	0.75	0.85	0.96
	Min-Min	0.52	0.65	0.77	0.92
	UDA	0.43	0.52	0.60	0.75
E×U	Max-Min	0.54	0.59	0.55	0.48
	Greedy	0.51	0.62	0.69	0.76
	Min-Min	0.49	0.60	0.73	0.89
	UDA	0.43	0.52	0.60	0.75
Makespan	Max-Min	381.8	719.8	1519.6	7013.6
	Greedy	414.7	680.8	1213.4	4427.1
	Min-Min	431.8	700.4	1146.3	3780.6
	UDA	496.3	832.1	1424.8	4570.3

The efficacy of UDA is always 1. The efficacy of Min-Min stays above 0.95 and actually is slightly higher with 256 tasks than with 16 tasks. The efficacy of Fast Greedy deteriorates slightly as the average number of tasks per processor increases, while the efficacy of Max-Min deteriorates considerably (from over 0.7 to under 0.5). Hence as the average number of tasks per processor increases, Min-Min and UDA perform better, relative to Max-Min. The results in Tables 2 suggest that UDA and Min-Min are schemes that generally achieve consistently high efficacy, but are somewhat deficient with respect to utilization. In contrast, Max-Min scores consistently high on utilization, but suffers from poor efficacy. This observation serves as the basis for an approach to develop an improved mapping heuristic, detailed in the next section.

4   Enhancing Mapping Heuristics

In this section, we show how the insights gained by use of the efficacy and utilization metrics can be used in enhancing existing mapping heuristics and/or developing new heuristics. First we provide the basic idea behind the approach and then demonstrate it by applying it to the UDA heuristic.

When mapping independent tasks in a heterogeneous environment, in order to achieve a low makespan, the product $E{\times}U$ must be maximized. From the results of the previous section, we observe that some existing mapping schemes achieve high efficacy, while some others achieve higher utilization. None of the schemes is consistently superior with respect to both efficacy and utilization. The basic idea behind our approach to enhancing mapping heuristics is to start with a base mapping scheme and attempt to improve the mapping with respect to the metric that the scheme fares poorly in. Thus, if we were to start with Min-Min or UDA, we find that efficacy is generally very good, but utilization is not as high as desired. On the other hand, Max-Min is consistently superior with respect to utilization, but often has very poor efficacy. If we start with an initial mapping that has high efficacy, we could attempt to incrementally improve utilization without suffering much of a loss in efficacy. In contrast, if we start with a high-utilization mapping, the goal would be to enhance efficacy through incremental changes to the mapping, without sacrificing much utilization.

The UDA heuristic stands at one end of the efficacy-utilization trade-off spectrum – it always has a maximal efficacy of 1.0, but usually has poor utilization. The idea behind the proposed enhancement is to start with the initial high-efficacy mapping produced by UDA, and progressively refine it by moving tasks, so that utilization is improved without sacrificing too much efficacy. This is done in two phases – a Coarse-Tuning phase, followed by a Fine-Tuning phase.

4.1 Coarse Tuning

Starting from an initial mapping that has very high efficacy, the objective of the Coarse Tuning phase (CT) is to move tasks from overloaded machines to underloaded machines in a way that attempts to minimize the penalty due to the moves. This penalty factor for moving task i from machine j to machine k is defined as:

$$move\_penalty(i,j,k) = \frac{ETC[i,k]}{ETC[i,j]} \qquad (4)$$

The algorithm uses an *Upper Bound Estimate* (UBE) for the optimal makespan. This is determined by use of any fast mapping heuristic. Here we use the Fast-Greedy algorithm to provide the estimate. Fast-Greedy has a running time of $O(N)$ for N tasks (whereas Min-Min and Max-Min are $O(N^2)$ algorithms). The CT phase attempts to move tasks from all overloaded processors (with total load greater than UBE) to underloaded processors (with total load under UBE) so that at the end of the phase, all processors have a total load of approximately UBE.

The moves are done by considering the penalties for tasks on the overloaded processors. The CT phase first considers the tasks mapped to the maximally loaded processor. For each task, the minimal move-penalty is computed, for a move from its currently mapped machine to the next best machine for the task. The tasks are sorted in decreasing order of the minimal move-penalty. The CT phase attempts to retain those tasks with the highest move-penalty and move tasks with lower move-penalties to other processors. Scanning the tasks in sorted move-penalty order, tasks are marked off as immovable until the total load from the immovable tasks is close to UBE. The remaining tasks are candidates for moving to underloaded processors.

For each movable task on an overloaded processor, a *move-diff penalty* (MDP) is then computed. Initially this is the difference between the task's execution times on the third-best machine and second-best machine for the task, normalized by the execution time on the best machine (i.e. the currently mapped machine, since UDA maps all tasks to their best machines):

$$MDP(i,1) = \frac{ETC[i,m3] - ETC[i,m2]}{ETC[i,m1]} \tag{5}$$

where $m1$, $m2$, m3 are respectively the best, second best and third best machines for task i.

As the algorithm progresses, and some of the machines get filled, the move-diff penalty is recalculated using the two next best (unfilled) machines. Thus in general :

$$MDP(i,k) = \frac{ETC[i,m3] - ETC[i,m2]}{ETC[i,m1]} \tag{6}$$

where $m1$, $m2$, m3 are respectively the best, k^{th} best and $(k+1)^{th}$ best machines for task i. In the extreme case when $k = m$, we set $ETC[i,m+1] = 0$.

The movable tasks are moved in decreasing order of their move-diff penalty. The rationale behind this criterion is as follows. All movable tasks are to be moved to some machine or the other. The lowest penalty for each task will result if it is moved over to its second-best machine. However, it is very likely that after several tasks are so moved, the more powerful of the underloaded machines will reach a load close to UBE, preventing additional moves onto them. Later moves will likely be forced onto less powerful underloaded machines, thereby suffering a larger move penalty. So it would be preferable to first move those tasks which would have the greatest penalty if they were forced to their third-best choice instead of their second-best choice, and so on. This is captured by the move-diff penalty. Due to space limitations, we omit the pseudo-code for the heuristic, and refer the reader to [11] for details.

4.2 Fine Tuning

After the Coarse Tuning phase, all processors are expected to have a load of approximately UBE. A fine-tuning load-balancing phase is then employed. It first attempts to move tasks from the most loaded machine to some under-loaded machine, so that the total makespan decreases maximally. Tasks in the most loaded machine are ordered by the penalty for moving the task to the worst machine. Task moves are attempted in order of increasing penalty, i.e. those tasks suffering the lowest penalty from moving are moved first. Note that the order of task selection for moves is exactly the opposite to that used in the Coarse Tuning phase - during Fine Tuning, the lowest penalty tasks are moved first, while the highest penalty tasks are first moved during Coarse Tuning. This difference is a consequence of the fact that with Coarse Tuning, due to the knowledge of UBE, all tasks to be moved are known a priori. In contrast, during Fine Tuning, we do not know how many tasks can be moved with decrease in makespan. When no more task moves from the most-loaded machine are possible that reduce makespan, pair-wise swaps are attempted between the most-loaded machine and any other machine. The reader is referred to [11] for details.

5   Performance Evaluation

In this section, we compare the four existing heuristics discussed earlier - Min-Min, Max-Min, Fast-Greedy and UDA, with the enhanced UDA-based schemes. The UDA-CFT version incorporates two stages of enhancements: coarse tuning and fine tuning. We also studied the effectiveness of both these enhancements independently.

Simulation experiments were performed for different heterogeneity parameters for tasks and machines, as well as for varying number of tasks per processor. It was observed that the relative effectiveness of the heuristics did not change much with respect to machine and task heterogeneity. Therefore we only present results for a single value of task heterogeneity (100) and machine heterogeneity (10). Fig. 1 shows results for the seven schemes: Min-Min, Max-Min, Fast-Greedy, UDA, UDA-CFT, UDA-Coarse, and UDA-Fine. The data displayed includes the efficacy, utilization, and makespan metrics for the heuristics. The metrics are shown for different number of tasks being mapped to 8 machines.

Let us first consider the data for the 16 tasks case. The graph displaying E vs. U shows that UDA has an efficacy of 1.0, but a utilization of only 0.42. The application of the Coarse Tuning and Fine Tuning (UDA-CFT) to UDA results in considerable improvement to the utilization (to almost 0.7) with a slight loss of efficacy (to a little under 0.9). The $E \times U$ product rises to around 0.6 from 0.42, with a corresponding decrease in makespan. The performance difference between the UDA-FT and UDA-CFT schemes is very small. This is because UDA produces a mapping with makespan close to the makespan of Fast-Greedy, i.e. there is very little scope for tasks to be moved from over-loaded to under-loaded machines. Like UDA, Min-Min also displays high efficacy (around 0.96) but low utilization (around 0.52). With Max-Min, we have higher inherent utilization (0.8) but lower efficacy (0.72).

The trends are very similar for the other three cases shown in Fig. 1. In all cases, the enhanced versions improve utilization at the price of a small decrease in efficacy. As the average number of tasks per processor increases, the utilization of the base schemes increases. As a consequence, the extent of improvement possible through the enhanced schemes diminishes. The reader is referred to [11] for results using different values of task/data heterogeneity and average number of tasks per processor, for consistent and inconsistent systems. The results for other cases are qualitatively similar to those seen in Fig. 1.

Table 3 presents data on the execution time for the various heuristics, i.e. the time needed to execute the heuristics to generate the mappings. The execution times were obtained by executing the heuristics on a Sun Ultra 4 Sparc machine. The number of tasks was varied from 256 to 2048, with the number of processors of the simulated system being 32. The execution time for the UDA-CFT heuristic can be seen to be about two orders of magnitude lower than the Min-Min or Max-Min heuristics.

Table 3. Execution times (seconds) for mapping heuristics

Heuristic	Number of tasks (32 machines)			
	256	512	1024	2048
Greedy	<0.01	<0.01	0.01	0.01
UDA	<0.01	<0.01	<0.01	0.01
UDA-Coarse	<0.01	0.01	0.01	0.03
UDA-CFT	<0.01	0.01	0.02	0.05
UDA-Fine	<0.01	<0.01	0.01	0.02
MinMin	0.19	1.00	3.16	12.46
MaxMin	0.20	1.05	3.33	13.07

6   Conclusions

The aim of this work was to obtain insights into different static mapping heuristics for heterogeneous environments by use of two metrics – efficacy and utilization. These metrics provide a clear and consistent characterization of several previously proposed mapping heuristics. The insights obtained from the characterization can be useful in improving existing mapping schemes and in developing new strategies. This was demonstrated by developing an enhancement to the UDA mapping scheme. Starting with a mapping that had maximum efficacy, a two-phase tuning heuristic was proposed to improve the utilization of the mapping. The enhanced scheme produces mappings that are comparable or better than Min-Min (reported to be the best heuristic in previous studies), with the total execution time for the enhanced UDA scheme being over an order of magnitude lower than Min-Min. We believe that these metrics will also be useful in understanding and characterizing task mapping strategies in the more complex context of online dynamic scheduling.

References

1. A.H. Alhusaini, V. K. Prasanna, and C. S. Raghavendra. A Unified Resource Scheduling Framework for Heterogeneous Computing Environments, 8th Heterogeneous Computing Workshop (HCW '99), Apr. 1999
2. R. Armstrong, D. Hensgen, and T. Kidd. The Relative Performance of Various Mapping Algorithms is Independent of Sizable Variances in Run-time Predictions, 7th IEEE Heterogeneous Computing Workshop (HCW'98), Mar. 1998, pp.79-87
3. B. Armstrong, R. Eigenmann. Performance Forecasting: Towards a Methodology for Characterizing Large Computational Applications. Proceedings of the 1998 International Conference on Parallel Processing, Aug. 1998, pp. 518-527
4. F. Berman, R. Wolski, S. Figueira, J. Schopf, and G. Shao. Aplication-Level Scheduling on Distributed Heterogeneous Networks, Proceedings of Supercomputing 1996

5. F. Berman and R. Wolski. Scheduling from the Perspective of the Application, from Proceedings of Symposium on High Performance Distributed Computing, 1996

6. T. D. Braun, H. J. Siegel, N. Beck, L. L. Bölöni, M. Maheswaran, A. I. Reuther, J. R. Robertson, M. D. Theys, Bin Yao, D. Hensgen, and R. F. Freund. A Comparison Study of Static Mapping Heuristics for a Class of Meta-tasks on Heterogeneous Computing Systems, 8th IEEE Heterogeneous Computing Workshop (HCW'99), Apr. 1999, pp.15-29

7. T.D. Braun, H.J. Siegel, N.Beck, L.L. Bölöni, M. Maheswaran, A.I. Reuther, J.P. Robertson, M.D. Theys, and B. Yao. A Taxonomy for Describing Matching and Scheduling Heuristics for Mixed-machine Heterogeneous Computing Systems, IEEE Workshop on Advances in Parallel and Distributed Systems, Oct. 1998, pp. 330-335

8. F. Chang, V. Karamcheti, Z. Kedem. Exploiting Application Tunability for Efficient, Predictable Parallel Resource Management. Proceedings of the 13th International Parallel Processing Symposium / Symposium 10th Symposium on Parallel and Distributed Processing, Apr. 1999, pp.749-758

9. D. Feitelson, A. Weil. Utilization and Predictability in Scheduling the IBM SP/2 with Backfilling. Proceedings of the 12th International Parallel Processing / Symposium 9th Symposium on Parallel and Distributed Processing, Apr. 1998, pp. 542-548

10. I. Foster and C. Kesselman. The Grid: Blueprint for a New Computing Infrastructure, Morgan Kaufmann Publishers, 1998

11. P. Holenarsipur, V. Yarmolenko, J. Duato, D.K. Panda, and P. Sadayappan, "Characterization and Enhancement of Static Mapping Heuristics for Heterogeneous Systems," Technical report OSU-CISRC-2/00-TR07, Department of Computer and Information Science, Ohio State University, 2000

12. O. H. Ibarra and C. E. Kim. Heuristic Algorithms for Scheduling Independent Tasks on Nonidentical Processors. Journal of the ACM, Vol.24, No.1, Jan. 1977, pp. 280-289

13. M. Kafil and I. Ahmad. Optimal Task Assignment in Heterogeneous Computing Systems. 6th IEEE Heterogeneous Computing Workshop (HCW'97), Apr. 1997, pp. 135-146

14. W. Leinberger, G. Karypis, and V. Kumar. Multi-Capacity Bin Packing Algorithms with Applications to Job Scheduling under Multiple Constraints. Proceedings of the 1999 International Conference on Parallel Processing, Aug. 1999, pp. 404-413

15. M. Maheswaran, Shoukat Ali, H. J. Siegel, D. Hensgen, and R. F. Freund. Dynamic Matching and Scheduling of a Class of Independent Tasks onto Heterogeneous Computing Systems. 8th IEEE Heterogeneous Computing Workshop (HCW'99), Apr. 1999, pp. 30-44

16. A. Radulescu and A. van Gemund. FLB: Fast Load Balancing for Distributed-Memory Machines. Proceedings of the 1999 International Conference on Parallel Processing, Aug. 1999, pp. 534-542

17. J. Schopf, F. Berman. Performance Prediction in Production Environments. Proceedings of the 12th International Parallel Processing/Symposium 9th Symposium on Parallel and Distributed Processing, Apr. 1998, pp. 647-654

18. G. Shao, R. Wolski and F. Berman. Performance Effects of Scheduling Strategies for Master/Slave Distributed Applications. UCSD Technical Report No. CS98-598

19. H. Singh and A. Youssef, Mapping and Scheduling Heterogeneous Task Graphs using Genetic Algorithms. 5th IEEE Heterogeneous Computing Workshop (HCW'96), Apr. 1996

20. H. Topcuoglu, S. Hariri, and Min-You Wu. Task Scheduling Algorithms for Heterogeneous Processors. 8th IEEE Heterogeneous Computing Workshop (HCW'99), Apr. 1999, pp. 3-14

21. L. Wang, H. J. Siegel and V. P. Roychowdhury. A Genetic-Algorithm-Based Approach for Task Matching and Scheduling in Heterogeneous Computing Environments. 5th IEEE Heterogeneous Computing Workshop (HCW'96), Apr. 1996

22. L. A. Yan, J. K. Antonio. Estimating the Execution Time Distribution for a Task Graph in a Heterogeneous Computing System. 6th IEEE Heterogeneous Computing Workshop (HCW'97), Apr. 1997, pp. 172-184

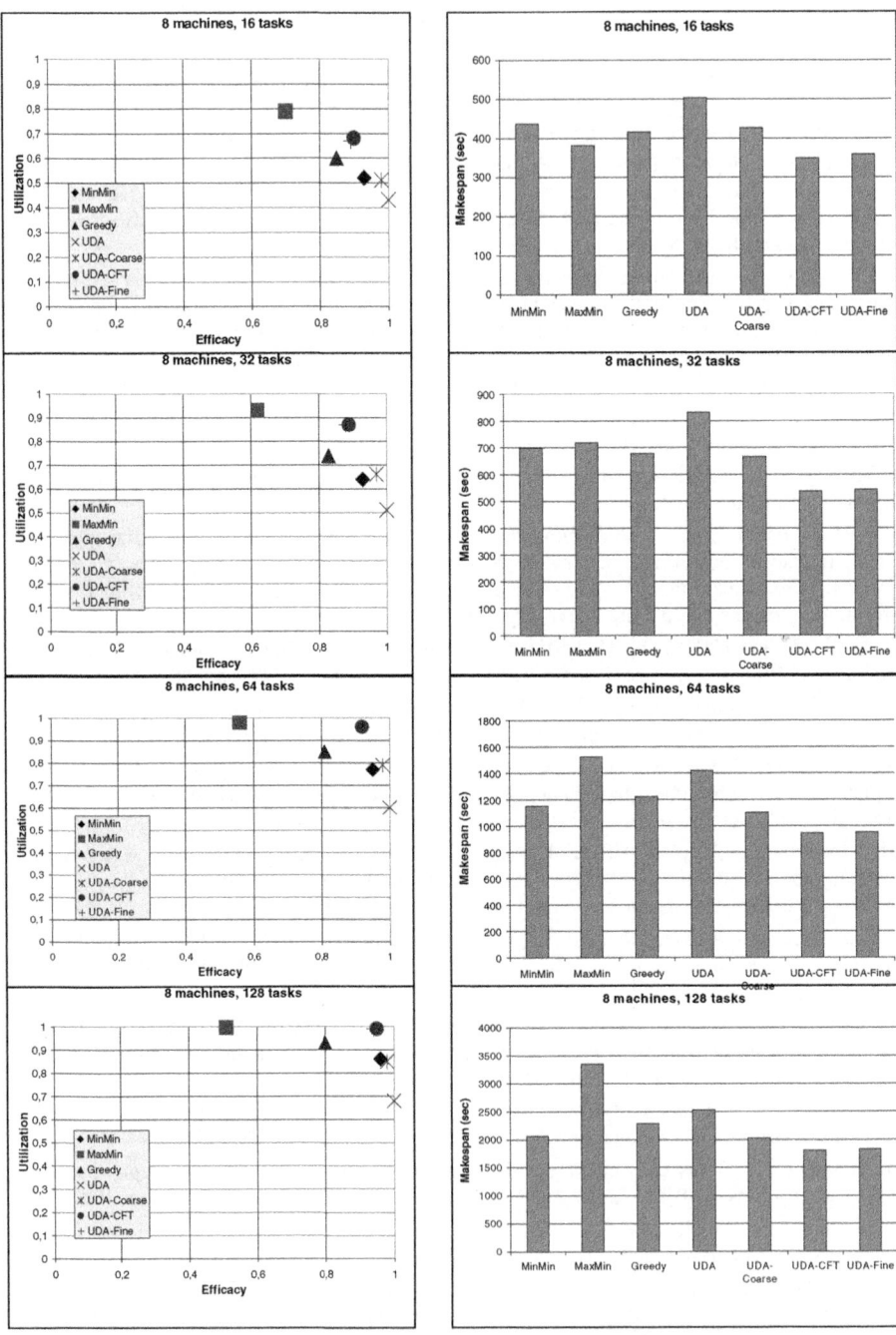

Fig. 1. Efficacy, utilization and makespan for mapping heuristics

Session I-B

Algorithms
Chair: Assaf Schuster
Israel Institute of Technology, Technion

Optimal Segmented Scan
and Simulation of Reconfigurable Architectures
on Fixed Connection Networks*

Alan A. Bertossi and Alessandro Mei

Department of Mathematics, University of Trento, Italy.
{bertossi,mei}@science.unitn.it

Abstract. Given n elements $x_0, \ldots, x_{n-1}$, and given n bits $b_0, \ldots, b_{n-1}$, with at least one zero, the segmented scan problem consists in finding the prefixes $s_i = x_i \otimes b_i s_{(i-1) \mod n}$, $i = 0, \ldots, n-1$, where $\otimes$ is an associative binary operation that can be computed in constant time by a processor. This paper presents:

(i) an $O(\log B)$ time optimal algorithm for the segmented scan problem on a $(2n-1)$-node toroidal X-tree, where B is the maximum distance of two successive zeroes in $b_0, \ldots, b_{n-1}$;

(ii) a novel definition of locally normal algorithms for trees and meshes of trees;

(iii) a constant slow-down, optimal, and locally normal simulation algorithm for a class of reconfigurable architectures on the mesh of toroidal X-trees, if the log-time delay model is assumed;

(iv) a constant slow-down optimal simulation of locally normal algorithms for meshes of toroidal X-trees on the hypercube.

1 Introduction

Given n elements $x_0, \ldots, x_{n-1}$, and an associative operation $\otimes$, a scan (also known as parallel prefix) operation consists in computing the prefixes $s_i = x_0 \otimes x_1 \otimes \cdots \otimes x_i$, $i = 0, \ldots, n-1$.

Scans are probably the most important parallel operations. They are so commonly used in computation that research has been done in order to endow sequential systems with scans as primitive parallel operations [2]. A lot of research has been done to find algorithms for this problem on most of the known parallel models of computation, from PRAMs to trees and hypercubes. For example, $O(\log n)$ time is needed to perform a scan operation on an $O(n/\log n)$-processor EREW PRAM, on an $O(n/\log n)$-node tree, and on an $O(n/\log n)$-node hypercube. All these results are clearly time and work optimal, within constant factors.

An important particular case of this problem arises when the scan operation is to be done separately on segments of the input vector $x_0, \ldots, x_{n-1}$. The

---

* This work has been supported by grants from the University of Trento and the Provincia Autonoma di Trento.

M. Valero, V.K. Prasanna, and S. Vajapeyam (Eds.): HiPC 2000, LNCS 1970, pp. 51–60, 2000.

segments are indicated by giving n bits $b_0, \ldots, b_{n-1}$, and the segmented scan problem thus consists in computing the prefixes:

$$s_0 = x_0$$
$$s_i = x_i \otimes b_i s_{(i-1)},$$

where $i = 1, \ldots, n-1$. Of course, this problem can still be solved in $O(\log n)$ time on the above parallel models, as this is just a particular scan operation. However, if there are enough b_is set to zero in the segmenting sequence, the dependencies of the output prefixes are so localized that sub-logarithmic algorithms could exist, and an $\Omega(\log n)$ time lower bound does not hold any more. Indeed, one expects the above operation to require just $O(\log B)$ time to be completed, where B is the size of the longest segment, which is clearly a lower bound on the time needed to perform a segmented scan on the above models. In this paper, it is shown that this is an upper bound too, for X-trees and hypercubes.

Strangely enough, this nice property of segmented scans has never been remarked in the literature. The main problem is probably that it is not clear what is the improvement. Indeed, $O(\log n)$ time is still needed to know B, and thus to predict the time needed by the operation. However, in some situations, it is possible to know in advance a sub-linear upper limit on B, and thus that the problem can be solved in sub-logarithmic time.

An important example of this situation arises when algorithms written for reconfigurable architectures are to be simulated on fixed-connection networks like trees and hypercubes. Indeed, the peculiar feature of a reconfigurable architecture of segmenting its buses in order to perform different broadcasting operations on different segments of buses, is a case in which, under proper assumptions, B can be exactly predicted. Exploiting the above result, it is thus possible to develop optimal simulation algorithms for a class of reconfigurable architectures.

This paper is organized as follows: Section 2 reviews the models of parallel computation used; Section 3 describes a time optimal algorithm for the segmented scan problem on the toroidal X-tree; as applications, Section 4 and Section 5 show how the mesh of toroidal X-trees an the hypercube optimally simulate a class of reconfigurable architectures.

2   Fixed-Connection and Reconfigurable Networks

A *fixed-connection network* [3] is a parallel model of computation consisting in a network of synchronous processors. The topology of a fixed-connection network is usually described by means of an undirected graph $\mathcal{G} = (V, E)$, where nodes stand for processors, and edges for links between processors.

Each processor of the network has a local control program and a local storage, and is allowed to perform a constant number of word operations. The size of a word is $O(\log n)$, where n is the size (i.e. number of nodes) of the network. We will also limit the complexity of the operations, by allowing only elementary arithmetic. The edges of the graph $\mathcal{G} = (V, E)$ describe the topology. If $uv \in E$, the processors represented by nodes u and v are connected by a physical link, which can be used to carry communications between the two processors.

Time is divided into *steps* by a *global clock*. At each step, each processor:

(1) reads messages coming from its neighbors,
(2) performs the computation described by its local control,
(3) sends a constant number of words to its neighbors.

A *reconfigurable network* is a network of processors operating synchronously. Differently from a fixed-connection network, each node can dynamically connect and disconnect its edges in various patterns by configuring a local switch.

Each switch has a number of I/O ports and each port is directly connected to at most one edge. While the edges outside the switch are fixed, the internal connections between the I/O ports of each switch can be locally configured in $O(1)$ time by the processor itself into any partition of the ports. In this way, during the execution of an algorithm, the edges of the network are dynamically partitioned into edge-disjoint *subgraphs*. Every such subgraph forms a *sub-bus*, and allows the processors of the sub-bus to broadcast a message to all the other processors sharing the same sub-bus.

Several variants of the model described above have been defined, depending on the kind of allowed partitions and of the sub-bus arbitration.

- *Linear reconfigurable network (LRN)*. Only partitions made of pairs and singletons are allowed (see Figure 1(a)). In this way, each sub-bus has the form of either a path or a cycle.
- *General reconfigurable network (RN)*. Any partition of the ports is allowed (see also Figure 1(b)). Thus, the possible configurations are any graph partition in edge-disjoint connected subgraphs.

Moreover, two main models are considered with respect to the kind of sub-bus arbitration: the *exclusive-write* model, where each port of the network must be reached by at most one message at any step of computation; and the *common-write* model, where more messages are allowed to reach the same port during the same step, provided that all of them contain the same value which in turn consists in a single bit only.

Finally, two models have been considered in the literature for delay along buses. According to a so called *unit-time delay* model [6], the time required by a broadcast operation to be completed is constant, regardless of the length of the sub-bus. Also considered has been a *log-time delay* model [6], which assume that the time required by a broadcast operation is logarithmic on the length (i.e. number of switches traversed) of the sub-bus.

A *reconfigurable mesh* is a reconfigurable network whose topology is a 2-dimensional grid. An important variant of this model is the *basic reconfigurable mesh*, where the connection patterns are limited in such a way that sub-buses can run only along rows and columns, thus no bent is allowed. More formally, the only patterns allowed are those shown in Figure 2. Clearly, all the variants

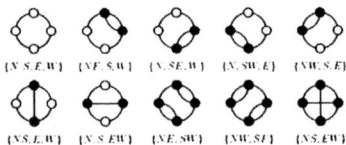

(a) LRN Model

(b) additional configura-
tions allowed in the RN
model

Fig. 1. All the configurations allowed on a four port switch.

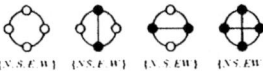

Fig. 2. All the configurations allowed on a basic reconfigurable mesh.

of the reconfigurable network model apply to the (basic) reconfigurable mesh, too.

In this paper, the log-time delay basic reconfigurable mesh is assumed. Moreover, we will endow the grid with toroidal links, which connect the first and last node on each row and each column (see Figure 3). This is useful in order to consider other well-known reconfigurable architectures as particular cases of the toroidal basic reconfigurable mesh. Namely, we refer to the HV reconfigurable mesh [1], the mesh with separable buses [5], and the polymorphic-torus network [4]. All the results for the basic reconfigurable mesh presented in this paper directly extend to all of them.

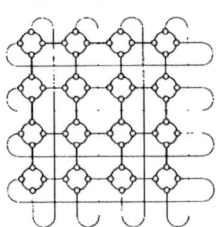

Fig. 3. A 4×4 basic reconfigurable mesh with toroidal links.

3   Optimal Segmented Scan Operations on Toroidal X-Trees

Let $\mathbb{A}$ be a set whose elements can be coded in a word, and let $\otimes$ be an associative binary operation over $\mathbb{A}$, such that $\otimes$ can be computed in $O(1)$ time by a processor.

Given an n-uple, $n = 2^k$, of elements $x_0, \ldots, x_{n-1} \in \mathbb{A}$, and given an n-uple of bits $b_0, \ldots, b_{n-1}$, with at least one zero, the *segmented scan* problem consists in finding the prefixes

$$s_i = x_i \otimes b_i s_{(i-1) \mod n},$$

where $i = 0, \ldots, n - 1$.

The bits $b_0, \ldots, b_{n-1}$ define a partition of $x_0, \ldots, x_{n-1}$ in the following way. Let $h > 0$ be the number of zeroes in $b_0, \ldots, b_{n-1}$, and let i_j, $j = 1, \ldots, h$, be the index of the j-th zero in the sequence. Naturally, a partition of the indices $0, \ldots, n-1$ is given by $B_1, \ldots, B_h$, where $B_j = \{i_j, i_j + 1, \ldots, i_{j+1} - 1\}$, $j = 1, \ldots, h-1$, and $B_h = \{i_h, i_h+1, \ldots, 0, 1, \ldots, i_1 - 1\}$. Clearly, the segmented scan problem consists in computing a scan operation independently inside each set of the partition.

This problem has a straightforward lower bound of $\Omega(\log \max_j |B_j|)$ time on a fixed-connection network. Unfortunately, the natural solution of letting a classical complete binary tree solve the problem, by providing $x_0, \ldots, x_{n-1}$ to its leaves, is not optimal. Indeed, assuming $n \geq 4$ to be a multiple of 4, if the partition is such that $B_j = \{2j - 1, 2j \mod n\}$, $j = 1, \ldots, n/2$, then $B_{n/4}$ has one element in the left subtree of the root, and the other element in the right one. The two leaves are thus $\Omega(\log n)$ far, and this is a lower bound on the time needed by the tree to solve the problem. In this case $\log \max_j |B_j| = 1$, hence, a binary tree cannot match the $\Omega(\log \max_j |B_j|)$ time lower bound.

In order to optimally solve this problem, a slightly more complex architecture is needed.

3.1   The Toroidal X-Tree

A toroidal X-tree (TX-tree) $T = (V, E)$ of size $m = 2^{k+1} - 1$ is a fixed-connection network composed by m nodes $v_1, \ldots, v_m \in V$, connected by bidirectional links in a tree-like fashion. The nodes are partitioned into $k + 1$ levels. The l-th level, $l = 0, \ldots, k$, is formed by nodes $v_{2^l}, \ldots, v_{2^{l+1}-1}$. The nodes on level l, except level 0, are connected in such a way that v_{2^l+i} has a bidirectional link to v_{2^l+i+1}, $i = 0, \ldots, 2^l - 2$, and $v_{2^{l+1}-1}$ to v_{2^l}. Moreover, each node v_i, $i = 2, \ldots, 2^l - 1$, is linked to its parent $v_{\lfloor i/2 \rfloor}$ in level $l - 1$ (see Figure 4). Intuitively, T is a complete binary tree of m nodes which has been extended by adding links in such a way to form a ring on each level of the tree.

For the sake of clarity, it is useful to define a new operation $\oplus 1$ over the set $\{1, \ldots, m\}$ of indices of the nodes of a toroidal X-tree. This operation will be

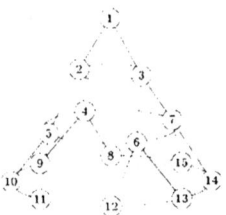

Fig. 4. A toroidal X-tree of 15 nodes.

used to find the index of the "next" node in the ring of nodes at any level of the TX-tree. More formally,

$$i \oplus 1 = \begin{cases} 2^l & \text{if } x = 2^{l+1} - 1, \text{ for some } l > 0, \\ i+1 & \text{otherwise.} \end{cases}$$

Similarly, a $\ominus 1$ operation can be defined which produces the index of the "previous" node in the ring. We also use a $\oplus c$ ($\ominus c$) operation, where c is a positive constant, obtained by iterating $\oplus 1$ ($\ominus 1$) c times.

3.2 An Optimal Algorithm for the Segmented Scan Problem

We start by giving a "weak" definition of the segmented scan problem, which will be useful in the second part of this paper.

Definition 1. *Given an n-uple, $n = 2^k$, of elements $x_0, \ldots, x_{n-1} \in \mathbb{A}$, and given h segments $B_j = \{\alpha_j, (\alpha_j + 1) \mod n, \ldots, \omega_j\} \subset \{0, n-1\}$ such that:*

(1) $\alpha_j < \alpha_{j+1}$, for all $j \in \{1, \ldots, h-1\}$;
(2) $|B_j \cap B_{j+1}| \leq 1$, for all $j \in \{1, \ldots, h-1\}$; and $|B_h \cap B_1| \leq 1$;

the overlapping segmented scan problem consists in computing a scan operation within each segment B_j, $j = 1, \ldots, h$.

It is easy to see that the segmented scan problem is a particular case of the overlapping version, where the segments B_j, $j = 1, \ldots, h$, form a partition of $\{0, \ldots, n-1\}$.

A toroidal X-tree $T = (V, E)$ of size $m = 2n - 1$ has $n = 2^k$ leaves and can be used to optimally solve an overlapping segmented scan problem of size n. Assume that $x_0, \ldots, x_{n-1}$ are stored in the leaves of the TX-tree in such a way that the node of index $i + 2^k$ stores x_i, $i = 0, \ldots, n-1$. Also assume that it stores a two bit register $c_i = c_{i,1} c_{i,2}$ such that $c_{i,1}$ is equal to zero if and only if there exists j such that $\alpha_j = i$, and $c_{i,2}$ is equal to zero if and only if there exists j such that $\omega_j = i$. Note that a segmented scan problem with parameters $b_0, \ldots, b_{n-1}$ can be converted into an overlapping one by setting $c_i = b_i b_{(i+1) \mod n}$.

For each segment B_j, $j = 1, \ldots, h$, define a sub-TX-tree $ST_j = (V_j, E_j)$ in the following way. $I_0^{(j)}$ is the set of indices of the nodes storing an integer x_i such that $i \in B_j$. Recursively define $I_{r+1}^{(j)}$, $r \geq 0$, to be the set of indices of the nodes having both sons in $I_r^{(j)}$. Let r_j be the minimum index such that $|I_{r_j}^{(j)}| \leq 2$. Then, let V_j, the set of nodes of ST_j, be equal to $I_0^{(j)} \cup \cdots \cup I_{r_j}^{(j)}$, and let E_j contain all the edges in E connecting two nodes of V_j (see Figure 5 for example). It is useful to state a few simple facts on the structure of ST_j.

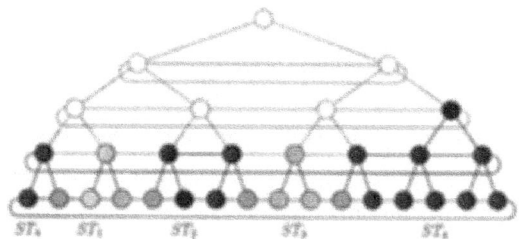

Fig. 5. In this example, $B_1 = \{1, \ldots, 4\}$, $B_2 = \{4, \ldots, 7\}$, $B_3 = \{7, \ldots, 10\}$, $B_4 = \{10, \ldots, 15, 0, 1\}$. Nodes in ST_2 and ST_4 are dark, while nodes ST_1 and ST_3 are light. Edges and nodes which do not belong to any ST_j are dashed.

Fact 1 *For all $j = 1, \ldots, h$, ST_j is connected.*

Fact 2 *For all $j = 1, \ldots, h$, r_j is $O(\log |B_j|)$.*

Fact 3 *For all $j = 1, \ldots, h$, and for all nodes $v_i \in ST_j \setminus I_{r_j}^{(j)}$, at least one of $v_{\lfloor i/2 \rfloor}$, $v_{\lfloor (i \ominus 1)/2 \rfloor}$, and $v_{\lfloor (i \oplus 1)/2 \rfloor}$, is in ST_j;*

ST_j is well-suited to compute the prefix operation inside B_j. Indeed, the topology of ST_j is close to the topology of a tree, and its low diameter allows to achieve optimal time.

Lemma 1. *For all $j = 1, \ldots, h$, the diameter of ST_j is $O(\log |B_j|)$.*

At this point, we can show how a sub-TX-tree ST_j can solve the problem in the segment B_j. Initially, each leaf v_{i+2^k}, $i = 0, \ldots, 2^k - 1$, stores x_i in a register y_{i+2^k}, and the two bits c_i in a two bit register d_{i+2^k}.

Lemma 2. *The scan operation on the values x_i, $i \in B_j$, can be performed in $O(\log |B_j|)$ time on the sub-TX-tree ST_j.*

Theorem 4. *A toroidal X-tree of size $2n - 1$ can time-optimally solve the overlapping segmented scan problem in $\Theta(\max_j \log |B_j|)$ time.*

Corollary 1. *A toroidal X-tree of size $2n - 1$ can time-optimally solve the segmented scan problem in $\Theta(\max_j \log |B_j|)$ time.*

4   The Mesh of Toroidal X-Trees

An $r \times c$ *mesh of toroidal X-trees*, with r and c powers of two, is a fixed-connection network composed by r row and c column TX-trees of size $2c - 1$ and $2r - 1$, respectively, sharing their leaves. Specifically, the leaf v_{j+c}, $j = 0, \dots, c - 1$, of the i-th row TX-tree, $i = 0, \dots, r - 1$, is also the leaf v_{i+r} of the j-th column TX-tree. Let this node be denoted by $P_{i,j}$. By construction, the nodes $P_{i,j}$, $i = 0, \dots, r - 1$ and $j = 0, \dots, c - 1$, form a grid of processors. As it can be easily seen, an $r \times c$ mesh of toroidal X-trees is very similar to an $r \times c$ mesh of trees, with toroidal X-trees in the place of classical trees.

4.1   Optimal Simulation of Basic Reconfigurable Meshes

With a few adaptations, the broadcast operation along each sub-bus of a basic reconfigurable mesh becomes a segmented scan operation. Let $\rightsquigarrow$, the associative operation used, be defined as follows:

$$
x \rightsquigarrow y = \begin{cases} x & \text{if } y = \bot, \\ y & \text{if } x = \bot \text{ and } y \neq \bot, \\ error & \text{otherwise}, \end{cases}
$$

where x and y are two messages sent by two processors, $\bot$ is a symbol indicating that the processor is not transmitting any message, and *error* is a symbol for detecting multiple writes on the bus.

Fact 5 $\rightsquigarrow$ *is associative.*

Consider a row of an $r \times c$ basic reconfigurable mesh, with $r = 2^{k_1}$ and $c = 2^{k_2}$. At the first step of the computation, each processor opens or closes the switch on the row bus. Let b_i be either zero, if the switch of the i-th processor is open, or one, if it is closed. A broadcast operation along the sub-buses created can be simulated by executing two instances of an overlapping segmented scan with $x_0, \dots, x_{n-1}$ being the messages, and $c_i = b_i b_i$, $i = 0, \dots, n - 1$, the bits defining the segments $B_1, \dots, B_h$. This can be done in the following way. First compute the prefixes $s_0^1, \dots, s_{n-1}^1$ resulting from the computation of $x_0 \rightsquigarrow x_1 \rightsquigarrow \cdots \rightsquigarrow x_{n-1}$. Then, compute the prefixes $s_0^2, \dots, s_{n-1}^2$ resulting from the computation of $x_{n-1} \rightsquigarrow x_{n-2} \rightsquigarrow \cdots \rightsquigarrow x_0$. Clearly, both results can be obtained by Theorem 4, the latter being a specular version of the former. Finally, the result of the broadcast operation is easily computed by the equation:

$$
s_i = \begin{cases} \bot & \text{if } s_i^1 = \bot \text{ and } s_i^2 = \bot, \\ s_i^1 & \text{if } s_i^1 \neq \bot \text{ and } s_i^2 = \bot, \\ s_i^2 & \text{if } s_i^1 = \bot \text{ and } s_i^2 \neq \bot, \\ error & \text{otherwise}. \end{cases}
$$

Theorem 6. *Given an algorithm $\mathcal{A}$ for an $r \times c$ basic reconfigurable mesh, $r = 2^{k_1}$ and $c = 2^{k_2}$, an $r \times c$ mesh of toroidal X-trees can optimally simulate $\mathcal{A}$ with constant slow-down, using the log-time delay model.*

5   The Hypercube

A *k-dimensional hypercube* is a fixed-connection network consisting of $n = 2^k$ nodes. Each node has a unique index in $\{0, \dots, n-1\}$, and two nodes are linked with an edge if and only if the binary representations of their indices differ in precisely one bit.

5.1   Optimal Simulation of Basic Reconfigurable Meshes

The problem of simulating a basic reconfigurable mesh on a hypercube is solved by giving a simulation scheme for meshes of TX-trees on hypercubes. Indeed, combining this result with Theorem 6, the simulation of basic reconfigurable meshes on hypercubes is obtained as a byproduct.

It is well-known that a k-dimensional hypercube easily simulates with constant slowdown normal algorithms for a $(2^{k+1} - 1)$-node complete binary tree [3]. However, the problems here are that the simulation algorithm of Theorem 6 is not normal, and that a toroidal X-tree has more wires than a simple tree. While the latter is solved as a simple application of Gray codes, the former involves a weaker and novel definition of normality.

Definition 2 ([3]). *An algorithm $\mathcal{A}$ for a tree is said to be* normal *if:*
(a) it uses only nodes of one level at each step of computation, and
(b) consecutive levels are used in consecutive steps.

As said above, the simulation algorithm of Theorem 6 is not normal, because nodes at different levels are used in different parts of the TX-tree, but it is intuitively *almost* normal. The following definition formalizes this intuition.

Definition 3. *An algorithm $\mathcal{A}$ for a tree is said to be* locally normal *if for each root-leaf path:*
(a) it uses only one node at each step of computation, and
(b) consecutive nodes are used in consecutive steps.

The above definition directly applies to a number of tree-based architectures like X-trees, toroidal X-trees, meshes of trees, and meshes of toroidal X-trees. It is immediate to see that a normal algorithm is also locally normal, while the opposite is not true in general. In particular, this is the case of the simulation algorithm of Theorem 6.

Lemma 3. *The simulation algorithm of Theorem 6 is locally normal.*

By using the reflected Gray code, it is possible to find a mapping from the nodes of a TX-tree into a hypercube which permits an optimal simulation of locally normal algorithms.

Let $G_s^{(w)}$, $w \geq 1$, $s = 0, \dots, 2^w - 1$, be the s-th element of the w-bit binary reflected Gray code. We map the i-th node of the l-th level of the $(2^{k+1} - 1)$-node TX-tree into the node of index $G_i^{(l)} 0^{k-l}$ of the 2^k node hypercube, where 0^x stands for a sequence of x zeroes. Note that, in this way, adjacent nodes of the TX-tree are mapped to either the same node or adjacent nodes of the hypercube. Indeed, $G_i^{(l)}0^{k-l}$ differs in one bit only from both $G_{i+1 \bmod 2^l}^{(l)}0^{k-l}$ and $G_{i-1 \bmod 2^l}^{(l)}0^{k-l}$, and, if $l > 0$, $G_i^{(l)}0^{k-l}$ differs in at most one bit from $G_{\lfloor i/2 \rfloor}^{(l-1)}0^{k-l+1}$.

Theorem 7. *Given a locally normal algorithm $\mathcal{A}$ for a toroidal X-tree of size $m = 2^{k+1} - 1$, a k-dimensional hypercube can simulate $\mathcal{A}$ with constant slow-down.*

Consequently, an n-node hypercube can also solve the segmented scan problem in $\Theta(\max_j \log |B_j|)$ time.

In order to extend the previous result to simulate an $r \times c$ mesh of toroidal X-trees, with $r = 2^{k_1}$ and $c = 2^{k_2}$, map the i-th node of the l-th level of the row r', $r' = 0, \dots, r-1$, into the node of index $G_{r'}^{(k_1)}G_i^{(l)}0^{k_2-l}$ of the $2^{(k_1+k_2)}$ node hypercube. Similarly, map the i-th node of the l-th level of the column c', $c' = 0, \dots, c-1$, into the node of index $G_i^{(l)}0^{k_1-l}G_{c'}^{(k_2)}$ of the hypercube. Note that, by using this mapping, a single node of the hypercube may have to simulate the action of two nodes of the mesh in the same step, even if the algorithm is locally normal. Of course, this fact causes a slow-down of 2, at most, which is still constant.

Corollary 2. *Given a locally normal algorithm $\mathcal{A}$ for an $r \times c$ mesh of toroidal X-trees, with $r = 2^{k_1}$ and $c = 2^{k_2}$, a (k_1+k_2)-dimensional hypercube can simulate $\mathcal{A}$ with constant slow-down.*

Finally, by combining Corollary 2 and Theorem 6, the following theorem is also proved.

Theorem 8. *Given an algorithm $\mathcal{A}$ for an $r \times c$ basic reconfigurable mesh, with $r = 2^{k_1}$ and $c = 2^{k_2}$, a $(k_1 + k_2)$-dimensional hypercube can optimally simulate $\mathcal{A}$ with constant slow-down, using the log-time delay model.*

References

1. Y. BEN-ASHER, D. GORDON, AND A. SCHUSTER, *Efficient self-simulation algorithms for reconfigurable arrays*, Journal of Parallel and Distributed Computing, 30 (1995), pp. 1–22.

2. G. E. BLELLOCH, *Scans as primitive parallel operations*, IEEE Transactions on Computers, 38 (1989), pp. 1526–1538.

3. F. T. LEIGHTON, *Introduction to parallel algorithms and architectures: arrays, trees, hypercubes*, Morgan Kaufmann, San Mateo, CA, 1992.

4. M. MARESCA, *Polymorphic-torus network*, IEEE Transactions on Computers, 38 (1989), pp. 1345–1351.

5. S. MATSUMAEAND AND N. TOKURA, *Simulating a mesh with separable buses*, Transactions of the Information Processing Society of Japan, 40 (1999), pp. 3706–3714.

6. R. MILLER, V. K. PRASANNA, D. I. REISIS, AND Q. F. STOUT, *Parallel computations on reconfigurable meshes*, IEEE Transactions on Computers, 42 (1993), pp. 678–692.

Reducing False Causality
in Causal Message Ordering

Pranav Gambhire and Ajay D. Kshemkalyani

Dept. of EECS, University of Illinois at Chicago, Chicago, IL 60607-7053, USA
{pgambhir, ajayk}@eecs.uic.edu

Abstract. A significant shortcoming of causal message ordering systems is their inefficiency because of false causality. False causality is the result of the inability of the "happens before" relation to model true causal relationships among events. The inefficiency of causal message ordering algorithms takes the form of additional delays in message delivery and requirements for large message buffers. This paper gives a lightweight causal message ordering algorithm based on a modified "happens before" relation. This lightweight algorithm greatly reduces the inefficiencies that traditional causal message ordering algorithms suffer from, by reducing the problem of false causality.

1 Introduction

In a distributed system, causal message ordering is valuable to the application programmer because it reduces the complexity of application logic and retains much of the concurrency of a FIFO communication system. Causal message ordering is defined using the "happens before" relation, also known as the causality relation and denoted $\longrightarrow$, on the events in the system execution [11]. For two events $e1$ and $e2$, $e1 \longrightarrow e2$ iff one of the following conditions is true: (i) $e1$ and $e2$ occur on the same process and $e1$ occurs before $e2$, (ii) $e1$ is the emission of a message and $e2$ is the reception of that message, or (iii) there exists an event $e3$ such that $e1 \longrightarrow e3$ and $e3 \longrightarrow e2$.

Let $Send(M)$ denote the event of a process handing over the message M to the communication subsystem. Let $Deliver(M)$ denote the event of M being *delivered* to a process after it is been received by its local communication subsystem. The system respects *causal message ordering* (CO) [3] iff for any two messages M_1 and M_2 sent to the same destination, $(Send(M_1) \longrightarrow Send(M_2))$ $\Longrightarrow (Deliver(M_1) \longrightarrow Deliver(M_2))$. In Figure 1, causal message ordering is respected if message M_1 is delivered to process P_3 before message M_3.

When a message arrives out of order with respect to the above definition, a causal message ordering system buffers it and delivers it only after all the messages that should be seen before it in causal order, have arrived and have been delivered.

Causal message ordering is very useful in several areas such as managing replicated database updates, consistency enforcement in distributed shared memory,

M. Valero, V.K. Prasanna, and S. Vajapeyam (Eds.): HiPC 2000, LNCS 1970, pp. 61–72, 2000.

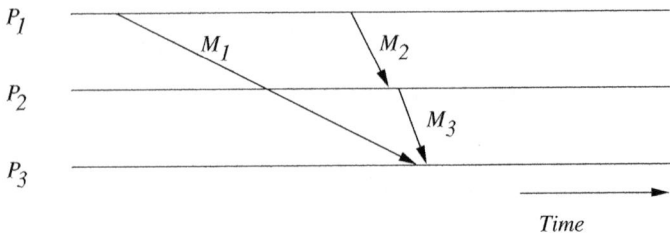

Fig. 1. Causal message ordering.

enforcing fair distributed mutual exclusion, efficient snapshot recording, global predicate evaluation, and data delivery in real-time multimedia systems. It has been implemented in systems such as Isis [3], Transis [1], Horus, Delta-4, Psync [12], and Amoeba [9]. The causal message ordering problem and various algorithms to provide such an ordering have been studied in several works such as [3,4,10,13,14,16] which also provide a survey of this area.

A causal message ordering abstraction and its implementation were given by Raynal, Schiper and Toueg (RST) [13]. For a system with n processes, the RST algorithm requires each process to maintain a $n \times n$ matrix - the $SENT$ matrix. $SENT[i,j]$ is the process's best knowledge of the number of messages sent by process P_i to process P_j. A process also maintains an array $DELIV$ of size n, where $DELIV[k]$ is the number of messages sent by process P_k that have already been delivered locally. Every message carries piggybacked on it, the $SENT$ matrix of the sender process. A process P_j that receives message M with the matrix SP piggybacked on it is delivered M only if, $\forall i$, $DELIV[i] \geq SENT[i,j]$. P_j then updates its local $SENT$ matrix $SENT_j$ as: $\forall i \forall j$, $SENT_j[k,l] = max(SENT_j[k,l], SP[k,l])$. Several optimizations that exploit topology, communication pattern, hardware broadcast, and underlying synchronous support are surveyed in [10] which also identifies and formulates the necessary and sufficient conditions on the information for causal message ordering and their optimal implementation.

Cheriton and Skeen [6] pointed out several drawbacks of the causal message ordering paradigm and these were further discussed in [2,5,15]. The most significant was that every implementation of a CO system has to deal with *false causality*. False causality is the insistence of the system to impose a particular causality ordering of events even though the application semantics do not require such an ordering.

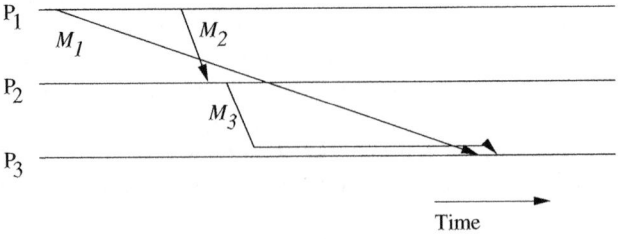

Fig. 2. False causality. No semantic causal dependence between $Send(M_1)$ and $Send(M_3)$.

In Figure 2, $Send(M_1) \longrightarrow Send(M_3)$. Assume that $Send(M_1)$ and $Send(M_3)$ are not causally related according to the application semantics. Now suppose M_3 reaches P_3 before M_1 does. In a causal message ordering system, it is required that $Deliver(M_1) \longrightarrow Deliver(M_3)$ and hence M_3 is buffered till M_1 is received and delivered. $Send(M_1)$ and $Send(M_3)$ are not semantically related and semantically P_3's behavior does not depend on the order in which it receives and is delivered M_1 and M_3. Hence, buffering of M_3 is really unnecessary. This buffering is wasteful of system resources and the withholding of message delivery unnecessarily delays the system execution.

This paper addresses the topic of reduction of false causality in causal message ordering systems. We propose an algorithm in which the incidence of false causality is much lower than in a conventional causal message ordering system that is based purely on the "happens before" relationship.

The notion of false causality arising from the "happens before" relation itself has been identified in several other contexts earlier, even before Cheriton and Skeen pointed out its drawbacks in the performance of causal message ordering algorithms. When Lamport defined the "happens before" relation $\longrightarrow$ [11], he had pointed out that "$e1 \longrightarrow e2$ means that it is *possible* for event $e1$ to causally affect event $e2$" but $e1$ and $e2$ need not necessarily have any semantic dependency. Fidge proposed a clock system to track true causality more accurately in a system with multithreaded processes [7]. However, this scheme and its variants are very expensive. Tarafdar and Garg pointed out the drawbacks of false causality in detecting predicates in distributed computations [17].

Section 2 gives the system model and a framework to design relations that have varying degrees of false causality. Section 3 presents a practical and easy to implement partial order relation based on the framework of Section 2, that reduces many of the false causal relationships modeled by the $\longrightarrow$ relation. Based on this new relation, the section then presents a lightweight algorithm that reduces false causality in causal message ordering. Section 4 concludes.

2   System Model

A distributed system is modeled as a finite set of n processes communicating with each other by asynchronous message passing over reliable logical channels. There is no shared memory and no common clock in the system. We assume that channels deliver messages in FIFO order. A process execution is modeled as a set of events, the time of occurrence of each of which is distinct. An event at a process could be a message send event, a message delivery event, or an internal event. A message can be multicast at a send event, in which case it is sent to multiple processes. A distributed computation or execution is the set of all events ordered by the "happens before" relation $\longrightarrow$ [11], also defined in Section 1. Define $e1 \stackrel{=}{\longrightarrow} e2$ as $(e1 \longrightarrow e2) \vee (e1 = e2)$. The jth event on process P_i is denoted as e_i^j. Each process P_i has a default initialization event e_i^0. The set of all events E forms a partial order $(E, \longrightarrow)$. The causal past of an event e is denoted $EC(e) = \{e' \mid e' \stackrel{=}{\longrightarrow} e\}$. The causal past of an event e_i projected on

P_j is denoted $EC_j(e_i) = \{e'_j \mid e'_j \xrightarrow{=} e_i\}$. For any event e_i^k, define a vector *event count* $ECV(e_i^k)$ of size n such that $ECV(e_i^k)[j] = |EC_j(e_i^k)| - 1$. $ECV(e_i^k)[j]$ gives the number of computation events at P_j in the causal past of e_i^k.

False causality is defined based on the true causality partial order relation $\xrightarrow{s}$ on events; the $\xrightarrow{s}$ relation is the analog of the $\longrightarrow$ relation, that accounts only for semantically required causality, and is defined similar to that in [17].

Definition 1. *Given two events* $e1$ *and* $e2$, $e1$ *semantically depends on* $e2$, *denoted as* $e1 \xrightarrow{s} e2$, *iff the action taken at* $e2$ *depends on the outcome of* $e1$.

We assume that a message delivery event is always semantically dependent on the corresponding message send event. Furthermore, if $e1$ and $e2$ are on different processes and $e1 \xrightarrow{s} e2$, then $\exists M \mid e1 \xrightarrow{s} Send(M) \xrightarrow{s} Delivery(M) \xrightarrow{s} e2$. With this interpretation, $\xrightarrow{s}$ is the transitive closure of a "local semantically depends on" relation and the "happens before" imposed by message send and corresponding delivery events.

By substituting $\xrightarrow{s}$ for $\longrightarrow$ in the traditional definition of causal message ordering, the resulting definition will be termed "semantic causal ordering". In contrast, the traditional causal message ordering will be termed the "happens before causal ordering". We will also refer to the causal message ordering problem as simply the causal ordering problem.

If the only ordering imposed on events is to respect the semantic causality relation defined above, then there is no performance degradation due to false causality. In other words, if messages M_1 and M_2 are sent to the same destination, then M_1 need not be delivered before M_2 if $Send(M_1) \xrightarrow{s}\!\!\!\!\!/ \;\; Send(M_2)$, but M_1 will be delivered before M_2 if $Send(M_1) \xrightarrow{s} Send(M_2)$. This model defines causality among events based on the application semantics. An implementation of this model requires that the semantic causal dependencies of each event be available. Existing programming language paradigms do not permit such specifications and neither does an API for such a specification seem practical. It is possible for a compiler to extract such information by analyzing data dependencies. Alternatively, analogous to Fidge [7] and Tarafdar-Garg [17], we can assume that this information is computable using techniques given in [7]. In fact, as we will show, the causal ordering algorithm we propose requires *only* that the following information is available: for any event, information about the most recent local event on which the event has a semantic dependency.

Definition 2. *Given two events* $e1$ *and* $e2$, $e1$ *weakly causes* $e2$, *denoted as* $e1 \xrightarrow{w} e2$, *iff* $e1 \xrightarrow{s}\!\!\!\!\!/ \;\; e2 \wedge e1 \longrightarrow e2$.

For a computation $(E, \longrightarrow)$ and a complete specification of the true causality $(E, \xrightarrow{s})$ in the computation, the amount of false causality is the size of the $\xrightarrow{w}$ relation, which is $\longrightarrow(E \times E) \setminus \xrightarrow{s}(E \times E)$.

Ideally, it is desired to implement the $\xrightarrow{s}$ relation, and have $\xrightarrow{w}$ be the empty relation on $E \times E$. The difficulties in having the programmer specify the

$\xrightarrow{s}$ relation are given above. Though a compiler or an alternate mechanism can identify the exact set of all local and nonlocal events that semantically precede the current event to implement true causality, the overhead of tracking such a set of events as the computation progresses is nontrivial. Therefore, our objective is to approximate $\xrightarrow{s}$ at a low cost to make an implementation practical, and minimize the size of the $\xrightarrow{w}$ relation on $E \times E$. To this end, we introduce the vector MCV to track the latest event at each process such that if the happens-before causal order for a message presently sent is enforced only with respect to messages sent in the computation prefix identified by such events, then semantic causal order is not violated. The vector MCV naturally identifies a computation prefix denoted MC. Thus, for a message sent at any event e, the vector $MCV(e)$ ensures that if happens-before causal ordering is guaranteed with respect to messages sent in $MC(e)$, then semantic causal ordering is guaranteed with respect to all messages sent in the causal past of the event e.

Definition 3. *For any event e, define vector $MCV(e)$ and set $MC(e)$ to have the following properties.*

- *The maximum causality vector $MCV(e)$ is a vector of length n with the following properties:*
  - *(Containment:) $\forall j$, $MCV(e)[j] \leq ECV(e)[j]$, and*
  - *(Semantic dependency satisfaction:) $e_j^l \xrightarrow{s} e_i^k \implies l \leq MCV(e_i^k)[j]$*
- *$MC(e) = \{e_j' \mid e_j' \longrightarrow e_j^{MCV(e)[j]}\}$, i.e., $MC(e)$ is the computation prefix such that the latest event of this prefix at each process P_j is $MCV(e)[j]$.*

We now make the following Proposition 1 which holds because there is no event that is not in $MCV(e)[j]$ that semantically precedes the event e.

Proposition 1. *For any event e, if every message sent by each P_j among its first $MCV(e)[j]$ events is delivered in happens-before causal order with respect to any messages sent at event e, then every message sent in $ECV(e)$ is delivered in semantic causal order with respect to any messages sent at event e.*

Observe that in general, there are multiple values of MCV that will satisfy Definition 3. Any formulation of MCV consistent with Definition 3 can be used by a causal ordering implementation to reduce false causality. Clearly, different formulations of MCV reduce the false causality to various degrees and can be implemented with varying degrees of ease. Two desirable properties of a good formulation of MCV are:

- It should eliminate as much of the false causality as possible.
- It should be implementable with low overhead.

Happens-before causal ordering of a message sent at event e needs to be enforced only with respect to messages sent in the computation prefix $MC(e)$. To implement this causal ordering, for each event e_i^k, process P_i needs to track the number of messages sent by each process P_j up to event $e_j^{MCV(e_i^k)[j]}$ to every other process. As observed by process P_i at e_i^k, the count of all such messages sent by each P_j up to $e_j^{MCV(e_i^k)[j]}$ to every other process P_l can be tracked by a matrix $SENT[1 \ldots n, 1 \ldots n]$, where $SENT[j, l]$ is the number of messages sent by P_j up to $e_j^{MCV(e_i^k)[j]}$ to P_l. Analogously, we track the count of all messages sent by each P_j up to $e_j^{ECV(e_i^k)[j]}$ to each other process P_l, by using the matrix $SENT\_ECV[1 \ldots n, 1 \ldots n]$.

For a traditional causal ordering system, $MCV(e) = ECV(e)$ and this implementation exhibits the negative effects caused by the maximum amount of false causality. Note that here $SENT = SENT\_ECV$ and the matrix $SENT$ as we have defined then degenerates to the matrix $SENT$ as defined in [13].

3   Algorithm

3.1   Preliminaries

We propose the following formulation of vector $MCV(e)$, that is easy to implement and gives a good lightweight solution to the false causality problem. The definition uses the *max* function on vectors, which gives the component-wise maximum of the vectors.

Definition 4. *1. Initially, $\forall i$, $MCV(e_i^0) = [0, \ldots, 0]$.*
2. For an internal event or a send event e_i^k,

$$MCV(e_i^k) = ECV(e_i^k) \quad \text{if } \exists\, e_q^p \mid MCV(e_i^{k-1})[q] < p \le ECV(e_i^k)[q] \;\wedge\; e_q^p \xrightarrow{s} e_i^k,$$
$$MCV(e_i^{k-1}) \; \text{otherwise}$$

3. For a delivery event e_i^k of a message sent at e_m^r,

$$MCV(e_i^k) = max(MCV(e_i^{k-1}), MCV(e_m^r))$$

Observe that the only way $e_q^p \xrightarrow{s} e_i^k$, where $q \ne i$, is if $\exists\, e_i^{k'}$, $e_q^{p'}$ such that $e_i^{k'} \xrightarrow{s} e_i^k$, $e_i^{k'} = Deliver(M)$, $e_q^{p'} = Send(M)$, and $e_q^p \xrightarrow{s} e_q^{p'}$. Also observe that the MCV of an event is the *max* of the ECVs of one or more events in its causal past, and therefore, analogous to EC, if $Deliver(M)$ belongs to the MC of some event, then $Send(M)$ also belongs to the MC of that event. Based on the above two observations, it is possible to simplify the condition test in Definition 4.2 as follows.

Definition 5. *1. Initially, $\forall i$, $MCV(e_i^0) = [0, \ldots, 0]$.*
2. For an internal event or a send event e_i^k,

$$MCV(e_i^k) = ECV(e_i^k) \quad \text{if } \exists\, e_i^p \mid MCV(e_i^{k-1})[i] < p \le k \;\wedge\; e_i^p \xrightarrow{s} e_i^k,$$
$$MCV(e_i^{k-1}) \quad \text{otherwise}$$

3. For a delivery event e_i^k of a message sent at e_m^r,

$$MCV(e_i^k) = max(MCV(e_i^{k-1}), MCV(e_m^r))$$

Informally, the above formulation of MCV identifies the following events at each process. *For any event e_i^k, (I) $MCV(e_i^k)[j]$, $j \neq i$, identifies the latest event e_j such that some event at P_i occurring causally after e_j and in $EC(e_i^k)$ depends semantically on e_j; (II) If e_i^k depends semantically on some event at P_i that occurs after $MCV(e_i^{k-1})[i]$, then $MCV(e_i^k)$ identifies e_i^k at P_i; otherwise it identifies $MCV(e_i^{k-1})[i]$ at P_i.* Causal ordering of a message sent at e_i^k needs to be enforced only with respect to messages sent in the computation prefix up to these identified events.

Lemma 1 shows that this formulation of MCV is an instantiation of Definition 3. Specifically, it states that each event that semantically happens before e_i^k belongs to the computation prefix up to the events indicated by $MCV(e_i^k)$.

Lemma 1. *Definition 5 satisfies the "Containment" property and the "Semantic Dependency Satisfaction" property of MCV described in Definition 3, i.e., $(\forall j, MCV(e)[j] \leq ECV(e)[j])$ and $e_j^l \xrightarrow{s} e_i^k \Rightarrow l \leq MCV(e_i^k)[j]$.*

The formulation of MCV in Definition 5 is easy to implement because we can observe the following property.

Property 1. From Definition 5, it follows that

1. At a send event e_i^k, determining $MCV(e_i^k)$ requires $MCV(e_i^{k-1})$ and identifying the most recent local event on which there is a semantic dependency of e_i^k. This information can be stored locally at a process.
2. At a delivery event e_i^k of a message sent at event e_m^r, determining $MCV(e_i^k)[j]$ requires $MCV(e_i^{k-1})[j]$ and $MCV(e_m^r)[j]$. $MCV(e_m^r)[j]$ can be piggybacked on the message sent at e_m^r.

The following property based on Definition 5 implies that the entries in the $SENT$ matrix at a process are monotonically nondecreasing in the computation.

Property 2. $MCV(e_i^k)[j] \geq MCV(e_i^{k-1})[j]$ i.e., $MCV(e)[j]$ is monotonically nondecreasing at any process.

3.2 Algorithm to Reduce False Causality

Definition 5 of MCV is realized by the algorithm presented in Figure 3. This algorithm is based on the causal ordering abstraction of Raynal, Schiper and Toueg [13] because it provides a convenient base to express the proposed ideas. The proposed ideas can be superimposed on more efficient causal ordering algorithms such as those proposed and surveyed in [10].

Recall that in [13], each send event and each delivery event updates the $SENT$ matrix to reflect the maximum available knowledge about the number of messages sent from each process to every other process in the causal past. This implies that any message M has to be delivered in causal order with respect to all the messages sent in the causal past of the event $Send(M)$, even though there may be no semantic dependency between M and these messages. This is a source of false causality which can be minimized by the lightweight algorithm proposed here.

Each process P_i maintains the data structures $DELIV_i$, $SENT\_CONC_i$ and $SENT\_PREV_i$, described below.

- $DELIV_i$: array[1..n] of integer.
  $DELIV_i[j]$ is P_i's knowledge of the number of messages from process P_j that have been delivered to P_i thus far. It is initialized to zeros.
- $SENT\_PREV_i$: array[1..n,1..n] of integer.
  $SENT\_PREV_i[j,l]$ at e_i^k is P_i's knowledge of the number of messages sent from P_j to P_l up to the event $e_j^{MCV(e_i^k)[j]}$. It is initialized to zeros.
- $SENT\_CONC_i$: array[1..n,1..n] of integer.
  $SENT\_CONC_i[j,l]$ at e_i^k is P_i's knowledge of the number of messages sent from P_j to P_l after the event $e_j^{MCV(e_i^k)[j]}$. It is initialized to zeros.

Recall that the $n \times n$ matrix $SENT\_ECV$ at e_i^k, defined in Section 2, gave the count of all messages sent by each P_j up to $e_j^{ECV(e_i^k)[j]}$ to each other process P_l, in $SENT\_ECV[j,l]$. The two matrices $SENT\_PREV$ and $SENT\_CONC$ at any process have the invariant property that $SENT\_PREV_i[j,l] + SENT\_CONC_i[j,l] = SENT\_ECV_i[j,l]$, as will be shown in Lemma 2. The row $SENT\_PREV_i[j,\cdot]$ reflects the row $SENT\_ECV_i[j,\cdot]$ up to the event $e_j^{MCV(e_i^k)[j]}$, whereas the row $SENT\_CONC_i[j,\cdot]$ reflects the row $SENT\_ECV_i[j,\cdot]$ after that event. The challenge is to maintain the $SENT\_PREV$ and $SENT\_CONC$ matrices at a low cost so as to retain the above property.

The causal ordering algorithm that minimizes false causality is given in Figure 3. At a message send event, steps (E1)-(E7) are executed atomically. Step (E1) determines based on the sender's semantics if the message being sent, and implicitly all future messages sent in the system, should be delivered in causal order with respect to all messages sent so far, i.e., whether $MCV(e_i^k) = ECV(e_i^k)$. Specifically, the test *should_Semantically_Precede* uses two inputs: $MCV(e_i^{k-1})$ and the latest local event on which there is a local dependency (Property 1). These two inputs are used to check for Definition 5.2, i.e., to determine whether there exists an event e_i^p such that $e_i^{MCV(e_i^{k-1})[i]} \longrightarrow e_i^p \wedge e_i^p \xrightarrow{s} e_i^k$, in which case $MCV(e_i^k)$ will be greater than $MCV(e_i^{k-1})$. If $MCV(e_i^k)$ is greater than $MCV(e_i^{k-1})$, then the condition *should_Semantically_Precede* becomes *true* and $MCV(e_i^k)$ should be set to $ECV(e_i^k)$. In this case, steps (E2)-(E5) update the matrices $SENT\_PREV$ and $SENT\_CONC$ to reflect this; otherwise $SENT\_PREV$ and $SENT\_CONC$ are left unchanged. Step (E6) sends the message with the two matrices piggybacked on it. Step (E7) updates $SENT\_CONC$ to reflect the message(s) just sent.

When a message M, along with the sender's $SENT\_PREV$ and $SENT\_CONC$ matrices SP and SC piggybacked on it, is received by a process, M can be delivered only if the number of messages delivered locally so far is greater than or equals the number of messages sent to this process as per SP (step (R1)). Steps (R2)-(R8) are executed atomically. The message gets deliv-

Data structures at P_i

D1. $DELIV_i$: array [1..n] of integer
D2. $SENT\_PREV_i$: array [1..n,1..n] of integer
D3. $SENT\_CONC_i$: array [1..n,1..n] of integer

Emission of message M from P_i to P_j

E1. if ($should\_Semantically\_Precede$) then
E2.   for $k = 1$ to n do
E3.   for $l = 1$ to n do
E4. $SENT\_PREV_i[k,l] = SENT\_PREV_i[k,l] + SENT\_CONC_i[k,l]$
E5. $SENT\_CONC_i[k,l] = 0$
E6. Send($M, SENT\_PREV_i, SENT\_CONC_i$)
E7. $SENT\_CONC_i[i,j] = SENT\_CONC_i[i,j] + 1$

Reception of (M, SP_M, SC_M) at P_j from P_i

R1. Wait until $(\forall k, SP_M[k,j] \leq DELIV_j[k])$
R2. Deliver M
R3. $SC_M[i,j] = SC_M[i,j] + 1$
R4. $DELIV_j[i] = DELIV_j[i] + 1$
R5. for $k = 1$ to n do
R6.   for $l = 1$ to n do
R7. $SENT\_CONC_j[k,l] = \max(SENT\_CONC_j[k,l] + SENT\_PREV_j[k,l],$
                $SP_M[k,l] + SC_M[k,l]) - \max(SENT\_PREV_j[k,l], SP_M[k,l])$
R8. $SENT\_PREV_j[k,l] = \max(SENT\_PREV_j[k,l], SP_M[k,l])$

Fig. 3. Causal message ordering algorithm to minimize false causality.

ered in step (R2). Steps (R3)-(R4) update the data structures SC and $DELIV$ to reflect that the message was sent and has now been delivered, respectively. Steps (R5)-(R8) update the data structures $SENT\_CONC$ and $SENT\_PREV$. These steps ensure that $SENT\_PREV$ reflects the maximum knowledge about the messages that were sent by events in the MC of the current event, while $SENT\_CONC$ reflects the maximum knowledge about the messages that were not sent by events in the MC of the current event.

3.3   Correctness Proof

We state some lemmas and the main theorem that prove the correctness of the algorithm. See the full paper for details [8].

Lemma 2 gives the invariant among $SENT\_PREV$, $SENT\_CONC$ and $SENT\_ECV$ at any event.

Lemma 2. $SENT\_PREV_i + SENT\_CONC_i = SENT\_ECV_i$

Lemma 3 states that the $SENT\_PREV$ matrix reflects exactly all the messages sent by various processes up to $MCV(e)$. This includes all the send events that semantically precede all local events up to the current event.

Lemma 3. *The messages sent by P_j until $e_j^{MCV(e_i^k)[j]}$ correspond exactly to the messages represented by $SENT\_PREV_i[j,\cdot]$ at e_i^k.*

Lemma 4 states that messages represented by the matrix $SENT\_CONC$ are the messages sent concurrently in terms of semantic dependency with respect to the current event. These are the messages with respect to which no "happens-before" causal ordering needs to be enforced in order to meet the semantic causal ordering requirements.

Lemma 4. *The messages represented by $SENT\_CONC_i[j, \cdot]$ at e_i^k correspond exactly to the messages sent in the left-open right-closed duration $(MCV(e_i^k)[j], ECV(e_i^k)[j]]$.*

Theorem 1. *The algorithm given in Figure 3 implements causal message ordering with respect to the relation $\xrightarrow{s}$.*

3.4 Algorithm Analysis

With the proposed approach, any message M multicast at event e_i should be delivered in causal order with respect to all the messages represented by $SENT\_PREV$, i.e., sent in the computation prefix $MC(e_i)$. Messages represented by $SENT\_CONC$ are those with respect to which no false causality is imposed (follows from Lemma 4 and Proposition 1). The amount of false causality in enforcing causal delivery of M is the size of $\xrightarrow{w}(MC(e_i) \times MC(e_i))$, in contrast to the traditional causal ordering where the amount of false causality is the size of $\xrightarrow{w}(EC(e_i) \times EC(e_i))$. While $\xrightarrow{w}(MC(e_i) \times MC(e_i))$ may still be large (although smaller than that for the traditional approach), indicating that much false causality may still exist, this is not so on close analysis. With the proposed approach, false causality is potentially imposed only with respect to some of the messages sent up to MCV. Each $MCV(e_i^k)[j]$ will usually be much less than $ECV(e_i^k)[j]$. Hence, false causality is potentially imposed only with respect to some of the messages sent in the more distant past. In practice, message delivery times tend to have an exponential distribution. Hence, messages sent in the distant past up to the events indicated by $MCV(e_i^k)$ would have been delivered with high probability and the present message could most likely be delivered as soon as it arrives and without any buffering. Only in the case that some message sent in the distant past has not been delivered, and a false causality exists on such a message, that the proposed algorithm will unnecessarily delay the present message and require some buffering. We expect that such a case will have a low probability of occurrence, and when it occurs, the extra delay incurred by the imposed false causality will be small.

The computational complexity of the proposed algorithm is the same as that of the RST algorithm. The $O(n^2)$ extra computation in steps (R5)-(R7) is the same order of magnitude as in the RST algorithm. The $O(n^2)$ extra computation in steps (E1)-(E5) is of the same order of magnitude as in the RST algorithm. In terms of space complexity, the RST algorithm requires $n^2 \times m$ bits of storage space and message overhead, where m is the size of the message counter, whereas the proposed algorithm requires $2n^2 \times m$ bits of storage space and message overhead, which are comparable.

The only additional overhead is to implement Definition 5. By Property 1, this requires the MCV vector of the previous local event and the identity of the latest local event on which there is a semantic dependency. The former information is already computed; the latter can be assumed to be available as in [7,17] or can be extracted from compiler data.

4   Concluding Remarks

False causality in causal message ordering reduces the performance of the system by unnecessarily delaying messages and requiring large buffers to hold the delayed messages. We presented an efficient algorithm for implementing causal ordering that eliminates much of the false causality. In particular, the algorithm eliminates the false causality in the near *past* of any message send event. Some false causality with respect to messages sent in the more distant past exists. It is expected that such false causality will have a minimal degradation on the performance of the ideal causal ordering implementation because messages sent in the more distant *past* will have been delivered with high probability and cause buffering of the present message with low probability. The implementation presented here is lightweight and requires about the same order of magnitude overhead as the baseline RST algorithm, with minimal additional support.

Acknowledgements

This work was supported by the U.S. National Science Foundation grants CCR-9875617 and EIA-9871345.

References

1. Y. Amir, D. Dolev, S. Kramer and D. Malki, Transis: A communication sub-system for high-availability, *Proc. 22nd International Symposium on Fault-tolerant Computing*, IEEE Computer Society Press, 337-346, 1991.
2. K. Birman, A response to Cheriton and Skeen's criticism of causal and totally ordered communication, *Operating Systems Review*, 28(1): 11-21, Jan. 1994.
3. K. Birman, T. Joseph, Reliable communication in the presence of failures, *ACM Transactions on Computer Systems*, 5(1): 47-76, Feb. 1987.
4. K. Birman, A. Schiper and P. Stephenson, Lightweight causal and atomic group multicast, *ACM Transactions on Computer Systems*, 9(3): 272-314, Aug. 1991.
5. J. Caroll, A. Borshchev, A deterministic model of time for distributed systems, *Proc. Eighth IEEE Symposium on Parallel and Distributed Processing*, 593-598, Oct. 1996.
6. D.R. Cheriton, D. Skeen, Understanding the limitations of causally and totally ordered communication, *Proc. 11th ACM Symposium on the Operating Systems Principles*, 44-57, Dec. 1993.
7. C. Fidge, Logical time in distributed computing systems, *IEEE Computer*, 24(8): 28-33, Aug. 1991.

8. P. Gambhire, Efficient Causal Message Ordering, M.S. Thesis, University of Illinois at Chicago, April 2000.

9. F. Kaashoek, A. Tanenbaum, Group communication in the Ameoba distributed operating system, *Proc. Fifth ACM Annual Symposium on Principles of Distributed Computing*, 125-136, 1986.

10. A. Kshemkalyani, M. Singhal, Necessary and sufficient conditions on information for causal message ordering and their optimal implementation, *Distributed Computing*, 11(2), 91-111, April 1998.

11. L. Lamport, Time, clocks, and the ordering of events in a distributed system, *Communications of the ACM*, 21(7): 558-565, July 1978.

12. L.L. Peterson, N.C. Bucholz and R.D. Schlichting, Preserving and using context information in interprocess communication, *ACM Transactions on Computer Systems* 7(3), 217-246, 1989.

13. M. Raynal, A. Schiper, S. Toueg, The causal ordering abstraction and a simple way to implement it, *Information Processing Letters* 39:343-350, 1991.

14. L. Rodrigues, P. Verissimo, Causal separators and topological timestamping: an approach to support causal multicast in large-scale systems, *Proc. 15th IEEE International Conf. on Distributed Computing Systems*, May 1995.

15. R. van Renesse, Causal controversy at Le Mont St. Michel, *Operating Systems Review*, 27(2):44-53, April 1993.

16. A. Schiper, A. Eggli, A. Sandoz, A new algorithm to implement causal ordering, *Proc. Third International Workshop on Distributed Systems*, Nice, France, LNCS 392, Springer-Verlag, 219-232, 1989.

17. A. Tarafdar, V. Garg, Addressing false causality while detecting predicates in distributed programs, *Proc. 18th IEEE International Conf. on Distributed Computing Systems*, 94-101, May 1998.

Working-Set Based Adaptive Protocol
for Software Distributed Shared Memory

Sung-Woo Lee and Kee-Young Yoo

Department of Computer Engineering, Kyungpook University, Korea
swlee@purple.knu.ac.kr, yook@bh.knu.ac.kr

Abstract. Recently, many different protocols have been proposed for software Distributed Shared Memory (DSM) that can provide a shared-memory programming model for distributed memory hardware. The adaptive protocols of these protocols attempt to allow the system to choose between different protocols based on the access patterns it observes in an application. This paper describes several problems that deteriorate the performance of a hybrid protocol[6], an adaptive invalidate/update protocol. To address these problems, this paper then presents a working-set based adaptive invalidate/update protocol that uses a working-set model as the criteria for determining whether to update or invalidate. The proposed protocol was implemented in CVM [7], a software DSM system, and evaluated using eight nodes of an IBM SP2. After experimenting with various working-set window sizes, it was confirmed that the proposed protocol could track an access pattern better than the hybrid protocol, plus with a very small window size the protocol was able to optimize the overall performance.

1 Introduction

Software distributed shared memory systems (DSM) provide programmers with the illusion of shared memory on top of message-passing hardware [2]. These systems provide a low-cost alternative for shared-memory computing, since they can be built using standard workstations and operating systems. Although many different protocols have been proposed for implementing a software DSM [1,4,9], the relative performance of these protocols is application-dependent: the memory access patterns of the application determine which protocol will produce a good performance. Accordingly, it would be interesting to build a system with multiple protocols, and allow the system to choose between the different protocols based on the access patterns it observes in a particular application [1,6,9,10,12].

The lazy hybrid (LH) protocol [5,6], an adaptive invalidate/update protocol, is a lazy protocol similar to the lazy release consistency (LRC) [5] using an invalidate protocol, however, instead of invalidating the modified pages, it updates some of the pages at the time of synchronization. The decision on updating a page depends on whether or not the target processor has accessed the page before.

This paper first describes three problems that deteriorate the performance of the LH protocol as follows: (i) the more diffs that are updated, the longer latency of the synchronization, as a result, the overall performance can be degraded, (ii) the protocol

M. Valero, V.K. Prasanna, and S. Vajapeyam (Eds.): HiPC 2000, LNCS 1970, pp. 73–82, 2000

continues to apply the update protocol unnecessarily to pages accessed once, yet not accessed later, and (iii) on lock synchronization, a release processor is unable to determine the access pattern that an acquire processor generated in the latest interval.

To cope with these problems, a working-set model is proposed as the criteria for determining whether to update or invalidate, and consequently, a Working-set based Adaptive invalidate/update Protocol (WAP) is presented. In a conventional operating system, the *working-set model* has been used for a demand paging subsystem that permits greater flexibility in mapping the virtual address space of a process into the physical memory of a machine [11]. For this model, Denning formalized the *working-set* of a process, which is the set of pages that the process has referenced in its last Δ memory references; the number Δ is called the *window* of the working-set.

WAP attempts to exploit spatial locality by assuming that a process tends to localize its references to the working-set and this set only changes slowly. Therefore, when sending the working-set with a lock acquire message, WAP allows the releasing processor to know the latest access pattern of the acquiring processor and only updates the pages in the working-set. As a result, WAP attempts to constrain the number of diffs via updates, thereby optimizing a tradeoff between the number of diffs via updates and the latency of the synchronization. In addition, WAP can propagate the access pattern of a processor earlier than LH, and therefore, has more chances to update.

2 Background

This paper focuses on page-based software DSM systems that use a multiple writer protocol and lazy release consistency (LRC) protocol, such as TreadMarks [4] and CVM [7]. In multiple writer protocols [1], write detection is performed by twinning and diffing. On the first write to a shared page, an identical copy of the page (a *twin*) is made. The twin is then compared with the modified copy of the page to generate a *diff*, a record of the modifications to the page. This diff is then returned in response to a page fault request.

The LRC [5] protocol, which is a release consistency [3] implementation, delays the propagation of shared memory modifications by processor p to processor q until q executes an acquire corresponding to a release by p. On an acquire operation, the last releaser can determine the set of *write notices* that the acquiring processor needs to receive, i.e. the set of *notices* that precede the current acquire operation in the partial order. A write notice is an indication that a page has been modified within a particular interval, however, it does not contain the actual modifications. Upon receiving the notices piggybacked on the lock grant message, the acquirer then causes the corresponding page to be invalidated. Access to an invalidated page causes a page fault. At this point, the faulting processor must then retrieve and apply to the page all the diffs that were created during the intervals that preceded the faulting interval in the partial order [3].

3 Lazy Hybrid Protocol and Its Limitations

3.1 Lazy Hybrid (LH) Protocol

The LH [6], an adaptive invalidate/update protocol, is a lazy protocol similar to LRC using an invalidate protocol, however instead of invalidating the modified pages, it updates some of the pages at the time of an acquire. LH attempts to exploit temporal locality by assuming that any page accessed by a processor in the past will probably be accessed by that processor again in the future. Therefore, all pages that are known to have been accessed by the acquiring processor are updated.

Each processor uses copysets [5] to track the page accesses by other processors. At a synchronization point, the copyset is used to determine whether or not a given diff must be sent to a remote location. For each write notice to be sent, if the releasing processor has a diff corresponding to the write notice and the acquiring processor is in the local copyset of that page, that diff will be appended to the lock grant message.

On arrival at a barrier, each processor creates a list describing any local write notices that may not have been seen by the other processors. A list for processor p_j at processor p_i consists of processors p_i's notion of all the local write notices that have not been seen by p_j. p_i sends an update message to p_j containing all the diffs corresponding to the write notices in this list.

3.2 Limitation of LH Protocol

After executing several applications using the LH protocol and tracing their behaviors, several problems were identified that deteriorate the performance of the LH protocol. First, under LH, the increment in the amount of diffs updated increases the latency of the synchronization, because LH updates the diffs through a synchronization message. Even when the all diffs sent via updates were used by the target processor, the increased excessively latency of the synchronization still tended to deteriorate the overall performance.

Second, LH updates unnecessary diffs in some applications that have a migratory access pattern. These applications access each page only once or twice for their entire life, and access the majority of entire pages. In this case, LH continues to unnecessarily apply the update protocol to the pages that are only accessed once and never accessed later.

Third, on lock synchronization, a releasing processor's copyset cannot reflect the access pattern that an acquiring processor has generated in the recent interval. Since LRC only requires that the acquiring processor knows (or receives) the events of the memory updates (or notices) that precede the current acquiring operation in the partial order, the releasing processor cannot know the latest access events of the acquire processor. As a result, LH cannot help but refer an access history of older intervals, even though the latest events are the most important from the viewpoint of the conventional operating system.

In light of these problems, a new protocol is proposed that uses a working-set model to track an access pattern efficiently, as described in the next section.

4 Working-Set Based Adaptive Protocol

The proposed protocol, the Working-set based Adaptive Protocol (WAP), still updates some of the pages at the time of an acquire, like LH. However, WAP uses the working-set of a processor to determine whether to update or invalidate the page, whereas LH uses the copysets of the pages.

Each process maintains the working-set using a circular queue, where the size of the queue, Δ, is the size of the working-set window. Whenever not only a write fault but also a read fault occur, the protocol inserts that page number into the queue only if the current page number is different from that of the queue's rear, that is, the latest page number inserted. The reason for this is to manage the queue effectively against a prevalent access pattern that will generate a write fault to a page immediately after generating a read fault. On sending an acquire message, the acquiring processor appends the content of the queue into the message. Although the system must remove any redundant page numbers to achieve an ideal working-set, the entire content of the queue is appended in order to minimize the time overhead to do it.

The releasing processor creates a list describing the local write notices that may not have been seen by the acquiring processor. For each write notice in the list, if the releasing processor has the diff and the page of the notice is included in the working-set delivered just previously, the diff is appended to the lock grant message.

On arrival at a barrier, each processor except for the barrier master creates a list describing the local write notices that may not have been seen by the barrier master processor, and then appends the diffs of a notice into a barrier arrival message only if the notice was created by oneself and the notice's page is included in the latest working-set sent before by the master (this former condition can eliminate any redundant diff transmissions.). Also, the processor's working-set is appended. After all the barrier arrival messages arrive at the master, the master then sends barrier start messages to all other processors in the same way with a lock release.

5 Implementation and Evaluation

5.1 Platform and Applications

The two protocols, LH and WAP, as described in the previous section, were implemented in a CVM DSM system. The experimental environment consisted of eight nodes on an IBM SP2 running AIX 4.1.4. Six applications were used in this paper: TSP from the CVM package, Barnes, Water-Nsquared, Water-Spatial, and Ocean from Splash-2 [8]. Table I summarizes the applications and their input sets. The values in the table were obtained by the current study. The lock value was the number of lock acquiring messages sent to remote processors, and not the total count of lock acquires. The number of page requests shows indirectly how many pages were used by the application.

Table 1. Application Suite

Program	Input	Page re-	Barriers	Locks
Tsp	19 cities	704	0	142
Barnes	4096 particle	1399	8	69200
Ocean	128 by 128	1100	448	2960
Water-Spatial	512 molecul	973	9	355
Water-Nsqaured	512 molecul	385	12	385

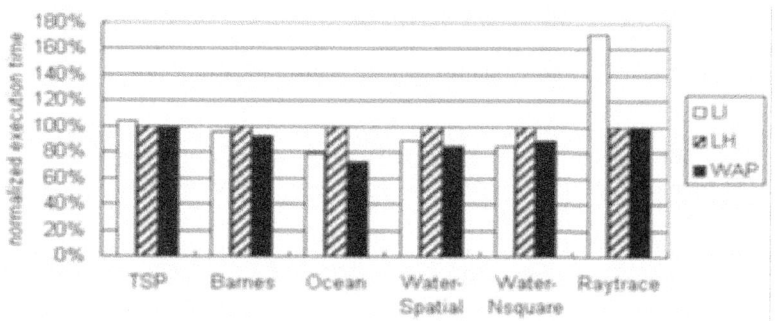

Fig. 1. Eight-processor normalized execution times for LI, LH, and WAP

5.2 Performance Evaluation

A performance comparison was acquired by executing application suite using WAP with various window sizes, LH, and the lazy invalidate (LI) protocol. The LI protocol is an LRC protocol using the invalidate protocol included in CVM package. Fig. 1 shows the normalized execution times for the six applications for each of the three protocols. In this figure, the execution times of WAP were obtained by selecting the optimal window size,Δ, which resulted in the minimal execution time among the other window sizes. The values of Δ was 300 for TSP, all other values were 20.

From Fig. 2 to Fig. 6, all figures present important elements of the performance with each application. For WAP, the graphs show the results of executing with the various values of Δ, that is, 20, 40, 60, 80, 100, 150, 200, 250, 300, 350, 400, 450, and 500. Graph (a) of each figure presents, in order, the execution time of the applications, the total time to create diffs, the total time waiting for diffs after sending a diff request, the total time to handle faults, and the total time to handle a lock acquire and barrier. Except for the execution time, all other values are the summation of the times of eight processors. Graph (b) presents the total number of message bytes delivered (msg_byte), the number of messages sent to remote processors in order to request a diff (remote_diff), the number of diffs created in the system (diff_creadted), the number of diffs created for updates (update_diff_creadted), the number of diffs delivered by updates (update_diffs_sent), and the ratio of diffs used by the system to diffs delivered by an update (hit_ratio).

TSP. With WAP, as the value of the window size, Δ, increases up to Δ=250, the number of diffs updated (update_diffs_sent) increases. Therefore, WAPs with Δ>=150 send more diffs via updates than LH and send fewer remote diff requests (remote_diff) than LH. Consequently, the number of remote diff requests with WAP when Δ=500 is less than half that of LH.

This result shows that since WAP causes the acquiring processor to send the latest access pattern to the releasing processor, WAP is able to adapt successfully to the access pattern of this application and thereby gains more chances to update than LH. However, although the overhead of WAP, related to the working-set transmission caused by an acquire, produces an increase in the total message bytes (msg_byte), fortunately, this overhead does not result in a longer lock latency. Nevertheless, with a large Δ value (Δ>300) the WAP performance is only marginally better than that of LH, since TSP has a very high computation to communication ratio.

Barnes. It is interesting to note that when Δ=500, although WAP generates a similar number of remote diff requests as LH, it sends about 50% fewer diffs via updates and achieves a 10% higher hit-ratio than LH. This result, also, shows that WAP exhibits a better adaptation ability to the access pattern than LH, as in TSP. However, in contrast to TSP, the overhead of WAP with a large Δ value results in much longer lock latency, and thereby a worse overall performance. The reason for this is the frequent lock synchronizations related to the very small size (close to zero) of the critical section, which then produces the WAP overhead. Consequently, when Δ=20, WAP performs 8% better than LH and 5% better than LI.

Ocean. For this application, although LH applies updates for the majority of the pages and reduces the remote diff requests by 85% compared with LI, only 66% of the diffs out of all the diffs updated by LH is used. As a result, updating unnecessary diffs results in longer synchronization latency. Consequently, the LH performance is much worse than that of LI. In contrast, with all values of Δ, WAP can constrain the number of diffs updated. Although WAP updates substantially fewer diffs than LH, the hit-ration of the updated diffs increases by about 30-36% and the synchronization latency is less than that of LH. In conclusion, when Δ=20, WAP performed 28% better than LH and 9% better than LI.

Water-Spatial. With WAP, as the value of the window size, Δ, increases up to Δ=250, the number of diffs updated also increases and the number of remote diff requests decreases. When Δ=250 WAP is able to update 50% more diffs than LH, and therefore generates 20% fewer remote diff requests. The fact that WAP can reduces the number of remote diff requests compared to LH even though LH achieves 100% hit-ratio, indicates the superior ability of WAP in tracking the access pattern of an application. When Δ=20, WAP performs 15% better than LH and 10% better than LI.

Water-Nsquare. With WAP, when Δ>60, the number of diffs updated does not increase. It is interesting to note that, when Δ=60, the number of remote diff requests of WAP is nearly equal to that of LH, even though the number of diffs updated by WAP is only 77% of that of LH. The reason for this is that WAP sends fewer unnecessary diffs via updates than LH. When Δ=20, WAP performs 11% better than LH and 5% worse than LI.

6 Conclusion

A new adaptive update/invalidate protocol based on a working-set model was presented in this paper. Although a unique optimal windows size value could not be selected for all applications, however, for most applications, with the use of a very small window size value (Δ=20 or 40), the proposed protocol (WAP) outperformed both LH and LI. The reason for this was that, even with such small window sizes, the proposed protocol was able to reduce the number of remote diff requests caused by page faults from LI, and reduce the latency of synchronization from LH through constraining the number of diffs via updates. Furthermore, for some applications, since the proposed protocol could to track the access pattern better than LH, WAP was able to either reduce the number of unnecessary diffs via updates, or create more chances for updating.

Experimental results highlighted a weakness in the proposed protocol, that is, the overhead caused by sending the working-set along with a lock acquire message. Due to this overhead, WAP with a larger working-set tend to make the overall execution time to be longer, even though many other elements of performance were improved. In particular, some applications generated excessive lock acquires or barriers, as a result, the overhead caused their performance to be degraded. However, in reality, DSM cannot help but produce very bad speedups with all of the protocols when using these applications. Therefore, it would appear that the applications need to be reconstructed for adjustment to DSM.

References

1. J. B. Carter, J. K. Bennett, and W. Zwaenepoel, "Implementation and Performance of Munin," *Proc. the 13th ACM Symposium on Operating Systems Principles*, pp. 152-164, Oct. 1991.
2. K. Li and P. Hudak, "Memory coherence in shared virtual memory systems," *ACM Transaction of Computer Systems* 7(4), 321-359, Nov 1989.
3. K. Gharachooloo, D. Lenoski, J. Laudon, P. Gibbons, A. Gupta and J. Hennessy, "Memory consistency and event ordering in scalable shared-memory multiprocessors," *Proceedings of the 17th Annual International Symposium on Computer Architecture*, pp. 15-26, May 1990.
4. P. Keleher et al., "TreadMarks: Shared Memory Computing on Networks of Workstations," *IEEE Computer*, pp. 18-28, Feb. 1996.
5. P. Keleher, "Distributed Shared Memory Using Lazy Release Consistency," PhD dissertation, Rice University, 1994.
6. P. Keleher, A. L. Cox, S. Dwarkadas, and W. Zwaenepoel, "An Evaluation of Software based Release Consistent Protocols", *Journal of Parallel and Distributed Computing*, Vol 29, pp. 126-141, October 1995.
7. P. Keleher, "CVM: The Coherent Virtual Machine", http://www.cs.umd.edu/ projects/cvm, November 1996.
8. S. Woo, M. Ohara, E. Torrie, J. P. Singh, and A. Gupta. "The SPLASH-2 Programs: Characterization and Methodological Considerations," *Proceedings of the 21st Annual International Symposium on Computer Architecture*, June 1995.
9. L.R. Monnerat and R. Bianchini. "Efficiently adapting to sharing patterns in software DSMs," *Proceedings. Of the 4th International Symposium on High-Performance Computer Architecture*, Feb. 1998.

80   S.-W. Lee and K.-Y. Yoo

10. C. Amza, A. Cox, S. Dwarkadas, and W. Zwaenepoel, " Software DSM protocols that Adapt Between and single Writer and Multiple Wirter," *Proceeding of he 3th International Symposium on High Performance Computer Architecture*, Feb 1997
11. P. B. Glavin, Operating System Concepts, Addison-Wesley Publishing Company, pp. 320-321, 1998
12. C. Amza, A. Cox, K. Rajamani, and W. Zwaenepoel, "Tradeoffs Between False Sharing and Aggregation in Software Distributed Shared Memory," *Proceedings of the 6th ACM SIGPLAN Symposium on Principles and Practice of Parallel Programming*, Jun 1997.

Appendix

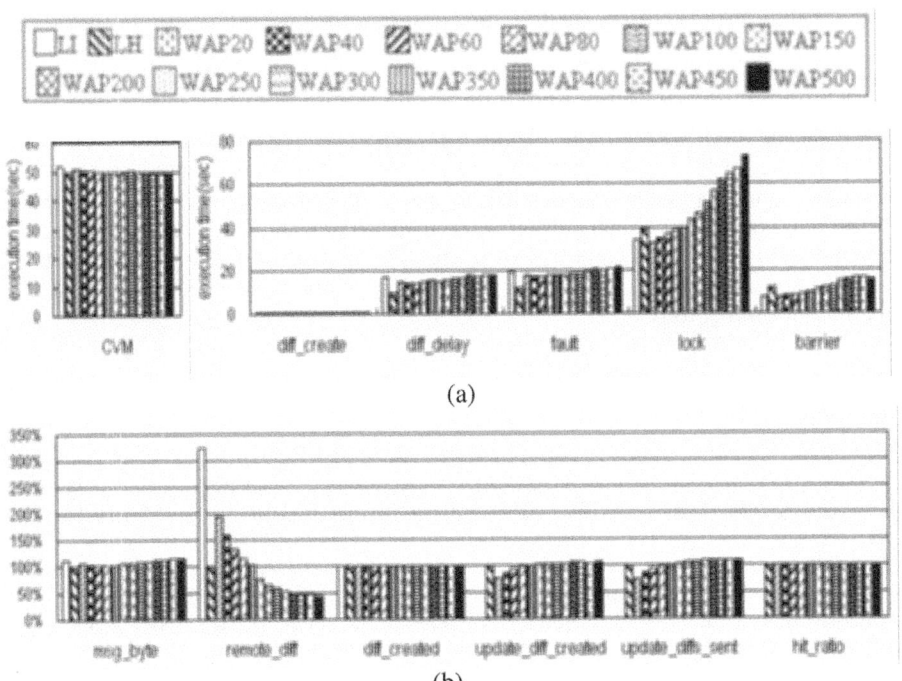

(a)

(b)

Fig. 2. Results for TSP

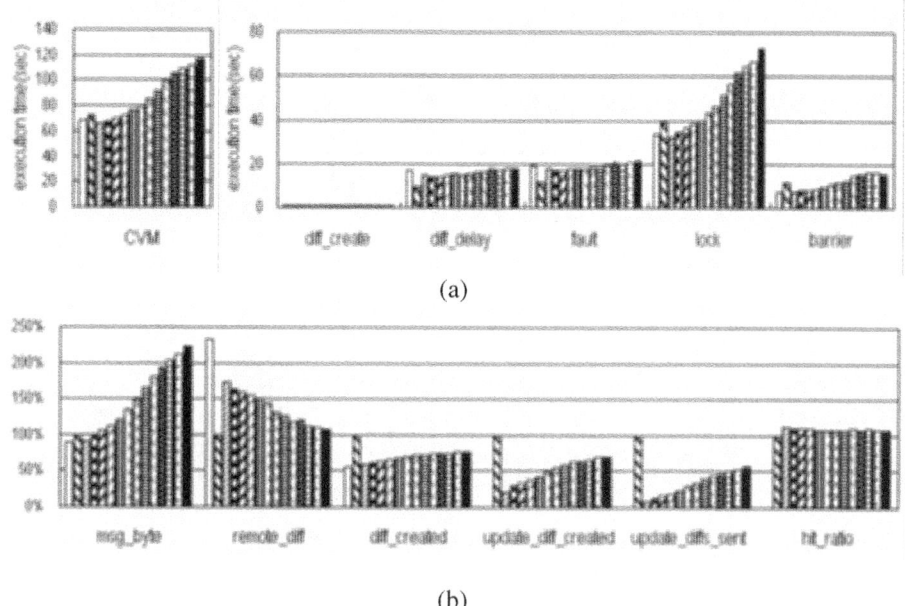

(a)

(b)

Fig. 3. Results for Barnes

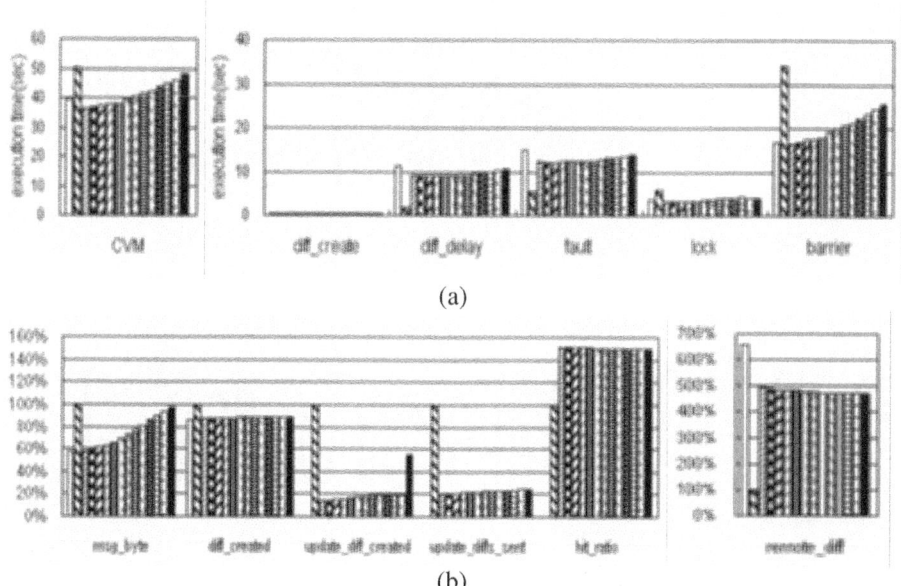

(a)

(b)

Fig. 4. Results for Ocean

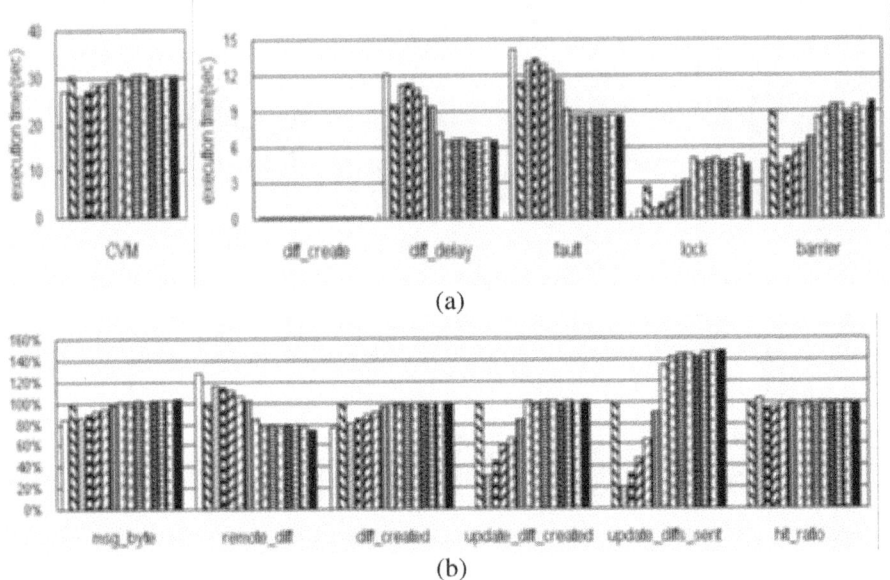

(a)

(b)

Fig. 5. Results for Water-Spatial

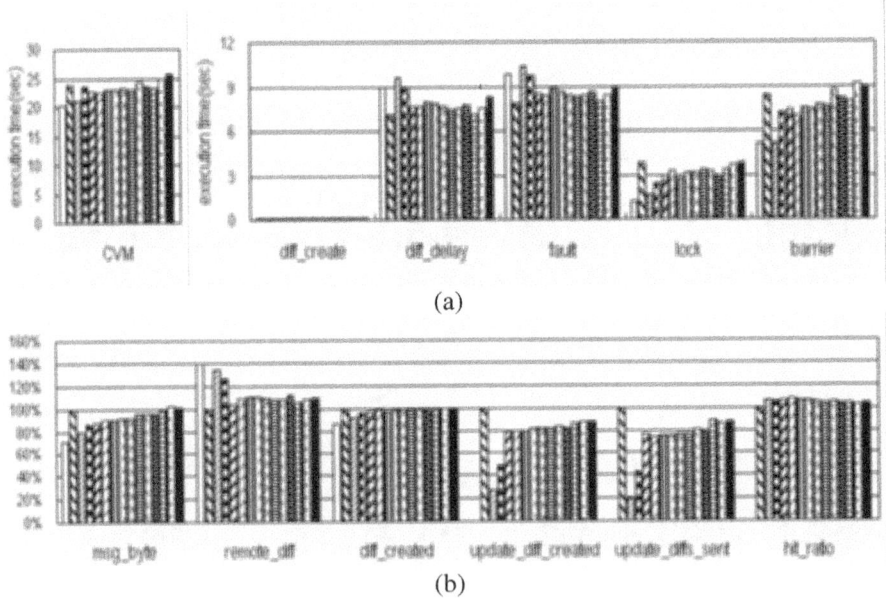

(a)

(b)

Fig.6. Results for Water-Nsquare

Evaluation of the Optimal Causal Message Ordering Algorithm

Pranav Gambhire and Ajay D. Kshemkalyani

Dept. of EECS, University of Illinois at Chicago, Chicago, IL 60607-7053, USA
{pgambhir, ajayk}@eecs.uic.edu

Abstract. An optimal causal message ordering algorithm was recently proposed by Kshemkalyani and Singhal, and its optimality was proved theoretically. For a system of n processes, although the space complexity of this algorithm was shown to be $O(n^2)$ integers, it was expected that the actual space overhead would be much less than n^2. In this paper, we determine the overhead of the optimal causal message ordering algorithm via simulation under a wide range of system conditions. The optimal algorithm is seen to display significantly less message overhead and log space overhead than the canonical Raynal-Schiper-Toueg algorithm.

1 Introduction

A distributed system consists of a number of processes communicating with each other by asynchronous message passing over reliable logical channels. There is no shared memory and no common clock in the system. A process execution is modeled as a set of events, the time of occurrence of each of which is distinct. A message can be multicast, in which case it is sent to multiple other processes. The ordering of events in a distributed system execution is given by the "happens before" or the causality relation [6], denoted by $\longrightarrow$. For two events $e1$ and $e2$, $e1 \longrightarrow e2$ iff one of the following conditions is true (i) $e1$ and $e2$ occur on the same process and $e1$ occurs before $e2$, (ii) $e1$ is the send of a message and $e2$ is the delivery of that message, or (iii) there exists an event $e3$ such that $e1 \longrightarrow e3$ and $e3 \longrightarrow e2$.

Let $Send(M)$ denote the event of a process handing over the message M to the communication subsystem. Let $Deliver(M)$ denote the event of M being *delivered* to a process after it is been received by its local communication subsystem. The system respects *causal message ordering* (CO) [2] iff for any pair of messages M_1 and M_2 sent to the same destination, $(Send(M_1) \longrightarrow Send(M_2)) \implies (Deliver(M_1) \longrightarrow Deliver(M_2))$.

Causal message ordering is valuable to the application programmer because it reduces the complexity of application logic and retains much of the concurrency of a FIFO communication system. Causal message ordering is useful in numerous areas such as managing replicated database updates, consistency enforcement in distributed shared memory, enforcing fair distributed mutual exclusion, efficient snapshot recording, and data delivery in real-time multimedia systems. Many

M. Valero, V.K. Prasanna, and S. Vajapeyam (Eds.): HiPC 2000, LNCS 1970, pp. 83–95, 2000.

causal message ordering algorithms have been proposed in the literature. See [2,3,5,8,9] for an extensive survey of applications and algorithms. Causal message ordering has been implemented in many systems such as Isis [2], Transis [1], Horus [3], Delta-4, Psync [7], and Amoeba [4].

Any causal message ordering algorithm implementation has two forms of space overheads, viz., the size of control information on each message and the size of memory buffer space at each process. It is important to have efficient implementations of causal message ordering protocols due to their wide applicability. The causal message ordering algorithm given by Raynal, Schiper and Toueg [8], hereafter referred to as the RST algorithm, is a canonical solution to the causal message ordering problem. It has a fixed message overhead and memory buffer space overhead of n^2 integers, where n (also denoted interchangeably as N) is the number of processes in the system. The Horus [3], Transis [1], and Amoeba [4] implementations of causal message ordering are essentially variants of the RST algorithm.

Recently, Kshemkalyani and Singhal identified and formulated the necessary and sufficient conditions on the information required for causal message ordering, and provided an optimal algorithm to realize these conditions [5]. This algorithm was proved to be optimal in space complexity under all network conditions and without making any simplifying system/communication assumptions. The authors also showed that the worst-case space complexity of the algorithm is $O(n^2)$ integers but argued that in real executions, the actual complexity was expected to be much less than n^2 integers, the overhead of the RST algorithm.

Although the Kshemkalyani-Singhal algorithm was proved to be optimal in space complexity by using a rigorous optimality proof, there are no experimental or simulation results about the quantitative improvement it offers over the canonical RST algorithm. The purpose of this paper is to quantitatively determine the performance improvement offered by the optimal Kshemkalyani-Singhal algorithm, hereafter referred to as the KS algorithm, over the RST algorithm. This is done by simulating the KS algorithm and comparing the amount of control information sent per message and the amount of the memory buffer space requirements, with the fixed overheads of the RST algorithm. The results over a wide range of parameters indicate that the KS algorithm performs significantly better than the RST algorithm, and as the network scales up, the performance benefits are magnified. With $N = 40$, the KS algorithm has about 10% of the overhead of the RST algorithm.

Note that the space overhead is the only metric of causal message ordering algorithms studied in this simulation because it was shown in [5] that the time (computational) overhead at each process for message send and delivery events was similar for the KS algorithm and for the canonical RST algorithm, namely $O(n^2)$.

Section 2 outlines the RST algorithm and the KS algorithm. Section 3 presents the model of the message passing distributed system in which the KS algorithm is simulated. Section 4 shows the simulation results of the KS algorithm in comparison to the results expected from the RST algorithm. Section 5 concludes.

2   Overview of the CO Algorithms

This section briefly introduces the RST algorithm [8] and the optimal KS algorithm [5] for causal message ordering. Both the algorithms assume FIFO communication channels and that processes fail by stopping.

2.1   The RST Algorithm

Every process in a system of n processes maintains a $n \times n$ matrix - the $SENT$ matrix. $SENT[i, j]$ is the process's best knowledge of the number of messages sent by process P_i to process P_j. A process also maintains an array $DELIV$ of size n, where $DELIV[k]$ is the number of messages sent by process P_k that have already been delivered locally. Every message carries piggybacked on it, the $SENT$ matrix of the sender process. A process P_j that receives message M with the matrix SP piggybacked on it is delivered M only if, $\forall i$, $DELIV[i]$ $\geq SENT[i,j]$. P_j then updates its local $SENT$ matrix $SENT_j$ as: $\forall k \forall l \in$ $\{1, \ldots, n\}$, $SENT_j[k, l] = max(SENT_j[k, l], SP[k, l])$. The space overhead on each message and in local storage at each process is the size of the matrix $SENT$, which is n^2 integers.

2.2   The KS Algorithm

Kshemkalyani and Singhal identified the necessary and sufficient conditions on the information required for causal message ordering, and proposed an algorithm that implements these conditions. To outline the algorithm, we first introduce some formalisms. The set of all events E in the distributed execution (computation) forms a partial order $(E, \longrightarrow)$ which can also be viewed as a *computation graph*: (i) there is a one-one mapping between the set of vertices in the graph and the set of events E, and (ii) there is a directed edge between two vertices iff either these vertices correspond to two consecutive events at a process or correspond to a message send event and a delivery event, respectively, for the same message. The causal past (resp., future) of an event e is the set $\{e' \mid e' \longrightarrow e\}$ (resp., $\{e' \mid e \longrightarrow e'\}$). A path in the computation graph is termed a *causal path*. $Deliver_d(M)$ denotes the event $Deliver(M)$ at process d.

The KS algorithm achieves optimality by storing in local message logs and propagating on messages, information of the form "d is a destination of M" about a message M sent in the causal past, *as long as* and *only as long as*

(*Propagation Constraint I:*) it is not known that the message M is delivered to d, and

(*Propagation Constraint II:*) it is not guaranteed that the message M will be delivered to d in CO.

In addition to the Propagation Constraints, the algorithm follows a *Delivery Condition* which states the following. A message M^* that carries information "d is a destination of M", where message M was sent to d in the causal past of $Send(M^*)$, is not delivered to d if M has not yet been delivered to d.

Constraint (I) and the Delivery Condition contribute to optimality as follows: To ensure that M is delivered to d in CO, the information "d is a destination of M" is stored/propagated *on* and *only on* all causal paths starting from $Send(M)$, but nowhere in the causal future of $Deliver_d(M)$.

Constraint (II) and the Delivery Condition contribute to optimality by the following transitive reasoning: Let messages M, M' and M'' be sent to d, where $Send(M) \longrightarrow Send(M') \longrightarrow Send(M'')$ and M' is the first message sent to d on all causal chains between the events $Send(M)$ and $Send(M')$. M will be delivered optimally in CO to d with respect to (w.r.t.) M'' if (i) M is guaranteed to be delivered optimally in CO to d w.r.t. M', and (ii) M' is guaranteed to be delivered optimally in CO to d w.r.t. M''. Condition (i) holds if the information "d is a destination of M" is stored/propagated *on* and *only on* all causal paths from $Send(M)$, but nowhere in the causal future of $Send(M')$ other than on message M' sent to d. This follows from the Delivery Condition. Condition (ii) can be shown to hold by applying a transitive argument comprising of conditions (II)(i) and (I). To achieve optimality, the information "d is a destination of M" must not be stored/propagated in the causal future of $Send(M')$ other than on message M' sent to d (follows from condition (II)(i)) or in the causal future of $Deliver_d(M)$ (condition (I)).

Information about a message (I) not known to be delivered to d and (II) not guaranteed to be delivered to d in CO, is explicitly tracked by the algorithm using the triple *(source, destination, scalar timestamp)*. This information is deleted as soon as either (I) or (II) becomes false. As the information "d is a destination of M" propagates along various causal paths, the earliest event(s) at which (I) becomes false, or (II) becomes false, are known as Propagation Constraint Points *PCP1* and *PCP2*, respectively, for that information. The information never propagates beyond its Propagation Constraint Points. With this approach, the space overhead on messages and in the local log at processes is less than the n^2 overhead of the RST algorithm, and is proved to be always optimal.

The information "d is a destination of M" is also denoted as "$d \in M.Dests$", where $M.Dests$ is the set of destinations of M for each of which (I) and (II) are true. In an implementation, $M.Dests$ can be represented in the local logs at processes and piggybacked on messages using the data structures shown in figure 1.

```
type LogStruct = record                 type MsgOvhdStruct = record
    sender : process_id;                     sender: process_id;
    clock: integer;                          clock: integer;
    numdests: integer;                       numdests: integer;
    dests: array[1..numdests] of process_id; numLogEntries: integer;
end                                          dests: array[1..numdests] of process_id;
                                             olog: array[1..numLogEntries] of LogStruct;
                                         end
```

Fig. 1. The log data structure and message overhead data structure.

The log is a variable length array of type `LogStruct`. Assuming that process_id is an integer, the size of a `LogStruct` structure is $3 + size(dests)$ integers, where $size(X)$ is the number of elements in the set X. The log space overhead is the sum of the sizes of all the entries in the log. The amount of overhead on a message required by the KS algorithm is the size of the `MsgOvhdStruct` structure sent on it. The size of the `MsgOvhdStruct` structure can be determined as $4 + size(dests) + SIZE(olog)$, where $SIZE(X)$ is the sum of the sizes of all the entries in the set X of `LogStruct`s. The message and log space overheads are determined in this manner in our simulation system.

3   Simulation System Model

A distributed system consists of asynchronous processes running on processors which are typically distributed over a wide area and are connected by a network. It can be assumed without any loss of generality that each processor runs a single process. Each process can access the communication network to communicate with any other process in the system using asynchronous message passing. The communication network is reliable and delivers messages in FIFO order between any pair of processes.

3.1   Process Model

A process is composed of two subsystems viz., the *application subsystem* and the *communication subsystem*. The application subsystem is responsible for the functionality of the process and the communication subsystem is responsible for providing it with causally ordered messaging service. The communication subsystem implements the causal message ordering algorithm in the simulation. The application subsystem generates message patterns that exercise the causal message ordering algorithm. The communication subsystem maintains a floating point clock, that is different from any clock in the causal message ordering algorithm. This clock is initialized to zero and tracks the elapsed run time of the process. Every process has a priority queue called the *in_queue* that holds incoming messages. This queue is always kept sorted in increasing order of the arrival times of messages in it.

Message structure: A message is the fundamental entity that transfers information from a sender process to one or more receiver processes. Each message M has a *causal_info* field, *time_stamp* field, and a *payload* field. The *causal_info* field is just a sequence of bytes on which a particular structure is imposed by the causal message ordering algorithm. The RST algorithm imposes a $N \times N$ matrix structure on the *causal_info* field. The KS algorithm imposes the structure given in figure 1. The communication subsystem uses the *time_stamp* field to simulate the message transmission times. The *in_queues* are kept sorted by the *time_stamp* field. The information that is contained in a message is referred to as its *payload*. In a real system, this would contain the application-specific packet of information according to the application-level protocol.

3.2   Simulation Parameters

The system parameters that are likely to affect the performance of the KS algorithm are discussed next.

- **Number of processes (N):** While most causal message ordering algorithms show good performance for a small number of processes, a good causal message ordering algorithm would continue to do so for a large number of processes. It is hence necessary to simulate any causal message ordering algorithm over a wide range of the number of processes. The number of processes in the system is limited only by the memory size and processor speed of the machine running the simulation. On an Intel Pentium III machine with 128 MB of RAM and the simulation framework being implemented in Java, we could simulate up to 40 processes.
- **Mean inter-message time (MIMT):** The mean inter-message time is the average period of time between two message send events at any process. It determines the frequency at which processes generate messages. The inter-message time is modeled as an exponential distribution about this parameter.
- **Multicast frequency (M/T):** The behavior of the KS algorithm may be sensitive to the number of multicasts. The ratio of multicasts to the total number of message sends (M/T) is the parameter on the basis of which the multicast sensitivity of the KS algorithm can be determined. Processes like distributed database updators have $M/T = 100\%$ and a collection of FTP clients have $M/T = 0$. We simulate the KS algorithm with M/T varying from 0 to 100%. The number of destinations of a multicast is best described by a uniform distribution ranging from 1 to N.
- **Mean transmission time (MTT):** The transmission time of a message here implicitly refers to the *msg. size/bandwidth + propagation delay*. We model this time as an exponential distribution about the mean, MTT. For the purpose of enforcing this mean, multicasts are treated as multiple unicasts and transmission time is independently determined for each unicast. When a process needs to send a message, it determines the transmission time according to the formula $Transmission\_time = -MTT * ln(R)$, where R is a perfect random number in the range [0,1]. This formulation of the transmission time can violate FIFO order. As most causal message ordering algorithms assume FIFO ordering, it is implemented explicitly in our system. Every process maintains an array LM of size n to track the arrival time of the last message sent to each other process. $LM[i]$ is the time at which the last message from the current process to process P_i will reach P_i. Should the transmission time determined be such that the arrival time for the next message at P_i is less than $LM[i]$, then the arrival time is fixed at $(LM[i] + 1)ms$. $LM[i]$ is updated after every message send to P_i.
  MTT is a measure of the speed of the network, with fast networks having small MTTs. We have varied MTT from $50ms$ to $5000ms$ in these simulations so as to model a wide range of networks.

3.3 Process Execution

All the processes in the system are symmetric and generate messages according to the same MIMT and M/T. The processes in a distributed system execute concurrently. But simulating each process as an independent process/thread involves inter-process/thread communication and the involved delays are not easy to control. Instead, a round-robin scheme was used to simulate the concurrent processes. Each simulated process is given control for a time slot of 500*ms*. A systemwide clock keeps track of the current time slot.

When a process is in control, it generates messages according to the MIMT. The sender of a message determines the transmission time using MTT, adds it to its current clock, and writes the result into the *time_stamp* field of the message. It then inserts this message into the *in_queue* of the destination process.

When a process gets control, it first invokes the communication subsystem. The communication subsystem looks at the head of its *in_queue* to determine if there are any messages whose *time_stamp* is lesser than or equal to the current value of the process clock. Such messages are the ones that must have already arrived and hence should have been processed before/during this time slot. All such messages are extracted from the queue and handed over to the causal message ordering delivery procedure in the order of their timestamps. The causal delivery procedure will buffer messages that arrived out of causal order. Note that this buffer is distinct from the *in_queue*. Messages in causal order are delivered immediately to the application subsystem. Blocked messages remain blocked till the messages that causally precede them have been delivered. The application subsystem then gets control and it generates messages according to the MIMT. The messages are handed over to the communication subsystem for delivery.

A process P_i stops generating messages once it has generated a sufficient number of messages (see Section 4) and flags its status as completed. The simulation stops when all the processes have their status flagged as completed.

4 Simulation Results

The KS algorithm was simulated in the framework presented in Section 3. The framework and the algorithm were implemented in Java using ObjectSpace JGL. The performance metrics used are the following.

- The average number of integers sent per message under various combinations of the system parameters, viz., N, MTT, $MIMT$, and M/T.
- The average size of the log in integers, under the same conditions.

Simulation experiments were conducted for different combinations of the parameters. For each combination, four runs was executed; the results of the four runs did not differ from each other by more than a percent. Hence, only the mean of the four runs is reported for each combination and the variance is not reported.

For each simulation run, data was collected for 25,000 messages after the first 5000 system-wide messages to eliminate the effects of startup. Every process P_i

in the system accumulates the sum of the number of integers I_i that it sends out on outgoing messages. After every message send event and every message delivery event, it determines the log size and accumulates it into a variable L_i. It also tracks m_i^s, the number of messages sent, and m_i^r, the number of messages delivered, during its lifetime. Once P_i has sent out $m_i^s = 30,000/N$ number of messages, it flags its status as complete and computes its mean message overhead $MMV_i = I_i/m_i^s$ and its log space overhead $LV_i = L_i/(m_i^r + m_i^s)$. These results are then sent to process P_0 which computes the systemwide average message overhead $\sum MMV_i/N$ and the systemwide average log space overhead $\sum L_i/N$. All the overheads are reported as a percentage of their corresponding deterministic overhead n^2 of the RST algorithm.

It is seen that the results for the log size overhead followed the same pattern as the results for the message size overhead in all the experiments. Hence, the log size overhead plots are not shown in this paper for space considerations.

4.1   Scalability with Increasing N

RST scales poorly to networks with a large number of processes because of its fixed overhead of n^2 integers. Although KS algorithm has $O(n^2)$ overhead, it is expected that the actual overhead will be much lower than n^2. We test the scalability of the KS algorithm by simulation.

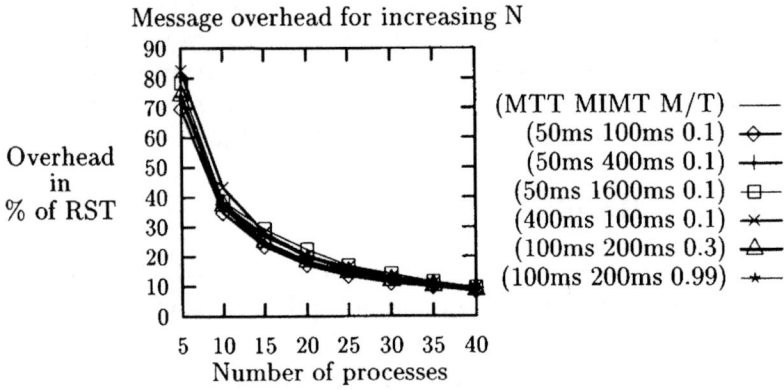

Fig. 2. Average message overhead as a function of N

The first three simulations were performed for (MTT, MIMT, M/T) fixed at $S_1(50ms, 100ms, 0.1)$, $S_2(50ms, 400ms, 0.1)$ and $S_3(50ms, 1600ms, 0.1)$. The number of processes was increased in steps of 5 starting from 5 up to 40. The results for the average message overhead are shown in figure 2. Observe that with increasing N, the message overhead rapidly decreases as a percentage of RST. Note that in all these simulations, the overhead is always significantly less

than that of RST. For the case of 40 processes, for all the simulations, the overhead is only 10% that of RST. For a small number of processes, the overheads
reported by KS are 80% of those of RST, but the overhead of RST itself is
low for such systems. Similar results are seen for the next three simulations:
(MTT, MIMT, M/T) fixed at $S_4(400ms, 100ms, 0.1)$, $S_5(100ms, 200ms, 0.3)$,
and $S_6(100ms, 200ms, 0.99)$ (the other three curves in figure 2). The latter two
simulations show that the improvement in overhead is unaffected by increasing
the traffic, modeled by increasing the multicast frequency to 30% and 99%.

It can be seen from figure 2 that the performance (overhead relative to the
RST algorithm) gets better when the number of processes is increased keeping
MTT, MIMT, and M/T constant. This is because increasing the number of processes implies an increase in the rate of generation of messages, given a constant
MIMT. As MTT is held constant, all these messages reach their destinations in
the same amount of time as with a lower number of processes. Hence, there is
greater dissemination of log information among the processes, thereby providing
impetus for the Propagation Constraints to work with more up to date information and purge more information from the logs. Thus as n increases, the logs get
purged more quickly and their size tends to be an increasingly smaller fraction
of n^2, the size of logs in the RST algorithm.

From all the simulations S_1 through S_6 and the above analysis, it can be
concluded that the KS algorithm has a better network capacity utilization and
hence better scalability when compared to RST.

4.2   Impact of Increasing Transmission Time

Increasing MTT is indicative of decrease in available bandwidth and increasing
network congestion. The space overheads of the RST algorithm are fixed at n^2,
irrespective of network congestion conditions. We ran simulations for systems
consisting of 10, 15, and 20 processes under varying MIMT and M/T to analyze
the impact of increasing MTT. The results for the average message overhead are
shown in figure 3.

The first three simulations fixed (N, MIMT, M/T) at $S_1(15, 400ms, 0.1)$,
$S_2(15, 800ms, 0.1)$ and $S_3(15, 1600ms, 0.1)$, respectively. The MTT was increased
from $200ms$ to $4800ms$ progressively in steps of 100 initially, 200 later, and
multiples of 2 finally. The fineness of the initial samples was necessary to see that
the overheads were growing fast initially but soon settled to a maximum. The
overhead of the algorithm as a % of the RST overhead first increases gradually
but soon reaches steady state despite further increases in MTT. This is explained
as follows. At low values of MTT, message transmission is very fast and hence
log sizes at the processes are small. However as MTT grows even slightly, the
message transmission rate falls and the log sizes begin increasing in size. Hence
a growth in overheads can be seen in the initial parts of the curves. However
once MTT becomes large, all the log sizes tend to a "steady-state" proportion of
n^2 (determined by other system parameters) but significantly less than n^2. This
trend is because the pruning of the logs by the Propagation Constraints is still
effective. Also recall that the sizes of the logs are bounded [5]; once a process P_i

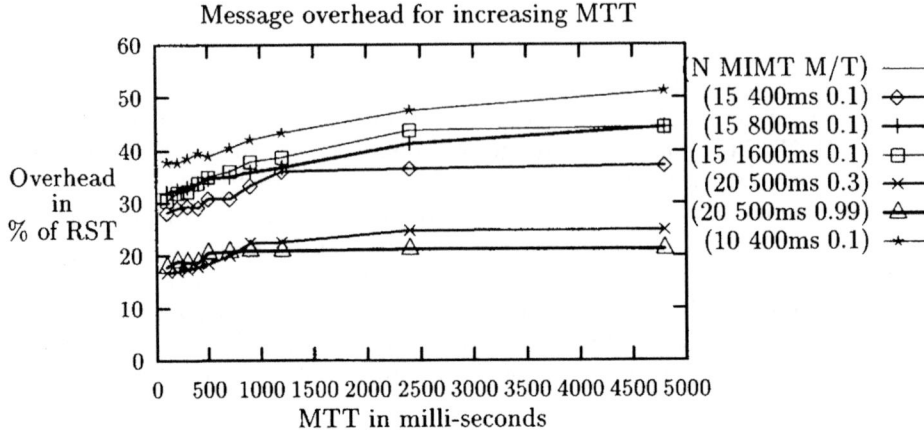

Fig. 3. Average message overhead as a function of MTT

has a log record of a message send to process P_j, a log record of a new message send to P_j can potentially erase all previous log records of messages sent to P_j. At lower MTT, because of faster propagation of log information, pruning logs using the Propagation Constraints is more effective and the logs are much smaller.

Note that despite an initial increase, the overhead is always significantly less than that of RST. For example, in simulations S_1, S_2, and S_3, the message overhead is never more than 40% that of RST. The next three simulations fixed (N, MIMT, M/T) at $S_4(20, 500ms, 0.3)$, $S_5(20, 500ms, 0.99)$, and $S_6(10, 400ms, 0.1)$. For simulations S_4 and S_5, the overhead is always less than 24% of that of RST.

The runs S_4 and S_5 show that increasing multicast frequency, thus increasing the network load, does not affect the overhead even under extreme network load conditions, i.e., under high MTT. This is because the log sizes have already reached a "steady-state" proportion of n^2 and multicasts cannot increase them much further. Besides, multicasts effectively distribute the log information faster into the system because they convey information to more number of processes. Thus when a multicast message is ultimately delivered, it can potentially cause a lot of log pruning at the destination. Thus we can conclude that the KS algorithm has better performance when compared to RST, even under high MTT.

4.3   Behavior under Decreasing Communication Load

The next set of simulations is aimed at determining the overhead behavior when the KS algorithm is used in applications that use communication sparingly. The values of (N, MTT, M/T) were fixed at $S_1(10, 100ms, 0.1)$, $S_2(15, 100ms, 0.1)$, $S_3(15, 800ms, 0.1)$, and $S_4(20, 100ms, 0.1)$ while varying MIMT from $100ms$ to $12800ms$, initially in steps of 100 and later in multiples of 2. The results for the average message overhead are shown in figure 4. As we were testing the system for behavior under light to moderate loads, we did not increase the traffic by increasing M/T.

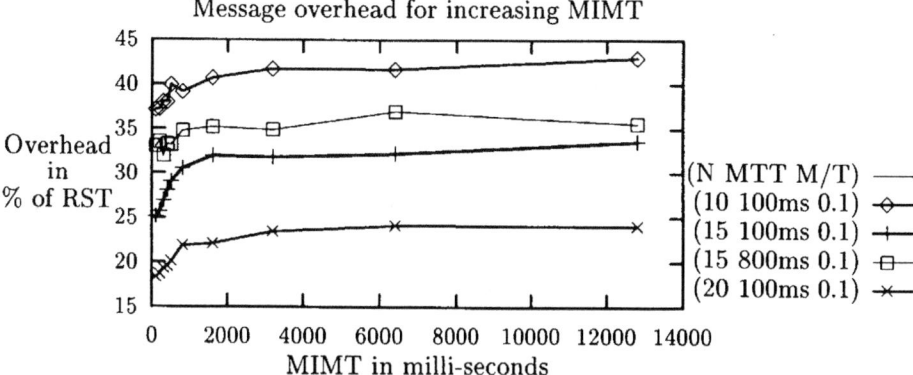

Fig. 4. Average message overhead as a function of $MIMT$

The results show a steep initial increase in overheads with increasing MIMT, followed by a leveling off to a steady overhead. Low MIMT means that messages are generated more frequently. As analyzed before, frequent message delivery disseminates log information faster and thus helps purge log entries. With increasing MIMT, message delivery information required by the Propagation Constraints to perform pruning of logs takes longer time in reaching all the processes that have the log record of a message send event. Hence the pruning of logs slows and log records grow in size with increasing MIMT. However as MIMT becomes very high, the generation of messages becomes infrequent. As new messages are generated very infrequently, the growth of a process's log is reduced. This causes the log growth rates to level off for high MIMTs.

Note that despite the steep initial increase, the overheads are always much less than those of RST. This is true even in the case of the 10 process simulation where, though the overhead is higher than for all other runs, it is still always lesser than 45% of that of RST.

4.4 Overhead for Increasing Multicast Frequency

The sensitivity of the KS algorithm to multicast frequency is of interest because multicasts seem to favor the pruning of logs.

We ran six simulation runs increasing M/T from 0.1 to 1.0 in steps of 0.1. The number of processes was varied starting from 25 and decreased to 12 across the simulations. MTT was progressively increased from $50ms$ to $500ms$ across the six runs. MIMT was varied from $400ms$ to $1000ms$. The results for the average message overhead are shown in figure 5.

For the two simulation runs (N, MTT, MIMT) = $S_1(25, 50ms, 400ms)$ and $S_2(20, 50ms, 400ms)$, the overheads are almost constant. For all the other runs,

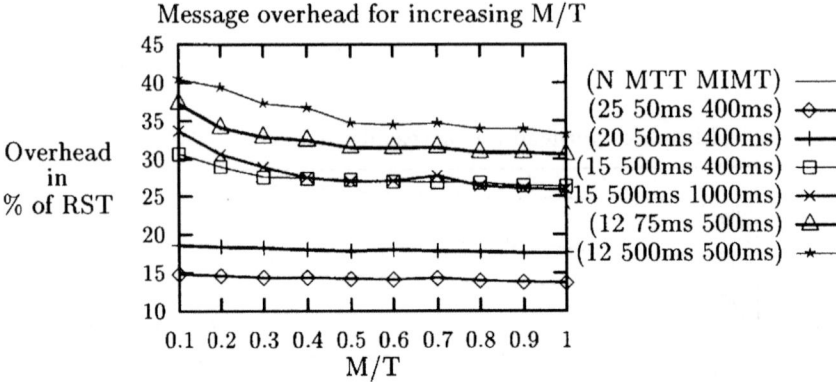

Fig. 5. Average message overhead as a function of M/T

they decrease with increase in M/T. The two simulation runs S_1 and S_2 represent networks of higher speed and more processes than the other runs. Because of the prompt delivery of all messages, increasing multicast frequency cannot decrease the overhead from the already existing minimal overhead. However for the other simulations, which have high MTT and/or high MIMT, increasing multicasts causes more efficient distribution of information which is useful to prune logs by effective application of the Propagation Constraints.

This experiment reaffirms our guess about the performance under high loads. Despite increasing network traffic by increasing M/T, the overheads decrease.

5 Concluding Remarks

This paper conducted a performance analysis of the space complexity of the optimal KS algorithm under a wide range of system conditions using simulations. The KS algorithm was seen to perform much better than the canonical RST algorithm under the wide range of network conditions simulated. In particular, as the size of the system increased, the KS algorithm performed very well and had an overhead rate of less than 10% of that for the canonical RST algorithm. The algorithm also performed very well under stressful network loads besides showing better scalability. As such, the KS algorithm which has been shown theoretically to be optimal in the space overhead does offer large savings over the standard canonical RST algorithm, and is thus an attractive and efficient way to implement the causal message ordering abstraction.

Acknowledgements

This work was supported by the U.S. National Science Foundation grants CCR-9875617 and EIA-9871345.

References

1. Y. Amir, D. Dolev, S. Kramer and D. Malki, Transis: A communication sub-system for high-availability, *Proceedings of the 22nd International Symposium on Fault-tolerant Computing*, IEEE Computer Society Press, 337-346, 1991.
2. K. Birman, T. Joseph, Reliable communication in the presence of failures, *ACM Transactions on Computer Systems*, 5(1): 47-76, Feb. 1987.
3. K. Birman, A. Schiper and P. Stephenson, Lightweight causal and atomic group multicast, *ACM Transactions on Computer Systems*, 9(3): 272-314, Aug. 1991.
4. M. F. Kaashoek and A. S. Tanenbaum, Group communication in the Ameoba distributed operating system, *Proceedings of the Fifth ACM Annual Symposium on Principles of Distributed Computing*, 125-136, 1986.
5. A. Kshemkalyani and M. Singhal, Necessary and sufficient conditions on information for causal message ordering and their optimal implementation, *Distributed Computing*, 11(2), 91-111, April 1998.
6. L. Lamport, Time, clocks, and the ordering of events in a distributed system, *Communications of the ACM*, 21(7): 558-565, July 1978.
7. L. L. Peterson, N. C. Bucholz and R. D. Schlichting, Preserving and using context information in interprocess communication, *ACM Transactions on Computer Systems*, 7(3), 217-246, 1989.
8. M. Raynal, A. Schiper, S. Toueg, The causal ordering abstraction and a simple way to implement it, *Information Processing Letters*, 39:343-350, 1991.
9. A. Schiper, A. Eggli, A. Sandoz, A new algorithm to implement causal ordering, *Proceedings of the Third International Workshop on Distributed Systems*, Nice, France, LNCS 392, Springer-Verlag, 219-232, 1989.

Register Efficient Mergesorting

Abhiram Ranade[1], Sonal Kothari[2], and Raghavendra Udupa[3]

[1] Indian Institute of Technology, Powai, Mumbai 400076, India
[2] 8343 Princeton Square Blvd East #906, Jacksonville FL 32256, USA
[3] IBM India Research Lab., IIT, New Delhi- 110 016, India

Abstract. We present a register efficient implementation of Mergesort which we call FAME (Finite Automaton MErgesort). FAME is a m-way Mergesort. The m streams are merged by organizing comparison tournaments among the elements at the heads of the streams. The winners of the tournament form the output stream. Many ideas are used to increase efficiency. First, the heads of the streams are maintained in the register file. Second, the tournaments are evaluated incrementally, i.e. after one winner is output the next tournament uses the results of the comparisons performed in the preceding tournaments and thus minimizes work. Third, to minimize register movement, the state of the tournament is encoded as a finite automaton. We experimented with 8-way and 4-way FAME on an Ultrasparc and a DEC Alpha and found that these algorithms were better than cache-cognizant Quicksort algorithms on the same machines.

1 Introduction

It has been recently noted that explicit management of the memory hierarchy is necessary to get high performance for many application problems. In this paper we consider the question for sorting. We are specifically concerned with the highest level of the memory hierarchy: registers, though we also pay attention to the cache.

Many sorting algorithms that attempt to adapt to the memory hierarchy have been proposed and studied in the literature. These studies have been concerned with the lower levels of the memory hierarchy, e.g. disk[1,2,6], or cache[4] or disk and cache[5]. To the best of our knowledge, little has been reported on how to exploit registers efficiently. Since modern processors have a large number of registers (typically 32), and these tend to have higher bandwidth to the ALU or lower access time or both, we believe there is an opportunity here to improve performance further.

Many of the ideas that have arisen while developing cache/disk cognizant sorting algorithms are also relevant while optimizing for register usage. One such idea is *tiling*: instead of making several passes over the entire dataset, it is better to break the dataset into *tiles* which can fit in the cache and then process each tile intensively as possible. Typically, recursive implementations of Quicksort and Mergesort implicitly employ tiling, while their common iterative formulations do

M. Valero, V.K. Prasanna, and S. Vajapeyam (Eds.): HiPC 2000, LNCS 1970, pp. 96–103, 2000.

not. Another idea is to use multiway merging – for m way merging the number of passes needed over data in Mergesort is $\log_m N$ for N keys. Clearly, large m is to be preferred. These techniques have been used by LaMarca and Ladner[4] as well as Nyberg et al[5].

While ideas such as multiway merging are useful for efficiently using registers, new ideas are also needed because unlike memory, registers cannot be addressed indirectly, i.e. most instruction sets do not provide ways to say "compare register i and register j, where i and j are themselves in registers k and l". Second, the number of registers is usually fairly small – this makes it possible to use special coding techniques.

In this paper we describe a register efficient sorting algorithm called FAME (Finite Automaton MErgesort). FAME is a m-way Mergesort. The m streams are merged by organizing comparison tournaments among the elements at the heads of the streams. The winners of the tournament form the output stream. FAME exploits registers by maintaining the keys at the head of the streams in the register file. In FAME the tournaments are evaluated incrementally, i.e. after one winner is output the next tournament uses the results of the comparisons performed in the preceding tournaments and thus minimizes work. This needs the state of the tournament to be somehow recorded. FAME does this in a novel manner: the state of the tournament is encoded in a finite automaton. The basic merging mechanism of FAME is driven by this finite automaton, so effectively the state of the finite automaton (and the tournament) is encoded in the program counter.

We describe 2 implementations of FAME and compare them with a *Memory tuned* Quicksort implemented along the lines described by LaMarca and Ladner[4]. We chose Quicksort as the basis for comparison mainly because it turns out to be the best algorithm for the range (4K to 4M keys) studied by them. We present comparisons on 2 machines – a Tandem(DEC) Alpha-250 and an Ultrasparc. On both these machines, FAME implementations perform better than Quicksort.

Our work is an extension of some of the work in [3].

Outline: In Section 2 we present the Finite Automaton MErgesort. Section 3 describes the main features of our benchmark Quicksort. In Section 4 we report our experiments. Section 5 discusses our conclusions and future work.

2   FAME

The basic merging iteration is as follows:

1. Conduct a tournament between the keys at the head of all (non-decreasing) streams to find the smallest key.
2. Append the winner to the output stream.
3. Advance the stream in which the winner was found.

Obviously, in each iteration the entire tournament need not be played out explicitly; after all, nearly the same keys participate in the ith tournament as well as the $i + 1$th (the only difference being the key that won the ith tournament). More precisely, let T_i denote the transcript for the ith tournament (assuming it is played out completely). If m is the number of streams being merged, then a transcript is simply a sequence of $m-1$ bits that record (in some fixed order) the results of the $m - 1$ comparisons performed in the tournament. The important observation is: Transcript T_i and T_{i+1} differ in at most $\log m$ bits. To see this, suppose that the key at the head of the sth stream won the ith tournament. Then the keys involved in the $i + 1$th tournament are the same as those in the ith tournament, except for the key from stream s. Thus, the transcript can only be different for the $\log m$ comparisons from the sth leaf to the root.

Thus, our basic action requires only $\log m$ comparisons to be performed. The important question, however, is how to keep track of the transcript conveniently so that at each point the comparisons to be peformed are known easily.

To do this, our code is organized as a finite automaton. In each state of the automaton one basic action takes place, and then the automaton transits to another appropriate state. States of the automaton corresponds to possible transcripts of the tournament: after completing the tournament for finding the ith winner, the automaton enters the state T_i (as defined above).

Action in each state of the automaton: In each state all we need to do is to *incrementally* compute the next transcript. From the description of the current state (transcript for the last tournament) we know which stream s contributed the previous winner, and hence the new key for the current tournament, and hence which $\log m$ comparisons need to performed to complete the current tournament. The code for these comparisons is "hard-wired" into the code for each state. When a state is visited, these comparisons get performed. As a result the winner is known and appeneded to the output stream. The stream from which the winner came is advanced. Finally, we transit to the state corresponding to the newly constructed transcript.

We give a detailed example. Consider the case of $m = 4$ streams, denoted s_0, s_1, s_2, s_3. Let $H(s)$ denote the key at the head of stream s. The tournament tree is a complete binary tree with 4 leaves, and has 3 internal nodes, i.e. 3 comparisons need be performed. The transcript for such a tournament thus has 3 bits $b_0 b_1 b_2$. The bits are interpreted as follows. $b_0 = 1$ if $H(s_0) < H(s_1)$, and 0 otherwise. $b_1 = 1$ if $H(s_2) < H(s_3)$ and 0 otherwise. $b_2 = 1$ if the smaller of $H(s_0), H(s_1)$ is smaller than the smaller of $H(s_2), H(s_3)$ and 0 otherwise. The Automaton has 8 states. Figure 1 shows the code for state 101 as an example.

The code is based on the following observations. First we know that for the previous tournament $H(s_0) < H(s_1)$, $H(s_2) \geq H(s_3)$ and $H(s_0) < H(s_3)$. Thus $H(s_0)$ was the smallest. Thus, $H(s_0)$ is new when the automaton reaches this state. Hence we need to compare the new $H(s_0)$ with $H(s_1)$, and the smaller among them with $H(s_3)$. Depending upon which way the comparisons go, this will cause either $H(s_0)$, $H(s_1)$ or $H(s_3)$ to be determined as the smallest. The smallest key thus determined is appended to the output stream; stream from which it came is advanced, and finally a transition is made to the appropriate next state.

State101:   If $H(s_0) < H(s_1)$ then
   If $H(s_0) < H(s_3)$ then
    Append $H(s_0)$ to output stream. Advance s_0. Go to State101.
   else
    Append $H(s_3)$ to output stream. Advance s_3. Go to State100.
  else
   If $H(s_1) < H(s_3)$ then
    Append $H(s_1)$ to output stream. Advance s_1. Go to State001.
   else
    Append $H(s_3)$ to output stream. Advance s_3. Go to State000.

Fig. 1. Code for State 101

Figure 1 omits some details, e.g. checking whether streams have ended. In the actual code, we use sentinels to terminate each stream. With this, it suffices to count the number of keys inserted into the output stream – when the requisite number are inserted, then we simply append the sentinel and the merging is declared complete. More importantly sentinels ensure that none of the streams terminates prematurely – at least the sentinel of the stream will stay till the end.

Use of Registers: The motivation for the above mechanism is of course the use of registers. The keys at the heads of all streams and pointers to the stream heads are maintained in registers. An additional register is needed to count the number of keys that have been appended to the output stream. Thus $2m + 1$ registers are needed.

Code length: Note that the code for any state must explicitly and separately deal with all the outcomes of $\log m$ comparisons. Thus there are m cases to consider, so the code will have length $O(m)$. Since there are 2^{m-1} states whose code must be written out explicitly, the total code length is thus $O(m2^m)$.

2.1   Comments

We note that the merging mechanism described above has several good features. First, it actually allows a large amount of the state to be held in registers during processing. Second, it completely eliminates register movements and in general unnecessary copying. An alternative to our method would be to organize the tournament as a heap – we coded this approach but it was hopeless due to the index calculations and register moves involved in navigating through the heap.

The main drawback of the scheme, in our opinion, is that the code for it is long. For example, the code for 8 way FAME exceeded the instruction cache length on one of the machines we experimented with (DEC 3000). If the instruction cache is long enough, as was the case for the two machines we studied extensively, this is not a serious problem.

It should be noted however, that while FAME attempts to minimize register moves (practically eliminates them), this may not be important on modern superscalar machines where the moves might be executed in parallel anyway.

2.2   Implementation Details

We did two implementations of FAME, one for $m = 4$ and the other for $m = 8$. The choice of m is governed by several factors. First, to keep the tournament tree simple, we would like it to be a power of 2. Second, to implement the merge mechanism well we need the processor to have $2m + 1$ registers at least. Since most modern processors have about 32 registers, this limits m to 8. Finally, the code length is proportional to $m2^{m-1}$. For $m = 16$ this would be too large, unlikely to fit in the instruction caches of most processors.

Our implementation was in the C language, and the code for the merging mechanism described above was generated by using C preprocessor macros. This is inevitable, especially for the case $m = 8$ which has 128 states. The code for each state is also considerably long, since it must handle each of the 8 possible ways in which 3 comparisons can get resolved. To ensure that the heads of the streams and the pointers to heads are kept in registers, we declared the associated variables to be of storage class *register*.[1]

The use of the sentinels can be expensive (excessive memory requirement) for merging very short sequences. For example, if we were to merge length 1 sequences, then half the memory would be taken up by sentinels. We avoided this by using Quicksort to sort very short sequences. The threshold for this was determined experimentally: it was about 16 for the DEC alpha, 128 for the Ultrasparc and 512 for the RS 6000.

Our merging code was organized recursively. This organization has the overhead of recursion, but it ends up exploiting locality and gives better cache performance.

3   Quicksort

Our Quicksort is based on the memory tuned Quicksort of LaMarca and Ladner[4] and includes several standard optimizations: (i) For sequences of length 32 we performed insertion sort. (ii) Instead of selecting the splitter at random we chose it to be the median of 3 randomly chosen elements. (iii) The algorithm was coded in a recursive manner which is known to exploit locality better and thus give better cache behaviour. The implementation was in the C language and was compiled with highest level of optimization.

4   Experiments

We experimented extensively on two machines: a DEC Alpha, and an Ultrasparc. The complete configurations are as follows

	Registers	L1 D-Cache	L1 I-Cache	L2 Cache	Clock Speed
Alpha-250	32	16 KB	16 KB	2 MB	266 MHz
Ultrasparc	24 (window)	16 KB	16 KB	512 KB	167 MHz

[1] We examined the generated code and found that the compiler had indeed allocated registers as we had directed.

We report the time to sort using various algorithms and various numbers of keys. In all cases the keys are 32 bit integers. The sorting time is reported on a per key basis, so that to calculate the actual sorting time it is necessary to multiply the plotted time by the number of keys. Three algorithms are considered (i) Quicksort (ii) 4-way FAME, (iii) 8-way FAME. The reported time is the median of the times taken for sorting 100 randomly generated data sets. For consistency, the same data sets were used for all algorithms. We have not shown the standard deviation, but it was less than 1 % for FAME for more than 4096 keys, and less than 3 % for Quicksort. The mean was very close to the median (easily within the standard deviation).

We also ran experiments on a DEC 3000 and an Intel Pentium. On the Pentium FAME performed badly. This is to be expected because the Pentium does not have many registers. The behaviour on the 3000 is discussed in Section 4.3.

4.1   Alpha-250 Results

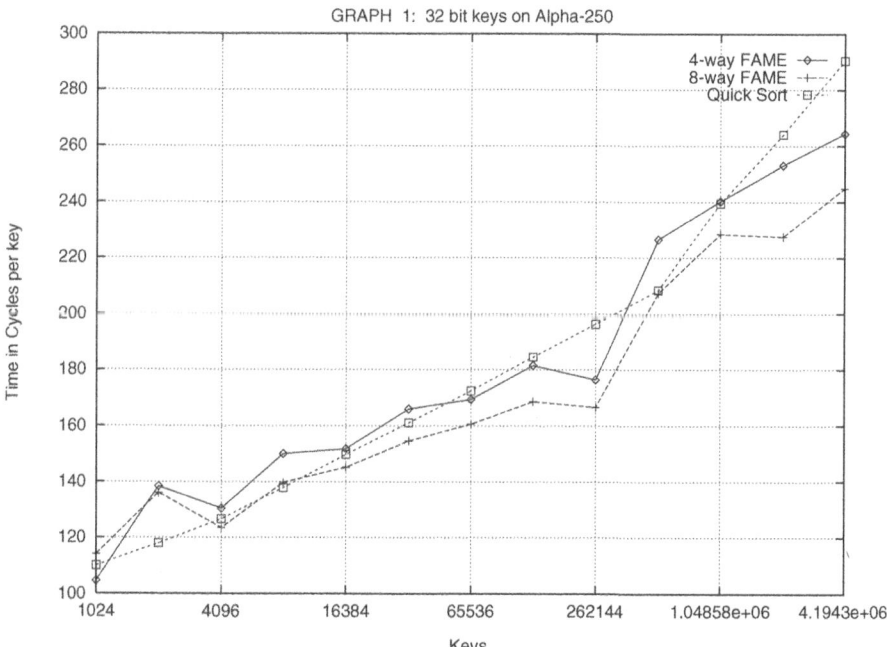

It is seen that the 8-way FAME outperforms Quicksort for more than 8192 keys. Beyond 65536 keys, even the 4-way FAME performs better, except for 524288 keys.

The plots for FAME show several minor bumps and one major bump at 524288 keys. At 524288 keys the data no longer fits the 2MB secondary cache (remember that the Mergesort requires about twice the memory of Quicksort), it is because of this that we think the time steeply rises. It should be noted that 8-way FAME performs better than Quicksort inspite of this rise. Quicksort also shows this steep rise, but it happens at 1 M keys. At 1 M keys and more, Quicksort performance is worse than 8 way as well as 4 way FAME.

The minor bumps in the plots for FAME arise probably because of the way in which FAME switches to Quicksort. Nominally the switch happens when the number of keys drop down below 32. But because the merging is 4 way or 8 way, the switch could happen at either 8 keys, or 16 keys or 32 keys.

4.2   Ultrasparc Results

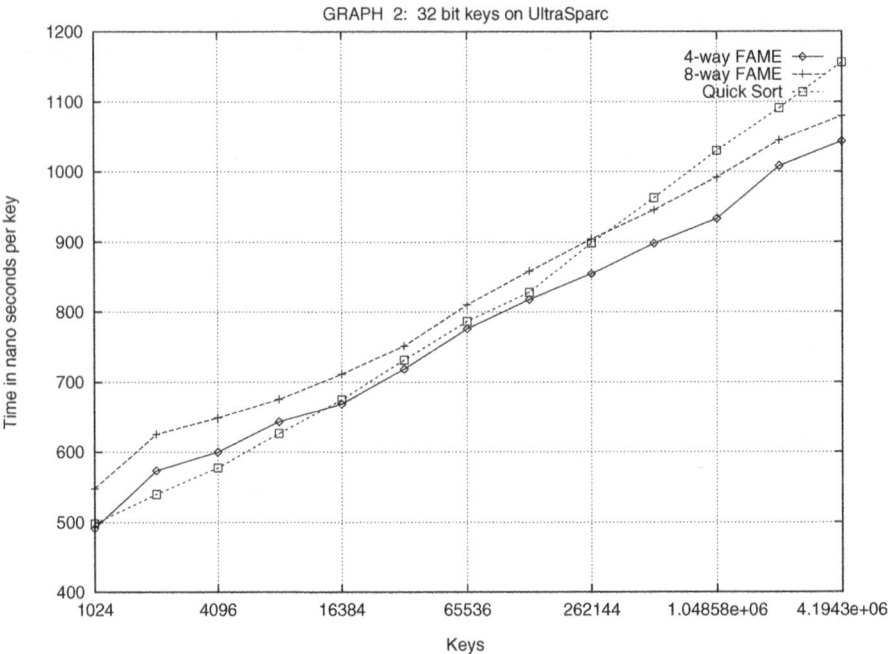

It is seen that 4-way FAME outperforms Quicksort for more than 16K keys. 8-way FAME outperforms Quicksort beyond 512K keys, but is itself worse than 4-way FAME. The plots here are much smoother than those for the Alpha-250, presumably because the processor is slower and thus better matches the memory speeds.

4.3   DEC 3000

The interesting observation here was that 4-way FAME was better than Quicksort, but 8 way FAME was not. Our DEC 3000 (MIPS processor) machine had a small instruction cache, which was easily seen to be incapable of holding the entire code for the 8 way FAME. Thus, we believe that the poor performance of 8 way FAME was due to instruction cache thrashing. It will be interesting to run the experiments on a MIPS architecture with a larger cache.

5   Concluding Remarks

We expected 8 way FAME to outperform both the 4 way FAME and the Quicksort. The number of comparisons made by all the algorithms are roughly similar, the difference is in the number of passes over memory. For N keys, Quicksort makes $\log_2 N$ passes over memory, 4 way FAME makes $\log_4 N$ passes and 8 way FAME $\log_8 N$ passes.

Our expectation held up for the Alpha-250, but not for the Ultrasparc. A possible explanation is that the code simplicity of 4 way FAME compensates for its additional passes over memory. But this needs more analysis, certainly. It should be noted that modern processors are very complex – with lots of mechanisms such as superscalar issue, pipeline interlocks, branch prediction mechanisms etc. coming into play. The final performance is a result of these complex interactions.

In summary, we have presented a strategy for exploiting registers in sorting programs. We have also presented a preliminary evaluation of the strategy on two architectures, where the strategy was seen to yield performance improvements. It will be interesting to evaluate the strategy on other processor architectures.

We suspect that much more work can be done in this area. Following LaMarca and Ladner, it might be interesting to seek register efficient implementations for all the standard sorting algorithms such as Selection sort, Heapsort, Radix sort and Quicksort, not just Mergesort. While we suspect that the only serious challenger to FAME will be a register efficient Quicksort (say a multiway Quicksort), studying the register efficient variants of even selection sort will be of interest, since that gets used when the number of keys drops down below 32. For this size, selection sort can be potentially extremely efficient because it could be made to run entirely inside the registers!

References

1. Alok Aggarwal and Jeffrey Scott Vitter. I/O-complexity of sorting and related problems. *Communications of the ACM*, 31:1116–1127, 1988.
2. D. E. Knuth. *The art of computer programming*, volume 3. Addison-Wesley, 1973.
3. Sonal (Sancheti) Kothari. Algorithm Optimization Techniques for the Memory Heirarchy. M.Tech thesis, Indian Institute of Technology, Bombay, 1999.
4. A. LaMarca and R. Ladner. The influence of caches on the performance of sorting. In *Proceedings of the ACM-SIAM Symposium on Discrete Algorithms*, 1997.
5. C. Nyberg, T. Barclay, Z. Cvetanovic, J. Gray, and D. Lomet. AlphaSort: a RISC machine sort. In *Proceedings of the 1994 ACM SIGMOD international conference on Management of data*, volume 23, pages 233–242, 1994.
6. Jeff Vitter. External Memory Algorithms and Data structures: Dealing with Massive Data, 1999. manuscript.

Session II-A

High-Performance Middleware
Co-Chairs:
Shikaresh Majumdar, Carleton University
Gabriel Kotsis, University of Vienna

Applying Patterns to Improve
the Performance of Fault Tolerant CORBA

Balachandran Natarajan[1], Aniruddha Gokhale[2],
Shalini Yajnik[2], and Douglas C. Schmidt[3]

[1] Dept. of Computer Science,
Washington University
One Brookings Drive
St. Louis, MO 63130
bala@cs.wustl.edu
[2] Bell Laboratories,
Lucent Technologies,
600 Mountain Avenue,
Murray Hill, NJ 07974
{agokhale,shalini}@lucent.com
[3] University of California,
616E Engineering Tower,
Irvine, CA 92697
schmidt@ece.uci.edu

Abstract. An increasing number of mission-critical systems are being developed using distributed object computing middleware, such as CORBA. Applications for these systems often require the underlying middleware, operating systems, and networks to provide end-to-end quality of service (QoS) support to enhance their efficiency, predictability, scalability, and fault tolerance. The Object Management Group (OMG), which standardizes CORBA, has addressed many of these QoS requirements the recent Real-time CORBA and Fault Tolerant CORBA (FT-CORBA) specifications. This paper describes the patterns we are incorporating into a FT-CORBA service called DOORS to eliminate performance bottlenecks caused by common implementation pitfalls.

1   Introduction

Emerging trends: Applications for next-generation distributed systems are increasingly being developed using standard services and protocols defined by distributed object computing middleware, such as the Common Object Request Broker Architecture (CORBA) [1]. CORBA is a distributed object computing middleware standard defined by the OMG that allows clients to invoke operations on remote objects without concern for where the object resides or what language the object is written in [2]. In addition, CORBA shields applications from non-portable details related to the OS/hardware platform they run on and the communication protocols and networks used to interconnect distributed objects. These features make CORBA ideally suited to provide the core communication infrastructure for distributed applications.

M. Valero, V.K. Prasanna, and S. Vajapeyam (Eds.): HiPC 2000, LNCS 1970, pp. 107–120, 2000.

A growing number of next-generation applications demand varying degrees and forms of quality of service (QoS) support from their middleware, including efficiency, predictability, scalability, and fault tolerance. In CORBA-based middleware, this QoS support is provided by Object Request Broker (ORB) endsystemsORB endsystems consist of network interfaces, operating system I/O subsystems, CORBA ORBs, and higher-level CORBA services.

Addressing middleware research challenges with patterns: Our prior research on CORBA middleware (`www.cs.wustl.edu/~schmidt/corba-research.html`) has explored many efficiency, predictability, and scalability aspects of ORB endsystem design, including static and dynamic scheduling, event processing, I/O subsystem and pluggable protocol integration, synchronous and asynchronous ORB Core architectures, systematic benchmarking of multiple ORBs, optimization principle patterns for ORB performance, and measuring performance of a CORBA fault-tolerant service. This paper focuses on another dimension in the ORB endsystem design space: *applying patterns to improve the performance of Fault Tolerant CORBA (FT-CORBA) implementations*.

A *pattern* names and describes a recurring solution to a software development problem within a particular context [3]. Patterns help to alleviate the continual re-discovery and re-invention of software concepts and components by documenting and teaching proven solutions to standard software development problems. For instance, patterns are useful for documenting the structure and participants in common communication software micro-architectures, such as *active objects* [4] and *brokers* [5]. These patterns are generalizations of object-structures that have been used successfully to build flexible, efficient, event-driven, and concurrent communication software, including ORB middleware.

In general, patterns can be categorized as follows:

Design patterns: A design pattern [3] captures the static and dynamic roles and relationships in solutions that occur repeatedly when developing software applications in a particular domain. The design patterns we apply to improve the performance of FT-CORBA include: *Abstract Factory, Active Object, Chain of Responsibility, Component Configurator, and Strategy*.

Architectural patterns: An architecture pattern [4] expresses a fundamental structural organization schema for software systems that provides a set of predefined subsystems, specifies their responsibilities, and includes rules and guidelines for organizing the relationships between them. The architectural patterns we apply to improve the performance of FT-CORBA include: *Leader/Followers and Reactor*.

Optimization principle patterns: An optimization principle pattern [6] documents rules for avoiding common design and implementation mistakes that degrade the performance, scalability, predictability, and reliability of complex systems. The optimization principle patterns we applied to improve performance of FT-CORBA include: *optimizing for the common case, eliminating gratuitous waste, and storing redundant state to speed up expensive operations*.

Paper organization: The remainder of this paper is organized as follows: Section 2 summarizes the recently adopted Fault Tolerant CORBA (FT-CORBA) specification. Section 3 describes the patterns we are using to improve the performance of our FT-CORBA service called DOORS; and Section 4 presents concluding remarks.

2   Overview of the Fault Tolerant CORBA Specification

The Fault Tolerant CORBA (FT-CORBA) [7] specification defines a standard set of interfaces, policies, and services that provide robust support for applications requiring high reliability. The fault tolerance mechanism used to detect and recover from failures is based on *entity redundancy*. Naturally, in FT-CORBA the redundant entities are replicated CORBA objects.

Replicas of a CORBA object are created and managed as a "logical singleton" [3] composite object. Figure 1 illustrates the key components in the FT-CORBA architecture. All components shown in the figure are implemented as standard CORBA objects, *i.e.*, they are defined using CORBA IDL interfaces and implemented using servants that can be written in standard programming languages, such as Java, C++, C, or Ada. The functionality of each component is described below.

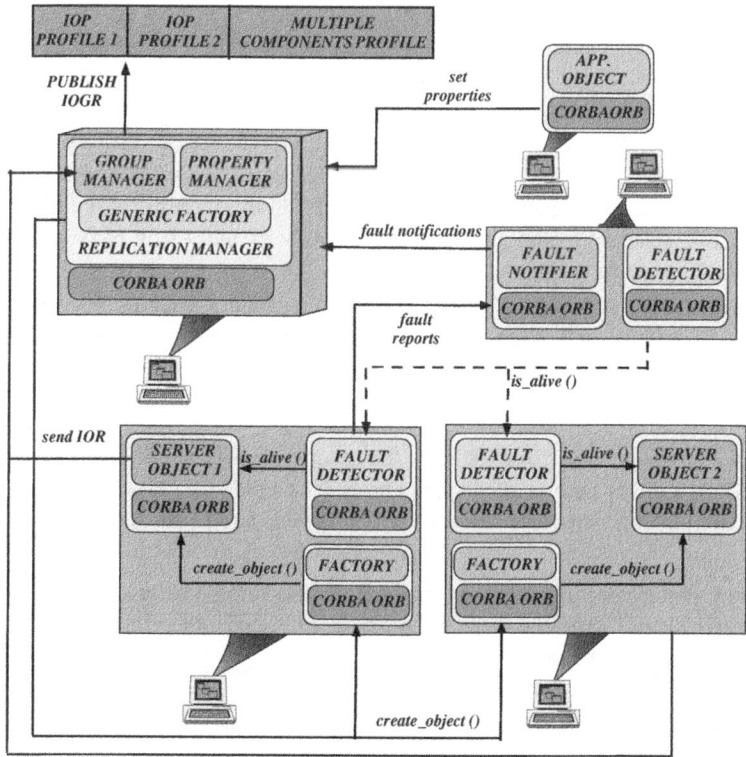

Fig. 1. The Architecture of Fault Tolerant CORBA

Interoperable object group references (IOGRs): FT-CORBA standardizes the format of interoperable object references (IOR) used for the individual replicas. An IOR is a flexible addressing mechanism that identifies a CORBA object uniquely [1]. In addition, it defines an IOR for composite objects called the *interoperable object group reference* (IOGR).

FT-CORBA servers can publish IOGRs to clients. Clients use these IOGRs to invoke operations on servers. The client-side ORB transmits the request to the appropriate server-side object that handles the request. The client application need not be aware of the existence of server object replicas. If a server object fails, the client-side ORB cycles through the object references contained in the IOGR until the request is handled successfully by a replica object. The references in the IOGR are considered invalid only if all server objects fail, in which case an exception is propagated to the client application.

ReplicationManager: This component is responsible for managing replicas and contains the following three components:

1. PropertyManager: This component allows properties of an object group to be selected. Common properties include the replication style, membership style, consistency style, and initial/minimum number of replicas. Example replication styles include the following:

- COLD_PASSIVE – In this replication style, the replica group contains a single primary replica that responds to client messages. If a primary fails, a backup replica is spawned on-demand to function as the new primary.
- WARM_PASSIVE – In the WARM_PASSIVE replication style, the replica group contains a single primary replica that responds to client messages. In addition, one or more backup replicas are pre-spawned to handle crash failures. If a primary fails, a backup replica is selected to function as the new primary and a new backup is created to maintain the replica group size constant.
- ACTIVE – In the ACTIVE replication style all replicas are primary and handle client requests independently of each other. To ensure a single reply sent to the client and to maintain consistent state amongst the replicas, a special group communication protocol is necessary.

Membership of a group and data consistency of the group members can be controlled either by the FT-CORBA infrastructure or by applications. FT-CORBA standardizes both application-controlled and infrastructure -controlled membership and consistency styles.

2. GenericFactory: For the infrastructure-controlled membership style, the `Gener-icFactory` is used by the `ReplicationManager` to create object groups and individual members of an object group.

3. ObjectGroupManager: For the application-controlled membership style, applications use the `ObjectGroupManager` interface to create, add, or delete members of an object group.

Fault Detector and Notifier: `FaultDetectors` are CORBA objects responsible for detecting faults via either a *pull-based* or a *push-based* mechanism. A *pull-based* monitoring mechanism periodically polls applications to determine if their objects are "alive." FT-CORBA requires application objects to implement a `PullMonitorable` interface that exports an `is_alive` operation. A *push-based* monitoring mechanism can also be implemented. In this scheme, which is also known as a "heartbeat monitor," applications implement a `PushMonitorable` interface and send periodic heartbeats to the `FaultDetector`.

FaultDetectors report faults to `FaultNotifiers`. In turn, the `FaultNotifiers` propagate these notifications to a `ReplicationManager`, which performs recovery actions.

Logging and Recovery: FT-CORBA defines a logging and recovery mechanism that is responsible for intercepting and logging CORBA GIOP messages from client objects to servers. Distributed applications can employ this mechanism via an infrastructure-controlled consistency style. If a failure occurs, a new replica is chosen to become the "primary." The recovery mechanism then re-invokes the operations that were made by the client, but which did not execute due to the primary replica's failure. In addition, it retrieves a consistent state for the new replica. The logging and recovery mechanism ensures that failovers are transparent to applications. For the application-controlled consistency style, applications are responsible for their own failure recovery.

FT-CORBA is designed to prevent single points of failure within a distributed object computing system. As a result, each component described above must itself be replicated. Moreover, mechanisms must be provided to deal with potential failures and recovery.

3   Applying Patterns to Improve DOORS Fault Tolerant CORBA Performance

Implementations of FT-CORBA, such as DOORS, are representative of complex communication software. Optimizing this type of software is hard since seemingly minor "mistakes," such as poor choice of concurrency architectures and data structures, lack of caching, and the inability to configure parameters dynamically, can adversely affect performance and availability. Therefore, developing high-performance, predictable, reliable, and robust software requires an iterative optimization process that involves (1) performance benchmarking to identify sources of overhead and (2) applying patterns to eliminate the identified sources of overhead. The patterns described in this section are shown in Table 1.

[8,9] describe a family of optimization principle patterns and illustrate how they have been applied in existing protocol implementations, such as `TCP/IP` and CORBA `IIOP`, to improve their performance. Likewise, our prior research on developing extensible real-time middleware [10,6] has enabled us to document the design, architectural, and optimization principle patterns used to improve performance and predictability.

This section focuses on the various design, architectural, and optimization principle patterns we are applying to systematically improve the performance of the DOORS FT-CORBA implementation. We focus on these patterns since they were the most strate-

Table 1. Patterns for Implementing FT-CORBA Efficiently

#	Problem	Pattern	Pattern Category
1	Missed Polls in `FaultDetector`	Leader/Followers	Architectural
2	Excessive overhead of recovery	Active Object	Design
3	Excessive overhead of service lookup	Optimize for the common case	Optimization
		Eliminate gratuitous waste	Optimization
		Store extra information	Optimization
4	Tight coupling of data structures	Strategy	Design
		Abstract Factory	Design
5	Inability for dynamic configuration	Component Configurator	Design
6	Property lookup	Chain of Responsibility	Design
		Perfect hash functions	Optimization

gic to eliminate sources of overhead in DOORS FT-CORBA that arose from common implementation pitfalls.

In the following discussion, we outline the forces underlying the key design challenges that arise when developing high-performance FT-CORBA middleware, such as DOORS. We also describe which patterns resolve these forces and explain how these patterns are used in DOORS. In general, the absence of these patterns leaves these forces unresolved.

3.1   Decoupling Polling and Recovery

Context: In DOORS, FT-CORBA objects operating under the PULL-based fault monitoring style are polled at specific intervals of time by a separate poller thread in the `FaultDetector`. In the event of failure, this poller thread identifies an application object crash and reports the failure to the `ReplicationManager`, which then performs the recovery.

Problem: If the polling thread polls and reports failures in the same thread of control, then in the event of failure it will block until the `ReplicationManager` has recovered from failure. This will cause missed polls to other application objects. This behavior is unacceptable for systems requiring high availability. A naive solution would be to create a separate polling thread for each application object. This strategy does not scale, however, as the number of objects polled by the fault monitor increase. Thus, the force that must be resolved involves ensuring that the `FaultDetector` polls all application objects at the specified intervals, even when the poller thread is blocked during failure recovery.

Solution → the Leader/Followers pattern: An effective way to avoid unnecessary blocking is to use the *Leader/Followers* pattern [4]. This pattern provides an efficient concurrency model where multiple threads take turns sharing a set of event sources in order to detect, demultiplex, dispatch, and process service requests that occur on these event sources.

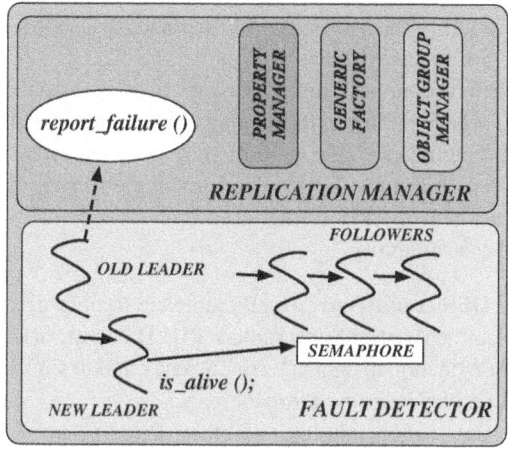

Fig. 2. Applying the Leader/Followers Pattern in DOORS

Figure 2 illustrates how this pattern is implemented in DOORS's `FaultDetector`. A pool of threads is allocated *a priori* to poll a set of application objects. One thread is elected as the leader to monitor the application objects. When a failure is detected, one of the follower threads is promoted to become the new leader, which then polls the remaining application objects. In contrast, the previous leader thread informs the `ReplicationManager` of the failure and blocks until recovery completes, at which point the previous leader thread becomes a follower. This pattern resolves the force of polling all application objects, even when the poller is blocked on the recovery of a failed object.

3.2   Decoupling Recovery Invocation and Execution

Context: When the PULL-based monitoring style is used in the DOORS implementation, the poller thread of the `FaultDetector` is responsible for polling application objects at constant time intervals. Whenever an application object fails to respond to the poll message, the `FaultDetector` must report the failure to the `ReplicationManager`. In contrast, the `ReplicationManager` can receive such reports from more than one `FaultDetector`. The DOORS's `ReplicationManager` serializes the failure report requests by handling them sequentially. For performance-sensitive applications with high availability requirements, it is imperative that the `ReplicationManager` be notified of failures and that recovery occur within a bounded amount of time.

Problem: The blocking behavior of the `FaultDetector`'s poller thread and missed polls discussed earlier precludes the propagation of failure reports from other application objects monitored by the same `FaultDetector` to the `ReplicationManager`. Also, since the `ReplicationManager` serializes all the failure reports, this degrades its responsiveness. In production systems, a `ReplicationManager` may receive many failure reports. Handling the failure reports sequentially incurs significant delay in the recovery process for queued requests.

A naive solution based on creating a thread per-report failure request scales poorly in a dynamic environment where failure requests may arrive in bursts. In addition, thread creation is expensive and inefficient programming may yield excessive synchronization overhead. Thus, the forces that must be resolved involve ensuring faster response to failure reports and faster recovery. Resolving these forces enables lower time to attain stability and hence higher availability.

Solution → the Active Object pattern: An efficient way to optimize system recovery and stabilization is to use the *Active Object* pattern [4]. This pattern decouples method execution from method invocation to enhance concurrency and to simplify synchronized access to an object that resides in its own thread.

In DOORS, the invocation thread of the `FaultDetector` calls the `report_failure` operation on the proxy object of the `ReplicationManager` which is exposed to it. Figure 3 shows how the `FaultDetector` can call the `report_failure` operation on the `ReplicationManager` proxy. This call is made in the `FaultDetector`'s thread of control. The proxy then hands off the call to the scheduler of the `ReplicationManager`, which enqueues this call and returns control to the `FaultDetector`. The call is then dispatched to the `ReplicationManager` servant, which executes this call in the `ReplicationManager`'s thread of control.

3.3   Caching Object References of FaultDetector in the ReplicationManager

Context: The FT-CORBA standard does not specify how the `ReplicationManager` informs the `FaultDetectors` to start monitoring the objects. A typical solution is to contact a CORBA Naming Service to obtain the `FaultDetector` object reference.

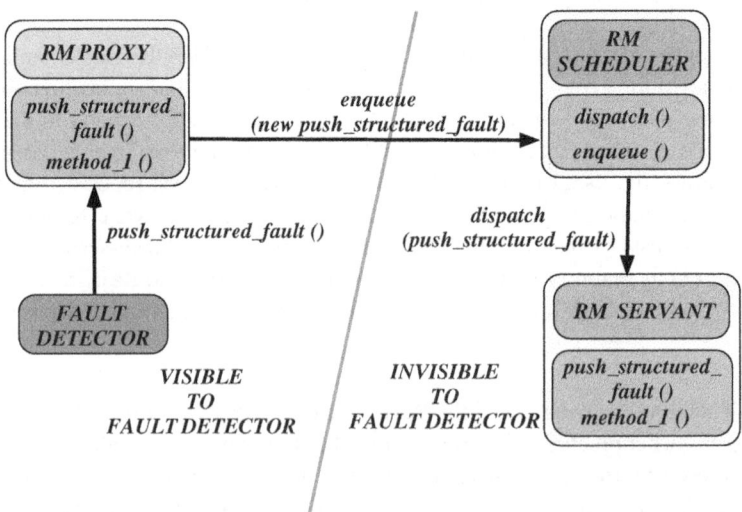

Fig. 3. Applying the Active Object Pattern in DOORS

Problem: Making remote calls to the Naming Service to obtain a `FaultDetector` object reference incurs a cost that affects the object group creation and recovery process. This cost becomes unnecessary when the `FaultDetector` remains unchanged, unless it has crashed and is rejuvenated. Hence, the force to be resolved involves minimizing the time spent in obtaining the object references of the `FaultDetectors`.

Solution → Optimize for the common case by storing redundant information and eliminating gratuitous waste: Unless the `FaultDetector` has itself crashed, there is no need for the `ReplicationManager` to obtain the object reference of the `Fault-Detector` each time it is needed. Instead, it can cache this information, which avoids the round-trip delay of invoking the `Naming Service` remotely.

During initialization, the `ReplicationManager` obtains the `FaultDetec-tor`'s object reference and stores it in an internal table, as shown in Figure 4. The

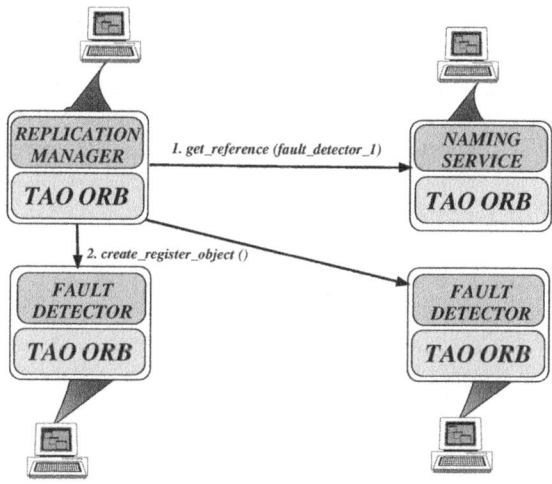

Fig. 4. Optimizing for the Common Case in DOORS

only time DOORS must obtain a new object reference is when the `FaultDetector` crashes, which happens infrequently in a properly configured system. This optimization can improve the time to recovery and system stabilization significantly, thereby enhancing the performance and availability of the application.

3.4   Support Interchangeable Behaviors

Context: As explained in Section 2, the FT-CORBA standard specifies several properties, such as replication styles and fault monitoring styles, and their values, which can be set on a per-object group, per-type, or per-domain basis. In addition, the FT-CORBA standard provides operations to override these properties or to retrieve their values. The `ReplicationManager` that inherits the `PropertyManager` interface implements these operations. Moreover, efficient implementations are possible only when efficient data structures are used to store and access these properties. The choice of data structures depends on (1) the number of properties supported by the `ReplicationManager` and (2) the maximum number of different types of object groups that are permitted.

Problem: One way to implement FT-CORBA is to provide only static, non-extensible strategies that are hard-coded into the implementation. This design is inflexible, however, since components that want to use these options must (1) know of their existence, (2) understand their range of values, and (3) provide an appropriate implementation for each value. These restrictions make it hard to develop highly extensible services that can be composed transparently from configurable strategies.

Solution → the Strategy pattern: An effective way to support multiple behaviors is to apply the *Strategy* pattern [3]. This pattern factors out similarities among algorithmic alternatives and explicitly associates the name of a strategy with its algorithm and state.

We are enhancing different components of DOORS, such as the `Replication-Manager` and `FaultDetector`, to use the Strategy pattern. These enhancements enable developers of FT-CORBA middleware to configure these components with implementations that are customized for their requirements. Figure 5 illustrates how the Strategy pattern is applied in DOORS. As shown in this figure, different replication styles

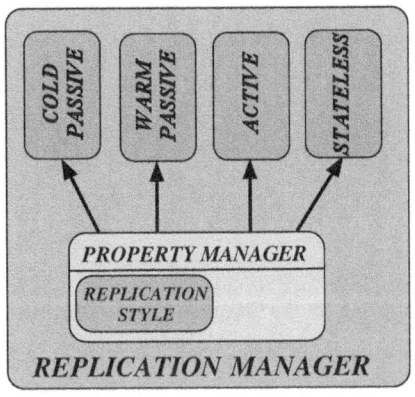

Fig. 5. Applying the Strategy Pattern in DOORS

can be configured as strategies that are selectable by applications at run-time. Moreover, new strategies, such as ACTIVE_WITH_VOTING, can be added without affecting existing strategies.

3.5 Consolidating Strategies

Context: Section 3.4 describes how the Strategy pattern can be applied to configure various requirements in the FT-CORBA service. There could be multiple strategies that offer various features, such as fault monitoring style or membership style. It is important to configure only semantically compatible strategies.

Problem: An undesirable side-effect from extensive use of the Strategy pattern in complex software is the maintenance problems posed by the possible semantic incompatibilities between different strategies. For instance, the FT-CORBA service cannot be configured with active replication style and application controlled membership style. In general, the forces that must be resolved to compose all such strategies correctly involve (1) ensuring the configuration of semantically compatible strategies and (2) simplifying the management of a large number of individual strategies.

Solution → the Abstract Factory pattern: An effective way to consolidate multiple strategies into semantically compatible configurations is to apply the *Abstract Factory* [3] pattern. This pattern provides a single access point that integrates all strategies used to configure the FT-CORBA middleware, such as DOORS. Concrete subclasses then aggregate compatible application-specific or domain-specific strategies, which can be replaced *en masse* in semantically meaningful ways.

In the DOORS FT-CORBA implementation, abstract factories are used to encapsulate internal data structure-specific strategies in components such as `Replication-Manager` and `FaultDetector`. Figure 6 depicts how the property list in the `Repli-cationManager` uses abstract factories. The property abstract factory encapsulates

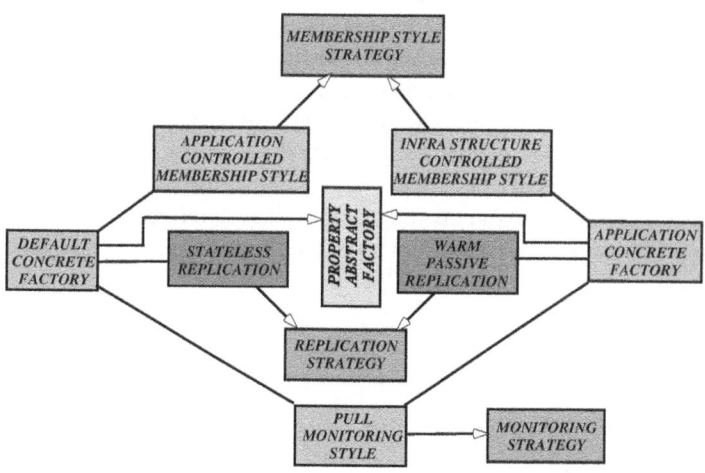

Fig. 6. Applying the Abstract Factory Pattern in DOORS

the different property strategies, such as the membership strategy, monitoring strategy, and replication strategy. By using a property abstract factory, DOORS can be configured to have different property sets conveniently and consistently.

3.6   Dynamically Configuring DOORS

Context: FT-CORBA implementations can benefit from the ability to extend their services *dynamically*, *i.e.*, by allowing their strategies to be configured at run-time. The FT-CORBA standard allows applications to dynamically set certain fault tolerance properties of the application's replica group registered with the `ReplicationManager`. These properties include the list of factories that create each replica object of the replica group or the minimum number of replicas required to maintain the replica group size above a threshold.

Problem: Although the Strategy and Abstract Factory patterns simplify the customization for specific applications, these patterns still require modifying, recompiling, and relinking the DOORS source code to enhance or add new strategies. Thus, the key force to resolve involves decoupling the behaviors of DOORS strategies from the time when they are actually configured into DOORS.

Solution → the Component Configurator pattern: An effective way to enhance the dynamism is to apply the *Component Configurator* pattern [4]. This pattern employs explicit dynamic linking mechanisms to obtain, install, and/or remove the run-time address bindings of custom Strategy and Abstract Factory objects into the service at installation-time and/or run-time.

DOORS's `ReplicationManager` and `FaultDetector` use the Component Configurator pattern in conjunction with the Strategy and Abstract Factory patterns to dynamically install the strategies they require without (1) recompiling or statically relinking existing code, or (2) terminating and restarting an existing `ReplicationManager` or `FaultDetector`. Applications can use this pattern to dynamically configure the appropriate replication style, monitoring style, polling interval, and membership style into the DOORS FT-CORBA service. Figure 7 shows how these properties are dynamically linked. The use of the Component Configurator pattern allows the behavior of DOORS's `ReplicationManager` and `FaultDetector` to be customized for specific application requirements without requiring access to, or modification of, the source code.

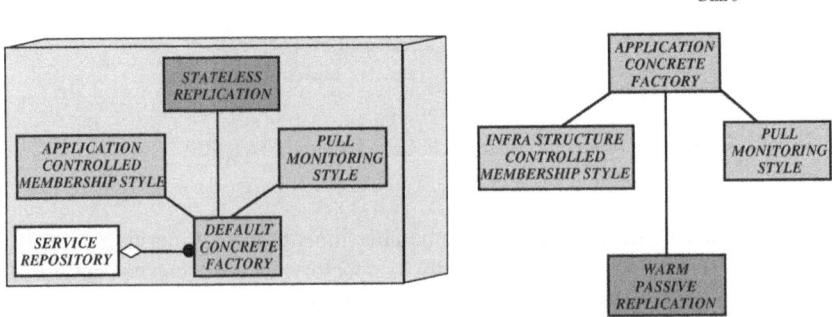

Fig. 7. Applying the Component Configurator Pattern in DOORS

3.7 Efficient Property Name-Value Lookups

Context: The `ReplicationManager` of a FT-CORBA service is required to lookup the fault-tolerant properties of the object groups registered with it during object group creation and recovery. Properties are also located when an application retrieves them or overrides previous values. The FT-CORBA standard defines a hierarchical order in which properties must be found. First, the properties must be located for the object group that is the target of the request. If it is not found, then a lookup is made on a repository that holds properties for all object groups of the same type. If that lookup also fails, another lookup is performed on the domain-specific repository that acts as the default for all the object groups that are registered with the `ReplicationManager`.

Problem: The hierarchical lookup ordering mandated by the FT-CORBA standard underscores the need for an efficient strategy to locate fault-tolerance properties. Thus, the force that must be resolved involves efficient lookups of fault-tolerance properties guided by the order specified in the FT-CORBA standard.

Solution → the Chain of Responsibility pattern and perfect hashing: An efficient way to perform hierarchical property lookups is to use the *Chain of Responsibility* pattern [3], which decouples the sender of a request from its receiver, in conjunction with *perfect hashing* [6] to perform optimal name lookups. The Chain of Responsibility pattern links the receiving objects and passes the request along the chain until an object handles the request. Perfect hashing is applicable because the number of properties supported by a `ReplicationManager` can be configured *a priori*.

Since the fault-tolerance properties supported by a `ReplicationManager` are determined *a priori*, the DOORS service uses a perfect hash function generated by GNU gperf [6] to perform an $O(1)$ lookup on the property name. The Chain of Responsibility pattern is applied by passing the request from one hash table to the other until the property is found or the search fails, as illustrated in Figure 8.

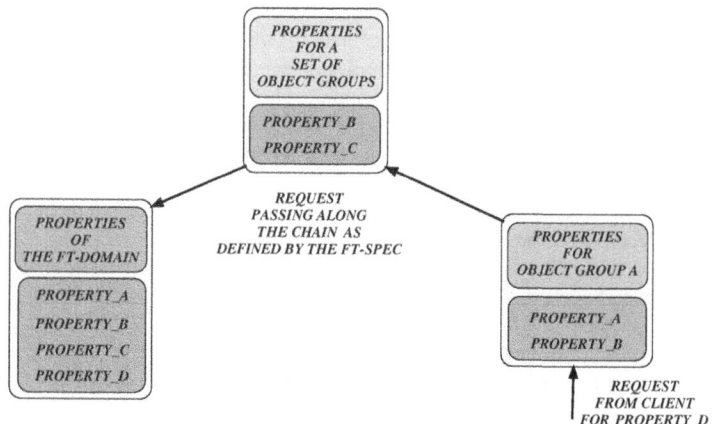

Fig. 8. Applying the Chain of Responsibility Pattern in DOORS

4   Concluding Remarks

A growing number of CORBA applications with stringent performance requirements also require fault tolerance support. To address the fault tolerance requirements, the OMG recently standardized the Fault Tolerant CORBA (FT-CORBA) specification. The most flexible strategy for providing fault tolerance to CORBA applications is via higher-level CORBA services.

To make FT-CORBA usable by performance-sensitive applications it must incur negligible overhead. To address these requirements, therefore, an FT-CORBA implementation should possess the following properties:

1. The fault detection and failovers incurred by servers should be transparent to clients.
2. Response time to the client should be bounded and predictable, irrespective of server failovers.
3. The overhead incurred by the fault tolerance framework should maintain application performance requirements, such as efficiency and scalability, within designated bounds end-to-end.

We have identified common pitfalls in FT-CORBA implementations that degrade performance. To eliminate these overheads, we are applying key design, architectural, and optimization principle patterns to improve the performance, extensibility, scalability, and robustness of the FT-CORBA implementation.

References

1. Object Management Group, *The Common Object Request Broker: Architecture and Specification*, 2.3 edition, June 1999.
2. Steve Vinoski, "CORBA: Integrating Diverse Applications Within Distributed Heterogeneous Environments," *IEEE Communications Magazine*, vol. 14, no. 2, February 1997.
3. Erich Gamma, Richard Helm, Ralph Johnson, and John Vlissides, *Design Patterns: Elements of Reusable Object-Oriented Software*, Addison-Wesley, Reading, MA, 1995.
4. Douglas C. Schmidt, Michael Stal, Hans Rohnert, and Frank Buschmann, *Pattern-Oriented Software Architecture: Patterns for Concurrency and Distributed Objects, Volume 2*, Wiley & Sons, New York, NY, 2000.
5. Frank Buschmann, Regine Meunier, Hans Rohnert, Peter Sommerlad, and Michael Stal, *Pattern-Oriented Software Architecture – A System of Patterns*, Wiley and Sons, 1996.
6. Irfan Pyarali, Carlos O'Ryan, Douglas C. Schmidt, Nanbor Wang, Vishal Kachroo, and Aniruddha Gokhale, "Applying Optimization Patterns to the Design of Real-time ORBs," in *Proceedings of the Conference on Object-Oriented Technologies and Systems*, San Diego, CA, May 1999, USENIX.
7. Object Management Group, *Fault Tolerant CORBA Specification*, OMG Document orbos/99-12-08 edition, December 1999.
8. George Varghese, "Algorithmic Techniques for Efficient Protocol Implementations ," in *SIGCOMM'96 Tutorial*, Stanford, CA, August 1996, ACM.
9. Aniruddha Gokhale and Douglas C. Schmidt, "Optimizing a CORBA IIOP Protocol Engine for Minimal Footprint Multimedia Systems," *Journal on Selected Areas in Communications special issue on Service Enabling Platforms for Networked Multimedia Systems*, vol. 17, no. 9, Sept. 1999.
10. Douglas C. Schmidt and Chris Cleeland, "Applying Patterns to Develop Extensible ORB Middleware," *IEEE Communications Magazine*, vol. 37, no. 4, April 1999.

Design, Implementation and Performance Evaluation of a High Performance CORBA Group Membership Protocol[*]

Shivakant Mishra[1] and Xiao Lin[2]

[1] Department of Computer Science, University of Colorado
Boulder, CO 80309-0430, USA.
mishras@cs.colorado.edu
http://www.cs.colorado.edu/_mishra
[2] Department of Computer Science, University of Wyoming
Laramie, WY 82071-3682, USA.

Abstract. This paper describes the design, implementation, and performance evaluation of a CORBA group membership protocol. Using CORBA to implement a group membership protocol enables that protocol to operate in a heterogeneous, distributed computing environment. To evaluate the effect of CORBA on the performance of a group membership protocol, this paper provides a detailed comparison of the performance measured from three implementations of a group membership protocol. One implementation uses UDP sockets, while the other two use CORBA for interprocess communication. The main conclusion is that CORBA can be used to implement high performance group membership protocols. There is some performance degradation due to CORBA, but this degradation can be reduced by carefully choosing an appropriate design.

1   Introduction

Group communication services have been proposed as mechanisms to construct high performance, highly available, dependable, and real-time applications [3,9,2]. At present, these services are mostly implemented on a homogeneous, distributed computing environment. This is a major limitation. Our goal is to address this problem of operating group communication services in a heterogeneous, distributed computing environment. In particular, we investigate the design, implementation, and performance of a group communication service implemented using the Common Object Request Broker Architecture (CORBA) [11].

In this paper, we investigate the design, implementation, and performance evaluation of a group membership protocol in CORBA. In particular, we describe the implementation and performance of a group membership protocol called the three-round, majority agreement group membership service (TRM) [4] using CORBA.

---

[*] This work was supported in part by an AFOSR grant F49620-98-1-0070.

M. Valero, V.K. Prasanna, and S. Vajapeyam (Eds.): HiPC 2000, LNCS 1970, pp. 121–130, 2000.
© Springer-Verlag Berlin Heidelberg 2000

Efforts in providing object replication support using a group communication service in CORBA include integration approach [7], alternate protocol approach [5,10], and service approach [6,8]. We have chosen a service approach in our endeavor to design and implement a CORBA group membership protocol. In particular, we address the following questions: (1) how practical is it to implement a group membership protocol using the service approach?, (2) what is the performance overhead of such an implementation?, (3) what are the sources of this performance overhead?, and (4) are the current CORBA specifications sufficient for implementing a group membership protocol using the service approach? We provide answers to these questions by describing three implementations of a group membership protocol called the three-round, majority agreement group membership protocol [4].

The first implementation uses the UDP socket interface, and runs in a homogeneous, distributed computing environment, on a network of SGI workstations running IRIX 6.2. This implementation follows the standard techniques that have been used in the past to implement group membership protocols. We will refer to this implementation as the *socket implementation*. The second and the third implementations use CORBA and run on a network of SGI workstations running IRIX 6.2, Sun Sparcstations running Solaris, PCs running Windows 95, and PCs running Windows NT 4.0. We will refer to these two implementations as the CORBA implementations. The second implementation uses IONA Orbix 2.3, while the third implementation uses IONA Orbix 2.3 and UDP socket interface. We will refer to the second implementation as the *pure CORBA implementation*, and the third implementation as the *hybrid CORBA implementation*.

We provide a performance comparison between the performance measured from the three implementations, identify the sources of performance overhead due to CORBA, and discuss some techniques to improve the performance of a CORBA group membership protocol. The main conclusion of this paper is that the current CORBA technology is suitable for implementing a high performance group membership protocol in a heterogeneous, distributed computing environment using the service approach. While there is some performance overhead due to CORBA, it is significantly lower than the performance overhead we observed in CORBA atomic broadcast protocol [8]. Furthermore, this performance overhead can be reduced to a certain extent by carefully choosing an appropriate design.

2   Group Communication Service

Figure 1 shows the relationship between application clients, application servers, and a group communication service. An application with high availability, dependability, and/or real-time responsiveness requirements is constructed by implementing application servers on multiple machines. These application servers replicate the application state and use a group communication service to coordinate their activities in the presence of concurrent event occurrences, asynchrony, and processor or communication failures. Application clients that need application services interact with one of the application servers.

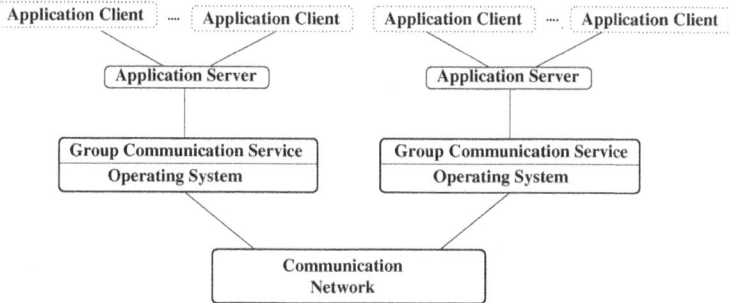

Fig. 1. Relationship between clients, servers, and group communication service.

We have chosen a group membership protocol called the three round, majority agreement group membership protocol (TRM) for implementation using CORBA. TRM is one of five group membership protocols presented in [4]. This protocol's high performance and simple design makes it an ideal candidate for real world implementations. In this section, we describe this protocol briefly. For a detailed description, see [4].

As the name implies, TRM consists of three rounds of message exchanges. In the beginning, a processor creates a new group and joins that group. After that, it starts sending *probe* messages to all processors. When a processor *p* receives a *probe* message from another processor whose ID is less than its own, it creates a new group and sends a *new group invitation* message to all other processors. Processor *p* is termed as the group creator. Receivers of this message reply with a *acceptance* message if they want to join the new group. When the group creator receives *acceptance* messages from other processors, it includes them as members in the new group and sends back a *join* message to inform them of successful join. If a processor receives more than one *new group invitation* message, it responds to the one sent by processor with a greater ID number. Eventually, the processor with the greatest ID among a set of processors that can communicate with one another forms a group that reaches a stable state. This stability is achieved by the leader of the group (member with smallest ID) periodically sending *I am alive* message to its right neighbor, who in turn passes that message to its right neighbor, and so on. When the *I am alive* message gets back to the leader, the group is stable. In addition to the *I am alive* message, the group leader also sends *probe* message to all processors that are not currently in the group.

In an asynchronous distributed system, there is no bound on communication delays. In TRM, a processor takes appropriate actions based on messages received with a fixed time interval. If a member doesn't receive *I am alive* message from its left neighbor with in a prescribed period, it declares a group failure and starts its own group. Similarly, if the group creator doesn't receive *acceptance* message from certain processors after sending a *new group invitation* message, it simply omits those processors from the group and sends *join* messages to only those processors that responded in time. On the other hand, if a processor responds to the group creator by sending an *acceptance* message, but doesn't receive a *join* message in time, it creates a new group.

3   Implementation

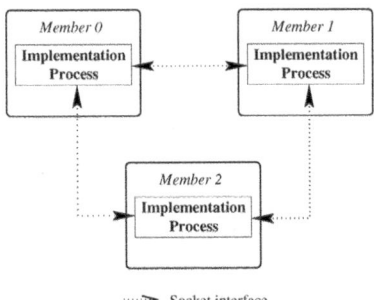

Fig. 2. Socket Implementation.

We have implemented the three round, majority agreement group membership protocol in three ways. The first implementation, called the socket implementation, uses UDP socket interface for communication. Each member is implemented by a single process that sends, receives and processes messages, and maintains the protocol state. Figure 2 shows an outline of this implementation.

The next two implementations, called the CORBA implementations, use CORBA. These implementations use two important properties of CORBA: standardized ORB and remote invocation. In accordance with the relationship between application clients, application servers, and a group communication service (see Figure 1), each group member provides two sets of IDL interfaces: *members_interface* and *application_interface*. The members_interface specifies interactions between different group members, and the application_interface specifies interactions between an application server and a group member. Together, these two interfaces allow different application servers and different group members to run on machines that have different architectures and run different operating systems. In addition, these interfaces allow the implementation of application servers and group members in different programming languages.

Group members running on different hosts communicate with one another via the ORB by using the members_interface. This interface allows different group members to run on machines of different architectures and use different programming languages for their implementation. Figure 3 shows the usefulness of this interface. The ORB makes remote invocation in CORBA as simple as a local function call. In our design, message transfer between different group members has been facilitated by passing messages as parameters of the remote invocation.

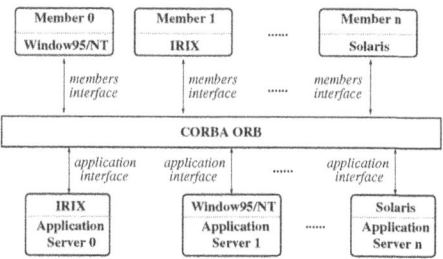

Fig. 3. Heterogeneity via the members and application interfaces.

An application server invokes various group communication service operations by using the application_interface. This allows application servers and group members to run on different machines of different architectures. In addition, this interface also enables an application server and a group member to be implemented in different programming languages. Figure 3 shows the usefulness of this interface.

The CORBA implementations use Basis Object Adapter as opposed to the Portable Object Adapter of the ORB. Each processor looks for other processors using their ID numbers. CORBA expresses network failures with predefined exception classes. A client captures these exceptions, and informs them to the processor (member) as communication failures. Failures can occur both at server lookup and at server method invocation.

The two CORBA implementations differ from each other in the way a member is implemented. In the pure CORBA implementation, there is no clearly defined role for server and client. Each group member implements both a server and a client. At the start, a member creates an object for its implementation. At initialization, this object exports to the ORB the newly created object by using instruction *boa.obj_is_ready(this)*. It then acts as a server by invoking *boa.impl_is_ready()*. When a group member needs to send messages to the other group members by invoking functions exported by the other members, it acts as a client. So, in the pure CORBA implementation, server code and client code are implemented in the same program. This implementation is similar to the member implementation in the socket implementation. Figure 4 shows this implementation for a group of size three. Each member is implemented by a single program that is responsible for sending messages, receiving messages, and processing messages.

A consequence of incorporating server and client codes in the same program is that there is a need for creating threads for performing various activities. In CORBA, an invocation from a client to a server function terminates only when either the server responds, or an exception occurs. If the main thread that responds to the clients invocations (acting as a server) invokes another members function (acting as a client), it may block for a significant period of time and will not be responsive to the client invocations. To avoid this, the main thread creates new threads to carry out each client duties.

In the hybrid CORBA implementation, each group member is implemented in two parts: one implements the server functions of a member and the other implements the client functions. The two parts are implemented by two separate processes called the *server process* and the *client process*. Since both processes run on the same processor, any form of interprocess communication provided by the underlying operating system may be used for communication between them. For example, in the Unix operating system, pipes or UDP sockets may be used. We have used UDP sockets in our implementation. Figure 5 shows this implementation for a group of size three.

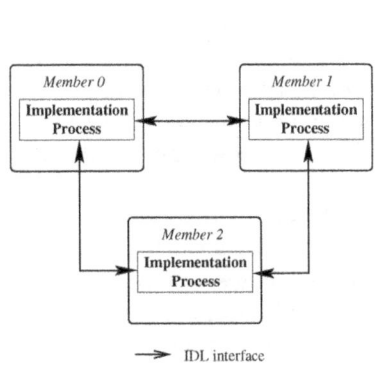

Fig. 4. Pure CORBA

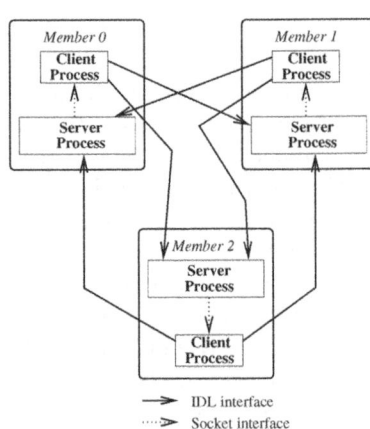

Fig. 5. Hybrid CORBA

The client process implements the application_interface and the server process implements the members_interface. The server process is responsible for receiving messages from other group members. As soon as a message is received, it passes that message to the client process. The client process is responsible for implementing all group membership functionalities. In particular, it processes messages sent by other members, services application server requests, implements atomicity, order, and termination semantics, maintains a consistent group membership, and so on. This implementation of a group member by two processes avoids the need for creating separate threads for various client activities.

4   Performance Evaluation

Implementation of the three-round, majority agreement group membership protocol using CORBA consists of about 3,000 lines of C++ code, not including the three automatically generated C++ source and header files from the IDL compiler. For CORBA, we used Orbix 2.3 from the Iona Technologies [1]. This implementation runs on a network of SGI workstations running IRIX 6.2, Sun Sparcstations running Solaris, PCs running Windows 95, and PCs running WindowsNT 4.0. To evaluate the effect of using CORBA on the performance of a group membership protocol, we measured its performance on a network of SGI workstations. The reason for this is that the socket implementation runs on a network of SGI workstations.

4.1   Performance Indices

There are three performance indices we have measured to evaluate the effect of CORBA on the performance of the three-round, majority agreement group membership protocol: *initialization stabilization time, failure stabilization time*, and *recovery stabilization time*. The initialization stabilization time measures the time to initialize the system. It is the time interval between the moment the last processor is started and the moment the group is stable, i.e. when the new group leader receives *I am alive* message for the first time. The failure stabilization time measures the time needed to construct a new group after a failure notification of a group member is first received. It is the time interval between the moment the failed member's right neighbor determines that its left neighbor has failed and the moment the new group is stable. Finally, the recovery stabilization time measures the time needed to incorporate a newly recovered processor in the group. It is the time interval between the moment the recovered processor is started and the moment the next group is stable.

Table 1. Membership Protocol Performance

δ (msec)	Implementation Technique	Initialization Stabilization Time	Failure Stabilization Time	Recovery Stabilization Time
1000	Socket	1117	2113	992
	Pure CORBA	2341	3034	1992
	Hybrid CORBA	1523	2193	1413
500	Socket	1157	1071	1002
	Pure CORBA	2123	2052	2042
	Hybrid CORBA	1506	1156	1379
200	Socket	1028	496	981
	Pure CORBA	2179	1518	2069
	Hybrid CORBA	1411	524	1302

4.2   Performance

Performance of the three round, majority agreement group membership protocol depends on the values of three constants: δ, π, and μ. δ is a time interval with in which any message sent by a processor will most likely be received by the intended receiver. Each processor sends 'I am alive' message to its right neighbor at regular intervals of size π time units, and μ denotes the period with which 'probe' messages are sent by a group member to incorporate any newly recovered processor in the group.

We have adopted the suggestions given in [4] to construct timeout intervals based on the values of these three constants: $SendProbeTimer = \mu$, $SendIamAliveTimer = \pi$, $MissingJoinTimer = 3*\delta$, $MissingAcceptTimer = 2*\delta$, and $MissingNeighborTimer = \pi + rank() * \delta$, where $rank()$ is denotes the position of a group member in the group. We have measured various performance indices from the three implementations for three different values of δ: 1000 msec, 500 msec, and 200 msec. The values of μ and π were fixed at 1000 msec in all experiments. Table 1 shows the values of the three performance indices measured.

4.3   Performance Analysis

It is clear form these measurements that the socket implementation provides the best performance, followed by the hybrid CORBA and pure CORBA implementations. In general, only the failure stabilization time changes with a change in the δ value in the three implementations. The failure stabilization time reduces with a reduction in the δ value. This is because a lower value of δ implies a lower value of $MissingAcceptTimer$. This results in the leader of the new group wait for a shorter duration before sending $Iamalive$ message. The initialization and recovery stabilization times are not affected significantly by a change in the δ value, because all processors most likely respond before the timer expires during initialization and recovery periods.

The initialization and recovery stabilization times in the pure CORBA implementation are about two times larger than the same times in the socket implementation, while they are about 1.5 times larger in the hybrid CORBA implementation compared to the same times in the socket implementation. The failure stabilization times in the hybrid CORBA implementation are only slightly larger than those in the socket implementation, while they are significantly larger in the pure CORBA implementation.

The only difference between the CORBA implementations and the socket implementation of the three round, majority agreement group membership protocol is the method used for interprocess communication. CORBA implementations use the CORBA remote object invocation via OrbixORB, while the socket implementation uses UDP sockets. So, in order to understand the reasons for extra performance overhead in the CORBA implementations, we measured the average one-way communication delay between two SGI workstations for UDP sockets and OrbixORB.

For a message size of 100 bytes (approximate size of the three round, majority agreement group membership protocol messages), the average one-way communication delay was measured to be about 0.6 milliseconds for UDP sockets and 2.5 milliseconds for OrbixORB. This indicates that the communication delay in UDP is nearly four times smaller than that in OrbixORB. As we increased the message size, this difference decreased. However, since messages in the three round, majority agreement group membership protocol are of smaller sizes (around 100 bytes), the one-way communication delay difference between the two implementations is expected to be about four times.

The extra performance overhead in all three performance indices in the CORBA implementations is due to this difference in one-way communication delay. An important point to note here ia that the performance degradation due to CORBA in the three-round, majority agreement group membership protocol is not as high as it was in the atomic broadcast protocol [8]. The main reason for this is that the performance of a group membership protocol is significantly affected by the values of various timers that the protocol uses. Generally, the values of these timers are much larger than the one-way communication delays of the network. For example, the one-way communication delay in our experiments was about 2.5 msec, while the values of various timers were more than 300 msec. As a result, the effect of the difference in one-way communication delays on the performance difference between different implementations is not significant.

There are two reasons for the poor performance of the pure CORBA implementation compared to the hybrid CORBA implementation. The first reason is that there is extra overhead in the pure CORBA implementation due extra thread creations and their synchronization. As mentioned in Section 3, a new thread is created for every client invocation to ensure that these invocations are serviced promptly. Some of these threads may run in parallel that gives rise to a need for thread synchronization. This thread creation and synchronization is avoided in the hybrid CORBA implementation by a clean separation of the client and the server functionalities. The second reason for the poor performance of the pure CORBA implementation compared to the hybrid CORBA implementation arises due to a need for name resolution. In both CORBA implementations, ID numbers are used to refer to a processor. So, a name resolution routine needs to be executed whenever a reference to a processor is not available. This overhead is smaller in the hybrid CORBA implementation than in the pure CORBA implementation because this name resolution needs to be done in the hybrid implementation only when the server process starts up. References to the server processes are always available to the client process, even during failure and recovery. This is not the case in the pure CORBA implementation.

5   Conclusion

We have implemented a CORBA group membership protocol using the service approach and measuring the performance overhead due to CORBA. Our goal was to find out the extent of this performance overhead. The performance degradation in the group membership protocol due to CORBA was observed to be between 1.5 to 2 times.

In addition to determining the extent of performance degradation due to CORBA in group membership protocols, we have proposed two design techniques using the service approach to implement group membership protocols. One technique uses only the CORBA ORB for interprocess communication, while the other uses CORBA ORB and UDP for interprocess communication. We observed that the second technique results in improving the protocol performance to some extent, because it avoids a need for thread creation or synchronization, and reduces the number of name resolutions performed.

The main conclusion we can draw from this work is that the current CORBA technology is suitable for implementing a high performance group membership protocol in a heterogeneous, distributed computing environment using the service approach. While there is some performance overhead due to CORBA, it is significantly lower than the performance overhead we observed in CORBA atomic broadcast protocol [8]. Furthermore, appropraite design can lead to further performance improvement in a CORBA group membership protocol.

References

1. Iona technologies, inc. URL: http://www.iona.com.
2. Y. Amir, L. Moser, P. Melliar-Smith, D. Agarwal, and P. Ciarfella. The totem single-ring ordering and membership protocol. *ACM Transactions on Computing Systems*, 13(4):311–342, 1995.
3. K. Birman, A. Schiper, and P. Stephenson. Lightweight causal and atomic group multicast. *ACM Transactions on Computer Systems*, 9(3):272–314, Aug 1991.
4. F. Cristian and F. Schmuck. Agreeing on processor group membership in asynchronous distributed systems. Technical Report CSE95-428, Department of Computer Science & Engineering, University of California, San Diego, CA, 1995.
5. M. Cukier, J. Ren, C. Sabnis, W. Sanders, D. Bakken, M. Berman, D. Karr, and R. Schantz. Aqua: An adaptive architecture that provides dependable distributed objects. In *Proceedings of the 17th Symposium on Reliable Distributed Systems*, West Lafayette, IN, Oct 1998.
6. P. A. Felber, B. Garbinato, and R. Guerraoui. The design of a corba group communication service. In *Proceedings of the 15th Symposium on Reliable Distributed Systems*, Niagara-on-the-Lake, Canada, Oct 1997.
7. IONA and Isis. *An Introduction to Orbix+Isis*. IONA Technologies Ltd. and Isis Distributed Systems, Inc., 1994.
8. S. Mishra, L. Fei, and G. Xing. Design, implementation, and performance evaluation of a corba group communication service. In *Proceedings of the 29th IEEE International Symposium on Fault-tolerant Computing*, Madison, WI, Jun 1999.
9. S. Mishra, L. Peterson, and R. Schlichting. Consul: A communication substrate for fault-tolerant distributed programs. *Distributed Systems Engineering*, 1(2):87–103, Dec 1993.
10. L. E. Moser, P. M. Melliar-Smith, and P. Narasimhn. Consistent object replication in the eternal system. *Theory and Practice of Object Systems*, 4(2), 1998.
11. J. Siegel. *CORBA Fundamentals and Programming*. John Wiley & Sons, Inc., 1996.

Analyzing the Behavior of Event Dispatching Systems through Simulation

Giovanni Bricconi[1], Elisabetta Di Nitto[2], Alfonso Fuggetta[2], and Emma Tracanella[1]

[1] CEFRIEL, via Fucini 2, 20133 Milano, Italy
{bricconi tracanel}@cefriel.it
[2] DEI, Politecnico di Milano, piazza L. da Vinci 32, 20133 Milano, Italy
{dinitto fuggetta}@elet.polimi.it

Abstract. The pervasive presence of portable electronic devices and the massive adoption of the Internet technology are changing the shape of modern computing systems. Applications are being broken down into smaller components that can be modified independently and that can be plugged in and out dynamically. All this drives the development of flexible, high scalable middleware. An interesting category of such middleware supports the event-based paradigm. In this paper we focus on the scalability issues we have faced in the development of an event-based middleware called JEDI. In particular, we focus on improving performances of such middleware when the number and distribution of components grows. In order to evaluate design alternatives, we have taken a simulative approach that has allowed us to analyze the design alternatives before actually implementing them.

1 Introduction

Modern computing systems are more and more oriented to serve scenarios in which any device is provided with a computational capability and is able to interact with the other devices it gets close to [7]. In this context, the need for scalable middleware that supports such interaction is growing. Such a middleware has to allow easy reconfiguration and plug in of new components. Moreover, it has to enable anonymous and multicast communication in order to support scenarios where components do not know what other components are around and which of them would be interested in theirs messages.

The kind of middleware that at the moment seems to address these requirements is based on the event-based approach, where applications are structured in autonomous components that communicate by generating and receiving *event notifications*. A component usually generates an event notification when it wants to let the "external world" know that some relevant event occurred in its internal state[1]. The *event dispatcher* provided by the middleware propagates the event to any component that has declared its interest by issuing a *subscription*. A subscription, therefore, can be seen as a constraint on the content of the events; when an event respects the condition

---

[1] In the following we will use the terms event notification and event indifferently since, from the viewpoint of event-based middleware, we do not need to distinguish between the occurrence of an event and the generation of the corresponding notification.

M. Valero, V.K. Prasanna, and S. Vajapeyam (Eds.): HiPC 2000, LNCS 1970, pp. 131–140, 2000.

stated by a subscription we say that the event is *compatible* with that subscription. The propagation of events is completely hidden to the components that generated them, thus the event dispatcher implements a multicasting mechanism that fully decouples event generators from event receivers. This provides two important effects. First, a component can operate in the system without being aware of the existence, number and location of other components. Second, it is always possible to plug a component in and out of the architecture without affecting the other components directly. These two effects guarantee a high compositionality and reconfigurability of a software system, and make event-based middleware particularly suited for the development of systems in which components operate autonomously and are loosely coupled. The underlying hypothesis is that an agreement exists between components on the structure of events. Several event-based platforms are currently available, either as research prototypes or as commercial products. [4] and [5] present the JEDI system and a taxonomy of some of the other existing platforms.

A main problem of all systems is scalability and performance. It can be noticed, in fact, that if the dispatcher has to manage every component subscription and notification, it can easily become a bottleneck for the whole system. To solve this problem in JEDI the event dispatcher is implemented as a distributed system composed of several *dispatching servers* organized in a hierarchy. Each of these servers shares with the others part of the received subscriptions and events in order to guarantee that connected components communicate properly.

In order to evaluate the performance and scalability of our solution and to identify possible improvements, we have taken a simulative approach. This allows us to determine the best alternative before actually implementing and deploying any possible solution. In this paper we compare through simulation the current implementation of JEDI with an alternative design, and we discuss advantages and disadvantages of both approaches. The rest of the paper is structured as follows. Section 2 presents an overview of the event-based middleware we are developing and of the alternative design we are facing with. Section 3 describes the simulation model of JEDI and Section 4 presents the results gathered from simulation. Section 5 discusses the related work and, finally, Section 6 provides some conclusions.

2 JEDI

JEDI stands for Java Event-based Distributed Infrastructure. It supports asynchronous and decoupled communication among distributed elements that are called *active objects* (AOs for short). Each active object interacts with other AOs by explicitly producing and consuming events. Event notifications in JEDI have a name and a number of parameters. For instance, SoftwareReleased(Editor, 1.3, WinNT) notifies that version 1.3 of a software called Editor has been released for WindowsNT. Subscriptions are syntactically similar to notifications; they have a name and some parameters. A subscription is *compatible* with every event that has the same values for the same fields. To allow the creation of more flexible subscriptions the operator "*" has been introduced, representing a kind of wildcard. For instance, subscription *(Editor, *, Win*) is compatible with all the notifications (including the one above) having any name, three parameters, and concerning all the Editor versions that run on WindowsNT, Windows98, Windows2000, ...

2.1   Subscriptions and Event Propagation

In JEDI subscriptions and notifications are managed by a hierarchy of dispatching servers (DS). Whenever an AO issues a subscription, the DS connected to that AO stores such subscription in its internal tables and forwards it to its parent, which, in turn repeats the procedure until the subscription arrives to the root DS. When a notification is generated, the receiving dispatching server forwards it to all its AOs and descendant dispatching servers that previously sent compatible subscriptions. Moreover, the dispatching server sends the notification to its parent that, in turn, acts in a similar way (without sending the notification back). Therefore, a notification can reach each AO that has issued a compatible subscription, regardless of the AO position in the hierarchy. The system guarantees that causally related notifications are received by AOs in the same order in which they are generated.

The hierarchical approach is certainly more scalable of a centralized one. However it requires DSs to get coordinated by propagating subscriptions and notifications. Such coordination traffic has to be kept as limited as possible in order to ensure good level of performance. In the next section we propose an improvement of the approach presented above.

2.2   Improving the Event Propagation Algorithm: Advertisements

As discussed in the previous section, in the current version of JEDI, events emitted by AOs always reach the root of the dispatching hierarchy even if there is no subscriber that can be reached through the root DS. So, when the number of events emitted in the time unit grows, the DSs located at the higher levels of the hierarchy may become overloaded. To relief this situation we introduce a new event propagation algorithm based on the hypothesis that AOs perform a new operation called *advertisement*. AOs use advertisement to specify which kind of events they will produce. The rules that define the matching between events and advertisements are the same holding between events and subscriptions (see previous section). Moreover, we say that an advertisement is *compatible* with a subscription if at least one event compatible with both of them exists. AOs can dynamically issue advertisements and withdraw them during their life cycle.

Each DS uses advertisements and subscriptions to create proper routing tables that cause events to be propagated only through paths leading to interested subscribers. In the new algorithm we propose, advertisements are routed toward the root of the dispatching hierarchy exactly as it was illustrated for subscriptions in the previous section. When receiving an advertisement, a DS looks for compatible subscriptions that are then forwarded to the sub-tree that has generated the advertisement.

As an example, let us consider the hierarchy shown in Fig. 1 where hexagons represent dispatching servers, while circles represent AOs. Let us suppose that AO β issues a subscription. According to the algorithm previously described, such subscription is notified to dispatchers D, B, and A. Now suppose that α issues an advertisement compatible with the subscription of β: F receives this advertisement and forwards it to B that detects the compatible subscription. Hence B propagates this subscription to F (notice that the subscription now is stored on every DS that connects α to β). Finally B propagates the advertisement to A that checks if the sub-tree rooted at C has communicated other compatible subscriptions. If not, it simply stores the advertisement in its internal tables. Thanks to this subscription propagation approach, all the events generated by α are routed to β through dispatcher B, and do not reach A unless a compatible subscription has been issued by any AO connected to A or to the subtree rooted by C.

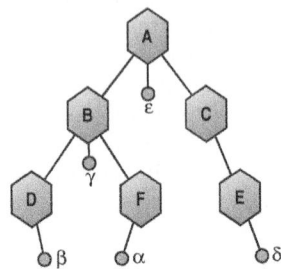

Fig. 1. Connections between dispatching servers and agents

Of course, new subscriptions compatible with an advertisement can be issued even after an advertisement has been propagated (or during the propagation). A dispatcher that receives a subscription, while routing it toward the root of the hierarchy, checks in its internal tables for compatible advertisements, and, if any, routes the subscription toward the senders of these advertisements.

Summarizing, advertisements and subscriptions are forwarded up to the root. In addition, subscriptions are sent toward advertisements so that they can be used to create virtual paths between the event producers and consumers.

To further limit the traffic caused by propagation of subscriptions and advertisements, we have exploited some optimization techniques introduced in [2] and [3]. In these papers it is shown that, given a pair of advertisements (or subscriptions) x and y, it may happen that all the events compatible to y are compatible to x too. In this case we say that *x covers y*. Whenever an advertisement (subscription), issued in a subtree of the dispatching hierarchy, is covered by another advertisement (subscription) issued in the same subtree, such advertisement (subscription) does not need to be propagated toward the root since it does not provide any additional information.

3 The Simulation Model

In order to define a simulation model that properly describes the interesting characteristics of JEDI, we have collected numeric data on the behavior of two existing applications that have been developed on top of JEDI in the past years. The analysis of such mass of data has allowed us to identify the main components of the simulation model and their principal characteristics.

The main components of the simulation model describe dispatching servers and active objects. Dispatching servers are described in terms of the operations they can perform, (i.e., manage subscriptions, advertisements, notifications, ...) and the distribution of the time spent in performing these operations as they have been determined from the measured data. Since the advertisement mechanism has not been developed so far in JEDI, the time needed to process advertisements has been estimated by considering that the algorithm that manages them is similar to the one used for subscriptions and exploits similar data structures.

Regardless of the application-dependent task they perform, AOs interact with the event-based middleware to issue subscriptions and advertisements, and to generate events. Based on this, for the purpose of our model, we have identified four elementary behaviors for AOs and we have associated them to proper simulation components called *agents*. Based on the behavior they embody, agents can be of four different types: *sinks*, *sources*, *proactive* and *reactive agents*. A source represents an AO that can only send events (and advertisements) during its life. A sink is able only to receive event notifications (and to send subscriptions). Proactive agents take the initiative by generating events and then wait for some reply events. Reactive agents wait for events and then generate new events.

In event-based middleware, inputs to the dispatching hierarchy (notifications, advertisements, and subscriptions) are correlated through the compatibility and covering relations defined in Section 2. This correlation captures the characteristics of the applications built on top of the middleware and it has an impact on the load of dispatching servers and on the network traffic. We have created a synthetic *correlation model* that, at simulation startup, automatically generates subscriptions, advertisements, and notifications to be assigned to each agent. This correlation model is based on few parameters that describe various application-dependent phenomena such as the distance covered by notifications in order to reach their subscribers, the coverage relationships defined on advertisements and subscriptions, etc.

The main parameter we have introduced is the *spreading coefficient* (*sc*). It provides an indication of the distance covered by events in the dispatching hierarchy. This coefficient is defined in the [0, 1] interval and occurs in the following formula:

$$R(n) = sc^n \qquad\qquad (1)$$

R(n) is the probability that an event issued by an agent directly connected to a dispatcher A has to be received by some agent directly connected to a dispatcher at a distance *n* from A. The distance between dispatchers is calculated on the basis of the topology of the dispatching hierarchy. For instance, in the topology of Fig. 1 dispatchers A and D are at a distance 2. Intuitively, the more *sc* (spreading coefficient) is close to 1, the more it is likely that events have to be spread across the whole hierarchy of DSs. Conversely, when *sc* is low, events and corresponding subscriptions are mostly localized in some dispatching hierarchy sub-tree. During simulation, we use the spreading coefficient as an input parameter, in order to test the behavior of the event-based system in different event propagation conditions. To capture the effect of covering relations for subscriptions and advertisements, we have introduced other simulation parameters that for space reasons are not be presented in this paper.

The environment we have selected to perform simulations is called OPNET [6]. In OPNET a model of the system to be simulated is defined by selecting components from proper libraries and by defining the way these components are connected together. OPNET provides a number of predefined libraries that model network components such as hubs, routers, and TCP/IP stacks. Using these libraries it is possible to easily define simulation models of local networks as well as WANs and wireless systems with mobile objects. In addition, OPNET allows users to define their own libraries to model the behavior of specific application-dependent elements. We have exploited this feature to define our simulation model and we have relied on the existing libraries as for modeling the underlying network infrastructure.

Before starting a simulation, the simulation model is customized by assigning values to the following parameters:

- The characteristics of the underlying physical network in terms of connectors bandwidth, latency, protocols, and topology.
- The topology of the system, i.e. the structure of the event dispatching hierarchy and their location on the physical nodes.
- The spreading coefficient and the parameters associated to the covering relations.
- The number and location of connected agents and their types.
- For each agent, the mean values of the distributions defining their life cycle (number of subscriptions, notification frequency, ...).

4 Simulation Results

The goal of simulation has been to understand when the usage of advertisements is advantageous in term of performances of the most critical parts of the system (root DS and network channels) and what happens when the bandwidth of network connections between dispatching servers decreases.

In order to set up the simulation scenarios, we have first identified the operational conditions of our system. For instance, we have defined the maximum number of events manageable by a dispatching server, we have analyzed the relations between the number of physically different nodes of computation and the load on the LANs, we have observed the bandwidth consumed by notifications to define the proper channel sizes to interconnect the various LANs, etc.

During simulation we have assumed that dispatchers are organized in a quaternary balanced tree. We have considered trees having 5, 21, or 85 dispatching servers organized in trees of 2, 3, or 4 levels respectively. Each dispatching server manages 52 agents located on 4 hosts connected to the same LAN. The number of agents connected to a dispatching server has been determined on the basis of some preliminary simulations in order to avoid saturation of the root DS. In all simulation scenarios the 52 agents connected to a dispatching server are categorized as follows: 16 proactive agents, 24 reactive agents, 8 sources, and 4 sinks. Proactive agents and sources send events every 8 seconds on average. The duration of simulation has been selected so that the initial transitory can be disregarded. The LANs connecting a dispatching server to its agents are 10Mbit/sec Ethernet networks. These are connected together through 1Mb/sec or 64 kb/sec communication channels, depending on the scenario being considered. We have always adopted the TCP protocol.

Fig. 2 and Fig. 3 compare the performance of the root DS when event propagation is exclusively based on subscriptions and when subscriptions and advertisements are used together to define the event routing tables. We call these two cases *subscriptions-based* and *advertisements-based*, respectively. The figures show the average percentage of time spent by the root dispatching server (i.e., the potential bottleneck of the system) in managing notifications (Fig. 2) and subscriptions (Fig. 3) plotted against the spreading coefficient introduced in Section 3. Intervals are drawn with a confidence level of 90%. The results are referred to a dispatching hierarchy of 21 dispatching servers connected through 1 Mb/sec links.

Fig. 2 shows that in the advertisements-based case the root DS spends much less time in handling notifications compared to subscriptions-based case. In fact, in the first case, the root DS has to handle only the events which actually need to be routed in subtrees different from their originators, while in the latter case, it handles all the events that are generated in the system. In the subscriptions-based case the growth in the spreading coefficient causes more events to be transmitted to other sub-trees thus resulting in more work on the side of the root dispatching server. Because of the growth in workload, in this case, the root dispatching server gets saturated for spreading coefficients higher than 0.4. In this case, therefore, the advantages of the advertisements-based approach against the subscriptions-based one is quite relevant. This result is also confirmed by the graphic of Fig. 3 showing that the advertisements-based approach does not result in an appreciable growth in processing time for handling subscriptions propagation for the root DS. We have also analyzed the processing time of advertisement, and we have noticed that this value is quite low (between 2.65% and 2.9%) and does not seem to be influenced by the spreading coefficient.

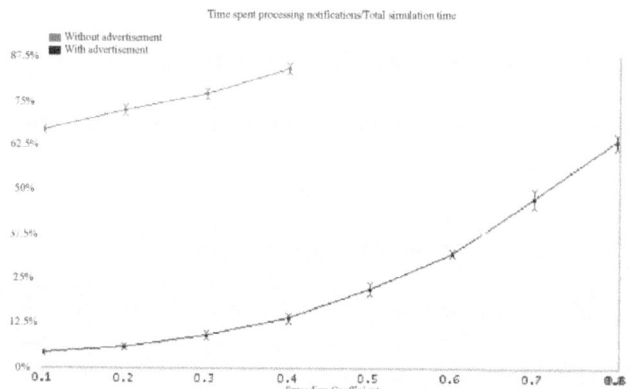

Fig. 2. Time spent by the root dispatching server in processing notifications

Similar simulations have been performed with dispatching hierarchies of 5 and 85 dispatchers. In the case of 85 dispatcher the subscriptions-based approach is not usable because, even for sc=0.1 the root DS gets saturated. The case of 5 dispatchers shows that for a small system the two approaches produce very similar results, with a difference of at most 5% in favor of the advertisement-based case.

The advantages of the advertisements-based algorithm increase when the bandwidth of the communication channels dedicated to the traffic of the event-based infrastructure decreases. This case applies when the event based middleware is deployed across the boundaries of a single organizations. Fig. 4 shows the results of a set of simulations performed on a model with 21 dispatchers connected among each other through 64 kb/sec links. The graphics show the utilization of communication links connecting dispatching servers at different levels in the hierarchy.The graphics on the left-hand side refer to the traffic directed toward a dispatching server and its controlled AOs (they reside on the same LAN), while those on the right-hand side refer to the traffic directed in the opposite direction. When the subscriptions-based approach is exploited, the 64 kbit/sec channels entering into the root DS get saturated for values of sc higher than 0.5. Conversely, the traffic outcoming from the root DS is more limited than in the advertisements-based approach. In this last case, in fact, subscriptions are sent downward to establish the routing between senders and subscribers.

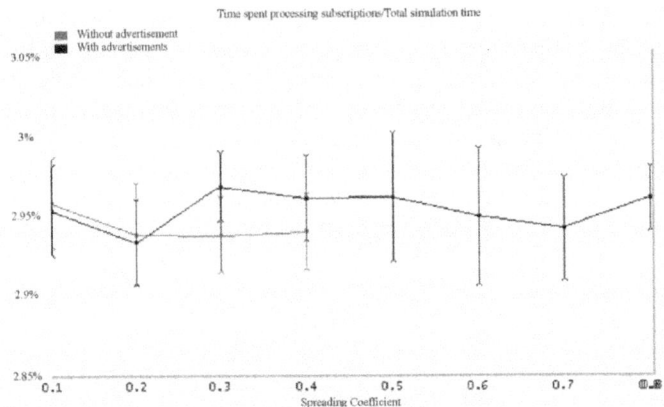

Fig. 3. Time spent by the root dispatching server in processing subscriptions

Concluding, in general the advertisements-based approach offers better performances in case of large systems where communications tend to be localized among groups of neighbor components. The advantages of the approach compared to the subscriptions-based approach are particularly evident when low speed communication channels are used. The subscriptions-based algorithm remains attractive for its simplicity in those cases where the dispatching infrastructure is not working under heavy load.

5 Related Work

While a number of researches and practitioners are focusing on the development of event based middleware, we know of few efforts devoted to understand performance and scalability of such middleware. A discussion of some preliminary requirements for scalable event-based middleware is presented in [7]. In particular, authors point out at the importance of having mechanisms for limiting network traffic among distributed dispatching servers located on wide-area network. With this respect, the advertisement algorithm we have developed for JEDI addresses, at least partially, some of these requirements. In the context of commercial systems, we are aware of two middleware, Smartsockets [8] and TIB/Rendezvouz [9], providing a distributed implementation of the event dispatcher. In both cases, however, distribution seems to be exploited to achieve reliability on a small size system more than scalability due to the massive distribution of components and efficiency of event propagation.

The problem of providing mechanisms for efficient multicast of events is tackled in [1]. In this paper authors present a new algorithm to verify the compatibility of notifications with subscriptions. Differently from what we do, they assume that subscriptions are known in advance at any node of the dispatching server network, and do not focus on how they are actually propagated. Based on this information, they can establish optimized paths for event notifications. The performances of the proposed algorithm are evaluated through simulation.

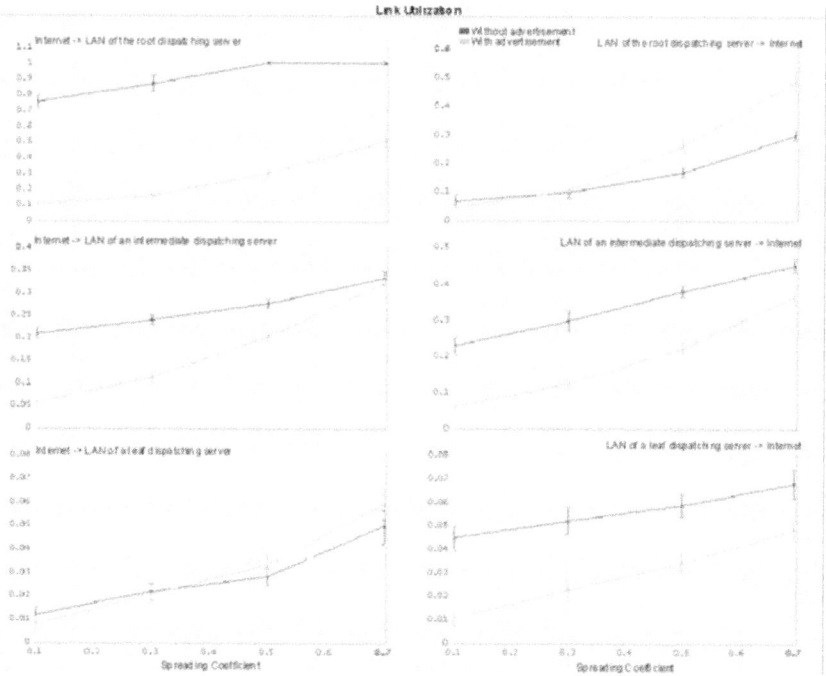

Fig. 4. Utilization of links between LANs at various levels in the hierarchy

Our work has its premises in a previous work of Carzaniga et al. ([2] and [3]), where an advertisement approach is presented and simulation is used for the first time in the context of the evaluation of event-based middleware. Differently from them, we have focused our effort on hierarchical systems. Moreover our algorithm avoids that advertisements and subscriptions floods the network of DSs. In the simulation model we have introduced two additional types of agents (proactive and reactive) and we have defined a model of locality based on the spreading coefficient. Also, we have tuned the simulation parameters by analyzing existing applications and relied on existing and proved models of network devices and protocols provided by a commercial simulator.

6 Conclusions

We have shown that the development of a scalable event-based middleware such JEDI requires special care in the definition of the mechanisms that allow dispatching components to distribute events regardless to the physical location of their originators and subscribers. We have exploited simulation with the purpose of validating the design alternatives we have defined. In particular, we have shown that the advertisements-based approach works well when event traffic is localized, while the subscription-based approach provides acceptable performances when events have to be dispatched to components distributed all over the dispatching hierarchy, assuming that the root dispatcher has been properly dimensioned. Based on the above observations we argue that the choice of the event propagation model depends on the purpose and structure of the application that is going to exploit it. Therefore, event-based infrastructures should not bundle themselves on a specific approach. Instead, they should be designed in such a way that the application designer is free to choose the approach that better suits his/her needs.

We are currently consolidating the simulation model and validating it through a proper analytical model. We aim at defining some load balancing mechanisms that avoid or contrast saturation in dispatching servers. Finally, we are extending the semantics of our event-based middleware by introducing new operations such as the possibility of generating events expecting a reply from their receivers.

Acknowledgements

We are grateful to Antonio Carzaniga, Gianpaolo Cugola, and Prof. Giuseppe Serazzi.

References

1. G. Banavar, T. Chandra, B. Mukherjee, J. Nagarajarao, R.E. Strom, and D.C. Sturman, "An Efficient Multicast Protocol for Content-Based Publish-Subscribe Systems". In Proceedings of ICDCS '99 -- Int'l Conference on Distributed Computing Systems.
2. A. Carzaniga, "Architectures for an Event Notification Service Scalable to Wide-area Networks". PhD Thesis. Politecnico di Milano. December, 1998.
3. A. Carzaniga, D.S. Rosenblum, and A.L. Wolf, "Achieving Expressiveness and Scalability in an Internet-Scale Event Notification Service.". Nineteenth ACM Symposium on Principles of Distributed Computing (PODC2000), Portland OR. July, 2000.
4. G. Cugola, E. Di Nitto, and A. Fuggetta, "Exploiting an Event-Based Infrastructure to Develop Complex Distributed Systems". Proceedings of the 20th International Conference on Software Engineering (ICSE 98), Kyoto (Japan), April 1998.
5. G. Cugola, E. Di Nitto, and A. Fuggetta, "The JEDI event-based infrastructure and its application to the development of the OPSS WFMS". To appear on IEEE Transactions on Software Engineering.
6. MIL3 Inc., "OPNET MODELER" reference guides. Vol. 2,3 and 11.
7. B. Segall and D. Arnold, "Elvin has left the building: A publish/subscribe notification service with quencing". Proceedings of AUUG97, September 1997.
8. Talarian, "Mission Critical Interprocess Communications - an Introduction to Smartsockets", White paper.
9. TIBCO, "TIB/Rendezvous", White Paper. http://www.rv.tibco.com/rvwhitepaper.html.

ParAgent: A Domain-Specific Semi-automatic Parallelization Tool

S. Mitra, S. C. Kothari, J. Cho, and A. Krishnaswamy

smitra@iastate.edu, kothari@iastate.edu, jeqcho@cs.iastate.edu, arvi@support.com

Abstract. Automatic parallelization is known to an intractable problem in general. This paper is about a new approach in which domain-specific knowledge is used to facilitate automatic parallelization. The research focuses on three widely used numerical methods: the finite difference method (FDM), the finite element method (FEM), and the boundary-element method (BEM). A prototype tool, called the ParAgent, has been developed to study the feasibility of the approach. The current version of the prototype can parallelize Fortran-77 programs based on the explicit time-marching FDM. The paper provides an overview of the new approach and some results of its application including the parallelization of the NCAR/Penn State Mesoscale Meteorology Model MM5. The manual parallelization of MM5 took about three years whereas the parallelization using ParAgent was done in about two weeks.

Introduction

Scientists have invested enormous time and effort into developing and refining legacy code for engineering and scientific applications. The solid pedigree associated with these legacy codes makes them very valuable. Computation and data intensive legacy code stand to benefit from application of parallel computing. For example, a 24-hour simulation using the mesoscale climate model MM5 [3] runs for close to two hours on a workstation. Long term climate studies to study global warming require 100-year climate simulations that would entail 8-year long runs! Clearly, parallel computing is imperative for such large-scale simulations.

The manual parallelization of legacy codes is time-consuming and prone to errors. To address this problem, considerable research effort has gone into developing automatic parallelization tools [6,12,16,20]. However, attempts to develop fully automatic parallelization tools have not yet succeeded [5,8,9]. So far the research has been mainly focused on parallelization of arbitrary sequential programs. But, the parallelization problem is too difficult at this level of generality and several sub-problems are known to be NP-complete. Another shortcoming of existing tools is that they depend mainly on syntactic analysis of programs and lack an effective way of dealing with the complex semantics of legacy code. Multiple factors including the physics of the problem, the mathematical model, the numerical technique contribute to the complex semantics.

We have developed a domain-specific interactive automatic parallelization tool called ParAgent [19,22]. The current system has the capability to process three-dimensional time-marching explicit finite difference codes written in Fortran-77 and

M. Valero, V.K. Prasanna, and S. Vajapeyam (Eds.): HiPC 2000, LNCS 1970, pp. 141–148, 2000
© Springer-Verlag Berlin Heidelberg 2000

produce parallel programs for distributed memory computers. This source-to-source translation tool is based on a new user-assisted approach in which domain-specific knowledge is used to facilitate automatic parallelization. Our approach to automatic parallelization is conceptually similar to the knowledge-based approach for program understanding adopted in software engineering [15,18,25].

Parallelization of the MM5 [3] mesoscale climate model code has been previously attempted using several different tools including the commercial tool FORGE [12]. As far as we know, none of the attempts has been successful. A team of four researchers at the Argonne National Laboratory worked for three years to produce a manual parallelization of MM5. Using ParAgent, a postdoctoral student parallelized MM5 in about two weeks.

This paper provides an overview of ParAgent and the results obtained by using it. The paper is organized as follows: the second section describes our approach to automatic parallelization; the third section describes ParAgent and its capabilities; the results and conclusions are presented in the fourth and fifth sections respectively. Details about the domain-specific approach and the architecture of ParAgent can be found in [22].

The Domain-Specific Approach

The main principle behind our approach is to design a structured parallelization process that incorporates knowledge about the underlying numerical method to facilitate automation. Note that a few numerical methods such as the finite-difference method (FDM), the finite-element method (FEM), and the boundary-element method (BEM) form the basis for a large majority of the legacy code for engineering and scientific applications.

Another important aspect of our approach is to blend automation and user assistance to provide a pragmatic solution. Parallelization involves many decisions and a lot of detailed work. Tedious and time-consuming tasks are the targets of automation and high-level decisions that can benefit from an expert advice are left to the user. For example, complex data dependency analysis is automated but the choice of parallelization strategy is left to the user. It is like a professor-student model where the professor may direct the parallelization and student works out the detail and implements the strategy suggested by the professor.

The analysis needed to arrive at a parallelizing scheme automatically is prohibitively complex and not practical. In our approach, the user suggests a high-level parallelization scheme, and ParAgent performs the data-dependency analysis to identify the communication patterns. This type of analysis is tedious, time-consuming and likely to cause errors if done manually; ParAgent provides critical help by automating it. For example, the user specifies that the MM5 be parallelized along I and J dimensions representing X and Y axes in the horizontal plane. ParAgent analyzes all variable that depend on the I and the J indexes to identify the communication patterns and their placement. This requires inter-procedural dataflow analysis [14] involving hundreds of variables and complex control structure spanning across subroutines.

Based on ten years of experience with parallelization of large scientific codes, we have designed an interactive and structured parallelization process that has three phases: diagnostics, communication analysis, and code generation. The first phase requires user assistance and the other two phases are completely automatic.

Diagnostic Phase

The parallelization process using ParAgent follows a strategy commonly used by expert programmers. A programmer first acquires knowledge about the numerical method and the specifics of how the method is applied in a given code. At this stage, consultation with the domain expert is often beneficial to resolve ambiguities and confirm the understanding about a given program. The diagnostic phase of ParAgent essentially mimics such consultation. During this phase, ParAgent helps the user by providing concise information about the code.

   To illustrate the diagnostic phase, consider the MM5 climate model. In this model, the user first specifies that it is based on the FDM and then processes each source file of the serial program through the ParAgent. The objective is to check if the program adheres to the finite differencing scheme. For example, ParAgent detected a problem with an array variable UJ1(I). A domain expert pointed out that UJ1 actually represents a slice of a 2 dimensional array U(I,J) with J set to 1. This sort of variable aliasing makes it hard to parallelize MM5. ParAgent identifies all uses of the variable UJ1 and helps the user modify the code to replace UJ1(I) with U(I,1).

Communication Phase

ParAgent uses a two-step process to identify communication. In the first step communication patterns are identified based on the indexing patterns of array variables. During this step, the objective is to ensure that all communication points and all variables to be communicated at each point are identified. The resulting number of communication points identified is often very large. The second step is for optimization of communication. This step analyzes communication to eliminate redundancy, performs grouping of messages, and finds optimal placements to reduce the number of communication points.

Code Generation Phase

The last phase of automatic parallelization primarily deals with two issues: global-to-local index transformation and insertion of communication primitives. Note that the hard task of identifying the communication pattern and synch/exchange points is done prior to this phase. This stage automates a lot of repetitive work such as changing the loop control statements to reflect the transformation to the local indices.

Advantages

This new approach has distinct advantages over existing approaches to compiler support for automatic parallelization [1,7,10,11,12,16,17,20,21]. It allows the user to provide high-level information such a parallelization strategy. Without such information, the parallelization can become an intractable search. The existing approaches do not include a diagnostic phase to analyze the code based on the characteristics of a numerical method. The diagnostic phase is needed to identify the quirks in the code as illustrated by an example above. Without the removal of such quirks, the existing tools are unable to perform the analysis necessary to identify communication. Once the code is diagnosed to ensure that it conforms and follows the characteristics of the numerical method, it is possible to simplify the parallelization process. For example, the communication analysis is based on a known template instead of an exhaustive data dependency analysis.

ParAgent

ParAgent is a parallelization tool built using the domain-specific approach described in the previous section. The current prototype addresses explicit time marching FDM. It takes serial Fortran-77 codes as input and produces parallel programs for distributed memory computers.

On launching the application, the user is presented with an inquiry screen. Here the user provides information about the class and the location of the code, the binding information (i.e. the loop indices that correspond to the grid dimensions), and the desired parallel mapping directions. Next, the user is presented with the diagnostics screen. The user can submit selected files for diagnosis. The system analyses the sequential program for inconsistencies and ambiguities based on the specific characteristics of the numerical method. Any inconsistency and/or ambiguity found are brought to the user's attention with help information about the possible reasons for the error. The user at this point can make changes to the code to fix the problem. When the code passes the diagnostics phase, the systems marks it as diagnosed.

When all the files have been successfully diagnosed, the user can move to the next screen, the parallelization screen. The user can select the underlying communication library (example RSL, MPI etc) and can select files for parallelization. Parallel code using embedded communication library calls is generated.

Capabilities

This section describes some of the important capabilities of ParAgent. These are grouped into diagnostic capabilities, visualization capabilities, and parallelization capabilities.

Diagnostic Capabilities
ParAgent automatically finds the correspondence (if any) between an array dimension and the grid dimension for each array variable after loop indices have been bound to grid dimensions. In some cases, user interaction is required to resolve array-space map conflicts such as for ambiguous uses of loop indices. This information is generated only once for each variable.

ParAgent allows files to be diagnosed one at a time. During this step, ParAgent verifies that the serial code conforms to the characteristics of the underlying numerical method. The system can identify conflicts such as data exchange patterns not consistent with the differencing scheme. Auxiliary information and specific references to the code are provided to help the user in resolving these problems. Each file needs to be diagnosed only once.

Visualization Capabilities

ParAgent traces the sequence of subroutine calls and displays the call-graph. This helps the user to understand the structure of the code. After parallelization, the call-graph contains annotations to indicate subroutines that will have communication.

After analyzing communication requirements, ParAgent displays the block structure of the code. The user can click on a block to view the serial code that corresponds to the block. Each block usually represents many lines of code that can be embedded in a parallel program as the code to be executed sequentially at each individual processor. The block structure also identifies the communication exchange points in the code.

For each variable, ParAgent displays the data exchange patterns at each of the communication points. The user can view these patterns in the form of stencils showing the communication that will occur in parallel processing. The stencil display shows the underlying differencing scheme at work.

Parallelization and Optimization Capabilities

ParAgent provides two alternatives for parallelizing a subroutine. In one alternative, the effects of subroutine calls from within the procedure are ignored. In the other alternative, inter-procedural analysis is performed to analyze the effects of all the subroutine calls. Subroutine calls may introduce additional communication and the selective process allows the user to view the effect of the different subroutine calls.

The diagnostic phase may take a couple of weeks for a large code consisting of several hundred files. However, the parallelization phase is automatic and takes a few minutes. During this phase, the actual parallel code is generated.

ParAgent performs inter-procedural analysis to determine global communication requirements and to perform communication optimization. Communication overhead is a function of the number of messages and the size of the messages. ParAgent reduces the number of total messages by: a) coalescing all messages going to the same processor at a given exchange point b) optimizing the number of data exchange points and c) by moving data exchange out of loops and subroutine calls whenever possible.

Results

ParAgent has been used to parallelize the following three climate and environmental modeling codes: MM5, RADM, RAMS. ParAgent runs on UNIX workstations or PCs running under LINUX. The input is Fortran-77 code and the output is a SPMD code using the MPI library for communication.

ParAgent has been demonstrated at several sites: the EPA/NOAA Atmospheric Sciences Modeling Division in North Carolina, a UNESCO-funded workshop on the project to Intercompare Regional Climate Simulations, the Center for Development of

Advanced Computing in India (CDAC), the National Center for Atmospheric Research (NCAR), and the Pacific Northwest National Laboratory (PNNL).

In this section, we present results of parallelization of two legacy codes.

Table 1. Performance of parallelized FDM benchmark (in seconds)

No of processors	N=61	N=122	N=244	N=488	N=976	N=1952
1	295.9	TL**	TL	TL	TL	TL
2	161.97	328.38	707.05	TL	TL	TL
4	97.68	182.69	369.12	823.88	TL	TL
8	62.76	108.99	200.72	424.61	1027.10	TL
16	36.76	64.95	120.12	228.61	524.06	1420.7
32	30.72	42.27	77.91	136.55	280.85	717.12
64	TS*	42.40	60.57	84.89	164.83	393.44

* problem size is too small
** problem size is too large to fit the memory

A FDM Benchmark

The first code is a FDM application that was obtained from Dayton University. This application is useful to assess the effect of problem size on performance of the parallel FDM program. This FDM application consists of 1443 lines of code distributed in 9 files. A team of two graduate students spent about 20 hrs each to manually parallelize this code. Using ParAgent, a different graduate student was able to parallelize this code in 2 hours.

The parallelized code was run on the Alice Cluster [2] in Ames Laboratory at ISU, which has 64 PCs connected by a fast Ethernet switch. Each PC in the cluster has a 200 MHz Pentium processor with 256KB cache and 256MB of main memory. The results of the parallelized code were compared to the serial version and verified to be accurate. The performance timings for the parallelized FDM benchmark are shown in Table 1.

MM5

The second code is the NCAR/Penn State Mesoscale Model MM5 [3]. This is a widely used mesoscale meteorology model. This is a complex Fortran 77 code with more than 200 files, several hundred variables with many levels of nested subroutines and complex control flow including loops spanning over several hundred lines of code. The parallelized code was run on a 64-node IBM SP1 machine. Each node had 64-MB memory.

The results of the parallelized code were compared to the serial version and verified to be accurate. The performance results for a 24-hour simulation of the parallelized MM5 code are shown in Table 2.

Table 2. Performance of parallelized MM5 (in seconds)

Processors	Grid Size	
	32x32x23	64x64x23
1	2760	18000
4	750	4520
16	225	1350
64	120	420

Conclusion

The prototype tool, ParAgent, validates the feasibility of our new approach. ParAgent has been used to successfully parallelize several large codes for climate and environmental modeling. Parallel code obtained from ParAgent has been verified to be accurate as well as efficient. Use of ParAgent can save considerable amount of time and effort in parallelization of large and complex legacy codes.

References

[1] V. Adve and C. Koelbel and J. Mellor- Crummey, Compiler Support for Analysis and Tuning of Data Parallel Programs, *Proceedings of the 1994 Workshop on Parallel Processing Tools and Environments*, 1994.
[2] Alice Cluster: http://www.scl.ameslab.gov.
[3] Anthes and Warner, Development of Hydrodynamic Models Suitable for Air Pollution and Other Mesometeorological Studies, *Mon. Weather Review*, #106, pp. 1045-1078, 1978.
[4] M. A. Ast and J. Labarta and H. Maz and A. Peruz and U. Schultz and J. Sole, A General Approach for an Automatic Parallelization Applied to the Finite Element Code PERMAS, *Proceedings of the HPCN conference*, 1995.
[5] P. Boulet and T. Brandes, Evaluation of Automatic Parallelization Strategies for HPF Compilers, *Lecture Notes in Computer Science*, 1067, pp 778, 1996.
[6] Cheng, A Survey of Parallel Programming Languages and Tools, Tech. Rep. RND-93-005, NASA Ames research center, Moffet Field CA 94035, 1993.
[7] J. Collard, Code Generation in automatic parallelizers, Proc. Of the Int. Conf. On Applications in Parallel and Distributed Computing, 1994.
[8] C. Cook and C. Pancake, What users need in parallel tool support: Survey results and analysis, *In proceedings of the scalable High performance Computing conference*, pp.40-47, 1994.
[9] M. Dion and J. L. Philippe and Y. Robert, Parallelizing compilers: what can be achieved?, *Lecture Notes in Computer Science*, 796, pp. 447-456, 1994.
[10] M. Dion and C. Randriamaro and Y. Robert, Compiling Affine Nested Loops: How to Optimize the Residual Communications after the Alignment Phase, *Journal of Parallel and Distributed Computing*, 38(2), pp. 176-187, 1996.

[11] R. Eigenmann and J. Hoeflinger and D. Padua, On the Automatic Parallelization of the Perfect Benchmarks, *IEEE Transactions on Parallel and Distributed Systems*, 9(1), pp. 5-23, 1998.

[12] R. Friedman, J. Levesque, G. Wagenbreth, Fortran Parallelization Handbook, Applied Parallel Research, 1995.

[13] M. Frumkin, A Comparison of Automatic Parallelization Tools/Compilers on the SGI Origin 2000, *SC'98: High Performance Networking and Computing: Proceedings of the 1998 ACM/IEEE SC98 Conference*, 1998.

[14] P. Havlak and K. Kennedy, Experience with interprocedural analysis of array side effects, *Proceedings, Supercomputing '90*, pp. 952-961, 1990.

[15] W. L. Johnson and E. Soloway, PROUST: Knowledge-based program understanding, *IEEE Trans. Softw. Eng.*, 11(3), 1985.

[16] P. Joisha and P. Banerjee, PARADIGM (version 2.0): A New HPF Compilation System, *Proc. 1999 International Parallel Processing Symposium*, 1999.

[17] M. Kandemir, A. Choudhary, N. Shenoy, P. Banerjee, J. Ramanujam, A Linear Algebra Framework for Automatic Determination of Optimal Data Layouts, *IEEE Transactions on Parallel and Distributed Systems*, 10(2), 1999.

[18] W. Kozaczynski and J. Q. Ning and A. Engberts, Program Concept Recognition and Transformation, *IEEE Transactions on Software Engineering*, 18(12), pp. 1065-1075, 1992.

[19] A. Krishnaswamy, Parallelization Agent - an interactive parallelizing environment, M.S. Thesis, 1999

[20] J. M. Levesque, Applied Parallel Research's xHPF system, IEEE parallel and distributed technology: systems and applications, 2(3), pp. 71-71, 1994.

[21] A. W. Lim and M. S. Lam, Maximizing parallelism and minimizing synchronization with affine partitions, *Parallel Computing*, 24, pp. 445-475, 1998.

[22] S. Mitra, A class based approach to parallelization of legacy codes, Ph.D. Thesis, 1997.

[23] B. D. Martino and H. P. Zima, Support of automatic parallelization with concept comprehension, Journal of Systems Architecture, 45, pp. 427-439, 1999.

[24] ParAgent Research: (Includes a tutorial) http://www3.ee.iastate.edu/~kothari

[25] C. Rich and R. C. Walters, The Programmer's Apprentice, ACM Press, 1990.

Practical Experiences with Java Compilation

Todd Smith[1], Suresh Srinivas[1], Philipp Tomsich[2], and Jinpyo Park[3]

[1] High Performance Programming Environments, Silicon Graphics, Inc.
MS 178, 1600 Amphitheatre Parkway, Mountain View, CA 94043
{tsmith, ssuresh}@sgi.com

[2] Institute of Software Technology, Vienna University of Technology
Favoritenstraße 9–11/E188, A–1040 Wien, Austria
phil@ifs.tuwien.ac.at

[3] School of Electrical Engineering, Seoul National University
Shinlim-Dong San 56-1, Kwanak-Gu, Seoul, Korea 151-742
jp@altair.snu.ac.kr

Abstract. The Java programming language and the underlying virtual machine model have introduced new complexities for compilation. Various approaches ranging from *just in time* (JIT) compilation to *ahead of time* (AOT) compilation are being explored with the aim of improving the performance of Java programs. The hurdles facing the achievement of high performance in Java and the strengths and weaknesses of different approaches to Java compilation are addressed in this paper, specifically within the context of SGI's effort to provide a high-performance Java execution environment for its computing platforms. The SGI JIT compiler and prototype AOT compiler are described, and performance results are presented and discussed.

1 Introduction

The Java language was designed to allow for fast and convenient development of very reliable and portable object-oriented software. Java applications are generally distributed in a platform-independent bytecode format and are tightly integrated with a number of libraries that foster software reuse and shorten development time. A Java execution environment thus typically consists of an interpreter or compiler accompanied by a large runtime environment. However, a number of these characteristics introduce additional complexities in the compilation process:

- **Portable executables.** Executables are not distributed natively compiled. Instead, a portable bytecode format is used, requiring either interpretation or a native compilation cycle before execution.
- **Strict error checking at run time.** A wide array of checks for exceptional conditions and erroneous code (e.g., out-of-bounds array accesses) are performed.
- **Lazy resolution of field and method accesses.** Java solves what is known as the *fragile base class* [9] problem by laying out objects and method tables at link time and resolving references to them lazily, as opposed to at compile-time as in C++.

M. Valero, V.K. Prasanna, and S. Vajapeyam (Eds.): HiPC 2000, LNCS 1970, pp. 149–157, 2000.
© Springer-Verlag Berlin Heidelberg 2000

– **Polymorphism and object-oriented coding styles.** Methods are virtual by default, leading to heavy use of expensive dynamic method dispatch. Typical object-oriented coding styles encourage the use of numerous small methods, making method invocation a performance bottleneck.

Consequently, the designers and implementors of commercial Java execution environments face design problems and choices that are rarely found in other languages. Multiple choices for executing Java bytecode need to be considered, including interpretation, just in time (JIT) and ahead of time (AOT) compilation, and an implementation may in fact combine multiple of these techniques.

The rest of this paper is organized as follows. Section 2 details some choices available for Java compilation and discusses their suitability for different application scenarios. Section 3 presents JIT and AOT compiler implementations for the SGI platform, and sections 4 and 5 compare and discuss their compile-time and run-time performance, respectively. Section 6 discusses related work and section 7 presents conclusions.

2   Choices for Java Compilation

Interpretation was the first choice available for Java implementations and helped accelerate early Java adoption due to its rapid retargetability. Its poor performance, however, quickly led to the proliferation of just in time (JIT) compilers, which translate bytecode to native code at run time, executing and caching it. A major benefit of JIT technology is that it integrates seamlessly into the Java execution environment and remains transparent to the application programmer and user. However, as the time required for compilation is added to the overall execution time of a program, time-consuming optimizations must be used sparingly or replaced by less costly algorithms. Code quality remains poor in comparison to traditional compilers due to this design requirement of fast compilation. Most modern systems with JIT compilers can be described as *mixed-mode*, in that they combine an interpreter with a JIT compiler: the interpreter runs initially and collects profiling information, and performance-critical methods are identified and compiled as execution progresses. This in turn allows the creation of JIT compilers that implement more costly optimizations, since they are invoked more selectively. Extending this approach further, multiple JIT compilers at different levels of sophistication can be employed within a single virtual machine [4].

With the growing importance of Java for server-side applications and the continuing demand for higher performance than that provided by early JIT compilers, implementors attempted to leverage existing mature compiler infrastructure for Java either by translating Java to C or by connecting a Java front end to a common optimization and compilation back end. The result of this approach is a system that compiles Java to native code ahead of time (AOT). Such compilers can produce completely static standalone executables, or they can work within the context of a traditional virtual machine which also supports interpretation or JIT compilation of dynamically loaded bytecode. Distributing applications in an AOT-compiled form, as opposed to distributing bytecode, gives developers more control over the environments in which their applications are deployed and more protection against decompilation and reverse engineering. However, AOT-compiled code obviously lacks the simple portability of bytecode, and AOT systems tend to have lower levels of conformance with the standard Java platform, particularly with its dynamic aspects. AOT systems offer the hope of higher performance than that available with traditional virtual machines and JIT compilers, but have yet to make that promise a clear reality.

3   Implementation of SGI JIT and AOT Compilation

SGI's Java execution environment for Irix/MIPS is currently shipping with a JIT compiler. In addition to these released products, SGI has developed a prototype AOT compiler which has been integrated with the virtual machine and JIT compiler to form a mixed-mode execution environment.

3.1   Implementation of the SGI JIT Compiler

The SGI JIT compiler is designed to work in conjunction with a virtual machine derived from Sun's reference implementation, communicating with it using the JIT API defined by Sun. Designed for very fast compilation, it has a simple structure and constructs no intermediate representation or control flow graph; bytecode is simply translated to machine code in a table-driven manner. This translation generally happens individually for each bytecode instruction, the only exception being two-instruction compare/branch sequences, which are translated together as a pair in order to generate code of reasonable quality. Complex bytecodes, such as those involving object allocation or complicated exception checks are translated into calls to runtime routines. Methods are translated lazily just prior to their first execution.

Each method begins with a prologue which sets up a Java stack frame as defined by the JVM, initializing it to the extent necessary for exception handling and garbage collection. Most of this prologue, as well as other common code sequences, are kept in a common code area for shared use by all methods, reducing code size.

The compiler uses a very fast, primitive form of global register allocation: frequently used method arguments, local variables and constants are mapped one-to-one to global registers for the entire method. This algorithm uses no live range information and makes no effort to allocate operand stack items to registers. Everything not chosen for global allocation is allocated within basic blocks to temporary registers and saved to memory at block boundaries, before method calls, and whenever an exception may be thrown. Additionally, register spilling may occur within a basic block due to register pressure.

Resolving a bytecode instruction's symbolic reference to a field or method is specified in detail in the Java virtual machine specification [20]. Although resolving these references eventually has to produce field or method table offsets to enable correct execution, that resolution cannot legally take place until the instruction is actually executed. For this reason the compiler defers generation of the final code for these references and instead emits a branch to runtime routines implementing the required resolution and offset calculation. Before control is returned to the JIT-compiled code, the branch is overwritten with the final code using the calculated offset.

Runtime exception checks are translated into explicit tests in the emitted code, branching to runtime routines to unwind the stack and transfer control to the appropriate exception handler.

3.2   Implementation of the SGI AOT Compiler

The SGI AOT compiler prototype has a novel design: rather than providing its own run-time system and producing native executables, it produces dynamically linked libraries which operate in the context of our Java virtual machine. The AOT system consists of a Java front end, a traditional multi-language compiler back end, and runtime support.

The front end, fej, reads in Java class files, converts the bytecodes to its own intermediate representation, constructs a control flow graph, and then performs high-level optimizations, including type inferencing and the associated removal of certain runtime checks. The methods are next translated into a representation suitable for the back end of the SGI MIPSPro compiler [15], a highly optimizing compiler originally developed for C, C++, and Fortran with recent modifications for Java. The MIPSPro compiler performs all of the traditional scalar compiler optimizations as well as advanced memory hierarchy optimizations. As it includes far more sophisticated optimizations than the JIT compiler, its compilation time is substantially higher. The output of AOT compilation is a dynamically linked library which contains one function symbol for each Java method in the given class.

Aside from the exceptions described below, the execution model for JIT-compiled and AOT-compiled code is very similar, allowing them to share a large amount of runtime support code. For each loaded class, the JIT compiler runtime support routines search for a AOT compiler-generated library corresponding to that class and then locate the compiled code for each method. If no compiled code is found for a given class or method, the JIT compiler is invoked instead. JIT-compiled and AOT-compiled methods share the same calling convention, allowing them to coexist without performance penalty.

Due to the nature of the SGI linker and operating system, code in a library is not rewritten at run time, so the AOT compiler does not use the same scheme as the JIT compiler for bytecode instructions which symbolically reference fields and methods. Instead, code is generated which causes a conditional trap to the operating system if a reference has not yet been resolved, and the trap handler performs the resolution. Following that, code is generated which simply accesses the class constant pool explicitly to load the offset necessary for the instruction. This results in larger and slower code than with the JIT compiler, where the rewritten code does not need to access the constant pool. Exception checking is also done by means of conditional traps. Management of exception handlers and stack unwinding is done using the standard DWARF [1] format for object file debugging information and the DWARF runtime support library. These mechanisms are far slower than the custom exception handling and stack unwinding in the JIT compiler runtime support.

4   Compile-Time Performance for Java Compilation Methods

This section and the following section present compile-time and run-time performance results, respectively, for the SGI JIT compiler, the SGI AOT compiler, and the third-party AOT system TowerJ [18] on the SGI platform. TowerJ is a clean-room, multi-platform Java AOT compiler whose emphasis is on server-side Java performance. Its current claim to fame is achieving the highest reported performance on VolanoMark 2.1 [19], measured on IA-32 Linux. Some compile-time experiments for non-SGI platforms are also included.

The benchmarks used were SPEC JVM98 1.03 [16], the jBYTEmark 0.9 suite of small Java kernels, Embedded CaffeineMark 3.0 [2], and VolanoMark 2.1. For our tests, SPEC JVM98 was modified to eliminate use of AWT, which is not supported by the TowerJ implementation. The benchmarking platforms included: an SGI Origin 2000 (300 Mhz R10000) with SGI's JDK 1.1.6, with and without the AOT compiler, SGI's Java2 v1.2.2, and TowerJ 3.3.b1; a Dell Pentium II (350 Mhz) running Red Hat Linux with IBM's JDK 1.1.8 [6]; and a Sun UltraSPARC (143 MHz) with Sun's Java2 v1.2.2 with the HotSpot 1.0 server compiler [17].

4.1   Comparison of SGI JIT and AOT Compilation Time for SPEC JVM98

JIT compilation of the entire SPEC JVM98 suite takes 2.4 seconds, which is a statistically insignificant addition to the SPEC JVM98 execution time of approximately 7 minutes. However, the SGI AOT compilation time is approximately 30 minutes and the TowerJ compilation time is in the same order of magnitude, both easily dwarfing the benchmark's run time. Obviously the technology in both of these compilers would be unsuitable in the context of a JIT compiler, and this increase in compile time should make possible an appreciable improvement in run time.

4.2   JIT Compile Time vs. Run Time

The SGI JIT was designed for fast compilation, as described in section 3.1 and illustrated by the results in section 4.1. However, it is interesting to examine other data points on the design spectrum of compile time vs. run time.

JIT compilers do not in general provide ways of reporting compilation time. Nevertheless, one way of measuring its effect is to run a small kernel repeatedly within one virtual machine invocation and to measure the run time of each iteration. Each iteration will include some combination of bytecode interpretation, JIT-compiled code execution, and JIT compilation, depending on the design of the JIT compiler and its interaction with the virtual machine. The run times of the iterations should be expected to decrease over time, as more time is spent in JIT-compiled code execution and less time is spent in bytecode interpretation or JIT compilation.

Figure 1 shows run times over three iterations of the jBYTEmark kernels with the SGI JIT compiler, and over four iterations for IBM's JIT compiler and Sun's HotSpot 1.0 server compiler. Since the SGI JIT compiler compiles every method before its first execution, run times with it have stabilized by the second iteration, with all time being spent in JIT-compiled code execution. The extra time in the first iteration is spent in JIT compilation, ranging up to 12% of the final JIT-compiled code execution time for these small kernels.

The IBM and HotSpot compilers are more sophisticated than the SGI JIT compiler, with more of the conventional compiler analyses and optimizations. However, the IBM JIT compiler does modify some of these traditional algorithms for shorter compilation time. Both systems delay compiling methods until their execution time or frequency have reached certain thresholds. In the results in figure 1 for both these compilers, the

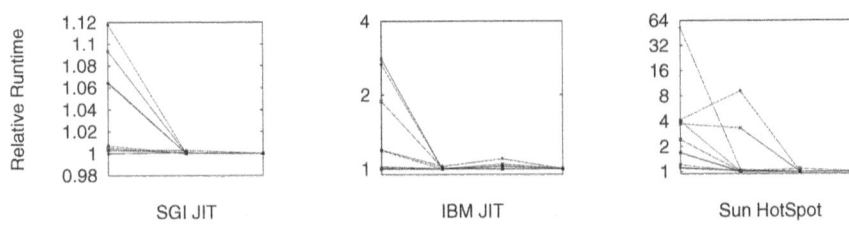

Fig. 1. Relative run times for multiple iterations of jBYTEmark kernels. Each line represents timings for an individual kernel, with successive iterations along the horizontal axis. All times are scaled such that the time of the final iteration for each kernel is 1.0. The graphs' vertical axes use three different scales, the last two being logarithmic

effects of longer compile time, higher-quality generated code, and delayed compilation can be seen in the higher first-iteration times and in the greater variance among later iterations, as more compilations are performed. Despite the longer compilation times of the IBM and HotSpot compilers, the quality of the code they produce and their selective application make them suitable and effective for use at run time as well.

5   Run-Time Performance for Java Compilation Methods

5.1   Comparision of SGI JIT and AOT Run Time for SPEC JVM98

Table 1 gives run times for several SPEC JVM98 tests using the SGI JIT compiler, SGI AOT compiler, and TowerJ compilers. These should not be considered official SPEC JVM98 results since they were not obtained according to the official SPEC JVM98 run rules. SGI AOT results were obtained using AOT-compiled SPEC JVM98 code at the -O3 optimization level but with JIT-compiled standard Java libraries.

The most performance-critical methods in mpegaudio supply the AOT compiler's global optimizer with large basic blocks of array computation and floating-point code to optimize; it does well, in particular removing many redundant memory operations. This

Table 1. SPEC JVM98 JIT vs. AOT Run-Time Performance

	Time (s)			Speedup over SGI JIT	
	SGI JIT	SGI AOT	TowerJ	SGI AOT	TowerJ
compress	62.2	61.1	69.7	1.03	.90
jess	61.1	77.2	63.9	.79	.96
raytrace	64.0	98.2	82.4	.65	.78
db	79.5	89.3	416.6	.89	.19
mpegaudio	50.2	39.5	68.7	1.27	.73
mtrt	70.0	106.2	84.7	.66	.83
jack	54.8	75.4	101.4	.82	.54

in general is the class of application for which the AOT compiler is superior. Exception-laden code, as in jack, tends to be slower with the AOT compiler because of the less efficient DWARF runtime support for exception handling and stack unwinding. Most of the run time in raytrace and mtrt is spent in small accessor methods which simply load and return a field from an object. In terms of instruction count, AOT-compiled invocations of these small methods are 32% longer than JIT-compiled invocations, due to longer field access and method invocation code, as described earlier, and longer method prologue/epilogue code, since the AOT compiler's generic back end is not optimally adapted for the shared JIT/AOT calling convention.

On the whole, the AOT compiler is unable to achieve significantly higher performance than the JIT compiler. While it is able to use the advanced compiler back-end technology SGI had already developed for other languages, that technology is in fact not currently well-suited for optimizing Java code. It does not contain the analyses necessary for removal of null pointer or array subscript checks, and its optimizer does not operate well in the presence of the traps used for runtime checking and resolution of field and method accesses. Finally, the inability to rewrite the code for those accesses hurts performance of AOT-compiled code for the vast majority of applications.

Without an understanding of the TowerJ implementation, it is difficult to explain its performance in detail. Profiling data indicates that at least for the worst cases, db and jack, its inefficiency may actually lie in the runtime support, e.g., synchronization, garbage collection, or threading support. The compiler itself is not the only important piece of the performance picture, particularly for languages like Java.

5.2   Comparison of SGI JIT and AOT Run Time for Other Benchmarks

Embedded CaffeineMark 3.0 [2] is a recent version of one of the earliest Java benchmarks. It is actually a suite of *micro-benchmarks*, very small pieces of code intended to measure specific aspects of a system's performance rather than to represent real applications.

Table 2 presents CaffeineMark results for the SGI JIT and AOT compilers. The AOT compiler is far superior to the JIT compiler for these tests, as their artificial code and tight loops are handled easily by the global optimizer and scheduler. While this provides further evidence that the AOT compiler does certain things well, the striking

Table 2. Embedded CaffeineMark 3.0 JIT vs. AOT Run-Time Performance

	SGI JIT	SGI AOT	Speedup with AOT
Sieve	3304	7693	2.328
Loop	8572	25572	2.983
Logic	4623	33232	7.188
String	7986	7525	0.942
Float	5230	14835	2.836
Method	3150	2733	0.867
Overall score	5081	11219	2.208

difference between these results and those for SPEC JVM98 illustrate why benchmarks like CaffeineMark have fallen out of favor in recent years; they do not tend to be good predictors of performance in real applications.

VolanoMark 2.1 [19] is a pure Java server benchmark measuring the average number of messages transferred per second by a server among multiple clients. Despite TowerJ's well-documented VolanoMark performance on IA-32 Linux, it is able to achieve only 592 messages per second on Irix, compared to 1701 messages per second with the SGI Java2 v1.2.2 and JIT compiler. This illustrates not only that an AOT system can be outperformed by a virtual machine with a simple JIT compiler, but also that high performance is not necessarily a portable phenomenon. Developers of AOT systems who want to claim portability across multiple platforms must pay attention to performance tuning across those platforms.

6   Related Work

The research literature appears to contain no work like that presented in this paper, comparing the JIT and AOT approaches to Java compilation and presenting experimental results, particularly within the context of a single Java execution environment. Significant JIT compilers that have been well-documented in the research literature are those from IBM [6], Intel [7], and the CACAO [11] and LaTTe projects [12]. Those that have not include Sun's HotSpot compiler [17] and Compaq's Fast JVM [3]. Appeal's JRockit [10] virtual machine implements JIT compilation at multiple levels of optimization. IBM Research is also developing a dynamic optimizing compiler written completely in Java called Jalapeno [8].

In addition to TowerJ [18], clean-room AOT systems that are currently available include NaturalBridge's BulletTrain [14] and IBM's HPCJ. All of these products omit support for some parts of the Java platform, such as JNI or the Reflection API. In particular, TowerJ is the only one which supports true dynamic class loading, as its runtime support does contain a bytecode interpreter. IBM Research [5] has adapted IBM compilers for high-performance numerical computation in Java by targeting the removal of null pointer and array subscript checking, and by improving array performance through new library classes and the introduction of semantic expansion techniques into the compiler. Marmot [13] is a static, highly optimizing AOT compiler, written almost entirely in Java, developed at Microsoft Research.

7   Conclusions

The Java programming language and the underlying virtual machine model have introduced new complexities for compilation that are currently being addressed in a variety of ways. This paper has discussed the JIT and AOT approaches to Java compilation, presented specific implementations and performance results within the context of a single Java execution environment, and discussed the trade-offs of compile time vs. run time in JIT compilers.

While AOT compilation currently does achieve higher performance in certain domains and has the potential for even further improvement, it has yet to prove itself over

JIT compilation technology in general. At present, there are still natural places in the Java implementation landscape both for AOT compilers and for JIT compilers ranging from the simple to the sophisticated, and the best implementations may be those that combine multiple approaches intelligently. It is clear that the task of compiling Java for high performance poses a new set of challenges which deserve continuing attention.

References

1. DWARF2 specification. http://reality.sgi.com/dehnert_engr/dwarf/.
2. CaffeineMark 3.0. http://www.pendragon-software.com/pendragon/cm3.
3. Compaq's Fast JVM for Alpha Processors. http://www.digital.com/java/alpha.
4. David Detlefs and Ole Agesen. The Case for Multiple Compilers. In *Proc. OOPSLA 1999 VM Workshop on Simplicity, Performance and Portability in Virtual Machine Design*, 1997.
5. Artigas et. al. High Performance Numerical Computing in Java: Language and Compiler Issues. In *12th Workshop on Language and Compilers for Parallel Computers*, Aug 1999.
6. Kazuaki Ishizaki et. al. Design, Implementation, and Evaluation of Optimizations in a Just-In-Time Compiler. In *Proc. ACM Java Grande Conference 1999*, June 1999.
7. M. Cierniak et. al. Fast, Effective Code Generation in a Just-In-Time Java Compiler. In *Proceedings of PLDI 1998*, June 1998.
8. Vivek Sarkar et. al. The Jalapeno Dynamic Optimizing Compiler for Java. In *Proc. ACM Java Grande Conference 1999*, June 1999.
9. James Gosling. Java Intermediate Bytecodes. In *ACM SIGPLAN Workshop on Intermediate Representation*, 1995. http://java.sun.com/people/jag/.
10. The JRockit Virtual Machine. http://www.appeal.se/machines/about.
11. Andreas Krall and Reinhard Grafl. CACAO: a 64bit JavaVM just-in-time compiler. In *Proc. PPoPP 1997 Workshop on Java for Science and Engineering Computation*, June 1997.
12. The LaTTe Virtual Machine. http://latte.snu.ac.kr.
13. Marmot: an Optimizing Compiler for Java.
    http://www.research.microsoft.com/apl.
14. NaturalBridge. *BulletTrainTM optimizing compiler and runtime for JVM bytecodes.* http://www.naturalbridge.com.
15. SGI MIPSPro compilers.
    http://www.sgi.com/developers/devtools/languages/mipspro.html.
16. SPEC JVM98 1.03. http://www.spec.org/osg/jvm98.
17. Sun HotSpot Virtual Machine. http://java.sun.com/products/hotspot.
18. Tower Technologies. *TowerJ3.0 A New Generation Native Java Compiler And Runtime Environment.* http://www.towerj.com.
19. VolanoMark 2.1. http://www.volano.com/benchmarks.html.
20. F. Yellin and T. Lindholm. *The Java Virtual Machine Specification.* Addison-Wesley, 1996.

Session II-B

Applications
Chair: C.P. Ravikumar, Indian Institute of Technology, Delhi

Performance Prediction and Analysis of Parallel Out-of-Core Matrix Factorization*

Eddy Caron, Dominique Lazure, and Gil Utard

LaRIA
Université de Picardie Jules Verne
5, rue du Moulin neuf
80000 Amiens, France
http://www.laria.u-picardie.fr/PALADIN

Abstract. In this paper, we present an analytical performance model of the parallel left-right looking out-of-core LU factorization algorithm. We show the accuracy of the performance prediction for a prototype implementation in the ScaLAPACK library. We will show that with a correct distribution of the matrix and with an overlap of IO by computation, we obtain performances similar to those of the in-core algorithm. To get such performances, the size of the physical main memory only need to be proportional to the product of the matrix order (not the matrix size) by the ratio of the IO bandwidth and the computation rate: There is no need of large main memory for the factorization of huge matrix!

1 Introduction

Many of important computational applications involve solving problems with very large data sets [7]. For example astronomical simulation, crash test simulation, global climate modelling, and many other scientific and engineering problems can involve data sets that are too large to fit in main memory. Using parallelism can reduce the computation time and increase the available memory size, but for challenging applications the memory is ever insufficient in size: for instance in a mesh decomposition of a mechanical problem, a scientist would like to increase accuracy by an increase of the mesh size. Those applications are referred as "parallel *out-of-core*" applications.

To increase the available memory size, a trivial solution is to use the *virtual memory* mechanism presents in modern operating system. Unfortunatly, in [2] we shown this solution is inefficient if standard *paging policies* are employed. To get the best performances, the algorithm must be generally *restructured* with explicit I/O calls. In this paper, we present a study of such a restructuration for the matrix LU factorization problem.

The LU factorization is the kernel of many applications. Thus, the importance of optimizing this routine has not to be proved because of the increasing demand

---

* This work is supported by a grant of the "Pôle de Modélisation de la Région Picardie".

M. Valero, V.K. Prasanna, and S. Vajapeyam (Eds.): HiPC 2000, LNCS 1970, pp. 161–172, 2000.
© Springer-Verlag Berlin Heidelberg 2000

of applications dealing with large matrices. In this paper we present an analytical performance model of the parallel left-right looking *out-of-core* algorithm which is used in ScaLAPACK [1]. The aim of this performance prediction model is to derive optimization for the algorithm.

In Section 2 and 3, we describe the LU factorization and the ScaLAPACK parallel version. In Section 4 we present the *out-of-core* LU factorization and the analytical performance model in Section 5. In Section 6 we analyze the overhead of the algorithm and show how to avoid it.

2   LU Factorization

The LU factorization of a matrix $A = (a_{ij})_{1 \le i,j \le N}$ is the decomposition of A as a product of two matrices $L = (l_{ij})_{1 \le i,j \le N}$ and $U = (u_{ij})_{1 \le i,j \le N}$, such that $A = LU$ where L is *lower triangular* (i.e. $l_{ij} = 0$ for $1 \le j < i \le N$) and U is *upper triangular* (i.e. $u_{ij} = 0$ for $1 \le i < j \le N$).

A well know method for parallelization of the LU factorization is based on the *blocked right-looking* algorithm. This algorithm is based on a *block decomposition* of matrices A, L and U:

$$\begin{pmatrix} A_{00} & A_{01} \\ A_{10} & A_{11} \end{pmatrix} = \begin{pmatrix} L_{00} & 0 \\ L_{10} & L_{11} \end{pmatrix} \begin{pmatrix} U_{00} & U_{01} \\ 0 & U_{11} \end{pmatrix}$$

This block decomposition gives the following equations:

$$A_{00} = L_{00}U_{00} \quad (1) \qquad A_{10} = L_{10}U_{00} \quad (3)$$
$$A_{01} = L_{00}U_{01} \quad (2) \qquad A_{11} = L_{10}U_{01} + L_{11}U_{11} \quad (4)$$

These equations lead to the following recursive algorithm:

1. Compute the factorization $A_{00} = L_{00}U_{00}$ in equation (1) (may be by another method).
2. Compute L_{01} (resp. U_{10}) from equation (2) (resp. (3)). This computation can be done by triangular solve (L_{00} and U_{00} are triangular).
3. Compute L_{11} and U_{11} from equation (4):
   (a) Compute the new matrix $A' = A_{11} - L_{10}U_{01}$.
   (b) Recursively factorize $A' = L_{11}U_{11}$.

This algorithm is called *right-looking* because once new matrix A' is computed, the left part (L_{00} and L_{01}) of the matrix is not used in the recursive computation. It is also true for the upper part (U_{00} and U_{10}). Moreover, it is easy to show that this computation can be done *data in place*: only one array is necessary to hold initial matrix A and resulting matrices L and U.

For numerical stability, *partial pivoting* (generally row pivoting) is introduced in the computation. Then, the result of the factorization is matrices L and U plus the permutation matrix P such that $PA = LU$.

In right looking algorithm with partial pivoting, the factorization of A_{00} and the computation of L_{01} are merged in the first step. For the sake of presentation, we present an algorithm with partial pivoting (data in place) where row interchanges are applied in two stages.

1a. Compute factorization $P \begin{pmatrix} A_{00} \\ A_{10} \end{pmatrix} = \begin{pmatrix} L_{00} \\ L_{10} \end{pmatrix} U_{00}$ where P is permutation matrix which represents partial pivoting: the left part of matrix A (i.e. $\begin{pmatrix} A_{00} \\ A_{10} \end{pmatrix}$) is factorized.

1b. Apply pivot P to the right part of matrix A (i.e. $\begin{pmatrix} A_{01} \\ A_{11} \end{pmatrix}$)

2. Compute U_{01} from equation (2).

3a. Compute new matrix $A' = A_{11} - L_{10}U_{01}$.

3b. Compute L_{11}, U_{11} and P' by a recursive call of factorization $P'A' = L_{11}U_{11}$ (P' is the permutation matrix.)

4. Apply pivot P' to the lower left part of matrix A (i.e. the L_{10} computed in the first step). Finally, return the composition of P and P'.

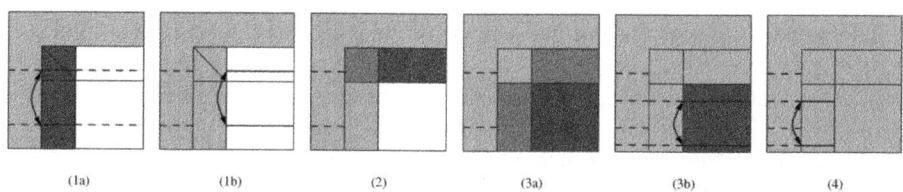

| (1a) | (1b) | (2) | (3a) | (3b) | (4) |

Fig. 1. A recursive call to the right-looking algorithm. Horizontal lines represent pivoting. Dashed lines represent part of rows which are not yet pivoted.

The Figure 1 shows the different steps for the second recursive call of the right-looking factorization.

3   Parallelization

In ScaLAPACK, the parallelization of the previous algorithm is based on a data-parallel approach: the matrix is distributed on processors and the computation is distributed according to the *owner compute rule*.

The matrix is decomposed in $k \times k$ blocks. As noticed above, at each recursive application of the right looking algorithm, the left and upper part of the matrix is factorized (modulo a permutation in the lower left part of the matrix). So, for *load balancing*, a cyclic distribution of the data is used.

The matrix is distributed *block cyclic* on a (virtual) grid of p rows and q columns of processors. The *block decomposition* of the algorithm (shown in Figure 1) corresponds to the *block distribution* of the matrix. So step 1a of the algorithm is computed by one column of p processors; step 2 is computed by one row of the q processors ; step 3a is computed by the whole grid. Pivoting step 1b (resp. 4) is executed concurrently with computation step 1a (resp. 3b).

Now let us describe more precisely the different steps of the algorithm. Step 1a is implemented by ScaLAPACK function `pdgetf2`, which factorize block of columns. For each diagonal element of the upper block (i.e. A_{00}) the following operations are applied:

1. determine pivot by a *reduce* communication primitive and *exchange* the pivot row with the current row;
2. *broadcast* the pivot row to columns of processors;
3. *scale*, i.e. divide, the column under pivot by the pivot value and update the matrix elements on the right of the column.

Step 2 of the algorithm is implemented by ScaLAPACK function `pdtrsm`: the left-upper block (i.e. A_{00}) is *broadcasted* to the processors row followed by a (BLAS) triangular solve.

Step 3a is implemented by ScaLAPACK function `pdgemm`: the blocks corresponding to U_{01} are broadcasted on columns (of processors); the blocks corresponding to L_{10} are broadcasted on rows (of processors); then the blocks are multiplied to update A'.

The performance of the parallel algorithm depends on the size of the blocks and the grid topology. The size of the block determines the degree and granularity of parallelism and also the performance of the BLAS-3 routines used by ScaLAPACK. The topology of the grid determines the cost of communications. In [3], it is shown that best performances are obtained with a grid with few rows: step 1a of the algorithm is fine grained and involves small communications (so a lot of communication latencies) for pivoting and for L_{10} computation.

4   Parallel Out-of-Core LU Factorization

Now, we consider the situation where matrix A is too large to fit in main memory. We present the parallel out-of-core left-right looking LU factorization algorithm used by the ScaLAPACK routine `pfdgetrf` for parallel out-of-core LU factorization [5]. Similar algorithms are also described in [8,6]. In the algorithm the matrix is divided in blocks of columns called *superblocks*. The width of the superblock is determined by the amount of physical available memory.

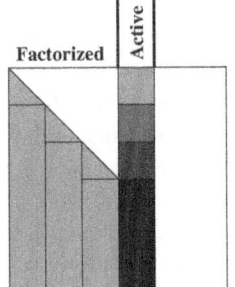

Fig. 2. Superblocks.

Like the previous parallel algorithm, the matrix is logically block cyclic distributed on the $p \times q$ grid of processors. But only blocks of the current superblock are in main memory, the other are on disks.

The parallel out-of-core algorithm is an extension of the parallel in-core algorithm. It factorizes the matrix from left to right, superblock per superblock. Each time a new superblock of the matrix is fetched in memory (called the *active* superblock), all previous pivoting and update of a *history of the right-looking algorithm* are applied to the active superblock. To do this update, superblocks lying on the left of the active superblock are read again. Once the update is finished, the right-looking algorithm resumes on the updated superblock, and the factorized active superblock is written on disk. Once the last superblock is factorized, the matrix is read again to apply the remaining row pivoting of the recursive phases (step 4).

The update of each active superblock is summarized in Figure 2. When a left superblock is considered (called the *current* superblock), the update consists in applying row pivoting to the active superblock and:

1'. read the under-diagonal part of current superblock;
2'. compute the U_{01} part of the active superblock by a triangular solve (function pdtrsm);
3'. update A_{11}, i.e. sub the product of U_{01} part of the active superblock by L_{10} of the current superblock (function pdgemm).

5   Performances Prediction

In this section we present an architectural model and a execution time prediction of the parallel out-of-core left-right looking algorithm.

---

pdgetf2:

$$\alpha_s \sum_{j=1}^{j=k}\left(\sum_{i=1}^{i=j-1}(2i-2)+\sum_{i=j+1}^{j=k}(2j-1)\right) \quad (5)$$

$$+\frac{\alpha_s}{p}\sum_{j=1}^{j=L}\left(k\times j+j\sum_{i=2}^{i=k}2(i-1)\right) \quad (6)$$

$$=\frac{\alpha_s}{p}\left(\frac{Nk^2}{6}+\frac{N^2k}{2}-\frac{Nk}{2}-\frac{N}{6}\right) \quad (7)$$

$$L\times\left(k\beta_1^p+\frac{k(k+1)}{2}\tau_1^p\right)=N\times(\beta_1^p+\frac{(k+1)}{2}\tau_1^p) \quad (8)$$

pdtrsm:

$$\frac{1}{q}\sum_{i=1}^{i=L-1}\alpha_t k^3\times i=\frac{\alpha_t}{2q}\times(N^2k-Nk^2) \quad (9)$$

$$(S(B-1)+\sum_{i=1}^{S-1}(i\times B))\times(\beta_p^q+k^2\tau_p^q)$$
$$=S(B-1)(\beta_p^q+k^2\tau_p^q)+ \quad (10)$$
$$\frac{S(S-1)}{2}(\beta_p^q+k^2\tau_p^q) \quad (11)$$

pdgemm:

$$\frac{1}{pq}\sum_{i=1}^{i=L-1}i^2\times 2\alpha_g k^3$$
$$=\frac{\alpha_g}{pq}\left(\frac{N^3+Nk^2}{3}-N^2k\right) \quad (12)$$

$$S((B-1)\beta_q^p+\frac{B(B-1)}{2}k^2\tau_q^p)$$
$$+\sum_{i=1}^{i=S}((B-1)\beta_p^q+\frac{B(B-1)}{2}k^2\tau_p^q$$
$$+(i-1)B(B-1)k^2\tau_p^q) \quad (13)$$

$$\sum_{i=1}^{i=S-1}i\times(B\beta_p^q+\frac{B(B-1)}{2}k^2\tau_p^q+(i-1)B^2k^2\tau_p^q)$$
$$=\frac{N(N-K)(6\beta_p^q+(Kk+4Nk-3k^2)\tau_p^q)}{12Kk} \quad (14)$$

$$\frac{S(S-1)}{2}(B\beta_q^p+K^2\tau_q^p) \quad (15)$$

IO:

$$2SNK\tau_{pq}^{io} \quad (16)$$

$$\sum_{i=1}^{i=S-1}i\times(\frac{B(B-1)}{2}k^2\tau_{pq}^{io}+(i-1)B^2k^2\tau_{pq}^{io})$$
$$=\frac{N(N-K)((Kk+4Nk-3k^2)\tau_{pq}^{io})}{12Kk} \quad (17)$$

Fig. 3. Costs of the different steps of the left-right looking algorithm for out-of-core LU factorization.

5.1 Architectural Model

The architectural model is a distributed memory machine with an interconnection network and one disk on each node, like a cluster. Each node stores its blocks on its own disk. Let characterize this kind of architecture by some constants representing the computation time, the communication time and the IO time.

Computation time. It is usually based on the time required for the computation of one floating point operation on one processor and is represented by a constant α. In fact, this time is not constant and depends on processor memory hierarchy and on the kind of computation. For instance a matrices multiply algorithm exhibits good cache reuse whereas product of a vector by a scale has poor temporal locality. So we distinguish three times for floating point operations which appear in the algorithm: α_g for matrix multiply , α_t for triangular solve, and α_s for scaling of vectors.

Communication time. As usual, the communication time is represented by the $\beta + V\tau$ model, where β is the startup time and τ is the time to transmit one unit of datum and V is the volume of data to be communicated. We consider only broadcast communication in our model. The constants β and τ are dependent on the topology of the virtual grid: β_p^q is the startup time for a column of p processors to broadcast data[1] on their rows, and $1/\tau_p^q$ represents the throughput. Similarly β_q^p and τ_q^p denote time for one row of q processors to broadcast data on their columns. These functions depend on communication network. For instance, for a cluster of workstations with a switch, the broadcast can be implemented by a tree diffusion. Then $\beta_p^q = \log_2 q \times \beta$ and $\tau_p^q = \log_2 q \times \frac{\tau}{p}$ where β is the startup communication time for one node and $1/\tau$ the throughput of the medium. With a hub (i.e. a bus), the model is: $\beta_p^q = p(q-1) \times \beta$ and $\tau_p^q = \tau$ if $q > 1$, $\tau_p^q = 0$ if $q = 1$.

IO time. The IO time is based on the throughput of a disk. Let τ^{io} be the time to read or write one word for one disk, then $\tau_p^{io} = \frac{\tau^{io}}{p}$ is the time to read or write p words in parallel for p independent disks.

5.2 Modelling

To model the algorithm, we estimate the time used by each function. For each function, we distinguish computation time and communication time, and we distinguish the intrinsic cost time of the parallel right-looking algorithm and cost time introduced by the out-of-core extension.

Let N be the matrix order, K be the column width of superblock, the block size is $k \times k$. The grid of processors is composed of p rows of q columns. We have the following constraints for the different constants: N is multiple of K, and K is multiple of k and q. Let $L = \frac{N}{k}$ be the block width of the matrix, $S = \frac{N}{K}$ the number of superblocks, and $B = \frac{K}{k}$ be the block width of a superblock.

---

[1] Data are equi-distributed on processors.

The Figure 3 collects costs of the different steps of the algorithm. For the sake of simplicity, we don't consider the pivoting cost in our analysis. This cost is mainly the cost of reduce operations for each element of the diagonal and the cost of row interchanges, plus the cost of re-read/write of the matrix. This time can be easily integrated in the analysis if necessary.

pdgetf2 cost. Step 1 of the algorithm (ScaLAPACK function `pdgetf2`) is applied on block columns (of width k) under the diagonal. There are L such blocks. This computation is independent of the superblock size. For the computation cost, we distinguish the computation of blocks on the diagonal (5) and the computation on the blocks under the diagonal (6). The total computation time for `pdgetf2` function for the whole $N \times N$ matrix is (7).

For communications in `pdgetf2`, for each block on the diagonal and for each element on the diagonal, the right part is broadcasted to the processor column (8).

pdtrsm cost. Step 2 of the algorithm (computation of U_{01}) is applied on each block of row lying on right of the diagonal: there is a triangular solve for each of them. The computation cost of a triangular solve between two blocks of size $k \times k$ is $\alpha_t k^3$. The total computation cost for every `pdtrsm` applied by the algorithm is (9).

The communication cost for `pdtrsm` is the broadcast of diagonal blocks to the processors row. One broadcast is done during the factorization of the active superblock (10), and another one is needed during the future updates (11).

pdgemm cost. Step 3 of the algorithm updates the trailing sub-matrix A'. The computation is mainly matrix multiply plus broadcast. For a trailing sub-matrix of order H, there is $(\frac{H}{k})^2$ block multiplications of size $k \times k$. The cost of such a multiplication is $2\alpha_g k^3$. The total computation cost is (12).

For the communication cost, we distinguish cost of factorization of the active superblock and cost of the update of the active superblock. For the factorization of the superblock, the cost is the broadcast of one row of blocks and the broadcast of one columns of blocks (13). The Figure 2 illustrates the successive updates for an active superblock. All blocks of column under diagonal in left superblock read are broadcasted (14). In the same time symmetric rows of blocks of the current superblock are broadcasted (15).

IO cost. The IO cost corresponds to the read/write of the active superblock (16) and the read of left superblocks (17).

5.3 Experimental Validation of the Analytical Model

To validate our prediction model, we ran the ScaLAPACK out-of-core factorization program on a cluster of 8 PC-Celeron running Linux and interconnected by a Fast-Ethernet switch. Each node has 96 Mb of physical memory. The model described in previous sections is instanced with the following constants (experimental measurements) : $1/\alpha_g = 237$ Mflops, $1/\alpha_t = 123$ Mflops, $1/\alpha_s = 16$ Mflops, $\beta = 1.7$ ms, $1/\tau = 11$ MB/s, $1/\tau_{io} = 1.8$ MB/s.

The Table 1 shows the comparison between the running time and the predicted time (in *italic*) of the program. We measured time for Input/Output, for Computation and we distinguished the Communication time during the factorization of active superblocks and the Communication time during the update of active superblocks. For computation and communication, running time was close to the predicted time. There was some differences for IO times. It is mainly due to our rough model of IO: IO performances are more difficult to model because access file performances depend on the layout of the file on the disk (fragmentation).

6   Out-of-Core Overhead Analysis

In comparison with the standard *in-core* algorithm, the overhead of the *out-of-core* algorithm is the extra IO cost and broadcast (of columns) cost for the update of the active superblock: for each active superblock, left superblocks must be read and broadcasted once again!

This overhead cost is represented by equations (11) and (14) for communications and (17) for IO. It is easy to show that if $K = N$ (i.e. $S = 1$) then this cost is equal to zero: it is the *in-core* algorithm execution time.

The overhead cost is $O(N^3)$, and is non negligible. In the following, we will show how to reduce this overhead cost. Let $O_C = (11) + (14)$ be the overhead communication cost and $O_{IO} = (17)$ be the overhead IO cost.

6.1   Reducing Overhead Communication Cost

As shown by the model and experimental results, the topology of the grid of processors has a great influence on the overhead communication cost:

Fact 1 *If the number of columns q is equal to 1, then $O_C = 0$!*

If there is only one column of processors, there is no broadcast of column during the update. If we consider a communication model where broadcast cost is increasing with the number of processors, then greater the number of columns is, greater is O_C. Figure 4 shows the influence of topology on the performances. In the same figure, there are plots for the predicted performances of the in-core right looking algorithm. We employed constants of our small PC-Celeron cluster. With a topology of one column of 16 processors (a ring) there is no extra communication cost. The difference with the *in-core* performances is due to the extra IO.

6.2   Overlapping IO and Computations

A trivial way to avoid the IO overhead is to overlap this IO by the computation. In the left-right looking algorithm, during updates of the active superblock, the

Table 1. Comparison of experimental and theoretical (in *italic*) running times and performances of the left-right looking out-of-core LU factorization. M is the matrix order, K the superblock wide, p the number of rows of the processor grid, q the number of columns, S is then number of superblocks. The size of the matrix in Gigabyte is given in the first column. Times are given in days (d), hours (h), minutes (m) and seconds (s).The last column shows the real and predicted performances in Mflops.

M	K	$p \times q$	IO	Computation	Comm. Active	Comm. Update	Execution time
12288	3072	1x8	5m 06s / *4m 19s*	16m 42s / *15m 54s*	3m 09s / *3m 00s*	4m 39s / *6m 28s*	29m 36s / *29m 41s* 693 Mflops / *714 Mflops*
1,2 Gb / S=4		2x4	4m 20s / *4m 19s*	13m 43s / *13m 27s*	0m 59s / *1m 16s*	1m 41s / *2m 21s*	20m 43s / *21m 23s* 983 Mflops / *970 Mflops*
		4x2	4m 44s / *4m 19s*	12m 23s / *12m 21s*	0m 34s / *0m 55s*	0m 58s / *1m 18s*	18m 39s / *18m 53s* 1088 Mflops / *1083 Mflops*
		8x1	4m 46s / *4m 19s*	12m 08s / *12m 01s*	1m 06s / *1m 22s*	1m 42s / *2m 16s*	19m 42s / *19m 58s* 1019 Mflops / *1023 Mflops*
20480	2048	1x8	31m 01s / *19m 52s*	1h 02m / *1h 04m*	10m 38s / *8m 12s*	52m 09s / *51m 51s*	2h 33m / *2h 23m* 610 Mflops / *687 Mflops*
3,3 Gb / S=10		2x4	33m 32s / *19m 52s*	1h 19m / *57m 31s*	2m 51s / *3m 10s*	18m 31s / *17m 59s*	2h 13m / *1h 38m* 706 Mflops / *985 Mflops*
		4x2	33m 10s / *19m 52s*	1h 04m / *54m 26s*	1m 07s / *1m 40s*	8m 14s / *6m 55s*	1h 46m / *1h 22m* 1208 Mflops / *1155 Mflops*
		8x1	30m 01s / *19m 52s*	1h 03m / *43m 32s*	1m 10s / *1m 52s*	7m 00s / *7m 37s*	1h 41m / *1h 12m* 917 Mflops / *1152 Mflops*
27648	1024	1x8	53m 03s / *1h 16m*	2h 45m / *2h 22m*	14m 42s / *14m 26s*	4h 03m / *4h 28m*	7h 55m / *8h 20m* 492 Mflops / *486 Mflops*
6,1 Gb / S=27		2x4	49m 48s / *1h 16m*	2h 17m / *2h 16m*	4m 25s / *5m 21s*	1h 21m / *1h 30m*	4h 32m / *5h 07m* 857 Mflops / *778 Mflops*
		4x2	50m 38s / *1h 16m*	2h 01m / *2h 11m*	1m 18s / *2m 20s*	25m 23s / *25m 36s*	3h 18m / *3h 57m* 1174 Mflops / *995 Mflops*
		8x1	51m 16s / *1h 16m*	1h 53m / *2h 11m*	0m 52s / *1m 58s*	13m 30s / *15m 00s*	2h 58m / *3h 43m* 1290 Mflops / *1054 Mflops*

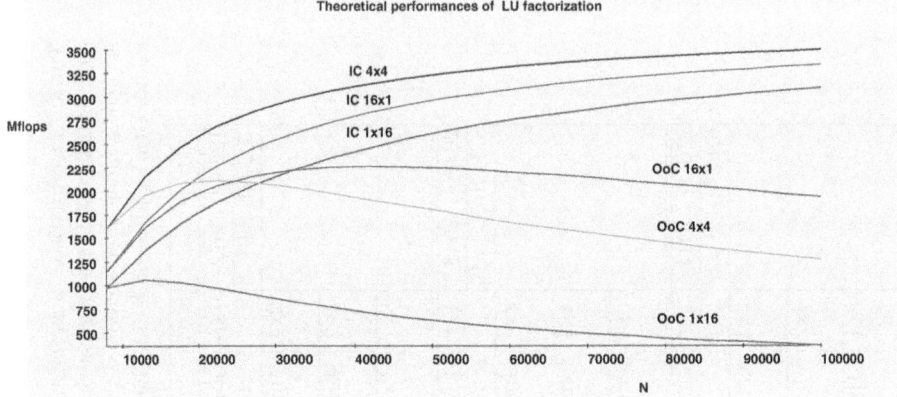

Fig. 4. Theoretical performances of the LU factorization on a cluster of 16 PC-Celeron/Linux (64Mb) interconnected by Fast-Ethernet switch: comparison of the parallel in-core (IC) right-looking algorithm and the parallel out-of-core (OoC) left-right looking algorithm, with 3 kinds of topology (1×16, 4×4, and 16×1). N is the matrix order.

left superblocks are read from left to right. An overlapping scheme is to read the next left superblock during the update of active superblock with the current one: if the time for this update is greater than the time for reading the next superblock, then the overhead IO cost is avoided.

Now, let consider the resource needed to achieve such a total overlapping. Let M be the amount of memory devoted to superblock in one processor. For a matrix order N the width of a superblock is then $K = \frac{pqM}{N}$. Let O^o_{IO} the overhead IO cost not overlapped in this new scheme.

Theorem 1. *If the number of column of processor q is equal to 1 and if $pM \geq N \frac{t_{io}}{2\alpha_g}$ then $O^o_{IO} = 0$.*

Proof. Let's go back over the update part of the algorithm (Figure 2). For the sake of simplicity, we underestimate the update computation time, and we consider only the main cost of this update: the `pdgemm` part. For each current superblock left the active superblock the update time is equal to the communication of B blocks of rows in the active superblock ($q = 1$ there is no communication of blocks of columns for the current superblock) plus the computation time of the update of the active superblock. The first part, i.e. the communication part, is equal to:

$$B^2 k^2 \tau^p_q + B \beta^p_q \tag{18}$$

Let H be the height of the current left superblock in the update of the active superblock (in number of $k \times k$ blocks). The computation time for the update of the active superblock with the current one is:

$$((H - B) \times B + \frac{B(B-1)}{2}) \times (2\frac{\alpha_g}{pq}k^3 B) \tag{19}$$

The reading time of the next left superblock is :

$$((H - 2B) \times B + \frac{B(B-1)}{2}) \times (k^2 \tau_{pq}^{io})) \tag{20}$$

Now consider the situation where the IO is overlapped by computation (i.e. $O_{IO}^o = 0$), that is $\frac{(19)+(18)}{(20)} \geq 1$. Note that if $\frac{(19)}{(20)} \geq 1$ then $\frac{(19)+(18)}{(20)} \geq 1$. So we restrict the problem to the following: determinate for which superblock width B

$$\frac{((H-B)B + \frac{B(B-1)}{2})}{((H-2B)B + \frac{B(B-1)}{2})} \times \frac{(2\frac{\alpha_g}{pq}k^3 B)}{k^2 \tau_{pq}^{io}} \geq 1$$

The first part of this expression is always greater than 1. We determinate for which superblock width the second part of the expression is greater than 1. By definition $K = k * B$ (K is the width of superblock in number of columns), and $\tau_p^{io}q = \frac{t_{io}}{pq}$. We have

$$\frac{(2\frac{\alpha_g}{pq}k^3 B)}{k^2 \tau_{pq}^{io}} \geq 1 \Leftrightarrow 2\frac{\alpha_g}{t_{io}}K \geq 1 \Leftrightarrow K \geq \frac{t_{io}}{2\alpha_g}$$

Since $K = pqM/N$ and $q = 1$, if $pM \geq N\frac{t_{io}}{2\alpha_g}$ then $\frac{(19)}{(20)} \geq 1$, i.e. $O_{IO}^o = 0$.

Note that in the algorithm, the width of the active and current superblock are equal. An idea to reduce the need for physical memory is to specify different width for the active and the current superblock during the update: increase the width of the active superblock (i.e. computation time) and reduce the width of current superblock (i.e. read time).

7   Conclusion

In this paper we presented a performance prediction model of the parallel out-of-core left-right looking LU factorization algorithm which can be found in ScaLA-PACK. This algorithm is mainly an extension of the parallel right-looking LU factorization algorithm. Thanks to these modelling, we isolated the overhead introduced by the out-of-core version. We observed that the best virtual topology to avoid the communication overhead is one column of processors. We shown that a straightforward scheme to overlap the IO by the computations allows us to reduce the IO overhead of the algorithm. We determined the memory size which is necessary to avoid the IO overhead. The memory size needed is proportional to the square root of the matrix size.

To see if this result is practicable, consider a small cluster of PC-Celeron with 16 nodes and with a Fast-Ethernet switch. To factorize a 80 Gigabytes matrix (a 100000 matrix order) we need 26 Megabytes (MB) of memory per superblock (active, current and prefetched) per node, i.e. 78 MB per node! The predicted execution time to factorize the matrix is 4.5 days without overlapping, and 2.5 days otherwise. If we substitute the Intel Celeron processors of 237 Mflops by Digital Alpha AXP processors of 757 Mflops, then the needed memory size per processor is 252 MB! The predicted computation time is about 36 hours without overlapping and about 21 hours otherwise (1.7 faster). This last time is the estimated time for the in-core algorithm with the same topology (i.e. one column of 16 processors). With a better topology for the in-core algorithm (4 columns of 4 processors), the in-core algorithm takes 18 hours to factorize the matrix, but the memory needed by node is 5 Gigabytes: 20 times greater than the memory necessary for the out-of-core version!

We plan to integrate this overlapping scheme in the ScaLAPACK parallel out-of-core left-right looking function, and experimentally validate the theoretical improvement. We plan also to study a general overlapping scheme where both IO and communication are overlapped by computation based on a extension of a previous work for the *in-core* case [4].

References

1. L. S. Blackford, J. Choi, A. Cleary, E. D'Azevedo, J. Demmel, I. Dhillon, J. Dongarra, S. Hammarling, G. Henry, A. Petitet, K. Stanley, D. Walker, and R. C. Whaley. *ScaLAPACK Users' Guide*. SIAM, Philadelphia, 1997.
2. Eddy Caron, Olivier Cozette, Dominique Lazure, and Gil Utard. Virtual Memory Management in Data Parallel Applications. In *HPCN'99, High Performance Computing and Networking Europe*, volume 1593 of *LNCS*. Springer, April 1999.
3. J. Choi, J. Demmel, I. Dhillon, J. Dongarra, S. Ostrouchov, A. Petitet, K. Stanley, D. Walker, and R. C. Whaley. LAPACK Working Note: ScaLAPACK: A Portable Linear Algebra Library for Distributed Memory Computers - Design Issues and Performances. Technical Report UT-CS-95, Department of Computer Science, University of Tennessee, 1995.
4. F. Desprez, S. Domas, and B. Tourancheau. Optimization of the ScaLAPACK LU factorization routine using Communication/Computation overlap. In *Europar'96 Parallel Processing*, volume 1124 of *LNCS*. Springer, August 1996.
5. Jack J. Dongarra, Sven Hammarling, and David W. Walker. Key Concepts for Parallel Out-Of-Core LU Factorization. *Parallel Computing*, 23, 1997.
6. Wesley C. Reiley and Robert A. van de Geijn. POOCLAPACK : Parallel Out-of-Core Linear Algebra Package. Technical report, Department of Computer Sciences, The University of Texas, Austin, October 1999.
7. J.M. Del Rosario and A. Choudhary. High performance I/O for massively parallel computers: Problems and Prospects. *IEEE Computer*, 27(3):59–68, 1994.
8. Sivan Toledo and Fred G. Gustavson. The design and implementation of SOLAR, a portable library for scalable out-of-core linear algebra computations. In *Proceedings of the Fourth Workshop on Input/Output in Parallel and Distributed Systems*, Philadelphia, May 1996. ACM Press.

Integration of Task and Data Parallelism: A Coordination-Based Approach*

Manuel Díaz, Bartolomé Rubio, Enrique Soler, and José M. Troya

Dpto. Lenguajes y Ciencias de la Computación. Málaga University
29071 Málaga, SPAIN
{mdr, tolo, esc, troya}@lcc.uma.es
http://www.lcc.uma.es

Abstract. This paper shows a new way of integrating task and data parallelism by means of a coordination language. Coordination and computational aspects are clearly separated. The former are established using the coordination language and the latter are coded using HPF (together with only a few extensions related to coordination). This way, we have a coordinator process that is in charge of both creating the different HPF tasks and establishing the communication and synchronization scheme among them. In the coordination part, processor and data layouts are also specified. The knowledge of data distribution belonging to the different HPF tasks at the coordination level is the key for an efficient implementation of the communication among them. Besides that, our system implementation requires no change to the runtime system support of the HPF compiler used. We also present some experimental results that show the efficiency of the model.

1 Introduction

High Performance Fortran (HPF) [1] has emerged as a standard data parallel, high level programming language for parallel computing. However, a disadvantage of using a parallel language like HPF is that the user is constrained by the model of parallelism supported by the language. It is widely accepted that many important parallel applications cannot be efficiently implemented following a pure data-parallel paradigm: pipelines of data parallel tasks [2], a common computation structure in image processing, signal processing or computer vision; multi-block codes containing irregularly structured regular meshes [3]; multidisciplinary optimization problems like aircraft design[4]. For these applications, rather than having a single data-parallel program, it is more appropriate to subdivide the whole computation into several data-parallel pieces, where these run concurrently and co-operate, thus exploiting task parallelism.

Integration of task and data parallelism is currently an active area of research and several approaches have been proposed [5][6][7]. Integrating the two forms of parallelism cleanly and within a coherent programming model is difficult [8].

---

* This work was supported by the Spanish project CICYT TIC-99-0754-C03-03

M. Valero, V.K. Prasanna, and S. Vajapeyam (Eds.): HiPC 2000, LNCS 1970, pp. 173–182, 2000.
© Springer-Verlag Berlin Heidelberg 2000

In general, compiler-based approaches are limited in terms of the forms of task parallelism structures they can support, and runtime solutions require that the programmer have to manage task parallelism at a lower level than data parallelism. The use of coordination models and languages to integrate task and data parallelism [4][9][10] is proving to be a good alternative, providing a high level mechanism and supporting different forms of task parallelism structures in a clear and elegant way. Coordination languages [11] are a class of programming languages that offer a solution to the problem of managing the interaction among concurrent programs. The purpose of a coordination model and the associated language is to provide a mean of integrating a number of possibly heterogeneous components in such a way that the collective set forms a single application that can execute on and take advantage of parallel and distributed systems.

BCL [12][13] is a Border-based Coordination Language focused on the solution of numerical problems, especially those with an irregular surface that can be decomposed into regular, block structured domains. It has been successfully used on the solution of domain decomposition-based problems and multi-block codes. Moreover, other kinds of problems with a communication pattern based on (sub)arrays interchange (2-D FFT, Convolution, solution of PDEs by means of the red-black ordering algorithm, etc.) may be defined and solved in an easy and clear way.

In this paper we describe the way BCL can be used to integrate task and data parallelism in a clear, elegant and efficient way. Computational tasks are coded in HPF. The fact that the syntax of BCL has a Fortran 90 / HPF style makes that both the coordination and the computational parts can be written using the same language, i.e., the application programmer does not need to learn different languages to describe different parts of the problem, in contrast with other approaches [7]. The coordinator process, besides of being in charge of creating the different tasks and establishing their coordination protocol, also specifies processor and data layouts. The knowledge of data distribution belonging to the different HPF tasks at the coordination level is the key for an efficient implementation of the communication and synchronization among them. In BCL, unlike in other proposals [6][10], the inter-task communication schedule is established at compilation time. Moreover, our approach requires no change to the runtime support of the HPF compiler used. The evaluation of an initial prototype has shown the efficiency of the model. We also present some experimental results.

The rest of the paper is structured as follows. In Sect. 2, by means of some examples, the use of BCL to integrate task and data parallelism is shown. In Sect. 3, some preliminary results are mentioned. Finally, in Sect. 4, some conclusions are sketched.

2   Integrating Task and Data Parallelism Using BCL

Using BCL, the computational and coordination aspects are clearly separated, as the coordination paradigm proclaims. In our approach, an application consists of a coordinator process and several worker processes. The following code shows the scheme of both a coordinator process (at the left hand side) and a worker process (at the right hand side).

```
program program_name           Subroutine subroutine_name (. . .)
DOMAIN declarations            DOMAIN declarations ! dummy
CONVERGENCE declarations       CONVERGENCE declarations ! dummy
PROCESSORS declarations        GRID declarations
. . .                          GRID distribution
DISTRIBUTION information        GRID initialization
DOMAINS definitions            do while .not. converge
BORDERS definitions               . . .
. . .                             PUT_BORDERS
Processes CREATION                . . .
end                               GET_BORDERS
                                  Local computation
                                  CONVERGENCE test
                               enddo
                               . . .
                               end subroutine subroutine_name
```

The coordinator process is coded using BCL and is in charge of:

- Defining the different blocks or domains that form the problem. Each one will be solved by a worker process, i.e., by an HPF task.
- Specifying processor and data layouts.
- Establishing the coordination scheme among worker processes:
  - Defining the borders among domains.
  - Establishing the way these borders will be updated.
  - Specifying the possible convergence criteria.
- Creating the different worker processes.

On the other hand, worker processes constitute the different HPF tasks that will solve the problem. Local computations are achieved by means of HPF sentences while the communication and synchronization among worker processes are carried out through some incorporated BCL primitives.

The different primitives and the way BCL is used are shown in the next sections by means of two examples. The explanation is self contained, i.e., no previous knowledge of BCL is required.

2.1   Example 1. Laplace's Equation

The following program shows the coordinator process for an irregular problem that solves Laplace's equation in two dimensions using Jacobi's finite differences method with 5 points.

$$\Delta u = 0 \;\; in \; \Omega \tag{1}$$

where u is a real function, Ω is the domain, a subset of R^2, and Dirichlet boundary conditions have been specified on $\partial\Omega$, the boundary of Ω:

$$u = g \;\; in \; \partial\Omega \tag{2}$$

```
1) program example1
2) DOMAIN2D u, v
3) CONVERGENCE c OF 2
4) PROCESSORS p1 (4,4), p2(2,2)
5) DISTRIBUTE u (BLOCK,BLOCK) ONTO p1
6) DISTRIBUTE v (BLOCK,BLOCK) ONTO p2
7) u = (/1,1,Nxu,Nyu/)
8) v = (/1,1,Nxv,Nyv/)
9) u (Nxu,Ny1,Nxu,Ny2) <- v (2,1,2,Nyv)
10) v (1,1,1,Nyv) <- u (Nxu-1,Ny1,Nxu-1,Ny2)
11) CREATE solve (u,c) ON p1
12) CREATE solve (v,c) ON p2
13) end
```

The domains in which the problem is divided are shown in Fig. 1 together with a possible data distribution and the border between domains. Dot lines represent the distribution into each HPF task. Line 2 in the coordinator process is used to declare two variables of type DOMAIN2D, which represent the two-dimensional domains. In general, the dimension ranges from 1 to 4. These variables take their values in lines 7 and 8. These values represent Cartesian coordinates, i.e. the domain assigned in line 7 is a rectangle that cover the region from point $(1,1)$ to (Nxu, Nyu). From the implementation point of view, a domain variable also stores the information related to its borders and the information needed from other(s) domain(s) (e.g. data distribution).

The border is defined by means of the operator <-. As it can be observed in the program, the border definition in line 9 causes that data from column 2 of domain v refresh part of the column Nxu of domain u. Symmetrically, the border definition in line 10, produces that data from column 1 of domain v are refreshed by part of the column Nxu-1 of domain u.

A border definition can be optionally labeled with a number that indicates the connection type in order to distinguish kinds of borders (or to group them using the same number). The language provides useful primitives in order to ease (or even automatically establish) the definition of domains and their (possibly

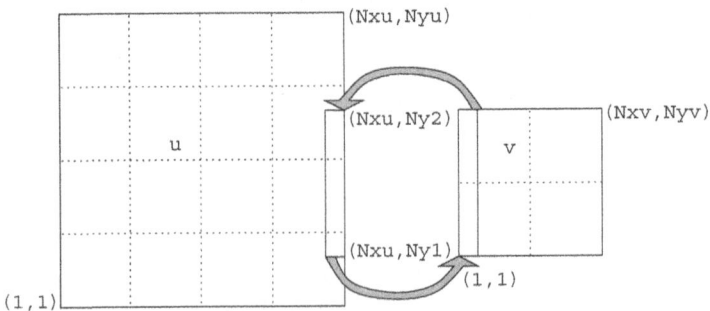

Fig. 1. Communication between two HPF tasks

overlapping) borders (e.g. intersection, shift, decompose, grow). The region sizes at both sides of the operator <- must be equal (although not their shapes). Optionally, a function can be used at the right hand side of the operator that can take as arguments different domains [12].

Line 4 declares subsets of HPF processors where the worker processes are executed. The data distribution into HPF processors is declared by means of instructions 5 and 6. The actual data distribution is done inside the different HPF tasks. The knowledge of the future data distribution at the coordination level allows a direct communication schedule, i.e., each HPF processor knows which part of its domain has to be sent to each processor of other tasks.

A CONVERGENCE type variable is declared in line 3, which is passed as an argument to the worker processes spawned by the coordinator. The clause OF 2 indicates the number of HPF tasks that will take part in the convergence criteria. The worker processes receive this variable as a dummy argument. However, when the type of the dummy argument is declared, the clause OF is not specified, as the worker processes do not need to know how many processes are solving the problem. This way, the reusability of the workers is improved (coordination aspects are specified in the coordinator process).

Lines 11 and 12 spawn the worker processes in an asynchronous way so that both HPF tasks are executed in parallel. The code for worker processes is shown in the following program:

```
1)  subroutine solve (u,c)
2)  DOMAIN2D u
3)  CONVERGENCE c
4)  double precision, GRID2D :: g, g_old
5)  !hpf$ distribute (BLOCK,BLOCK) :: g, g_old
6)  g%DOMAIN = u
7)  g_old%DOMAIN = u
8)  call initGrid (g)
9)  do i=1, niters
10)     g_old = g
11)     PUT_BORDERS (g)
12)     GET_BORDERS (g)
13)     call computeLocal (g,g_old)
14)     error = computeNorm (g,g_old)
15)     CONVERGE (c,error,maxim)
16)     Print *, "Max norm: ", error
17) enddo
18) end subroutine solve
```

Lines 2 and 3 declare dummy arguments u and c, which are passed from the coordinator. The GRID attribute appears in line 4. This attribute is used to declare a record with two fields, the data array and an associated domain. Therefore, the variable g contains a domain, g%DOMAIN, and an array of double precision numbers, g%DATA, which will be dynamically created when a value is assigned to the domain field in line 6. This is an extension of our language since a dynamic array can not be a field of a standard Fortran 90 record.

Note that line 5 is a special kind of distribution since it produces the distribution of the field DATA and the replication of the field DOMAIN.

Statement 10 produces the assignment of two variables with GRID attribute. Since g_old has its domain already defined, this instruction will just produce a copy of the values of field g%DATA to g_old%DATA. In general, a variable with GRID attribute can be assigned to another variable of the same type if they have the same domain size or if the assigned variable has no DOMAIN defined yet. In this case, before copying the data stored in the DATA field, a dynamic allocation of the field DATA of the receiving variable is carried out.

Lines 11 and 12 are the first where communication is achieved. The instruction PUT_BORDERS(g) in line 11 causes that the data from g%DATA needed by the other task (see instructions 9 and 10 in the coordinator process) are sent. This is an asynchronous operation. In order to receive the data needed to update the border associated to the domain belonging to g, the instruction GET_BORDERS(g) is used in line 12. The worker process will suspend its execution until the data needed to update its border are received.

In this example, there is only one border for each domain. In general, if several borders are defined for a domain, PUT_BORDERS and GET_BORDERS will affect all of them. However, both instructions may optionally have a second argument, an integer number that represents the kind of border that is desired to be "sent" or "received".

Local computation is accomplished by the subroutines called in lines 13 and 14 while the convergence method is tested in line 15. The instruction CONVERGE causes a communication between the two tasks that share the variable c. In general, this instruction is used when an application needs a reduction of a scalar value.

In order to stress the way our approach achieves the code reusability, Fig. 2 shows another irregular problem that is solved by the following program:

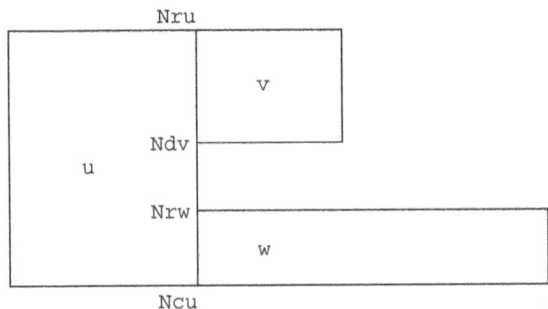

Fig. 2. Another irregular problem

```
 1) program example1_bis
 2) DOMAIN2D u, v, w
 3) CONVERGENCE c OF 3
 4) PROCESSORS p1 (4,4), p2(2,2), p3(2,2)
 5) DISTRIBUTE u (BLOCK,BLOCK) ONTO p1
 6) DISTRIBUTE v (BLOCK,BLOCK) ONTO p2
 7) DISTRIBUTE w (BLOCK,BLOCK) ONTO p3
 8) u = (/1,1,Ncu,Nru/)
 9) v = (/1,1,Ncv,Nrv/)
10) w = (/1,1,Ncw,Nrw/)
11) u (Ncu,1,Ncu,Nrw) <- w (2,1,2,Nrw)
12) u (Ncu,Ndv,Ncu,Nru) <- v (2,1,2,Nrv)
13) v (1,1,1,Nrv) <- u (Ncu-1,Ndv,Ncu-1,Nru)
14) w (1,1,1,Nrw) <- u (Ncu-1,1,Ncu-1,Nrw)
15) CREATE solve (u,c) ON p1
16) CREATE solve (v,c) ON p2
17) CREATE solve (w,c) ON p3
18) end
```

The most relevant aspect of this example is that subroutine `solve` does not need to be modified, it is the same one than in the example before. This is due to the separation that has been done between the definition of the domains (and their relations) and the computational part. Lines 15, 16 and 17 are instantiations of the same process for different domains.

2.2   Example 2. 2-D Fast Fourier Transform

2-D FFT transform is probably the application most widely used to demonstrate the usefulness of exploiting a mixture of both task and data parallelism [6][10]. Given an N×N array of complex values, a 2-D FFT entails performing N independent 1-D FFTs on the columns of the input array, followed by N independent 1-D FFTs on its rows.

```
 1) program example2
 2) DOMAIN2D a, b
 3) PROCESSORS p1 (Np), p2(Np)
 4) DISTRIBUTE a (*,BLOCK) ONTO p1
 5) DISTRIBUTE b (BLOCK,*) ONTO p2
 6) a = (/1,1,N,N/)
 7) b = (/1,1,N,N/)
 8) a <- b
 9) CREATE stage1 (a) ON p1
10) CREATE stage2 (b) ON p2
11) end
```

In order to increase the solution performance and scalability, a pipeline solution scheme is preferred as proved in [6] and [10]. This mixed task and data parallelism scheme can be easily codified using BCL. The code above shows the coordinator process, which simply declares the domain sizes and distributions, defines the border (in this case, the whole array) and creates both tasks. For this kind of problems there is no convergence criteria.

The worker processes are coded as follows. The stage 1 reads an input element, performs the 1-D transformations and calls PUT_BORDERS(a). The stage 2 calls GET_BORDERS(b) to receive the array, performs the 1-D transformations and writes the result. The communication schedule is known by both tasks, so that a point to point communication between the different HPF processors can be carried out.

```
subroutine stage1 (d)              subroutine stage2 (d)
DOMAIN2D d                         DOMAIN2D d
complex, GRID2D :: a               complex, GRID2D :: b
!hpf$ distribute a(*,block)        !hpf$ distribute b(block,*)
a%DOMAIN = d                       b%DOMAIN = d
do i= 1, n_images                  do i= 1, n_images
 ! a new input stream element       GET_BORDERS (b)
 call read_stream (a%DATA)          !hpf$ independent
 !hpf$ independent                  do irow = 1, N
 do icol = 1, N                      call fftSlice(b%DATA(irow,:))
  call fftSlice(a%DATA(:,icol))     enddo
 enddo                              ! a new output stream element
 PUT_BORDERS (a)                    call write_stream (b%DATA)
enddo                              enddo
end                                end
```

3   Preliminary Results

In order to evaluate the performance of BCL, a prototype has been developed. Several examples have been used to test it and the obtained preliminary results have successfully proved the efficiency of the model [13]. Here, we show the results for the two problems explained above.

A cluster of 4 nodes DEC AlphaServer 4100 interconnected by means of Memory Channel has been used. Each node has 4 processors Alpha 22164 (300 MHz) sharing a 256 MB RAM memory. The operating system is Digital Unix V4.0D (Rev. 878). The implementation is based on source-to-source transformations together with the necessary libraries and it has been realized on top of the MPI communication layer and the public domain HPF compilation system ADAPTOR [14]. No change to the HPF compiler has been needed.

Table 1 compares the results obtained for Jacobi's method in HPF and in BCL considering 2, 4 and 8 domains with a 128×128 grid each one. The program has been executed for 20000 iterations. BCL offers a better performance than HPF due to the advantage of integrating task and data parallelism. When the number of processors is equal to the number of domains (only task parallelism

Table 1. Computational time (in seconds) and HPF/BCL ratio for Jacobi's method

Domains	Sequential	HPF vs. BCL (ratio)		
		4 Processors	8 Processors	16 Processors
2	97.05	42.40/41.27 (1.03)	35.05/27.66 (1.27)	33.73/22.67 (1.49)
4	188.88	93.90/90.06 (1.04)	70.75/45.06 (1.57)	69.61/29.28 (2.38)
8	412.48	185.62/199.66 (0.93)	150.54/95.85 (1.57)	163.67/56.43 (2.90)

is achieved) BCL has also shown better results. Only when there are more domains than available processors, BCL has shown less performance because of the context change overhead among weight processes.

Table 2 shows the execution time per input array for HPF and BCL implementations of the 2-D FFT application. Results are given for different problem sizes. Again, the performance of BCL is generally better. However, HPF performance is near BCL as the problem size becomes larger and the number of processors decreases, as it also happens in other approaches [6]. In this situation HPF performance is quite good and so, the integration of task parallelism does not contribute so much.

4   Conclusions

BCL, a Border-based Coordination Language, has been used for the integration of task and data parallelism. By means of some examples, we have shown the suitability and expressiveness of the language. The clear separation of computational and coordination aspects increases the code reusability. This way, the coordinator code can be re-used to solve other problems with the same geom-

Table 2. Computational time (in milliseconds) and HPF/BCL ratio for the 2-D FFT problem

Array Size	Sequential	HPF vs. BCL (ratio)		
		4 Processors	8 Processors	16 Processors
32×32	1.507	0.947/0.595 (1.59)	0.987/0.475 (2.08)	1.601/1.092 (1.47)
64×64	5.165	2.189/1.995 (1.09)	1.778/1.238 (1.44)	2.003/1.095 (1.83)
128×128	20.536	7.238/7.010 (1.03)	5.056/4.665 (1.08)	4.565/3.647 (1.25)

etry, independently of the physics of the problem and the numerical methods employed. On the other hand, the worker processes can also be re-used with independence of the geometry. The evaluation of an initial prototype by means of some examples has proved the efficiency of the model. Two of them have been presented in this paper.

References

1. Koelbel, C., Loveman, D., Schreiber, R., Steele, G., Zosel, M.: The High Performance Fortran Handbook. MIT Press (1994)
2. Dinda, P., Gross, T., O'Hallaron, D., Segall, E., Stichnoth, J., Subhlok, J., Webb, J., Yang, B.: The CMU task parallel program suite. Technical Report CMU-CS-94-131, School of Computer Science, Carnegie Mellon University, (1994)
3. Agrawal, G., Sussman, A., Saltz, J.: An integrated runtime and compile-time approach for parallelizing structured and block structured applications. IEEE Transactions on Parallel and Distributed Systems, **6(7)** (1995) 747–754
4. Chapman, B., Haines, M., Mehrotra, P., Zima, H., Rosendale, J.: Opus: A Coordination Language for Multidisciplinary Applications. Scientific Programming, **6(2)** (1997)345–362
5. High Performance Fortran Forum: High Performance Fortran Language Specification version 2.0 (1997)
6. Foster, I., Kohr, D., Krishnaiyer, R., Choudhary, A.: A library-based approach to task parallelism in a data-parallel language. J. of Parallel and Distributed Computing, **45(2)** (1997)148–158
7. Merlin, J.H., Baden, S. B., Fink, S. J. and Chapman, B. M.: Multiple data parallelism with HPF and KeLP. In: Sloot, P., Bubak, M., Hertzberger, R. (eds.): HPCN'98. Lecture Notes in Computer Science, Vol. 1401. Springer-Verlag (1998) 828–839
8. Bal, H.E., Haines, M.: Approaches for Integrating Task and Data Parallelism. IEEE Concurrency, **6(3)** (1998) 74–84
9. Rauber, T., Rünger, G.: A Coordination Language for Mixed Task and Data Parallel Programs. 14th Annual ACM Symposium on Applied Computing (SAC'99). Special Track on Coordination Models. ACM Press, San Antonio, Texas. (1999) 146–155
10. Orlando S., Perego, R.: $COLT_{HPF}$ A Run-Time Support for the High-Level Coordination of HPF Tasks. Concurrency: Practice and experience, **11(8)** (1999) 407–434
11. Carriero, N., Gelernter, D.: Coordination Languages and their Significance. Communications of the ACM, **35(2)** (1992) 97–107
12. Díaz, M., Rubio, B., Soler, E., Troya, J.M.: Using Coordination for Solving Domain Decomposition-based Problems. Technical Report LCC-ITI 99/14. Departamento de Lenguajes y Ciencias de la Computacion. University of Málaga, (1999) http://www.lcc.uma.es/~tolo/publications.html
13. Díaz, M., Rubio, B., Soler, E., Troya, J.M.: BCL: A Border-based Coordination Language. International Conference on Parallel and Distributed Processing Techniques and Applications (PDPTA'2000), Las Vegas, Nevada. (2000) 753–760
14. Brandes, T.: ADAPTOR Programmer's Guide (Version 7.0). Technical documentation, GMD-SCAI, Germany. (1999) ftp://ftp.gmd.de/GMD/adaptor/docs/pguide.ps

Parallel and Distributed Computational Fluid Dynamics: Experimental Results and Challenges

M.J. Djomehri, R. Biswas, R.F. Van der Wijngaart, and M. Yarrow

Computer Sciences Corporation
NASA Ames Research Center, Moffett Field, CA 94035, USA
{djomehri,rbiswas,wijngaar,yarrow}@nas.nasa.gov

Abstract. This paper describes several results of parallel and distributed computing using a production flow solver program. A coarse grained parallelization based on clustering of discretization grids, combined with partitioning of large grids, for load balancing is presented. An assessment is given of its performance on tightly-coupled distributed and distributed-shared memory platforms using large-scale scientific problems. An experiment with this solver, adapted to a Wide Area Network environment, is also presented.

1 Introduction

Recent improvements in high-performance computing hardware have made the simulation of complex flow models a viable analysis tool. However, further efficiency increases are needed to integrate this tool into the design process, which requires large numbers of separate flow analyses. The only way in which drastic improvements in performance can be obtained in the short term is to utilize parallel and distributed computing techniques. In recent years, significant strides have been made towards this goal through parallelization of flow solver programs of importance to NASA. The objective of this paper is to assess the performance of one such code, identify potential bottlenecks, and determine their impact in the more demanding environment of distributed computing using widely-separated resources.

Two realistic cases are used to evaluate performance of the solver. They concern viscous calculations about complex configurations. The computations are done using the flow code OVERFLOW [2], which resolves the geometrical complexity of solution domains by letting sets of separately generated and updated *structured* discretization grids exchange information through interpolation. We consider both single, tightly-coupled platforms, and geographically separated machines connected using the Globus metacomputing toolkit [4].

The remainder of this paper includes a brief overview of the numerical method (Section 2) and parallelization strategy (Section 3) used. Section 4 describes the parallel implementation for a geographically distributed environment, and compares it with that used in our earlier work [1]. Results are presented in Section 5, and summary remarks and a discussion of future directions, including a strategy for improving the load balance, are given in Section 6.

M. Valero, V.K. Prasanna, and S. Vajapeyam (Eds.): HiPC 2000, LNCS 1970, pp. 183–193, 2000.
© Springer-Verlag Berlin Heidelberg 2000

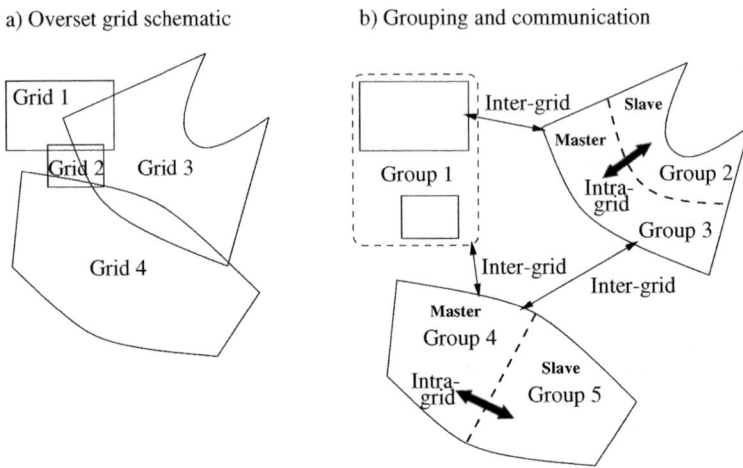

a) Overset grid schematic   b) Grouping and communication

Fig. 1. Basic partitioning strategy.

2   Numerical Method

OVERFLOW [2] uses *Chimera* [13] overset grid systems to solve the thin layer Navier-Stokes equations, augmented with a turbulence model. It is widely used in the aerodynamics community and is the most popular flow solver in use at NASA Ames Research Center. It uses finite differences in space, and implicit time-stepping. For steady-state problems such as those studied in this paper, fast relaxation to the final solution is achieved by a combination of a sophisticated multigrid convergence acceleration scheme and a local time-stepping method, which updates the solution based on a spatially varying virtual time increment.

OVERFLOW's speed and accuracy depend on the special properties of structured discretization grids. Such grids individually are not well suited for geometrically complex domains, so several are used, each of which covers only part of the domain. The resulting configuration is an overset grid system. The solution proceeds by updating, at each time step, the inter-grid boundaries with interpolated data from overlapping grids (Fig. 1a). The coordinates of the Chimera interpolation points are fixed in time and can be determined prior to the run.

3   Parallel Implementation Basics

Overset grid systems feature at least one level of exploitable parallelism: the solution on individual grids can be carried out independently by different processors, as described in our previous work [1,3]. However, that version of the code [12,14] did not allow individual grids to be distributed across several processors, creating a load balancing problem in case of large disparities between grid sizes. The new version [3,7,8] described here solves part of this problem by parallelizing the solution *within* each grid; thus, it provides a second level of exploitable parallelism. While this allows a better load balance in principle, three important sources of overall load imbalance remain.

3.1   Sources of Load Imbalance

First, all grid points are currently given equal weight in terms of associated work. However, some of the near-body grids require more work per point, because they need to solve the turbulence model in addition to the flow equations. For better load balance, we need to give larger weights to points inside turbulence regions. Second, processors may either solve part of one large grid, or one or more whole grids, but not both. This means that once a processor is assigned part of a large grid, it cannot further reduce any load imbalance by receiving more work. Third, even if the computational work is divided evenly among the processors, load imbalances may still result from disparities in communication volumes. The reason for this is twofold. First, if a large grid is distributed across multiple processors, the implicit solution process within this grid must be parallelized, which requires a significant amount of communication (indicated by the heavy arrows in Fig. 1b). This is in addition to any communications required to interpolate data from different neighbor grids. Second, the grid grouping and splitting strategy currently does not explicitly take into account the magnitude of the data volume incurred by the decomposition. In [1], we described a way of minimizing the maximum communication volume between processors. This strategy will have to be further refined to reflect the possibility of individual grids being distributed. We discuss a promising new load balancing strategy in Section 6.1. Some basic features of the parallel strategy implemented in OVERFLOW, used for the experiments reported in this paper, are discussed below.

3.2   Grouping Strategy and Grid Splitting

Load balancing the parallel algorithm involves a bin-packing strategy that forms a number of groups, each consisting of a grid and/or a cluster of grids. If no further work division takes place, the total number of grid points per group is limited by the memory allotted to each processor. As reported in [1], this strategy may produce a poor load balance, depending on the total number and size distribution of grids, and the number of processors. To alleviate this problem, the current grouping strategy in OVERFLOW allows us to divide large grids evenly across multiple processors while maintaining the implicitness of the numerical scheme within the grid; however, because a processor receiving part of a grid needs to exchange information with the other parts *during* the implicit solution process, it effectively needs to execute in lockstep with the other processors working on the grid. This has led to the requirement that a processor receiving a partial grid not receive any other work, to avoid starvation on the other processors working on the grid. As a consequence, each part of an equi-partitioned large grid is in a group by itself (Fig. 1b). The total number of processors equals the total number of groups.

The grouping algorithm is as follows. First, specify the maximum number N_{max} of grid points that each processor may receive. This number is often related to the amount of local memory available on each processor. Since the total number of points N_{tot} for the problem is given, we can now determine the minimum number of processors P_{min} required, i.e. $P_{min} = \lceil N_{tot}/N_{max} \rceil$. Next, sort the grids by size in descending order, and break up any grids whose size N exceeds the maximum into $\lceil N/N_{max} \rceil$ pieces. Each such piece is assigned its own group, and hence its own processor, as indicated above. The remaining whole grids whose sizes are below the maximum are assigned as follows. The largest is assigned to the next available processor, until all P_{min} processors own at least one grid. Each subsequent remaining grid is assigned to the processor that is responsible for the smallest total number of points thus far. When no more grids can be placed without exceeding the maximum number of points per processor, the remaining number of points is determined, and the minimum corresponding number of additional processors is computed. This process is repeated for assigning whole grids to processors until all can be placed.

OVERFLOW uses explicit message passing (MPI) for data transfers, which is suitable for both distributed and distributed-shared memory architectures. With grid splitting, message passing is also required between processors working on the same grid (Fig. 1b). Communication between groups of a partitioned grid follows the master-slave paradigm. One of the processors working within a subpartition of a grid is selected as the master of all processors in that grid. Upon completion of a time-step, all inter-group Chimera exchanges between the partitioned grid and other groups are achieved via the masters.

Most communication in the implicit flow solution occurs during the evaluation of the nonlinear forcing term (a data parallel stencil operation), and during the so-called line solves. The solves, which take place in all three coordinate directions, feature a data dependence in the active direction, which is resolved using pipelining. A detailed analysis shows that an intra-grid interface point involves the communication of 76 words per time step, whereas a Chimera interpolation point "consumes" only 5 words (without the turbulence model). This disparity in communication sizes per interface point is indicated in Fig. 1b by the relative thickness of the arrows that symbolize inter-group communications.

Intra-grid communications take place through updates of overlap points [7]. While this is a well-understood process, it adds significantly to the complexity of the implicit solution process, and requires many changes to the serial version of the code. It is also interesting to note that the parallel updates of Chimera boundaries is different from the serial updates, a fact that may impact solution stability. The difference may be thought of as block Gauss-Seidel versus block Jacobi iteration [3].

4   Parallel Distributed Computing

The distributed computing methodology used in this work is based on NASA's Information Power Grid (IPG) project [9]. It is one of several infrastructural approaches to *Grid computing* [5] (not to be confused with computations on discretization grids). IPG provides an environment for resource management with the ability to unify multiple physically separated computational resources into a single virtual machine. Large-scale problems whose memory requirements exceed those of individual resources can potentially be solved by employing the IPG; a large data set can be decomposed into several smaller, more manageable sets, and then be distributed across a collection of resources, resulting in a *parallel distributed* computation.

The parallel distributed methodology is similar to that of the parallel code. It uses the same grouping strategy and processor assignment, but processes are now distributed across distinct resources. Grid splitting is only allowed within a single resource to avoid voluminous communications between geographically separated machines. Data is transferred between processors by the MPICH-G [6] message passing library, in conjunction with the Globus metacomputing toolkit [4]. Functionally, the entire application is run as a single message passing program, and the application programmer need not be aware of any distinction between the multiple machines.

5   Results

Several experiments are presented to demonstrate the performance of OVER-FLOW. Parallel performance results on single, tightly-coupled machines: SGI Origin2000 (O2K) and Cray T3E, are discussed in Section 5.1, using two large-scale, geometrically complex, realistic test problems, consisting of grid systems totaling 9 (MEDIUM) and 33 (LARGE) million grid points, respectively. Parallel distributed performance on NASA IPG testbed systems — O2Ks at Ames, Langley and Glenn Research Centers — using the MEDIUM grid case above, is discussed in Section 5.2.

5.1   Parallel Performance

The current version of OVERFLOW has been run on a 128-processor O2K (R12000 300 MHz), and on a 512-processor Cray T3E (300 MHz). The test cases are as follows:

- MEDIUM consists of a wing-body configuration mounted on the splitter plate of the NASA Ames 12-foot Pressure Wind Tunnel (PWT), where internal flow of the tunnel and about the model has been simulated. The flow domain is discretized with 32 overset grids.
- LARGE consists of a complex configuration of a high-wing transport vehicle with nacelles and deployed flaps, discretized with 153 overset grids. No tunnel walls were modeled.

Fig. 2 shows an isometric view of some overset grids for a wing-body test object mounted in the PWT. The simulation of the tunnel's internal flow about the model takes into account the effects of the tunnel wall and other support equipment interferences. For all test cases, the following options were selected: one-equation Spalart-Allmaras turbulence model, Roe upwind scheme, ARC3D 3-factor diagonal scheme, and second and fourth order smoothing (all common choices for practical applications).

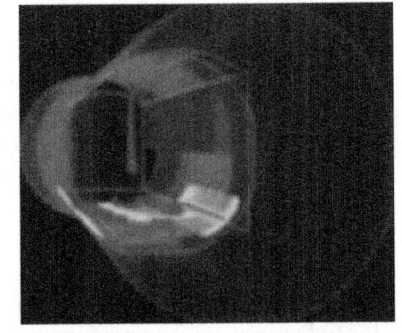

Fig. 2. Isometric view of wing-body overset grids in 12-foot PWT.

Performance statistics for the MEDIUM test case in Table 1 consist of *Wall-time* (the gross execution time in sec/time step, which includes computation, communication, and synchronization idle times), millions of floating point operations per second (Mflops), and the average and maximum inter-group (Chimera boundary) communication times. Data transfer times for the intra-grid communications are not listed separately. The Mflops reported on the O2K and the T3E are calculated relative to single-CPU Cray C90 Mflops for the same problem.

The T3E is a purely distributed-memory machine, and the size of MPI processes run on a node is limited by the physical memory located on that node. The smallest number of nodes on which test case MEDIUM can be run is 88, while LARGE requires a minimum of 203 nodes. By contrast, the O2K allows MPI processes to use as much memory as is physically available on the whole machine, so no minimum number of processors is required to solve either test case. However, we do try to maintain a reasonable balance between number of processors requested and maximum amount of memory used, so that the interference with jobs run by other users is minimized.

Table 1. Parallel performance for MEDIUM test case: 9 million grid points.

Machine	CPUs	Walltime (sec/step)	Mflops	Comm. time (sec/step)	
				Avg	Max
O2K	4	68.	540	3.8	11.
	16	18.	2055	3.0	8.8
	63	7.0	5257	1.6	3.8
	124	4.3	7667	1.1	2.7
T3E	88	17.	2127	3.5	12.
	271	7.2	5111	1.2	3.8
	370	7.9	4658	1.6	4.1
	492	6.8	5411	1.5	3.7

We conclude from Table 1 that the MEDIUM grid configuration allows reasonable scalability up to 124 and 271 processors on the O2K and T3E, respectively. Parallel efficiency on the O2K (based on 4 CPUs) is about 40% for processor sets in the range of 60 to 124. The code achieves a maximum of 7600 Mflops on the O2K with 124 CPUs, as compared with approximately 4000 Mflops on the Cray C90 with 16 processors using the serial code with multitasking directives. Similarly, the LARGE grid case results, shown in Table 2, indicate that this configuration scales well up to 96 and 510 processors on the O2K and T3E, respectively.

We note that scalability on the O2K tapers off sooner for the LARGE grid system than for the MEDIUM size configuration. This counter-intuitive result is due to the different distribution of grid sizes and inter-grid communications, and the concomitant poorer load balance. Performance on the T3E for the MEDIUM test case deteriorates beyond 271 CPUs due to communication overhead, in addition to the poor load balance.

Table 2. Parallel performance for LARGE test case: 33 million grid points.

Machine	CPUs	Walltime (sec/step)	Mflops	Comm. time (sec/step)	
				Avg	Max
O2K	16	52.0	2650	12.	24.6
	48	28.9	4768	5.0	12.5
	96	19.7	6994	4.5	9.30
	124	20.1	6855	5.5	10.9
T3E	203	34.2	4025	7.2	20.4
	299	31.0	4432	12.	25.5
	400	22.4	6043	5.2	13.3
	510	18.1	7613	4.8	12.2

5.2   Parallel Distributed Computing Performance

The NASA IPG testbed currently consists mostly of O2K systems, and this is the platform we used for the parallel distributed computing experiments. This was done to eliminate heterogeneity as a possible additional source of load imbalance. The machines used are at NASA research centers in California (Ames), Virginia (Langley), and Ohio (Glenn), respectively.

Since intra-grid interfaces are much more communication intensive than inter-grid Chimera interpolation, partitioned grids are never split across geographically

Table 3. Parallel distributed performance for MEDIUM test case, using IPG-Globus.

Number of Processors			Walltime (sec/step)	Comm. Time (sec/step)	
Ames (CA)	Langley (VA)	Glenn (OH)		Min	Max
1	1	0	195	15	48
2	0	0	175	1	27
2	1	1	98	5	30
4	0	0	91	0.7	29
2	5	1	52	1	18
8	0	0	43	0.3	9
2	12	2	32	2	16
0	16	0	23	0.1	13
8	8	8	26	2	13
0	0	24	15	0.2	8

separated machines. Moreover, the method for latency hiding, described in our previous work [1], is applicable only to Chimera updates, not to pipelining of the line solves in OVERFLOW. Performance results are summarized in Table 3 for the MEDIUM size configuration only, because the IPG testbed systems used are relatively small. Several runs were done on various numbers of processors selected from each machine. Listed are the results for totals of 2, 4, 8, 16, and 24 processors on multiple resources. For comparison, we also list single-resource results for the same number of processors. The last two columns of Table 3 refer to the minimum and maximum communication time per step, over all processors.

It follows that the scalability of the code on up to 24 processors is reasonable, although it is clear that the communication overhead becomes increasingly significant beyond 8 processors. The comparison of results for multiple versus single resources shows that both minimum and maximum communication times are significantly increased. While the maximum value is influenced by many factors (including a bad load balance), the minimum is governed virtually exclusively by the communication speed. The increase of this minimum by about an order of magnitude on multiple, geographically separated resources points to the necessity of latency hiding, as argued in [1], and in Section 6 below. Future plans also include algorithmic enhancements for better load balancing.

6   Challenges

The performance results discussed above demonstrate the feasibility of parallel and distributed computing on homogeneous IPG testbeds, although performance is significantly affected by an increase in communication time. In a realistic IPG environment, poorer connectivity and larger latencies due to geographical separation of the computers used, could further impact performance. Modifications must be made that minimize synchronization idle time and that hide latency by overlapping communication with computation. Currently, none of the aforementioned techniques have been implemented in OVERFLOW.

6.1   Improved Load Balancing Strategy

It is evident from Tables 1, 2, and 3 that there is a significant imbalance in the amounts of time spent on communications between the processors. The computational load imbalance is not listed explicitly, but it is on the order of 20% or more for most cases. Consequently, significant room for improvement is present. We propose the following strategy for better balancing the load.

Introduce the concept of *effective grid points*. These are weighted such that each effective point takes the same amount of computational work. The weight will be deduced from the presence or absence of a turbulence model.

Introduce the concept of *effective communication volume*, to be associated with exterior grid boundaries (related to Chimera interpolation updates), and with internal grid boundaries (related to partitionings of individual grids and the ensuing communications required by the implicit solution process). This volume equals the number of interface points, times the relative weight of the points (five for Chimera points, 76 for internal boundary points).

The first step in improving the current algorithm is to allow processors to work on job mixes that contain both partial and whole grids. Since subdividing single grids incurs a significant communication cost, we use the heuristic that it is best to limit that partitioning process to the smallest number of subdivisions possible. Hence, we use the method outlined in Section 3 for distributing those grids that exceed the maximum number of points allowed per processor, regardless of the number of *effective* points involved.

Subsequently, we construct a graph whose nodes consist of the numbers of effective grid points per grid or — for distributed grids — per grid subdivision, and whose edges consist of the effective communication volumes between grids and/or subdivision. This graph can then be partitioned using any of a number of efficient graph partitioners (for example, MeTiS [10], as proposed in [1]) that are capable of balancing total node weight per partition while minimizing total weight of the cut edges. The only constraint is that no partition receive more than one grid subdivision. However, the way that subdivisions are created guarantees that no more than one of them will fit on any processor anyway. The reason why processors should not receive more than one partial grid is that all processors that cooperate on a particular grid need to synchronize. If some participate in multiple such synchronized operations, it quickly becomes impossible to balance the load. But if they only participate in one during each time step, this can be scheduled as the first computational task, thus avoiding synchronization penalty.

When the partitioning is complete, several processors will generally be oversubscribed, having received too many grid points. The number of excess points is totaled, and a new number of processors is extrapolated from it, after which the assignment process is repeated until no processor exceeds the maximum allowable number of points. Implementation of this strategy requires only little coding.

Extension of this method to a widely distributed computing environment is relatively simple; the grouping strategy takes place in two stages. The first stage only assigns collections of whole grids to individual, geographically separated platforms, whose computational resources are listed as aggregate quantities. Again, we use a graph partitioner to balance computational loads and minimize communication volumes. The second stage balances the load within each platform — potentially partitioning large grids — using the method described above.

6.2   Latency Tolerance

The second ingredient for improving performance of distributed computing consists of techniques that hide communication. One such approach was implemented and tested in our previous work [1]. In this approach, named, Deferred Strategy, the numerical time-advancement procedures of the solution scheme were altered. In the original *synchronous* time-stepping method within OVER-FLOW, the Chimera boundary data is exchanged and updated prior to the start of each time step. In the deferred, *asynchronous* scheme, the Chimera boundary data exchange is initiated at the beginning of each time step, but completion of the exchange and the subsequent boundary value update is postponed to the end of the step, thus creating substantial opportunity for overlapping communication with computation.

In the deferred scheme, the Chimera updates lag one time step behind as compared with the original time-stepping method, introducing further explicitness (in a numerical sense) into the iteration procedure that might affect stability. Experiments conducted with this approach, however, show no convergence degradation for a rather large-scale, steady-state application. Unfortunately, neither do they show improvement in performance on two geographically separated O2K machines (total of up to eight processors). The reason is that the asynchronous algorithm was not supported by asynchronous hardware. A successful approach would require one or more co-processors allocated on each separate machine for data exchange alone, freeing other processors for computation. A software emulation of this concept, MPIHIDE [11], was initiated as a research project at the time of this work at Argonne National Laboratory, but has not yet been tested in conjunction with OVERFLOW. It is a subject of future research in this area.

References

1. S. Barnard, R. Biswas, S. Saini, R. Van der Wijngaart, M. Yarrow, L. Zechtzer, I. Foster, O. Larsson. Large-scale distributed computational fluid dynamics on the Information Power Grid using Globus. In *7th Symp. on the Frontiers of Massively Parallel Computation*, Annapolis, MD (1999) 60–67.
2. P. Buning, W. Chan, K. Renze, D. Sondak, I.-T. Chiu, J. Slotnick, R. Gomez, D. Jespersen. Overflow user's manual, version 1.6au. NASA Ames Research Center, 1995.
3. M.J. Djomehri, Y. Rizk. Performance and application of parallel OVERFLOW codes on distributed and shared memory platforms. In *NASA HPCCP/CAS Workshop*, Moffett Field, CA, 1998.
4. I. Foster, C. Kesselman. Globus: A metacomputing infrastructure toolkit. *International Journal of Supercomputer Applications*, 11 (1997) 115–128.
5. I. Foster, C. Kesselman (editors). *The Grid: Blueprint for a New Computing Infrastructure*. Morgan Kaufmann, San Francisco, 1999.
6. W. Gropp, E. Lusk, N. Doss, A. Skjellum. A high-performance, portable implementation of the MPI Message Passing Interface Standard. *Parallel Computing*, 22 (1996) 789–828.
7. G.P. Guruswamy, Y.M. Rizk, C. Byun, K. Gee, F.F. Hatay, D.C. Jespersen. A multilevel parallelization concept for high-fidelity multi-block solvers. In *SC97: High Performance Networking and Computing*, San Jose, CA, 1997.
8. D.C. Jespersen. Parallelizing overflow: Experiences, lessons, results. In *NASA HPCCP/CAS Workshop*, Moffett Field, CA, 1998.
9. W. Johnston, D. Gannon, W. Nitzberg. Information Power Grid implementation plan. *Working Draft*, NASA Ames Research Center, 1999.

10. G. Karypis, V. Kumar. A fast and high quality multi-level scheme for partitioning irregular graphs. Department of Computer Science Tech. Rep. 95-035, University of Minnesota, 1995.
11. O. Larsson, M. Feig, L. Johnsson. Some metacomputing experiences for scientific applications. *Parallel Processing Letters*, 9 (1999) 243–252.
12. R. Meakin. On adaptive refinement and overset structured grids. In *13th AIAA Computational Fluid Dynamics Conf.*, AIAA-97-1858, 1997.
13. J. Steger, F. Dougherty, J. Benek. A Chimera grid scheme. *ASME FED*, 5 (1983).
14. A. Wissink, R. Meakin. Computational fluid dynamics with adaptive overset grids on parallel and distributed computer platforms. In *Intl. Conf. on Parallel and Distributed Processing Techniques and Applications*, Las Vegas, NV (1998) 1628–1634.

Parallel Congruent Regions
on a Mesh-Connected Computer[*]

Chang-Sung Jeong, Sun-Chul Hwang,
Young-Je Woo, and Sun-Mi Kim

Department of Electronics Engineering, Korea University
Anamdong 5-ka, Sungbuk-ku, Seoul 136-701, Korea
Tel: 82-2-3290-3229, Fax: 921-0544, csjeong@charlie.korea.ac.kr

Abstract. In this paper, we present a parallel algorithm for solving the congruent region problem of locating all the regions congruent to a test region in a planar figure on a mesh-connected computer(MCC). Given a test region with k edges and a planar figure with n edges, it can be executed in $O(\sqrt{n})$ time if each edge in the test region has unique length; otherwise in $O(k\sqrt{n})$ time on MCC with n processing elements(PE's), and in $O(\sqrt{n})$ time for both cases using kn PE's, which is optimal on MCC within constant factor. We shall show that this can be achieved by deriving a new property for checking congruency between two regions which can be implemented efficiently using RAR and RAW operations. We also show that our parallel algorithm can be directly used to solve point set pattern matching by simple reduction to the congruent region problem, and it can be generally implemented on other distributed memory models.

1 Introduction

A mesh-connected computer(MCC) consists of several processing elements (PE's) each of which is connected to its four neighbors. Due to its architectural simplicity and easiness of exploiting massive parallelism, MCC has long been used as models of parallel computation in diverse application areas. In this paper, we deal with the *congruent region problem*(CRP) on MCC.

Given a planar figure G and a test region R, CRP is to locate all the regions in G which are congruent to R[1]. (See figure 1.) The problem arises frequently in several fields such as pattern matching, computer vision, and CAD systems. There have been a lot of sequential algorithms related to the congruity problems[4,5,6]. However, there are very few parallel algorithms for solving CRP. Shih, et. al.[1] gave an $O(n^2)$ parallel algorithm on a linearly-connected computer(LCC) with k PE's and $O(kn)$ memory, where n and k are the number of edges in G and R respectively, and based on their algorithm Boxer[2] presented $O(nlogn)$ and $O(logn)$ parallel algorithms on CREW model with k and kn PEs respectively using $O(kn)$ space. In this paper we present a parallel algorithm

---

[*] This work has been supported by KOSEF and Brain Korea 21 projects

M. Valero, V.K. Prasanna, and S. Vajapeyam (Eds.): HiPC 2000, LNCS 1970, pp. 194–203, 2000.
© Springer-Verlag Berlin Heidelberg 2000

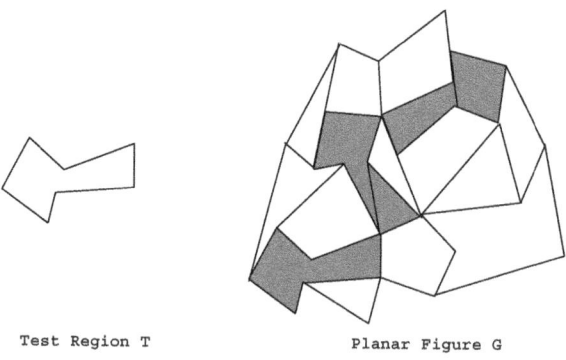

Test Region T   Planar Figure G

Fig. 1. Congruent region problem(CRP)

which can be executed in $O(\sqrt{n})$ time using $O(n)$ space if each edge in the test region has unique length; otherwise in $O(k\sqrt{n})$ time using $O(kn)$ space on a MCC with n PEs, and in $O(\sqrt{n})$ time using $O(kn)$ space for both cases on MCC with kn PEs, which is optimal within constant factor. As far as we know, this is the first parallel algorithm for CRP proposed on MCC. The previous works on LCC and CREW try to check if two regions are congruent using the relation of two adjacent edges in G and R. However, with such method, they can hardly be implemented with the same time complexity as ours on MCC even when using the same number of PE's as ours. In our parallel algorithm, we derive a new property which checks the congruent relation using the relative relation of each edge with respect to one distinct edge in R. This property enables us to design the parallel algorithm with the proposed time complexity on MCC by using RAR(random access read) and RAW(random access write) operations. Given a point set S of n points and a pattern set R of k points, point set pattern matching(PSPM) is to find all subsets $P \subset S$ such that R and P are congruent. PSPM is similar to CRP, but differs in that it deals with point sets instead of edge sets in planar figure. We shall show that our parallel algorithm can be directly used to solve PSPM by reducing it to the proper congruent region problem. The previous sequential or parallel algorithms [3,4] for PSPM can not be implemented with the same time complexity on MCC.

The outline of our paper is as follows: In section 2 we explain basic operations, some definitions and property of the congruent region. In section 3 we present parallel algorithms for solving CRP and in section 4 show how they can be used for solving PSPM. In section 5 we give a conclusion.

2   Preliminaries

In this section we shall explain basic operations on MCC, define the *congruent region problem* more formally by introducing some notations and terminologies, and then describe an important property which shall be used for the efficient computation of congruent regions in our algorithm.

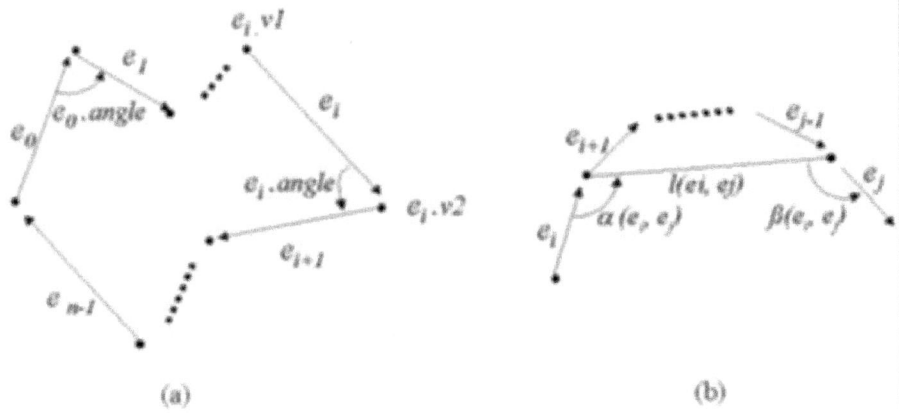

Fig. 2. Illustration of Definitions

We shall use RAR and RAW as the basic operations on MCC arranged in $\sqrt{n} \times \sqrt{n}$ array. In a RAR, each PE specifies the key of data it wishes to receive, or it specifies a null key, in which case it receives nothing. Each PE reads a data from the PE which contains its specified key. Several PE's can request the same key and read the same data. In a RAW, each PE specifies the key of data and sends its data to the PE which contains its specified key. If two or more PE's send data with the same specified key, then the PE with that key will receive the value obtained by summing all the data sent to it. RAR and RAW take $O(\sqrt{n})$ time on MCC [7,8]. In MCC, $O(\sqrt{n})$ time complexity is optimal within constant factor.

Let G be a graph with a vertex set V and an edge set E. A graph G is called a *planar figure* if it is *planar* and all the edges in E are straight lines.

Definition 1: Let $e_i.v_1$ and $e_i.v_2$ be the end vertices of an edge e_i. A *region* P is a cyclic chain of directed edges $\{e_0, e_1, .., e_{k-1}\}$ such that $e_i.v_2 = e_{i+1}.v_1$ for all i, $0 \leq i < k$ and each edge e_i is a straight line directed from $e_i.v_1$ to $e_i.v_2$ so that the interior of the region lies to the right side of each edge in P. Let $e_i.length$ and $e_i.angle$ denote the length of e_i and the interior angle between e_i and e_{i+1} respectively. (See figure 2.a.)

We assume that all the subscripts are modulo k. Let $S = \{s_0, s_1, .., s_{k-1}\}$ and $R = \{r_0, r_1, ..., r_{k-1}\}$ be the regions in the plane.

Definition 2: A pair of edges (s_i, s_{i+1}) in S is said to be *e-congruent* to a pair of edges (r_j, r_{j+1}) in R for i and j, $0 \leq i, j < k$ if and only if $s_i.length = r_j.length$, $s_{i+1}.length = r_{j+1}.length$, and $s_i.angle = r_j.angle$. S is said to be *r-congruent* to R if there exist i, $0 \leq i < k$ such that (s_{i+t}, s_{i+t+1}) is *e-congruent* to (r_t, r_{t+1}) for all t, $0 \leq t < k$.

Therefore, we can easily see that S is *r-congruent* to R if S can coincide with R by proper geometrical translation, rotation and reflection. CRP can be described formally using the above definition as follows: Given a planar figure G and a test region R, CRP is to find all the regions of G each of which is

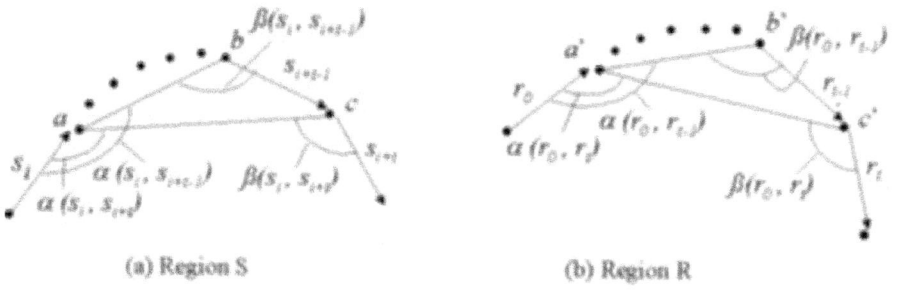

(a) Region S   (b) Region R

Fig. 3. Illustration of Lemma 1

r-congruent to R. From now on we shall refer to the region which is r-congruent to R simply by the *congruent region*.

Definition 3: If S is r-congruent to R, each edge s_i in S is one-to-one correspondence to an edge r_j in R such that (s_i, s_{i+1}) is *e-congruent* to (r_j, r_{j+1}), and s_i is said to *correspond* to r_j. We define the *starting edge* of S to be the one which corresponds to r_0.

In the following we derive another condition for S to be r-congruent to R, which can be used more efficiently for the computation of congruent regions on MCC.

Definition 4: For two edges e_i and e_j in a region $P = \{e_0, e_1, .., e_{k-1}\}$, the line connecting $e_i.v_2$ and $e_j.v_1$ is denoted $\ell(e_i, e_j)$ and its length by $\ell(e_i, e_j).length$, and the interior angle between e_i (resp. e_j) and $\ell(e_i, e_j)$ by $\alpha(e_i, e_j)$ (resp. $\beta(e_i, e_j)$). (See figure 5.b.) If e_j is identical to e_{i+1}, then $\alpha(e_i, e_j)$ and $\beta(c_i, c_j)$ are all equal to $e_i.angle$. Then, the *relative position* of e_j with respect to e_i is defined by a record $RP(e_i, e_j) = (e_i.length, e_j.length, \alpha(e_i, e_j), \beta(e_i, e_j), \ell(e_i, e_j).length)$.

Definition 5: A pair of edges (s_i, s_{i+t}) in S is said to be *relative e-congruent* to a pair of edges (r_0, r_t) in R for i, $0 \le i < k$ if $RP(s_i, s_{i+t}) = RP(r_0, r_t)$.

Lemma 1: S is r-congruent to R if and only if there exist i, $0 \le i < k$ such that (s_i, s_{i+t}) is *relative e-congruent* to (r_0, r_t) for all t, $1 \le t < k$.

Proof: Suppose that (s_i, s_{i+t-1}) and (s_i, s_{i+t}) are *relative e-congruent* to (r_0, r_{t-1}) and (r_0, r_t) respectively for $1 \le t < k$. We want to show that (s_{i+t-1}, s_{i+t}) is *e-congruent* to (r_{t-1}, r_t). Since $s_{i+t-1}.length$ and $s_{i+t}.length$ are equal to $r_{t-1}.length$ and $r_t.length$ respectively, all we have to prove is that $s_{i+t-1}.angle$ is equal to $r_{t-1}.angle$. Let T (resp. T') be a triangle determined by the vertices a, b, c (resp. a', b', c'), where a, b, c (resp. a', b', c') are $s_i.v_2$, $s_{i+t-1}.v_1$, $s_{i+t}.v_1$ (resp. $r_0.v_2$, $r_{t-1}.v_1$, $r_t.v_1$) respectively. (See figure 3.) Since $\ell(s_i, s_{i+t-1}).length = \ell(r_0, r_{t-1}).length$, $s_{i+t-1}.length = r_{t-1}.length$, and $\beta(s_i, s_{i+t-1}) = \beta(r_0, r_{t-1})$, T is congruent to T'. Therefore, $\angle bca = \angle b'c'a'$, and hence $s_{i+t-1}.angle = r_{t-1}.angle$, since $\beta(s_i, s_{i+t}) = \beta(r_0, r_t)$. We shall prove the other direction by induction. Suppose that (s_{i+t}, s_{i+t+1}) is *e-congruent* to (r_t, r_{t+1}) for all t, $0 \le t < k$ without loss of generality. Clearly, (s_i, s_{i+1}) is *relative e-congruent* to (r_0, r_1). Suppose that (s_i, s_{i+t-1}) is *relative e-congruent* to (r_0, r_{t-1}). We shall show that (s_i, s_{i+t}) is *relative e-congruent* to (r_0, r_t). Since S is *e-congruent* to R, $s_i.length = r_0.length$ and $s_{i+t}.length = r_t.length$. Since $\ell(s_i, s_{i+t-1}).length = \ell(r_0, r_{t-1}).length$, $s_{i+t-1}.length = r_{t-1}.length$, and $\beta(s_i, s_{i+t-1}) = \beta(r_0, r_{t-1})$, T is congruent to T'. Therefore, $\ell(s_i, s_{i+t}).length = \ell(r_0, r_t).length$, $\angle bac = \angle b'a'c'$ and $\angle bca = \angle b'c'a'$, and hence $\alpha(s_i, s_{i+t}) = \alpha(r_0, r_t)$ and $\beta(s_i, s_{i+t}) = \beta(r_0, r_t)$, since $\alpha(s_i, s_{i+t-1}) = \alpha(r_0, r_{t-1})$, and $s_{i+t-1}.angle = r_{t-1}.angle$, and it follows that (s_i, s_{i+t}) is *relative e-congruent* to (r_0, r_t). $\blacksquare$

3   Parallel Algorithm

In this section we shall explain how to find all the regions in a planar figure which are *r-congruent* to a test region. First we shall describe a parallel algorithm for the unique case that each edge in the test region has distinct length, and then for the general case with no restriction using the lemma 1. Each edge e in a planar figure G is duplicated into one directed from $e.v_1$ to $e.v_2$ and the other from $e.v_2$ to $e.v_1$ in order to take care of the geometrical reflection with respect to e. Let $E = \{e_0, e_1, .., e_{n-1}\}$ be a set of all the duplicated edges in G and let $R = \{r_0, r_1, .., r_{k-1}\}$ be a test region. Initially, all the edges in E and R are evenly distributed in local memory on MCC. Each congruent region S is represented by the index of the starting edge in S which corresponds to r_0.

1). Unique Case: There exists, for each edge e_i in E, unique edge r_{i_c} with the same length as e_i if it does. The detailed description of the algorithm is given below.

Parallel Algorithm Congruent-Region-I
Input: A set $E=\{e_0, e_1, ..., e_{n-1}\}$ of edges for a planar figure G and a set $R=\{r_0, r_1, ...,r_{k-1}\}$ of edges for a test region such that the length of each edge in R is unique.
Output: Each edge e_i in a congruent region S sets $e_i.flag$ to *true* and stores into $e_i.reg$ and $e_i.corr$ respectively the index of the starting edge of S which corresponds to r_0 and the index of edge in R to which e_i corresponds.

1). Find, for each edge r_i, its relative position $RP(r_0, r_i)$ with respect to r_0, and store it into $r_i.rel$.

2). Find, for each edge e_i in E, an edge r_{i_c} in R with the same length as e_i if it exists, and store the index i_c of r_{i_c} and the relative position $r_{i_c}.rel$ into $e_i.corr$ and $e_i.rel$ respectively, and set $e_i.flag$ to *true*. This can be done using RAR with the length as key. Note that e_i should correspond to r_{i_c} if there exists a congruent region to which e_i belongs.

3). Find, for each edge e_i with $e_i.flag$ set to *true*, an edge e_{i_0} in G such that (e_{i_0}, e_i) is *relative e-congruent* to (r_0, r_{i_c}). If there exists such e_{i_0}, store i_0 to $e_i.reg$; otherwise set $e_i.flag$ to *false*. This can be done by first determining, for each edge e_i, the two end vertices of e_{i_0} from e_i and $e_i.rel$, and then finding e_{i_0} in G by executing RAR with those two end vertices as key.

Comment: At step 3) we have computed, for each edge e_i with the same length as r_{i_c}, e_{i_0} which may be the starting edge of the congruent region where e_i

corresponds to r_{i_c}. However, we have not determined yet whether there exists a congruent region with e_{i_0} as the starting edge or not. The following lemma is used to check whether there exists a congruent region with e_{i_0} as the starting edge.

Lemma 2: For a set S of k edges, S is congruent to R if and only if e_{i_0} is identical for every edge e_i in S.

Proof: Suppose that e_{i_0} is identical for every edge e_i in S. For two distinct edges e_i and e_j in S, r_{i_c} should be different from r_{j_c}, since if r_{i_c} is identical to r_{j_c}, then (e_{i_0}, e_i) should be *relative e-congruent* to (e_{i_0}, e_j), making contradiction. Therefore, for each edge e_i in S there must exist unique r_{i_c} in R such that (e_{i_0}, e_i) is *relative e-congruent* to (r_0, r_{i_c}), and hence S is *r-congruent* to R by lemma 1. The other direction follows directly from lemma 1. ∎

Therefore, we can determine, for each edge e_p in G, whether there exists a congruent region with e_p as the starting edge by checking whether there is a set S of k edges such that for each edge e_i in S, e_{i_0} is e_p. Using this property we can find all the edges in E which are the starting edges of the congruent regions in G, and hence we can determine, for each edge e_i, whether it belongs to a congruent region or not by checking whether e_{i_0} is a starting edge.

4). Check, for each edge in E with its flag set to *true*, whether it may be the starting edge of a congruent region. This can be done as follows: Send, for each edge e_i with $e_i.flag$ set to *true*, 1 to the PE containing $e_{i_0}.sum$ using RAW. Then $e_{i_0}.sum$ stores the sum of all the 1's sent to it, and if it is k, there exists a congruent region with e_{i_0} as the starting edge by lemma 2, and set $e_{i_0}.flag$ to *true*; otherwise *false*.

5). Check, for each edge e_i with $e_i.flag$ set to *true*, whether it may belong to the congruent region with e_{i_0} as the starting edge. Since we have determined whether there exists a congruent region with e_{i_0} as the starting edge at step 4), this step can be done simply by checking whether $e_{i_0}.flag$ is true or not using RAR. Unless $e_{i_0}.flag$ is not true, there does not exist a congruent region with e_{i_0} as the starting edge, and set $e_i.flag$ to *false*. After this step, all the edges in the congruent regions have their flags to *true*, and each edge e_i in the congruent region S stores into $e_i.reg$ and $e_i.corr$ respectively the index of the starting edge in S and the index of the edge in R to which e_i corresponds.

End Algorithm.

The correctness of the parallel algorithm follows from the discussion in the algorithm, and the following theorem holds.

Theorem 1: Given a planar figure G with n edges and R with k edges, all the congruent regions can be computed in $O(\sqrt{n})$ time on MCC using n PE's and $O(n)$ space.

Proof: Step 1) takes $O(\sqrt{n})$ time for distribution of r_0. Step 2) and 3) can be done in $O(\sqrt{n})$ time for RAR respectively. Step 4) can be executed in $O(\sqrt{n})$ time for RAW and step 5) in $O(\sqrt{n})$ for RAR. Therefore, the overall time complexity is $O(\sqrt{n})$. Since we need $O(1)$ space for each edge in E and R, the overall space complexity is $O(n)$. ∎

2). General Case: Each edge e_i in E may belong to more than one congruent regions and correspond to different edges of R in each congruent region to which it belongs. Therefore, in the general case we need to check, for each edge e_i, whether there is a congruent region such that e_i corresponds to r_j, for $j = 0, 1, 2, .., k-1$. At the jth iteration, the computation of a congruent region where e_i corresponds to r_j is carried out similarly as in the unique case. The detailed description of the parallel algorithm is given below.

Parallel Algorithm Congruent-Region

Input: A set $E=\{e_0, e_1, ..., e_{n-1}\}$ of duplicated edges for a planar figure G and a set $R=\{r_0, r_1, ..., r_{k-1}\}$ of edges for a test region.

Output: Each edge e_i in E is associated with an array A_i of size k, each component of which consists of three fields: *flag, reg,* and *rel.* If $A_i[j].flag$ is set to *true*, e_i belongs to the congruent region S where e_i corresponds to r_j, and $A_i[j].reg$ stores the index of the starting edge of S. $A_i[j].rel$ stores the relative position $RP(r_0, r_j)$.

Comment: For each edge e_i in E there may exist more than one edges of R whose lengths are equal to e_i. Therefore, e_i may correspond to more than one edges of R in different congruent regions. In order to distinguish the congruent regions where e_i belongs but corresponds to different edges of R, e_i is associated with an array A of size k such that $A_i[j].flag$ is *true* if e_i belongs to a congruent region where e_i corresponds to r_j and $A_i[j].reg$ stores the index of the starting edge of the congruent region if it exists.

1).   Check whether each edge in R has unique length or not. If it is, find all the congruent regions using the parallel algorithm Congruent-Region-I.

2).   For each edge e_i do the following for $j = 0, 1, .., k-1$.

2.1).   Find the relative position $RP(r_0, r_j)$ and store it into $A_i[j].rel$.

2.2).   Set $A_i[j].flag$ to *true* if $e_i.length$ is equal to $r_j.length$.

3).   For each edge e_i with $A_i[j].flag$ set to *true* do the following for $j = 0, 1, .., k-1$.

3.1).   Find an edge e_{i_0} in E such that (e_{i_0}, e_i) is *relative e-congruent* to (r_0, r_j) if it exists. This can be done similarly as in step 3 of the previous algorithm for the unique case. If there exists such e_{i_0}, set $A_i[j].flag$ to *true*, and store i_0 to $A_i[j].reg$.

Comment: At the jth iteration of step 3.1 we have computed, for each edge e_i, e_{i_0} which may be the starting edge of the congruent region where e_i corresponds to r_j. However, we have not known yet whether there exists a congruent region with e_{i_0} as the starting edge. In the next step we check, for each edge e_{i_0} in E, whether there exists a congruent region with it as the starting edge.

3.2).   Send 1 to the PE containing $e_{i_0}.sum$ using RAW, where $i_0 = A_i[j].reg$. Then each PE containing $e_{i_0}.sum$ receives 1 and add it to $e_{i_0}.sum$.

4).   After step 3) $e_{i_0}.sum$ in each PE stores the sum of all the 1's sent to it. If the sum is k, there exists a congruent region with e_{i_0} as the starting edge by lemma 2, and set $A_{i_0}[0].flag$ to *true*; otherwise *false*.

Comment: If $A_{i_0}[0].flag$ is set to *true*, then there exists a congruent region S with e_{i_0} as the starting edge, and every edge e_i with $A_i[j].reg$ equal to i_0

becomes an edge of S which corresponds to r_j. Therefore, we can identify all the edges in the congruent regions as follows.

5). For each edge e_i with $A_i[j].flag$ set to *true* do the following for $j = 0, 1, .., k - 1$: Check whether $A_{i_0}[0].flag$ is *true* or not by RAR, where $i_0 = A_i[j].reg$. If it is true, e_i is an edge of the congruent region with e_{i_0} as the starting edge; otherwise set $A_i[j].flag$ to *false*. After this step if e_i corresponds to r_j in the congruent region with the starting edge e_{i_0}, $A_i[j].flag$ is set to *true* and $A_i[j].reg$ stores the index i_0 of the starting edge of S.

End Algorithm.

Theorem 2: Given a planar figure G with n edges and R with k edges, all the congruent regions can be computed in $O(k\sqrt{n})$ time on MCC using n PE's, and in $O(\sqrt{n})$ time using kn PE's and $O(kn)$ space.

Proof: The correctness of the parallel algorithm follows directly from the discussion given in the algorithm.

4   Point Set Pattern Matching

In this section we shall show that the parallel algorithm for finding congruent regions can be used in solving PSPM by transforming it to the proper CRP. Given a point set $S = \{s_0, s_1, .., s_{n-1}\}$ of n points and a pattern set $R=\{r_0, r_1, .., r_{k-1}\}$ of k points in Euclidean 2D space, PSPM is to find all subsets $P \subset S$ such that R and P are congruent. The parallel algorithm works by executing, for each s_j in S, a subprocedure *Pattern_Matching* for finding all subsets $P \subset S$ such that R and P are congruent while anchoring r_0 at s_j. (See figure 4.) The subprocedure is essentially identical to the parallel algorithm Congruent -Region described in the previous section. It can be described briefly as follows: Let $P = \{p_0, p_1, .., p_{k-1}\}$ be a subset of S with p_0 identical to s_j. Figure 4a illustrates five points $r_0 \sim r_4$ in a pattern set R, and figure 4b and c respectively show eight points $s_0 \sim s_7$ in S and five points $p_0 \sim p_4$ in $P \subset S$ congruent to R while anchoring r_0 at s_5. Suppose each point r_i (resp. p_i) in R (resp. P) is connected to r_0 (resp. p_0) by an edge e_i (resp. d_i). (See figure 5.) For explanation purpose, we assume that all the points in R and S are sorted in circular order. (In the real algorithm, the circular sorting is not necessary, since we use the relative position of each edge.) For each edge e_i, its relative position $RP(e_i, e_j)$ with respect to e_j is defined by a record $(e_i.length, e_j.length, \alpha(e_i, e_j))$, where $\alpha(e_i, e_j)$ is the interior angle between e_i and e_j. Similarly, for each edge d_i, its relative position $RP(d_i, d_j)$ with respect to d_j is defined by a record $(d_i.length, d_j.length, \alpha(d_i, d_j))$, where $\alpha(d_i, d_j)$ is the interior angle between d_i and d_j. We assume that for subscript $i, i + t$ is $(i + t \text{ modulo } k) + 1$ if $i + t \geq k$.

Definition 6: A pair of points (p_i, p_j) in P is said to be *relative p-congruent* to a pair of points (r_a, r_b) if $RP(d_i, d_j) = RP(e_a, e_b)$.

Definition 7: P is said to be *m-congruent* to R if and only if there exists i, $1 \leq i < k$ such that (p_{t+i}, p_i) is *relative p-congruent* to (r_{t+1}, r_1) for all t, $1 \leq t < k - 1$.

We can easily see that P is *m-congruent* to R if P coincides with R by proper geometrical translation, rotation and reflection, since the relative position of two

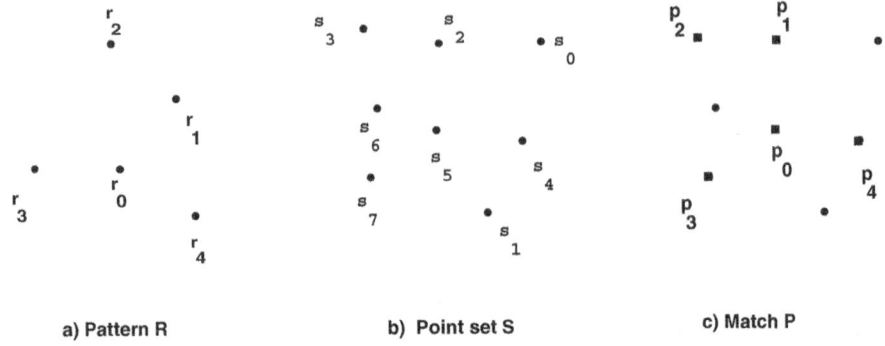

a) Pattern R   b) Point set S   c) Match P

Fig. 4. Point Set Pattern Matching

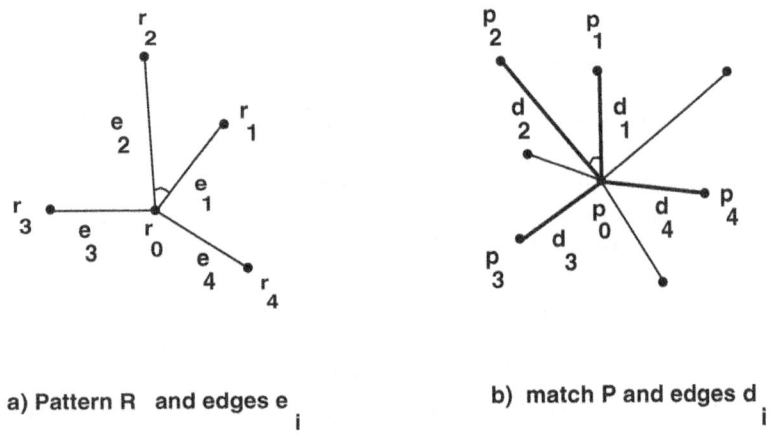

a) Pattern R   and edges e$_i$ b) match P and edges d$_i$

Fig. 5. Illustration of Definitions

corresponding edges in P and R is identical. Therefore, the subprocedure *Pattern_Matching* is reduced to the problem to find all the subsets P of S each of which is *m_congruent* to R, which is identical to the congruent region problem with minor change in relative position. Thus it can be computed in parallel with the same time and space complexity as the parallel algorithm *Congruent_Region* shown in the previous section. Since we have to iterate the subprocedure *Pattern_Matching* n times for PSPM, the following theorem follows:

Theorem 3: Given point sets S with n points and R with k points in Euclidean 2 dimensional space, PSPM can be computed in $O(n^{1.5})$ time using $O(n)$ space if the length of each edge e_i for r_i in R is unique; otherwise in $O(kn^{1.5})$ on MCC with n PE's using $O(kn)$ space, and in $O(n^{1.5})$ time for both cases on MCC with kn PE's using $O(kn)$ space.

5   Conclusion

Our parallel algorithm for CRP on MCC has several research contributions: First, as far as we know, it is the first parallel algorithm proposed on MCC, which takes $O(\sqrt{n})$ time if all the edges in R have unique length; otherwise $O(k\sqrt{n})$ time using n PEs, and $O(\sqrt{n})$ time for both cases using kn PEs, which is optimal within constant factor. Second, it can be exploited efficiently for solving more complex PSPM with simple reduction to CRP. Note that the previous parallel algorithms for solving CRP or PSPM on other parallel models such as CREW and LCC[1,2,?] can not be directly implemented with the same time complexity as ours on MCC even when using the same number of PEs as ours. Investigating the congruence relation by checking the relative position between edges as shown in lemma 1 and 2 enabled us to exploit RAR and RAW operations for the efficient location of congruent regions and hence develope the parallel algorithm with the proposed time complexity on MCC for CRP and PSPM. Third, the parallel algorithm for CRP and PSPM can be generally implemented on any parallel distributed memory model such as CCC(cube-connected computer), TC(tree-connected computer), PSC(perfect shuffled computer), etc without having to design new parallel algorithm if they are provided with RAR and RAW operations, since our parallel algorithm makes use of them as basic operations. If it takes $O(T(n))$ time for RAR and RAW, CRP can be executed in $O(T(n))$ if all the edges in R have unique length; otherwise $O(kT(n))$ time using n PEs, and $O(T(n))$ for both cases using kn PEs. For example, $T(n)$ is $O(log^2 n)$ for CCC, TC, and PSC in distributed memory model. It still remains as an open problem to find an optimal parallel algorithm for PSPM on MCC with n PEs for higher dimensions.

References

1. Z. C. Shih, R. C. T. Lee and S. N. Yang, A Parallel Algorithm for Finding Congruent Regions, *Parallel Computing 13* (1990) 135-142.
2. L. Boxer, Finding congruent Regions in Parallel, *Parallel Computing* (1992) 807-810.
3. L. Boxer, Scalable Parallel Algorithms for Geometric Pattern Recognition, *Journal of Parallel and Distributed Computing* (1999) 466-486.
4. P. J. de Renzende and D. T. Lee, Point Set Pattern Matching in d-dimensions, *Algorithmica* 13 (1995), 387-404.
5. K. Sugihara, An *nlogn* Algorithm for Determining the Congruity of Polyhedra, *J. Computer and System Sciences 29* (1984) 36-47.
6. M. D. Atkinson, An Optimal Algorithm for Geometrical Congruence, *J. Algorithms 8* (1987) 159-172.
7. D. Nassimi and S. Sahni, Data Broadcasting in SIMD Computers, *IEEE Trans. on Computer* (1981) 101-106.
8. M. Lu, Constructing the Voronoi Diagram on a Mesh-connected Computer, *Proc. of 1986 Int. Conf. on Parallel Processing* (1986) 806-811.

Can Scatter Communication Take Advantage of Multidestination Message Passing?*

Mohammad Banikazemi and Dhabaleswar K. Panda

Ohio State University, Columbus OH 43210, USA,
{banikaze,panda}@cis.ohio-state.edu

Abstract. Collective communications compose a significant portion of the communications in many high performance applications. The *multidestination* message passing mechanism has been proposed recently for efficient implementation of collective operations by using a fewer number of communication phases. However, this mechanism has been only employed for improving the performance of non-personalized collective operations such as multicast. In this paper, we investigate whether multidestination message passing can also help personalized communications such as one-to-all scatter. We propose a new scheme, called *Sequential Multidestination Tree-based* (SMT) scheme, which takes advantage of the multidestination message passing mechanism and provide a better performance. For a range of system sizes and parameters, it is shown that the SMT scheme outperforms other known schemes for a wide range of message lengths.

1 Introduction

Collective communications such as *broadcast, multicast, global reduction, scatter, gather, complete exchange,* and *barrier synchronization* have shown to compose a significant portion of the communications in many high performance applications [4]. Furthermore, these operations are used by the underlying system for operations such as resource management and maintenance of cache coherency in Distributed Shared Memory systems [2]. Thus, it is crucial to develop efficient mechanisms and algorithms for implementing these operations. Collective operations can be classified into two major groups: *non-personalized* and *personalized* operations. In non-personalized operations, the same message is exchanged among the participating nodes. In personalized operations, different messages are exchanged between the participating nodes.

A new *multidestination* message passing mechanism [6] has been proposed recently for efficient implementation of collective operations by using a fewer number of communication phases. The multidestination message passing schemes

* This research is supported in part by an IBM Cooperative Fellowship award, an OSU Presidential Fellowship, an NSF Career Award MIP-9502294, NSF Grant CCR-9704512, an Ameritech Faculty Fellowship award, and grants from the Ohio Board of Regents.

M. Valero, V.K. Prasanna, and S. Vajapeyam (Eds.): HiPC 2000, LNCS 1970, pp. 204–211, 2000.

have been used to implement many non-personalized collective communications such as multicast, broadcast, and barrier synchronization. However, the problem of improving the performance of personalized operations (such as scatter, gather and complete exchange) by using multidestination message passing mechanism has remained unsolved.

In this paper, we take on such a challenge and demonstrate that the performance of personalized collective communications can also be improved by using the multidestination message passing mechanism. In particular, we focus on the *one-to-all personalized* communication simply known as *scatter* [4]. We focus on systems supporting wormhole switching. As a starting point, we consider only systems with k-ary n-cube regular topologies and router-based multidestination message passing. However, the framework can be generalized to other systems with switch-based and network interface-based multidestination message passing.

The paper is organized as follows. The previously known schemes for scatter communication are discussed in Section 2. In Section 3, we present a new scheme (SMT) for implementing scatter. Performance evaluation results are presented in Section 4. Finally, we conclude the paper with conclusions and future research directions.

2 Current Approaches for Implementing Scatter Communication

In this section, we define the scatter collective communication operation and derive its lower bound for wormhole-routed systems. Then, we present two unicast-based scatter schemes which are used on current systems. We also investigate the conditions under which these schemes can meet the lower bound.

2.1 Scatter Communication and Its Lower Bound

Scatter is one of the personalized communication operations in which, one processor (source node) sends a different message to every other processor in a user-defined group [3]. For a system with P processors and message size of l bytes, the volume of the data which has to be sent out from the source node to the network is $(P-1)l$, and at least one message should be sent out. Therefore, the lower bound (LB) of the scatter operation latency for wormhole routed systems (in the absence of congestion) [1] can be derived as:

$$T_{LB} = t_s + (P-1)lt_p \tag{1}$$

where t_s is the communication start-up overhead and t_p is the transmission time per byte.

---

[1] We ignore the impact of node delay for simplicity.

2.2   Sequential Tree (ST) Scheme

The ST scheme is based on *sequential trees*. In this scheme the source node sends a separate message to each of the other processors participating in scatter. Therefore, the number of required communication steps is $P - 1$. Figure 1.a shows the sequential tree for a system with 8 processors. The algorithm for implementing the ST scheme can be found in [1].

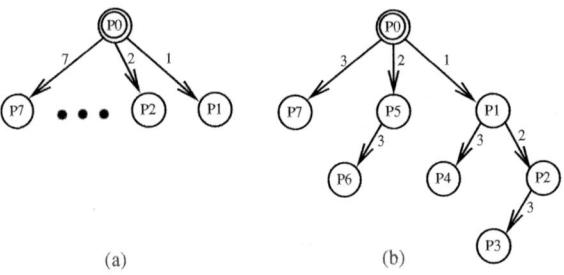

(a)   (b)

Fig. 1. Sequential tree (ST) (a) and Binomial Tree (BT) (b) schemes for an 8-processor system.

Let us analyze the latency of the scatter communication under the ST scheme. Depending on the system parameters, two different cases can be recognized: $t_s \leq lt_p$ and $t_s > lt_p$. Thus, the scatter latency can be written as:

$$T_{ST} = \begin{cases} t_s + (P-1)lt_p \text{ if } t_s \leq lt_p \\ (P-1)t_s + lt_p \text{ if } t_s > lt_p \end{cases} \tag{2}$$

It is obvious that only if $t_s \leq lt_p$ then the lower bound can be met. If this condition is not satisfied the latency of the scatter operation will exceed the lower bound. Because of the $(P-1)t_s$ term, the latency of scatter can increase rapidly when the system size increases or the start-up overhead is significant. This leads to the following observations:

Observation 1 *The ST scheme can achieve the lower bound only when $l \geq \frac{t_s}{t_p}$.*

Observation 2 *With $l < \frac{t_s}{t_p}$, the latency of scatter increases rapidly with increase in system size.*

2.3   Binomial Tree (BT) Scheme

The BT scheme is based on the well known *binomial trees*. Such a scheme is used in the MPL library for scatter implementation. In this approach, the number of processors which receive the message is doubled in each communication step. Figure 1.b illustrates the BT scheme on an 8-processor system. The algorithm for implementing the BT scheme is presented in [1].

Let us analyze the latency of the scatter communication under the BT scheme. Since the number of processors which receive their messages is doubled in each step, the number of required communication steps for the entire operation is $\log_2 P$. On the other hand, the data size decreases by a factor of two in every successive step. Therefore, the latency of scatter can be written as:

$$T_{BT} = \sum_{i=1}^{\log_2 P} t_s + \frac{Pl}{2^i} t_p$$
$$= \log_2 P t_s + (P-1) l t_p \qquad (3)$$

It can be seen that the BT scheme can never meet the lower bound due to additional number of start-ups. This leads to:

Observation 3 *The BT scheme is not capable of achieving the lower bound.*

3 Sequential Multidestination Tree (SMT) Scheme

In this section, we use multidestination message passing to develop better schemes for scatter. (A brief overview of the multidestination mechanism is presented in [1].) First, we provide the motivation behind such schemes. Then, we propose a generalized scheme. Finally, we present schemes for 2D tori/meshes and k-ary n-cubes.

3.1 Motivation

As presented in observations 1 and 3 only ST scheme can achieve the lower bound. Nevertheless, observation 1 shows that, the ST scheme can achieve the lower bound just for a limited variations of system parameters and message sizes. This lower bound is achieved only when the start-up overhead of sending messages are overlapped with the transmission time of other messages. To achieve this requirement, the $t_s \leq l t_p$ condition should hold. In other words, the size of the messages being transmitted should be large enough so that the next message can become ready for transmission not later than when the previous message has been completely injected to the interconnection network ($l \geq \frac{t_s}{t_p}$). In this section, we propose a new scheme that extends the cases in which the lower bound latency can be met by increasing the size of the messages which are being transmitted. The multidestination message passing mechanism helps us to achieve such increase of message length.

3.2 The General Scheme

In the SMT scheme, all processors which are participating in the scatter communication operation are grouped into smaller groups. The messages of all processors in each group are combined to form a multidestination message. Then, these different multidestination messages are sent out to the members of the corresponding groups. Depending on the routing scheme being used in the system and the arrangement of the data at the source node different groupings can be used. For example, as illustrated in Fig. 2, nodes in a 12-processor system are grouped into four groups (G1-G4). With such grouping, the source node uses four messages to cover these groups. If the smallest group of processors has g ($g \geq 1$) members then the size of the shortest message that is going to be transmitted is gl. Therefore, the condition under which the lower bound is met can be written as: $gl \geq \frac{t_s}{t_p}$. The minimum value of g to achieve the lower bound can be found from the following equation:

$$g_{min} = \frac{t_s}{lt_p} \tag{4}$$

Obviously, for bigger g's there are less constraints on system parameters and message sizes under which the lower bound can be achieved. However, the need for rearrangement of the original data and the routing constraints don't allow selecting any arbitrary g's. Assuming that all groups have g members, the execution time of the SMT scatter scheme can be derived as:

$$T_{SMT} = \begin{cases} t_s + (P-1)lt_p & \text{if } ts \leq lgt_p \\ \frac{P}{g}t_s + (g-1)lt_p & \text{if } ts > lgt_p \end{cases} \tag{5}$$

Observation 4 *The SMT scheme can achieve the lower bound when $l \geq \frac{t_s}{gt_p}$.*

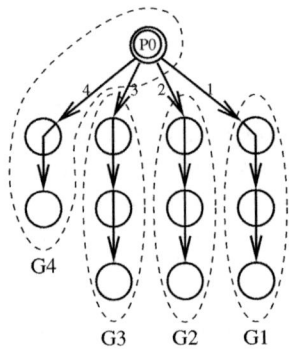

Fig. 2. Illustration of the Sequential Multidestination Tree (SMT) scheme.

3.3   The SMT Scheme for 2D Tori, Meshes, and k-ary n-cubes

Using this generalized approach we have show how this scheme can be used to implement scatter on 2D meshes/tori and higher dimensional systems with dimension-ordered routing [1]. In a $k \times k$ torus, it can be easily observed that the processors of each column can form a group. A multidestination message can deliver the data for all members of each group under dimension-ordered routing. The SMT scheme for 2D tori can be easily generalized for k-ary n-cubes. To implement the SMT scheme on k-ary n-cubes with dimension-ordered routing, it is enough to partition the processors along the lowest dimension and select g equal to k.

4   Performance Evaluation

In this section, We compare the latency of ST, BT, and SMT schemes for different system parameters and message sizes by using a wormhole-routed simulator called WORMulSim [5]. WORMulSim is a detailed flit level simulator which is built by using CSIM and takes care of flit-level contention. We consider an 8×8 torus with $t_s = 5.0$ μsec and $t_p = 10.0$ nsec as our base system and investigate the effect of the variations of system parameters on the latency of the scatter schemes. We also present the factor of improvements obtained by using SMT instead of ST and BT for systems with different characteristics.

4.1   Effect of Communication Start-up

Figure 3 shows the comparison results for three different start-ups ($t_s = 1.0, 5.0,$ and 10.0 μsec). It can be seen that the SMT scheme performs better than the ST scheme for all different message sizes. It can also be verified that the SMT performs better than BT for a wide range of message sizes. As t_s decreases the minimum message length for which the lower bound can be achieved decreases ($l \geq \frac{t_s}{g t_p}$). With $t_s = 10.0$ μsec, the SMT scheme achieves the lower bound for messages of size 128 bytes and more. With $t_s = 5.0$ and 1.0 μsec, lower bound is achievable for messages of size 64 and 16 bytes, respectively.

Figure 4 shows the factor of improvement of the SMT scheme over the ST scheme for three different startups. It can be observed that the ST scheme can never perform better than the SMT scheme. For a 1-byte message size, the SMT scheme can reduce the scatter latency up to a factor of 8 compared to the ST scheme. As t_s increases, the SMT scheme performs better than the ST scheme for a given message length. Similarly, Fig. 5 shows the factor of improvement of the SMT scheme over the BT scheme. It can be observed that the BT scheme can perform better than the SMT only for very short messages. For other message sizes, the SMT scheme can show up to 60% improvement compared to the BT scheme. It can be seen that as t_s reduces the maximum message size for which the BT scheme can perform better than the SMT scheme decreases.

As t_s keeps on reducing for new generation systems, these results indicate that the SMT scheme can outperform the unicast-based schemes (ST and BT) for a wider range of system parameters and message sizes.

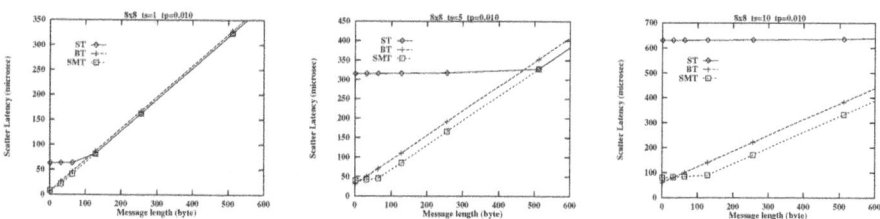

Fig. 3. Latency of three different schemes for different start-up times.

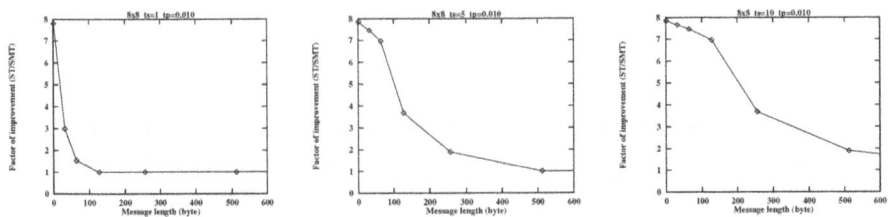

Fig. 4. Factor of improvement (T_{ST}/T_{SMT}) for different start-up times.

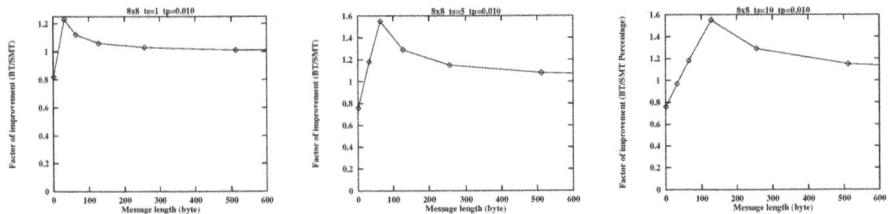

Fig. 5. Factor of improvement (T_{BT}/T_{SMT}) for different start-up times.

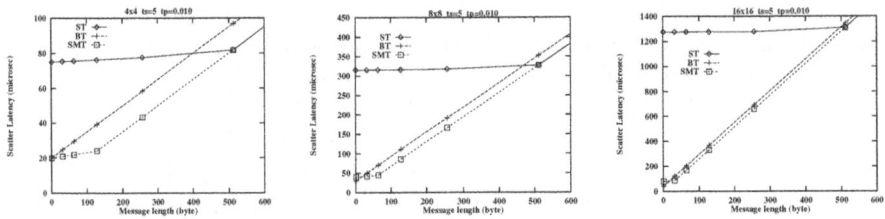

Fig. 6. Latency of three different schemes for different system sizes.

4.2 Effect of System Size

The effect of system size on the scatter schemes is shown in Fig. 6. It is interesting to observe that by increasing the system size the minimum length requirement for achieving the lower bound in the SMT scheme decreases. It can be seen that while the SMT scheme can not achieve the lower bound for messages shorter than 128 bytes on a 4×4 system, the lower bound is achieved even for 64-byte and 16-byte messages on 8×8 and 16×16 systems, respectively. Such performance improvement demonstrates the scalability of the SMT scheme with respect to increase in system size.

4.3 Effect of Propagation Cost

Figure 7 shows the latency of the three scatter schemes for different propagation times ($t_p = 5.0, 10.0,$ and 20.0 nsec). It can be observed that as propagation time increases, the performance of the ST scheme becomes worse. On the other hand, it can be seen that as t_p reduces, the factor of improvement obtained by using

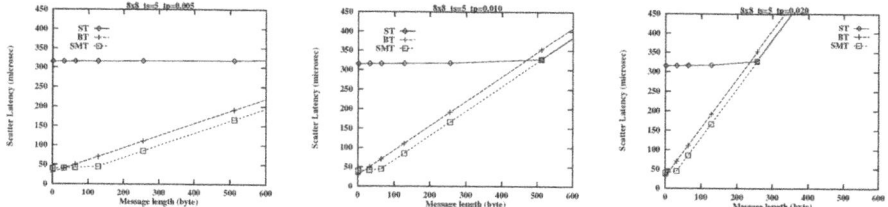

Fig. 7. Latency of three different schemes for different propagation times.

the SMT scheme over the BT scheme increases. The factor of improvement for message size of 128 bytes and $t_p = 5.0$ is 1.15. For a similar system with $t_p = 10.0$ and 5.0 nsec, the factors of improvement are 1.29 and 1.55, respectively.

The minimum message size above which the SMT scheme outperforms the BT scheme for a large set of system parameters is presented in [1]. It is shown that for a wide range of system parameters, system sizes, and message sizes the SMT scheme performs better than the BT scheme.

5 Conclusions

In this paper, we have presented a new approach to implement scatter communication in k-ary n-cube wormhole systems using multidestination message passing mechanism. Compared to the current unicast-based schemes, the proposed framework can meet the lower bound latency for a much wider range of system, technological, and application parameters. We are extending our work to irregular and tree-based topologies and switch-based and network interface-based multidestination message passing mechanisms.

References

1. M. Banikazemi and D.K. Panda. Can Scatter Communication Take Advantage of Multidestination Message Passing?. Technical Report OSU-CISRC-8/00-TR19, Dept. of Computer Science, The Ohio State University, 2000.
2. D. Dai and D. K. Panda. Reducing Cache Invalidation Overheads in Wormhole DSMs Using Multidestination Message Passing. In *International Conference on Parallel Processing*, pages I:138–145, Chicago, IL, Aug 1996.
3. L. Ni and P. K. McKinley. A Survey of Wormhole Routing Techniques in Direct Networks. *IEEE Computer*, pages 62–76, Feb. 1993.
4. D. K. Panda. Issues in Designing Efficient and Practical Algorithms for Collective Communication in Wormhole-Routed Systems. In *ICPP Workshop on Challenges for Parallel Processing*, pages 8–15, 1995.
5. D. K. Panda, D. Basak, D. Dai, R. Kesavan, R. Sivaram, M. Banikazemi, and V. Moorthy. Simulation of Modern Parallel Systems: A CSIM-based approach. In *Proceedings of the 1997 Winter Simulation Conference (WSC'97)*, pages 1013–1020, December 1997.
6. D. K. Panda, S. Singal, and R. Kesavan. Multidestination Message Passing in Wormhole k-ary n-cube Networks with Base Routing Conformed Paths. *IEEE Transactions on Parallel and Distributed Systems*, 10(1):76–96, Jan 1999.

Future Processors:
Invited Session

Co-Chair:
Sriram Vajapeyam, Indian Institute of Science
Mateo Valero, Technical University of Catalonia

Power: A First Class Design Constraint for Future Architectures

Trevor Mudge

Computer Science and Electrical Engineering
University of Michigan, Ann Arbor[1]

Abstract. In many mobile and embedded environments power is already the leading design constraint. This paper argues that power will also be a limiting factor in general purpose high-performance computers too. It should therefore be considered a "first class" design constraint on a par with performance. A corollary of this view is that the impact of architectural design decisions on power consumption must be considered early in the design cycle — at the same time that their performance impact is considered. In this paper we summarize the key equations governing power and performance, and use them to illustrate some simple architectural ideas for power savings. The paper then presents two contrasting research directions where power is important. We conclude with a discussion of the tools needed to conduct research into architecture-power trade-offs.

1 Introduction

The limits imposed by power consumption are becoming an issue in most areas of computing. The need to limit power consumption is readily apparent in the case of portable and mobile computer platforms — the laptop and the cell phone being the most common examples. But the need to limit power in other computer settings is becoming important too. A good example is the case of server farms. They are the warehouse-sized buildings that internet service providers fill with servers. A recent analysis presented in [1] has shown that a 25,000 sq. ft. server farm with about 8,000 servers will consume 2MW. Further it was shown that about 25% of the total running cost of such a facility is attributable to power consumption, either directly or indirectly.

It is frequently reported that the net is "growing exponentially," it follows then that the server farms will match this growth and with them the demand for power. A recent article in the Financial Times noted that 8% of power consumption in the US is for information technology (IT). If this component is set to grow exponentially without check it will not be long before the power for IT will be greater than for all other uses combined.

---

1. *Author's address:* Dept. EECS, The University of Michigan, 1301 Beal Avenue, Ann Arbor, MI 48109-2122, USA. *Tel:* +1 (734) 764.0203. *Fax:* +1 (734) 468.0152. *E-mail:* tnm@eecs.umich.edu. *WWW:* http://www.eecs.umich.edu/~tnm.

M. Valero, V.K. Prasanna, and S. Vajapeyam (Eds.): HiPC 2000, LNCS 1970, pp. 215-224, 2000.

To get an idea of the trends in power consumption of today's processors consider the following table. taken from [2]. The rapid growth in power consumption is obvious.

Table 1: Power Trends for the Compaq Alpha

Alpha model	Power (W)	Frequency (MHz)	Die size (mm^2)	Supply voltage
21064	30	200	234	3.3
21164	50	300	299	3.3
21264	90	575	314	2.2
21364	>100	>1000	340	1.5

The growth in power density of the die is equally alarming. It is growing linearly, so that the power density of the 21364 has reached about 30 W/cm^2 — three times that of a typical hot plate. This growth has occurred in spite of process and circuit improvements. Clearly, trading high power for high performance cannot continue, and architectural improvements will also have to be added to process and circuit improvements if the growth in power is to be contained. The only exceptions will be one-of-a-kind supercomputers built for special tasks like weather modelling.

In this introductory discussion we have briefly illustrated that power will be a limitation for most types of computers, not just ones that are battery powered. Furthermore, we have argued that system and architectural improvements will also have to be added to process and circuit improvements to contain the growth in power. It therefore seems reasonable that power should be dealt with at the same stage in the design process as architectural trade-offs

In the remainder of the paper we will expand on this theme starting, in the next section, with a simple model of power consumption for CMOS logic.

2   Power Equations for CMOS Logic

There are three equations that provide a model of the power-performance trade-offs for CMOS logic circuits. The equations are frequently discussed in the low-power literature and are simplifications that capture the essentials for logic designers, architects, and systems builders. We focus on CMOS because it will likely remain the dominant technology for the next 5-7 years. In addition to applying directly to processor logic and caches, the equations are also relevant for some aspects of DRAM chips. The first equation defines power consumption:

$$P = ACV^2f + \tau A VI_{short}f + VI_{leak} \qquad (1)$$

There are three components to this equation. The first is perhaps the most familiar. It measures the dynamic power consumption caused by the charging and discharging of the capacitive load on the output of each gate. It is proportional to the frequency of the operation of the system, f, the activity of the gates in the system, A (some gates may not switch every clock), the total capacitance seen by the gate outputs, C, and the square of the supply voltage, V. The second term captures the power expended due to the short-circuit current, I_{short}, that momentarily, τ, flows between the supply voltage and ground when the output of a CMOS logic gate switches. The third term measures the power lost due to leakage current that is present regardless of the state of the gate.

In today's circuits the first term dominates, immediately suggesting that the most effective way to reduce power consumption is to reduce the supply voltage, V. In fact, the quadratic dependence on V means that the savings can be significant: halving the voltage will reduce the power consumption to one quarter of it original value. Unfortunately, this comes at the expense of performance, or more accurately maximum operating frequency, as the next equations shows:

$$f_{max} \propto (V - V_{threshold})^2 / V \qquad (2)$$

In other words, the maximum frequency of operation is roughly linear in V. Reducing it will limit the circuit to a lower frequency. But reducing the power to quarter of its original value will only cut the maximum frequency in half. There is an important corollary to equations (1) and (2) that has been widely noticed: If a computation can be split in two and run as two parallel independent tasks, this form of parallel processing has the potential to cut the power in half without slowing the computation.

We can lessen the effect of reducing V in (2) by reducing $V_{threshold}$. In fact this must occur to allow proper operation of low voltage logic circuits. Unfortunately, reducing $V_{threshold}$ increases the leakage current, as the third equation shows:

$$I_{leak} \propto \exp(-V_{threshold}/(35mV)) \qquad (3)$$

Thus, this is a limited option for countering the effect of reducing V. It makes the leakage term in the first equation appreciable.

Summary: The are three important points that can be taken away from the above model. They are:

1. Reducing voltage has a significant effect on power consumption — $P \propto V^2$.
2. Reducing activity can too — in the simplest case this means turning off parts of the computer that are not being used.
3. Parallel processing is good if it can be done efficiently — independent tasks are ideal.

2.1   Other Figures of Merit Related to Power

The equation for power consumption given in (1) is an average value. There are two important cases where more information is needed. The first is peak power. Typically, systems have an upper limit, which if exceeded will lead to some form of damage — average power does not account for this. The second is dynamic power. Sharp changes in power consumption can result in inductive effects that can result in circuit malfunction. The effect is seen in "di/dt noise"— again, average power does not account for this.

The term "power" is often used quite loosely to refer to quantities that are not really power. For example, in the case of portable devices the amount of energy used to perform a computation may be a more useful measure, because a battery stores a quantity of energy, not a quantity of power. To refer to a processor as being "lower power" than another may be misleading if the computation takes longer to perform. The total energy expended may be the same in both cases — the battery will be run down by the same amount in both cases[1]. This leads to the idea of energy/operation. A processor with a low energy/operation is sometimes incorrectly referred to as "low power." In fact, the inverse of this measure, MIPS/W is frequently used as a figure of merit for processors intended for the mobile applications [3].

Although the MIPS/W is widely used as a figure of merit (higher numbers are better), it too can be misleading because cutting the power consumption in half reduces the frequency of operation by much less, because of the quadratic term in (1), as we have discussed. This has lead Gonzalez and Horowitz to propose a third figure of merit: energy*delay. This measure takes into account that, in systems whose power is modelled by (1), it is possible to trade a decrease in speed for higher MIPS/W. Unfortunately, the bulk of the literature uses MIPS/W or simply Watts, so we will continue this convention recognizing that occasionally it may suggest misleading trade-offs where "quadratic" devices like CMOS are concerned. Finally, it should be noted that if the computation under consideration must finish by a deadline, slowing the operation may not be an option. In these cases a measure that combines total energy with a deadline is more appropriate.

3   Techniques for Reducing Power Consumption

In this section we will survey some of the most common techniques that systems designers have proposed to save power. The scope of this brief overview includes logic, architecture and operating systems.

3.1   Logic

There are a number of techniques at the logic level for saving power. The clock tree can consume 30% of the power of a processor — the early Alpha 21064 exceeded this. Therefore, it is not surprising that this is an item where a number of power saving techniques have been developed.

3.1.1   Clock Gating

This is a technique that has been widely employed. The idea is to turn off those parts of the clock tree to latches or flip-flops that are not being used. Until a few years ago gated clocks were considered poor design practice, because the gates in the clock tree can exacerbate clock skew. However, more accurate timing analyzers and more flexible design tools have made it possible to produce reliable designs with gated clocks, and this technique is no longer frowned upon.

---

1. This is a simplification because the total energy that can be drawn from a battery after it has been charged depends, to some extent, on the rate at which the energy is drawn out. We will ignore this effect for our simplified analysis.

3.1.2   Half-Frequency/Half-Swing Clocks

The idea with the half-frequency clock is to use both edges of the clock to synchronized events. The clock can then run at half the frequency of a conventional clock. The drawbacks are that the latches are much more complex and occupy more area, and the requirements on the clock are more stringent.

The half-swing clock also increases the requirements on latch design, and it is difficult to employ in systems where V is low in the first place. However, the gains from lowering clock swing are usually greater than for clocking on both edges.

3.1.3   Asynchronous Logic

The proponents of asynchronous logic have pointed out that their systems do not have a clock and therefore stand to save the considerable power that goes into the clock tree. However, there are a drawbacks with asynchronous logic design. The most notable is the need to generate completion signals. This means that additional logic must be employed at each register transfer — in some case a double rail implementation is employed. Other drawbacks include difficulty of testing and absence of design tools.

There have been several projects to demonstrate the power saving of asynchronous systems. The Amulet, an asynchronous implementation of the ARM instruction set architecture, is one of the most successful [5]. It is difficult to draw definitive conclusions because it is important to compare designs that are realized in the same technologies. Furthermore, the asynchronous designer is at a disadvantage, because, as noted, today's design tools are geared for synchronous design. In any case, asynchronous design does not appear to offer sufficient advantages for there to be a wholesale switch to it from synchronous designs.

The area where asynchronous techniques are likely to prove important is in globally asynchronous, locally synchronous systems (GALS). These reduce clock power and help with the growing problem of clock skew across large chips, while still allowing conventional design techniques for most of the chip.

3.2   Architecture

The focus of computer architecture research, typified by the work presented at International Symposia on Computer Architecture and the International Symposia on Microarchitecture, has been on high performance. There have been two important themes pursued by this research. One has been to exploit parallelism, which we have seen can help reduce power. The other is to employ speculation — computations are allowed to proceed beyond dependent instructions that may not have completed. Clearly, if the speculation is wrong, energy has been wasted executing useless instructions. Branch prediction is perhaps the best known example of speculation. If there is a high degree of confidence that the speculation will be correct, then it can provide an increase in the MIPS/W figure. However, for this to be the case, the confidence level must often be so high that speculation is rarely employed as a means to reduce MIPS/W or power [7].

The area where new architectural ideas can most profitably contribute to reducing power is in reducing the dynamic power consumption term, specifically the activity factor, A, in (1).

3.2.1   Memory Systems

The memory system is a significant source of power consumption. For systems with relatively unsophisticated processors, cache memory can dominate the chip area. There are two sources of power loss in memory systems. First there is the dynamic power loss, due to the frequency of memory access. This is modelled by the first term[1] in (1). The second is the leakage current — the third term in (1).

There have been several proposals for limiting the dynamic power loss in memories by organizing memory so that only parts of it are activated on a memory access. Two examples are the filter cache and memory banking [8]. The filter cache is a small cache placed in front of the L1 cache. Its purpose is to intercept signals intended for the main cache. Its hit rate does not have to be very high, and its access time need be no faster than the L1 cache. Even if it is hit only 50% of the time, then the power saved is half the difference between activating the main cache and the filter cache. This can be significant. The second example, memory banking, is currently employed in some low power designs. The idea is to split the memory into banks and activate only the bank presently in use. It relies on the reference pattern having a lot of spatial locality, and thus is more suitable for instruction cache organization.

There is not much that can be done by the architect or systems designer to limit leakage, except to shut the memory down. This is only practical if the memory is going to be unused for a relatively long time because it will lose state and therefore must be backed up to disk. This type of shut down (often referred to as sleep mode) is usually handled by the operating system.

3.2.2   Buses

Buses are a significant source power loss. This is especially true for inter-chip buses. These are often very wide — the standard PC memory bus includes 64 data line and 32 address lines. Each require substantial drivers. It is not unusual for a chip to expend 15-20% of its power on these inter-chip drivers.

There have been several proposal for limiting this swing. One idea is to encode the the address lines into a Gray code. The reasoning is that address changes (particularly from cache refills) are often sequential, and counting in Gray code switches the least number of signals [10].

It is straightforward to adapt other ideas to this problem. Transmitting the difference between successive address values achieves a similar result to the Gray code. More generally, it has been observed that the address lines can be reduced by compressing the information in them [9]. These techniques are best suited to inter-chip signalling, because the encoding can be integrated into the memory controllers.

Continuing with the code compression concept, it has been shown that significant instruction memory savings results if the program is stored in compressed form and decompressed on the fly (typically on a cache miss) [11]. The reduction in memory size can translate to power savings. It also reduces the frequency of code overlays, another

---

1. The second term in *(1)* can also be lumped together with this.

source of power loss, and a technique still used in many digital signal processing (DSP) systems.

3.2.3   Parallel Processing and Pipelining

As we noted above, a corollary of our power model is that parallel processing can be a important technique for reducing power consumption in CMOS systems. Pipelining does not share this advantage, because the concurrency in pipelining it achieved through increasing the clock frequency which limits the ability to scale the voltage (2). This is an interesting reversal, because the microarchitecture for pipelining is simpler than for parallel instruction issue and therefore it has traditionally been the more common of the two techniques employed to speed up execution.

The degree to which computations can be parallelized varies widely. Some are "embarrassingly parallel." They are usually characterized by identical operations on array data structures. However, for general purpose computations typified by the SPEC benchmark suite there has been little progress on discovering parallelism. This is reflected in the fact that successful general purpose microprocessors rarely issue more than three or four instructions at once. Increasing instruction level parallelism is not likely to offset the loss due to hazards inhibiting efficient parallel execution. However, it is likely that future desktop architectures will have shorter pipes.

In contrast, common signal processing algorithms often possess a significant degree of parallelism. This is reflected in the architecture of DSP chips, which is notably different from desktop or workstation architectures. DSPs typically run at much lower frequencies and exploit a much higher degree of parallelism. Parallelism and direct support for a multiply-accumulate (MAC) operation, which occurs with considerable frequency in signal processing algorithms, means that in spite of their lower clock rates they can achieve high MIPS ratings. An example is given by Analog Devices 21160 SHARC DSP. It can achieve 600 Mflops using only 2W on some DSP kernels.

3.3   Operating System

The quadratic voltage term in (1) means that reducing voltage has great benefit for power savings, as we have noted several times. A processor does not have to run at it maximum frequency all the time to get its work done. If the deadline of a computation is known it may be possible to adjust the frequency of the processor and reduce the supply voltage. For example, a simple MPEG decode runs at a fixed rate determined by the screen refresh (usually once every 1/30th of a second). A processor responsible for this work could be adjusted to run so that it does not finish ahead of schedule and waste power.

It is very difficult to automatically detect periods where voltage can be scaled back, so current proposals are to provide an interface to the operating system that allows it to control the voltage. The idea is for the application to use these operating system functions to "schedule" its voltage needs [12]. This is a very effect way to save power because it works directly on the quadratic term in equation (1). Support for voltage scaling has already found its way into the next generation of StrongARM microprocessor from Intel, the XScale.

4 What Can We Do with a High MIPS/W Device?

The obvious applications for processors with a high MIPS/W are in mobile computing. The so-called 3G mobile phones that we can expect in the near future will have to possess remarkable processing capabilities. The 3G phones will communicate over a packet-switched wireless link at up to 2 Mbs. The link will support both voice and data and will be always be connected within areas that have 3G service. There are plans to support MPEG4 video transmission as well as other data intensive applications.

Today's cell phones are built around a two processors: a general purpose computer and a DSP engine. Both have to be low power, the lower the better. A common solution is to use an ARM processor for the general purpose machine and a Texas Instruments DSP chip. For 3G systems both of these processors will have to be much more powerful without sacrificing battery life. In fact, the processing requirements, given the power constraints, are beyond the present state-of-the-art. The design of such systems where power is a first class design constraint will be one of the next major challenge for computer architects. Furthermore, the immense number of units that will be sold — there are hundreds of millions of cell phone in use today — means that this platform will take over from the desktop as the defining application environment for computers as a whole.

The two-processor configuration of the cell platform has arisen out of the need to have a low power system that can perform significant amounts of signal processing and possess general purpose functionality for low-resolution display support, simple data base functions, and the protocols associated with cell-to-phone communications. From an architectural point of view this is not a particularly elegant solution and a "convergent" architecture that can handle the requirements of both signal processing and general purpose computing may be a cleaner solution. However, from a power perspective it may be easier to manage the power to separate components; either one can be easily turned off when not required. There are many more trade-offs to be studied.

While the cell phone and its derivatives will become the leading user of power efficient systems, it is by no means the only place where power is key, as we saw in the introduction. To go back to the server farm example, consider one of the servers: It has a workload of independent programs. Thus parallelism can be used without the inefficiencies often introduced by the need for intra-program communication and synchronization — multiprocessors are an attractive solution. A typical front-end servers that handles mail, web pages, and news, has a Intel compatible processor, 32M bytes of memory, an 8G byte disk and requires about 30W of power. Assume the processor is an AMD Mobile K6 with a total of 64K bytes of cache running at 400MHz. It is rated at 12W (typical). Compare this to the recently announced Intel XScale, which is Intel's next generation of StrongARM processor. It has the same total cache size but consumes only 450mW at 600MHz. (It can run from about 1GHz to below 100MHz; at 150Mhz is consumes just 40mW.) If we replace the K6 with 24 XScales we have not increased power consumption. For the K6 to process as many jobs as the 24-headed multiprocessor, it will have to have an architectural efficiency (e.g., SPECmarks) that is about 24 times that of an XScale.

There is a lot to disagree with in the above analysis. For example, the much more complex processor-memory interconnect required by the multiprocessor is not accounted for, nor has any consideration been given to the fact that the individual jobs may have an unacceptable response time. However, the point is to show that if power is the chief design constraint, then a low power but non-trivial processor like the XScale can introduce a new perspective into computer architecture. Consider the above analysis if we replaced the K6 with a 100W Compaq 21364: it would need to be 200 times as efficient.

5 Conclusion

We have made an argument for power to be a first class design constraint. We have also listed a number of ways that systems designers can contribute to satisfying this constraint. There is much more research to be done. Power aware design is no longer exclusively the province of the process engineer and circuit designer, although their contributions are crucial. By one account (see the talk by Deo Singh in [1]) we need architects and systems level designers to contribute a 2x improvement in power consumption per generation. This improvement is required in addition to the gains from process and circuit improvements.

To elevate power to a first class constraint it needs to be dealt with early on in the design flow, at the same point that architectural trade-offs are being made. This is the point where cycle accurate simulation is performed. This is problematic because accurate power determination can only be made after chip layout has been performed. However, very approximate values are usually acceptable early on in the design flow provided they accurately reflect trends. In other words, if a change is made in the architecture the approximate power figure should reflect a change in power that is in the correct direction.

Several research efforts are under way to insert power estimators into cycle level simulators. They typically employ event counters to obtain frequency measures for components of the architecture. These components are items such as adders, caches, decoders, and buses, for which an approximate model of power can be obtained. An early example was developed by researchers at Intel [13]. Others are Wattch [14] and SimplePower [15]. All three are based on the SimpleScalar simulator that is widely used in academe [16]. A fourth effort, PowerAnalyzer, under development by the author, T. Austin and D. Grunwald is expanding on the work in [13]. PowerAnalyzer will also provide estimates for di/dt noise and peak power [17].

Acknowledgment. The author was supported by contract from the Power Aware Computing and Communications program at DARPA. #F33615-00-C-1678

References

[1]   D. Singh and V. Tiwari. Power Challenges in the internet World. *Cool Chips Tutorial: An Industrial Perspective on Low Power Processor Design*, Eds. T. Mudge, S. Manne, D. Grunwald, held in conjunction with MICRO 32, Haifa Israel, Nov. 1999, pp.8-15. (http://www.eecs.umich.edu/~tnm/cool.html)

[2]   K.Wilcox and S. Manne. Alpha processors: A history of power issues and a look to the future. *Cool Chips Tutorial: An Industrial Perspective on Low Power Processor Design*, Eds. T. Mudge, S. Manne, D. Grunwald, held in conjunction with MICRO 32, Haifa, Israel, Nov. 1999, pp.16-37. (http://www.eecs.umich.edu/~tnm/cool.html)

[3]   Chart Watch: Mobile Processors. Microprocessor Report, vol. 14, archive 3, Mar. 2000, p.43

[4]   R. Gonzalez and M. Horowitz. Energy dissipation in general purpose microprocessors. *IEEE Jour. of Solid-State Circuits*, Sep. 1996, pp. 1277-1284.

[5]   S. Furber, J. Garside, and S. Temple. *Power Saving Features in Amulet2e. Power-Driven Microarchitecure Workshop*, Eds. T. Mudge, S. Manne, D. Grunwald, held in conjunction with ISCA 98, Barcelona, Spain, June 1998. (http://www.cs.colorado.edu/~grunwald/LowPowerWorkshop/agenda.html)

[6]   S. Moore, et al. Self Calibrating Clocks for Globally Asynchronous Locally Synchronous Systems. *Int. Conf. on Computer Design*, Austin, Texas, Sep. 2000.

[7]   S. Manne, A. Klauser and D. Grunwald. Pipeline gating: Speculation control for energy reduction. *Proc. 25th Int. Symp. Computer Architecture,* Barcelona, Spain, June 1998, pp. 132-141.

[8]   M. Johnson, M. Gupta and W. Mangione-Smith. Filtering memory references to increase energy efficiency. *IEEE Trans. on Computers*, vol. 49, no. 1, Jan. 2000, pp. 1-15.

[9]   A. Park, M. Farrens and G. Tyson,. Modifying VM hardware to reduce address pin requirements. *Proc. 25th Int. Symp. Computer Architecture,* Portland, Oregon, Dec. 1992, pp. 1-4.

[10]   L. Benini, et al. Address bus encoding techniques for system-level power optimization. *Proc. 1998 Design Automation and Test in Europe (DATE '98)*, pp. 861-866.

[11]   C. Lefurgy, E. Piccininni, and T. Mudge. Reducing code size with run-time decompression. *Proc. 6th Int. Symp. on High-Performance Computer Architecture*, Jan. 2000, pp. 218-227.

[12]   T. Pering, T. Burd, and R. Brodersen, Voltage scheduling in the lpARM microprocessor System, *Proc. 2000 Int. Symp. on Low Power Electronics and Desig*n, July 2000.

[13]   G. Cai and C. Lim. Architectural level power/performance optimization and dynamic power estimation. *Cool Chips Tutorial: An Industrial Perspective on Low Power Processor Design*, Eds. T. Mudge, S. Manne, D. Grunwald, held in conjunction with MICRO 32, Haifa Israel, Nov. 1999, pp.90-113. (http://www.eecs.umich.edu/~tnm/cool.html)

[14]   D. Brooks, V. Tiwari, and M. Martonosi. Wattch: A framework for architectural-level power analysis and optimizations. *Proc. 27th Int. Symp. Computer Architecture*, Vancouver, Canada, pp. 83-94.

[15]   N. Vijaykrishnan, et al. Energy-driven integrated hardware-software optimizations using SimplePower. *Proc. 27th Int. Symp. Computer Architecture*, Vancouver, British Columbia, Canada, June 2000, pp. 95-106.

[16]   D. Burger and T. Austin. *The SimpleScalar toolset, version 2.0.* Tech Rept. Computer Science Dept., Univ. Wisconsin, June 1997. (see also http://www.simplescalar.org)

[17]   http://www.eecs.umich.edu/~tnm/power/power.html

Embedded Computing:
New Directions in Architecture and Automation

B. Ramakrishna Rau and Michael S. Schlansker

Hewlett-Packard Laboratories,
1501 Page Mill Road, Palo Alto, California, U.S.A.
rau@hpl.hp.com, schlansk@exch.hpl.hp.com

Abstract. With the advent of system level integration (SLI) and system-on-chip (SOC), the center of gravity of the computer industry is moving from personal computing into embedded computing. The opportunities, needs and constraints of this next generation of computing are somewhat different from those to which we have got accustomed in general-purpose computing. This will lead to significantly different computer architectures, at both the system and the processor levels, and a rich diversity of off-the-shelf and custom designs. Furthermore, we predict that embedded computing will introduce a new theme into computer architecture: automation of architecture. In this paper, we elaborate on these claims and provide, as an example, an overview of PICO, the architecture synthesis system that the authors and their colleagues have been developing over the past five years.

1   Introduction

Over the past few decades, driven by ever increasing levels of semiconductor integration, the center of gravity of the computer industry has steadily moved down from the mainframe price bracket to the personal computer price bracket. Now, with the advent of system level integration (SLI) and system-on-chip (SOC), the center of gravity is moving into embedded computing.

Embedded computers hide within products designed for everyday use as they assist our world in an invisible manner. We refer to such products as *smart products*. Embedded computers have been incorporated into a broad variety of smart products such as video games, digital cameras, personal stereos, televisions, cellular phones, and network routers. A digital camcorder, for instance, uses a high-performance embedded processor to record or playback a digital video stream of data. These embedded processors achieve supercomputer levels of performance on highly specific tasks needed for recording and playback. It is the availability of high-density VLSI integration that makes it practical to provide such product-defining performance at an affordable cost.

As VLSI density increases embedded processors continue to provide more compute power at even lower cost. This is stimulating the rapid introduction of a vast

M. Valero, V.K. Prasanna, and S. Vajapeyam (Eds.): HiPC 2000, LNCS 1970, pp. 225–244, 2000

array of innovative smart products. Newly defined digital solutions, capable of inexpensively performing complex data manipulations provide revolutionary improvements to product functionality. Embedded processors are often used in products that previously relied on analog circuitry. When analog signals are digitally represented, digital processing performance increases can be used to provide higher speed, higher accuracy signal representation, more efficient storage, and more sophisticated processing uniquely available in the digital domain.

1.1   Product Requirements

Demanding product requirements often constrain the design of embedded systems from many sides. A smart product may simultaneously need, higher performance, lower cost, lower power, and higher memory capacity. High throughput is especially important in data-intensive tasks such as imaging, video, signal processing, etc. Products often have demanding real time requirements that further exaggerate processing needs. In these cases, embedded processors are required to perform complex processing steps in a limited and precisely known amount of time. Compute-intensive smart products use special-purpose processing engines to deliver very high performance that cannot be achieved with a general-purpose processor and software. Such products are often enabled by *non-programmable accelerators* (*NPAs*) that accelerate performance critical functions.

Often, cost is more important than performance, with budgets that allow only a few cents for critical chips. Power is obviously important for battery-operated devices, but can be equally important in office environments where densely stacked equipment requires expensive cooling. In such settings, long term operating costs can easily exceed the purchase price of equipment. Lack of memory storage is a serious problem and compression techniques are used to conserve storage especially for high-volume video, image, and audio data. The storage of large computer programs is often too costly and processors are valued for their small code size. In some systems, programs are compressed to reduce their size and later decompressed for execution.

While the speed of general-purpose processors may exceed 1GHz, embedded processors often execute at modest clock speeds relying instead on parallelism to achieve needed performance. The use of parallel execution, rather than high clock speed, allows for designs with cheaper, lower power circuits. These low cost and power-efficient designs are also less dependent on precisely tuned low level circuitry. Additional power management features, such as gated or variable speed clocks, may also be used.

1.2   Market Requirements

Smart products require newly developed, and highly specialized, embedded systems in order to provide the novel and product-defining features necessary in a competitive marketplace. Often, the introduction of a new smart product depends upon the successful design and fabrication of a new chip that gives the product its distinguishing, high value features. Time-to-market determines the success or failure of products and businesses. While one *smart product vendor* (*SPV*) awaits a tardy chip design before he can ship product, a competing SPV's product takes advantage of the higher profit margins and visible press available during early product introduction. The SPVs who deliver product early capture the lion's share of the profit and visibility. Those whose products are late, miss the market window and must settle for what is left. The ability to design rapidly provides a distinct market advantage.

There is a proliferation in the number of smart products being introduced, and an escalating number of embedded system designs are needed to support them. The problem is exacerbated by the shorter life cycles experienced by smart products in the consumer space, since each generation of a rapidly improving smart product requires new embedded systems in order to improve functionality.

The need for complex new chip designs cannot currently be satisfied. Product innovation is limited by our ability to perform expensive and time-consuming design tasks. A design bottleneck results from the use of a limited pool of highly talented engineers who must design a greater variety of complex chips at ever increasing rates.

1.3   Application Characteristics

Embedded applications are characterized by small well-defined workloads. Embedded systems often are used within single function products. Such products have product-specific and portable forms with simple and intuitive user interfaces. Single function products are based on far simpler software than general-purpose computers. They do not have the system crashes and reboots that we commonly associate with general-purpose systems. Their administration is simple and computer-illiterate users can enjoy their use. Single-function products with complex, high-performance, embedded systems are increasingly popular as the cost of their electronic content decreases. Many users do not, and cannot, write programs for their products. To them, programs are invisible logic which is used to provide product-defining functions.

Often, key kernels within applications represent the vast majority of required compute cycles. Kernels consist of small amounts of code which must run at very high performance levels to enable a smart product's functionality. In video, image and digital signal processing, the applications' execution times are often dominated by simple loop nests with a high degree of parallelism. These applications are typically characterized by a greater degree of determinacy then general-purpose applications. Not only is the nature of the application fixed, but key parameters such as loop trip counts or array sizes are pre-determined by physical product parameters such as the size of an imaging sensor. For such demanding application-specific products, high performance and high efficiency is often obtained using deeply pipelined function units which have been crafted into custom architectures designed to solve kernel codes.

Products may also have real-time constraints where time-critical computing steps must be completed before well understood deadlines to prevent system failure. Real-time systems require highly predictable execution performance. This often requires the use of a real-time operating system which provides task scheduling mechanisms necessary to guarantee predictable performance. Complex dynamic techniques, such as the use of virtual memory, caches, or dynamic instruction scheduling may not be allowed. When these techniques make the accurate prediction of performance excessively difficult, real-time constraints cannot be verified.

1.4   Overview of this Paper

The movement of the center of gravity of computing, from general-purpose personal computers to special-purpose embedded computers, will cause a major upheaval in the computer industry. Successful computer architecture has always resulted from a judicious melding of the opportunities afforded by the latest technologies with the requirements of the market, product and application. These are significantly different for embedded computing leading to substantially different computer architectures, at both the system and the processor levels, as well as a rich diversity of *off-the-shelf* (*OTS*) and custom designs. Furthermore, we believe that embedded computing will introduce a new theme into computer architecture: the automation of computer architecture.

In the rest of this paper, we elaborate on these points. In Section 2, we look at embedded computer architectures, and explain the increased emphasis on special-purpose architectures over general-purpose ones. Section 3 considers the issue of customization and the circumstances that force a SPV to resort to it instead of using an OTS solution. In Section 4 we argue for the use of automation in architecting and designing embedded systems, and we articulate our philosophy for so doing. As an example of this philosophy, Section 5 provides an overview of PICO, the architecture synthesis system that the authors and their colleagues have been developing over the past five years.

2   Embedded Computer System Architecture

To achieve the requisite performance, high-performance embedded systems often take a hierarchical form. The system consists of a network of processors; each processor is devoted to its specialized computing task. Processors communicate as required by the application and network connections among processors support only the required communications. Each processor also provides parallelism and is constructed using networks of arithmetic units, memory units, registers, and control logic. Hierarchical systems jointly offer both the process-level parallelism achieved with multiple processors as well the instruction-level parallelism achieved by using multiple function units within each processor.

For efficient implementation, and to provide product features such as highest performance, lowest cost, and lowest power, embedded computer system designs are often irregular at both levels of the design hierarchy. The entire system, as well as each of the processors, represents a highly special-purpose architecture that shows a strong irregularity that closely mirrors application requirements and ideally supports kernel needs.

Whereas specialization can be used to provide the very highest performance at the very lowest cost, embedded systems employ architectures which span a spectrum. At one end, a special-purpose architecture provides very high performance with very good cost-performance, but little flexibility. At the other end of the spectrum, a general-purpose architecture provides much lower performance and poorer cost-performance, but with all of the flexibility associated with software programming. A third choice, the OTS customizable architecture, provides an increasingly important compromise between these two extremes.

2.1   Special-Purpose Architectures

Smart products often incorporate special-purpose embedded processing systems in order to provide high performance with low cost and power. For example, a printer needs to perform multiple processing steps on its input data e.g.: error correction, decompression, image enhancement, coordinate transformation, color correction, and final rendering. Enormous computation is needed to perform these steps. To achieve the needed performance, printers often use a pipeline consisting of multiple processors; data is streamed through this pipeline of concurrently executing special-purpose processors. This style of design produces relatively inexpensive and irregular system architectures that are specialized to an application's needs.

In performance- and cost-critical situations, these embedded computer systems are specialized to simplify circuitry. Enormous savings can be achieved by specializing high-bandwidth connections among processors. Connections of appropriate bandwidth are provided between processors, exactly as needed, to accommodate very specific communication needs. While high bandwidth data paths are provided among processors that must exchange a large volume of data, no data paths are provided among processors that never communicate.

Likewise, the design of each processor is also specialized to the specific needs of the task that it performs. If we look within the design of each processor, we again see that specialization greatly simplifies needed circuitry. Each processor's performance, memory, and arithmetic capability are all adjusted to exactly match its dedicated task needs. Arithmetic units and registers are connected with a network of data paths that is specific to task needs. Each arithmetic unit, data path, or register is optimized to exact width requirements dictated by the statically known arithmetic precision of the operations and operands that they support. This specialization process again eliminates substantial amounts of unnecessary circuitry.

The control circuitry for each special-purpose processor can also be specialized to exact task requirements. Control circuits often degenerate to simple state machines which are highly-efficient in executing simple dedicated tasks. RAM structures within each processor are distributed according to need. Special table look up RAMs may be connected directly to the arithmetic units which use their operands. Each RAM is minimal in size in both number and width of its words. Rather precise information about the application is used to squeeze out unnecessary arithmetic, communication, and storage circuitry. Chained sequences of arithmetic, logical, and data-transfer operations are statically optimized to squeeze out additional circuitry. These optimized circuits perform multiple operations using far less logic than would be required if cascaded units were designed to execute each operation separately.

2.2   General-Purpose Architectures

The widespread use of special-purpose architectures sharply increases the number of distinct architectures that must be designed. If satisfactory programmable general-purpose system architectures existed, they would eliminate the need to specialize architectures to specific applications. Such general-purpose and domain-specific systems would offer the hope that new smart products could be designed using OTS parts that are reusable across many new smart products. General-purpose systems are highly flexible and are, in fact, reusable across almost all low-performance applications. General-purpose systems are designed in a number of ways. A general-purpose RISC processor can be re-programmed for a large variety of tasks. Domain-specific systems are customized to specific application areas (e.g. digital signal processing) but not to a specific smart product or application. Domain-specific systems often incorporate a control processor to run an operating system and a digital signal processor (DSP) to accelerate signal processing kernel code. Too often, however, general-purpose and domain-specific systems are not are not able to meet demanding embedded computing needs. RISC, superscalar, VLIW, and DSP architectures do not efficiently scale beyond their respective architectural limits. They have limits in clock speed and ability to exploit parallelism that make them costly and impractical at the highest levels of performance.

Symmetric multiprocessors are commonly used to extend system performance through the addition of more processors. Because general-purpose multiprocessors are designed without application knowledge, they provide uniform communication among processors. Identical processors are connected in a symmetric manner. General-purpose interconnection networks and multiprocessor shared memories provide the required general and parallel access. However, such highly connected and symmetric hardware scales poorly as the number of processors is increased. The communication hardware is either over-designed, permitting high-volume data transfers that never occur, or under-designed and unable to handle those that do. For large numbers of processors, the interconnect requires large amounts of chip area and the long transmission delays adversely affect the cycle time.

General-purpose systems rely on general-purpose processors for their computing horsepower. Each general-purpose processor uses flexible, general-purpose, function units (FUs) that can execute any common operation that might be needed in an arbitrary program. The general-purpose FU, however, is very expensive when compared to specialized function units (within custom processors) that execute only a single operation type. General-purpose processors often require symmetric access between function units and registers or RAMs. They use expensive and slow multi-ported register files and multi-ported RAMs to allow general (and parallel) communication and storage. Multi-ported register files and multi-ported RAMs do not scale and processors retain their efficiency at only modest levels of parallelism. Since the general-purpose processor is expected to be capable of executing a broad range of applications, it ends up being both over-designed and, often, under-designed for the specific aplication it is called upon to execute in the embedded system.

The control for a general-purpose processor is based on instructions that require wide access, complex shifting, and long distance transfer. The unpacking of the complex instruction formats needed to reduce code size can be very complex. This problem is especially difficult for wide-issue superscalar or VLIW machines. Their instruction units are well designed for supporting arbitrary and large programs on machines of modest issue width. But, for simple and highly parallel tasks, they are too expensive when compared to state-machine based controllers found in special-purpose systems.

2.3   Off-the-Shelf, Customizable Systems

In many settings, general-purpose and domain-specific OTS processor chips cannot deliver adequate computing power. Product designers seek other OTS chip architectures to meet high-performance needs. Field Programmable Gate Arrays (FPGAs) [12] provide one alternative approach. FPGAs use programmable logic cells interconnected with a network of wires and programmable switches. Rather than relying on normal sequential programming, FPGAs are programmed using hardware design techniques. FPGAs have traditionally been used to implement simple control logic and "glue logic"—the left over logic needed to glue key components together. An FPGA allows such logic to be collected within a single chip to reduce system cost. The density of FPGAs has grown to the point where complex data paths are now possible on a single FPGA. The architecture of FPGAs continues to improve as features are added to support wider data paths, wider arithmetic, and substantial amounts of on-chip local memory. With these improvements, FPGAs are increasingly used to implement special-purpose, high-performance processors.

The data path and control of an embedded architecture can be specified as a circuit or as network of hardware functions. The circuit can execute operations from a repertoire of memory and logic operations that are supported directly, or through libraries, in OTS FPGA hardware. The embedded system circuit can be carefully specialized to application needs. After completing logic design, the circuit is then mapped onto an FPGA. The simplified logic diagram is mapped onto logic cells and placed and routed within OTS FPGA hardware.

While not as easy as software programming, embedded system design using FPGAs eliminates expensive, risky, and slow chip design efforts and substantially decreases product risk and time-to-market. Because they are programmable, the cost of fixing bugs in FPGA-based systems is small. A fix can be quickly tried and shipped. Firmware downloads can be used to fix FPGA related problems. FPGAs offer an increasingly important programming paradigm for delivering high-performance processing and rapid time to market.

Due to inherent hardware costs for supporting programmable logic, FPGAs cannot hope to be as efficient as customized hardware. Though FPGAs may be an order of magnitude less efficient than customized hardware, FPGA-based designs of high-performance processors can be far more efficient than designs that rely solely on general-purpose processors for computing horsepower.

FPGA vendors now provide OTS chips that contain a general-purpose processor, FPGA, and RAM hardware in a single SOC. These FPGA-based architectures permit the design of complex special-purpose processing systems. The general-purpose processor provides flexible low-performance computing while the FPGA, along with its associated configurable RAM blocks, allows high-performance special-purpose processors to be implemented as programmable logic. FPGA libraries now also provide processor cores that are programmed as FPGA logic. These processors can be modified or enhanced for specific application needs. Other functionality, such as support for peripheral interfaces and programmable I/Os, further increase FPGA utility.

3   System Customization

By customization we mean the process of taking an OTS system and modifying it to meet one's requirements. In its extreme form, it entails designing the system from scratch. At one level, customization is quite commonplace. For instance when one buys a personal computer, one typically configures the amount of memory, disk, and the set of devices connected to the peripheral bus. But one never modifies the processor or the system architecture (e.g., its bus structure). That task is viewed as the domain of the semiconductor or computer manufacturer from whom we expect to buy an OTS system. Our discussion of customization is focused on this part of the overall embedded system, what we will refer to as the *central computing complex* (*CCC*), which is the set of processors, memories and interconnect involved in executing the embedded application.

Customization of the CCC increases the design cost of smart products and often delays product introduction. Whereas it can greatly simplify the system, allowing high performance at low cost, it requires a complicated design process. A customized system involves complex tasks, including architecture, logic, circuit, and physical design. Designs must be verified and masks fabricated before chip production can begin. The SPV is well advised to use an OTS CCC when possible; from the viewpoints of time-to-market, engineering effort and project risk, this is clearly the preferable approach.

And yet, SPVs routinely design custom CCCs, processors and accelerators. What are the circumstances that force them to do so? Why is it that what the SPV needs is not available OTS, and that using something OTS would fall far short of his needs?

3.1   Why Customize?

Our view is that three conditions, in conjunction, create the situation that forces a SPV to have to customize his CCC. Firstly, the smart product must have challenging requirements which can only be met by specializing the system or processor architecture, as described in Section 2. Else, the SPV could just use an OTS general-purpose design. Secondly, for the given application, the performance, perhaps even the usability, of designs within the space of meaningful, special-purpose designs must be very disparate. For this particular application, all the OTS designs must fall short to such an extent that it is worth the SPV's while to pay the costs of customization: longer time-to-market, greater engineering effort and increased project risk. Lastly, the space of worthwhile special-purpose designs must be so large, that it is not possible, or not economical, for someone to make them all available, on an OTS basis.

   This last criterion raises a further question. If it was worth the SPV's while to create a custom design, why was it not so for a manufacturer of OTS designs to have designed and offered the same thing? There are at least three reasons that serve as explanations. The most frequent reason is that the application, or a portion of it, represents the product-defining functionality that provides competitive differentiation to the smart product, i.e., it incorporates algorithms that are proprietary. If the performance requirements on these algorithms are sufficiently demanding, the CCC architecture must be specialized to reflect these proprietary algorithms; it must be customized. Secondly, the unit volume represented by a given smart product may be too low to make it worthwhile for a supplier to provide an OTS system. A final possibility, is that the diversity of special-purpose solutions demanded by SPVs is just too large for every one of them to be provided as OTS solutions, even though neither of the other two reasons is applicable. The SPV must fend for himself.

3.2   Customization Strategies

Customization incurs two types of design costs: architectural design cost and physical design cost. The former includes the design costs associated with architecting the custom system and any custom processors that it may contain, performing hardware-software partitioning, logic synthesis, design verification and, subsequently, system integration. Physical design cost includes the design costs associated with the floor-planning, placement and routing, as well as the cost of creating a mask set.

   While customization can be used to provide the very highest performance at the very lowest cost, SPVs employ a variety of strategies which span a spectrum. At one end, a custom architecture provides very high performance, but at very high architectural and physical design costs. At the other end of the spectrum, an OTS architecture provides much lower performance with the low design costs associated with software programming. A third choice, the OTS customizable system, provides an increasingly important compromise: the ability to use OTS parts, but requiring FPGA programming that is more complex and similar to physical design. Although the architectural cost and a part of the physical design cost must be borne, the mask set cost is avoided.

Minimizing architectural design cost. The key to minimizing architectural design cost is to reuse pre-existing designs to the extent possible. One strategy is to fix the system-level architecture, but to customize at the subsystem level. For instance, the system architecture may consist of a some specific interconnect topology, containing one or more specific OTS microprocessors plus one or more unspecified accelerators. The accelerators are defined by the application and must, therefore, be custom designs. However, only the architectural design cost of the accelerators is incurred. This strategy can be applied one level down, to the processors. Most of a processor's architecture can be kept fixed, but certain of the FUs, for instance, may be customized [5].

Minimizing physical design cost. An embedded system may either fit on a single chip, or be spread over multiple chips. For every part of a chip that has been customized, one must necessarily bear the cost of placement and routing. One can avoid this cost for the rest of the chip by using what is known as *hard IP*, i.e., subsystems that have been taken through physical design. Although the floorplanning, placement and routing for the chip as a whole must still be performed, the hard IP blocks are treated as atomic components, greatly reducing the complexity of this step. However for every chip that has been customized, even to a small extent, the entire cost of creating the mask sets for the chip must be borne.

At lower levels of VLSI integration, a system used to consist of multiple chips, of which only a few might have needed to be custom. Furthermore, the cost of creating a mask set was relatively low. The advent of SOC greatly reduces the cost of a complex embedded system. However, it comes with a disadvantage; *any* customization of the system, however tiny, requires a new mask set. Worse yet, the cost of creating a mask set is now in the hundreds of thousands of dollars, and rising.

Avoiding mask set costs. This is a powerful incentive to avoid VLSI design completely, motivating the notion of OTS, customizable SOCs or "reconfigurable hardware". The basic idea, as before, is to fix certain aspects of the system's and processors' architectures, and to allow the rest to be custom. The difference is that instead of implementing the custom accelerators and FUs using standard cells, they are implemented by mapping them, after performing logic synthesis, on to FPGAs. Accordingly, the OTS SOC contains the fixed portions of the system architecture, implemented as standard cells, but provides FPGAs instead of the custom processors, accelerators and FUs. By programming the FPGAs appropriately, the OTS SOC can be customized to implement a number of different custom system architectures. The entire architectural design cost for the custom subsystems is incurred, as are the costs of programming the FPGAs (similar in many ways to placement and routing), but the VLSI design costs are eliminated.

This is an extremely attractive approach when the desired custom system fits the system-level architecture of the OTS, customizable SOC. If not, the SPV must design a custom SOC. For high volume products, where application needs are well understood, and where high design cost and design time can be tolerated, customized SOCs provide higher performance at a lower cost than do FPGAs. However, its program-mability allows an OTS FPGA to serve a far greater variety of immediate product needs and allows complex products to get to market more quickly and to evolve with changing application requirements. Further benefits of this approach are quick prototyping and field programmability in the event that the nature of the customization needs to be changed.

Note that our discussion of OTS customizable system here and in Section 2.3 are two views of the same thing. There, our view of it was as an alternative style of architecture. Here, our view of it is the more traditional one, as a way of implementing a hardware design.

4   Automation of Computer Architecture

Embedded computing, with its distinct set of requirements and constraints, is gener-ating the need for large numbers of custom embedded systems. Whether they are implemented as custom SOCs or by using OTS, customizable SOCs, the architectural costs must be incurred. We believe that the need for mass customization, and its asso-ciated architectural costs, have brought into existence a new theme in computer ar-chitecture—the automation of computer architecture. We call this architecture synthe-sis in order to distinguish it from other forms of high-level synthesis such as behav-ioral synthesis [4, 7] and, to distinguish it from low-level synthesis such as logic synthesis [8].

4.1   When Is Automation Important?

One could argue that for the foreseeable future, an automatically architected computer system can be expected to be less well-designed than a manually architected and tuned design. This statement is not as obviously true as it might sound; the ability of an automated system to evaluate thousands of disparate designs could quite conceiva-bly yield a superior design. But if we accept this statement as true, the question that arises is under what circumstances it is desirable to use automation. We believe that there are at least three sets of circumstances that argue for automation.

The first one is when the desired volume of custom designs, stimulated by an ex-plosion in the number of smart products, exceeds the available design manpower. The demand might be due to a large number of either application-specific or domain-specific designs. It might also be due to shortened product life cycles, either because the relevant standards are evolving, or because of competitive pressures in a con-sumer business. Automation addresses the problem by sharply increasing the aggre-gate design bandwidth.

Automation is also useful when time-to-market or time-to-prototype is crucial. Often, the product definition could not be anticipated far enough ahead of time to use manual design methodologies, perhaps because a relevant standard had not yet converged, or perhaps the functionality of a new product was not yet clearly understood. In such cases, the speed of automated techniques is of great value.

A third motivation for automation occurs when the expected volume of the custom design is too small to permit the product to be economical using a manual design process. Automation reduces the design cost (which must be amortized over the small volume), and can make such a product viable.

4.2   A Philosophy of Automation

A typical reaction to the notion of automating computer system design is that it is a completely unrealistic endeavor. Typically, the assumption underlying this reaction is that the automatic design system would emulate the human design methodology. That would, indeed, be a very hard problem in artificial intelligence, since human designers tend to invent new solutions to problems they encounter during design. We do not believe that one should try to build an architecture synthesis system that does this. Instead, our approach picks the most suitable design out of a large, possibly unbounded, denumerable design space. This space has to be large and diverse enough to ensure that there is a sufficient repertoire of good designs so that a best design, selected from the space, will closely match application needs.

Framework and Parameters. Since it is impractical to explicitly enumerate every feasible design, the space of designs is defined by a set of rules and constraints that must be honored by each design in the space. We call this a *framework*. Within a framework, some aspects of the design, such as the presence of certain modules and the manner in which they are connected are predetermined. Other aspects of the design are left unspecified. Of these, some can be derived once the rest have been specified. We refer to the latter as *parameters*. The *specification* of a design consists of binding the values of the parameters. From these, the derived aspects of the design are computed. Together, and in the context of the framework, they constitute a completely specified design. We define *construction* to be the process of deriving the detailed, completely specified design once the parameters are given.

Our philosophy for deciding what is a parameter, and what is not, is determined operationally. When we believe that we have an algorithmic way of determining certain design details in an optimal or near-optimal manner, we view their definition as part of implementation. When we have no clear way of determining important attributes of a design, and we use heuristic search to determine well-chosen values, we view them as parameters. After all design parameters are bound, a design is completely specified and the design can then be constructed. In the specification of a given design, it is often the case that not every combination of parameters is valid. For a system design to be valid, certain parameters of one subsystem must match the corresponding parameters of another subsystem. These are expressed as validity constraints involving the parameters that must match [1].

Components. Designs are constructed by assembling lower-level components, picked from a component library. Sometimes, these components are parameterized with respect to certain of their attributes; once the parameters are specified, a component constructor can be used to instantiate the corresponding component. The components, or their constructors, are designed and optimized manually. The components must fit into the framework and, must collectively provide all of the building blocks needed to construct any design.

In addition to its detailed design, each component must have associated information needed during the construction of the system design. Information needed to properly interface components into a broader design context includes a description of a component's functional capability, a description of a component's input and output wiring needs, and a description of a component's externally visible timing and resource requirements. Components are often described as a network of lower-level components forming a design hierarchy.

A Paradigm for Automation. Typically, one has multiple evaluation metrics in mind (such as cost, performance and power) when picking a good design. Thus, finding an optimal design involves a multi-objective optimization task. A design is said to be a *Pareto-optimal design*, or a Pareto design for short, if there is no other design that is at least as good as it with respect to every evaluation metric, and better than it with respect to at least one evaluation metric. The set of all Pareto designs is the *Pareto set*, or the Pareto for short.

We automatically find a Pareto set using three interacting modules. The *spacewalker* explores the space of possible designs, looking for the Pareto-optimal ones. The space of possible designs is specified to the spacewalker by the user, who provides a range of values for each parameter. The *design space* is the Cartesian product of the sets of values for the various parameters. It defines the space of designs that the spacewalker must explore. At each step in the search, the spacewalker specifies a design by binding parameters. A *constructor* can take a design, as specified by the spacewalker, and construct a hardware realization of the chosen design. The effects of a component binding are evaluated using an *evaluator* that determines the suitability of the spacewalker's choice. Evaluation is most accurately performed by first executing the constructor for a design, with appropriately bound parameters, to produce a detailed design. Then the evaluators use the detailed design to compute the evaluation metrics. When the cost of constructing candidate detailed designs is excessive, approximate evaluation metrics can sometimes be quickly estimated directly from design parameters. The evaluation process uses multiple tools including compilers, simulators, and gate count estimators.

At each step in the search, the spacewalker invokes the constructor and evaluators to determine whether, in the context of the designs evaluated thus far, the latest design is Pareto-optimal. If the design space is small, the spacewalker may use an exhaustive search. Otherwise, at each step, it uses the evaluation metrics, possibly other statistics relating to the design, and appropriate heuristics to guide it in taking the next step in its search. The goal in this case is to find all, or most, of the Pareto-optimal designs while having examined a very small fraction of the design space. Spacewalking is hierarchical when an optimized system is designed using a spacewalking search and components of the system are treated as sub-systems that are in turn optimized using lower-level spacewalking searches.

A framework restricts the generated design to a subset of all possible designs that a human might have created. However, it is precisely from this that the power of a framework arises. It is difficult to conceive of how one could create constructors and evaluators capable of constructing and evaluating any possible design. Yet, without constructors and evaluators, automation would be impossible. The limits placed by a framework, on the types of designs that have to be evaluated and constructed, are crucial to making automation possible. The challenge, when designing a framework, is to choose one that is large and diverse enough to contain good designs, while at the same time retaining the ability to evaluate and construct every design in that framework.

5   The PICO Architecture Synthesis System

In order to illustrate our automation philosophy, we briefly describe PICO (Program In, Chip Out), our research prototype of an architecture synthesis system for automatically designing custom, embedded computer systems. It employs our paradigm for automation, in a hierarchical fashion, four times over. PICO takes an application written in C, automatically architects a set of Pareto-optimal system designs, and emits the structural VHDL for them. Currently, optimality is defined by two evaluation metrics: cost (gate count or chip area) and performance (execution time). PICO explores trade-offs between the ways in which silicon area can be utilized in such a system, presenting to the user a set of Pareto-optimal system designs. In the process, PICO does hardware-software co-design—partitioning the given application between hardware (one or more custom NPAs) and software (on a custom EPIC/VLIW processor[1] [10]). PICO also retargets a compiler to each custom VLIW processor; we call this processor-compiler co-design.

5.1   System Synthesis

At the system level, PICO's task is to identify the Pareto-optimal set of custom, application-specific, embedded system designs for a given application. Each system that PICO designs consists of a custom VLIW processor and a custom, two-level cache hierarchy. The cache hierarchy consists of a first-level data cache (Dcache), a first-level instruction cache (Icache), and a unified second-level cache (Ucache). In addition, the system may contain one or more custom NPAs that work directly out of the second-level cache. PICO exploits the hierarchical structure of this design space. Within PICO's framework, a system design consists of a VLIW processor, one or

---

[1] EPIC (Explicitly Parallel Instruction Computing) is a generalization of VLIW. For convenience, in the rest of this paper we use the term VLIW to include EPIC as well.

more NPAs, and a cache hierarchy. Accordingly, PICO decomposes the system design space into smaller design spaces, one for each of these major subsystems. The components that PICO uses to create a system-level design are the custom, application-specific VLIW processors, NPAs and cache hierarchies that are yielded by the spacewalkers and constructors for the various subsystems, as discussed below. The system-level parameters are the union of the parameters for the VLIW processor, the NPAs, and the cache hierarchy.

The system design space typically contains millions of designs, each requiring an hour or more of profiling, compilation, synthesis and simulation time. An exhaustive search of the design space is infeasible. Instead, PICO exploits the hierarchical structure of the design space. The basic intuition is that Pareto-optimal systems are composed out of Pareto-optimal component subsystems [1]. Accordingly, the system-level spacewalker invokes the subsystem-level spacewalkers to get the Pareto-optimal sets of subsystem designs. The set of all combinations of Pareto-optimal subsystems is far smaller that the original design space. In the simplest case, the system-level spacewalker would consider all these combinations, evaluate them, and discard all that are not Pareto-optimal system designs. But due to validity constraints, not all combinations are valid. Therefore, the system-level spacewalker requires that each subsystem-level spacewalker return not just a single Pareto set, but rather a set of parameterized Pareto sets. When composing subsystems, the system-level spacewalker enforces validity constraints by only combining designs from compatible, parameterized Pareto sets. For such compatible Pareto sets, it considers all combinations of subsystems, evaluates them, and discards all that are not Pareto-optimal system designs.

For each loop nest in the application, that is a candidate to be implemented in hardware, the spacewalker examines both options. If there are N such loop nests, the spacewalker explores 2^N system architectures, from those that have no NPAs at all, to those in which every loop nest has been implemented as a NPA. For each system architecture, it comes up with the Pareto set as described above. It then forms the union of these Pareto sets and finds the Pareto-optimal designs within this union. This final Pareto set contains all designs of interest for the application. Typically, at low performance levels, the Pareto-optimal designs contain no NPAs. Conversely, at sufficiently high levels of performance, all of the loop nests may be implemented as NPAs. In this manner, the system-level spacewalker makes different hardware-software partitioning decisions for Pareto optimal systems with varying cost and performance.

The system-level constructor utilizes the constructors for the VLIW, NPA and cache hierarchy subsystems to construct the subsystem designs. It then glues these subsystem designs together by synthesizing the appropriate hardware and software interfaces between the VLIW processor and the NPAs. Likewise, the system-level evaluators make use of the subsystem-level evaluators. System designs are evaluated by adding the costs and the execution times, respectively, of the component subsystems.

5.2   VLIW Synthesis

PICO-VLIW is the PICO subsystem that designs custom, application-specific VLIW processors and generates a parameterized set of Pareto sets [3]. In addition, it retargets Elcor (PICO's VLIW compiler) to each new processor so that it can compile the C application to that processor. The processors currently included within PICO-VLIW's framework encompass a broad class of VLIW processors with a number of sophisticated architectural and micro-architectural features [6, 9]. A complete specification of a VLIW processor within this framework involves hundreds of detailed decisions. If all of these were parameters, it would result in an extremely unwieldy design space exploration task. Our choice of the interface between the spacewalker and the VLIW constructor involves a delicate balance between giving the spacewalker adequate control over the architecture, without bogging it down by requiring it to specify all details. Our compromise is that the parameters that the spacewalker must specify are limited to the sizes and types of register files, the operation repertoire, and the requisite level of ILP concurrency. Thereafter, it is the job of the VLIW constructor to make the remaining detailed design decisions.

The number of parameters needed to specify a VLIW design is still relatively large. Consequently, even if the range for each parameter is small, the size of the design space can be extremely large. Furthermore, the evaluation of the performance of a VLIW design is time-consuming, since it involves compiling a potentially large application. The spacewalker, therefore, explores the design space using sophisticated search strategies and heuristics to prune the design space based on previously evaluated processor designs.

For each set of parameters generated by the spacewalker, the VLIW constructor designs the architecture and micro-architecture of the specified VLIW processor, including the execution datapaths, the instruction format and the instruction unit, and emits structural VHDL. It also automatically extracts that part of the *machine-description database* (*mdes*) [9] that drives Elcor during scheduling and register allocation. The VLIW constructor uses RTL components from PICO's macrocell library (such as adders, multiplexers and register files) to synthesize the VLIW processor. In addition to the gate-level design, each component has associated with it information regarding its area, gate count, and degree of pipelining. Functional unit macrocells also are annotated with the set of opcodes that they can execute.

The cost evaluator estimates the chip area and gate count for the design using parameterized formulae for area and gate count that are attached to each component in the macrocell library. These formulae are calibrated against actual designs. The performance evaluator estimates the execution time of the given application on the newly designed processor using Elcor and PICO's retargetable assembler. Both are automatically retargeted by supplying them with the mdes for the target processor. The schedule created by Elcor, along with the profiled frequency of execution of each basic block, suffices to estimate the execution time. The object code generated by the assembler serves two objectives during design space exploration. One is to evaluate the code size and its impact upon the cost of main memory. The other is to permit an estimate of its effect upon the Icache and Ucache miss rates and the resulting impact on execution time.

5.3   NPA Synthesis

PICO-N is the PICO subsystem for designing NPAs customized to a given loop nest and for obtaining a parameterized set of Paretos for such NPAs [11]. A design in the NPA framework consists of a synchronous, one- or two-dimensional array of customized processing elements (PEs) along with their local memories and interfaces to global memory, a controller, and a control and data interface to a host processor. Each PE is a datapath with a distributed register file structure. Each register file is a FIFO with random read access. Interconnections between the FIFOs and FUs exist only as needed by the computation, resulting in a sparse and irregular interconnect structure with connections to only some of the FIFOs' elements. PICO-N also synthesizes the code required to make use of the NPA.

The design parameters are the number of PEs, the initiation interval (II) between starting successive iterations on any single PE, and the amount of memory bandwidth to the second-level cache. Since the number of parameters is small, the spacewalker performs an exhaustive search through the design space defined by the ranges of the three parameters. Because the precise geometry of the array of PEs is not specified as a parameter, the spacewalker steps through all one- and two-dimensional array geometries which have the specified number of PEs. This serves as an additional parameter for the constructor. The results of this search are used to create a set of Paretos, each one parameterized by the number of memory ports that the NPA has to the second level cache.

The NPA constructor starts off with a loop nest expressed as a sequential computation working out of global memory. It first tiles the loop nest, creating a new set of sequential outer loops, that will run on the VLIW processor, along with an inner loop nest with fewer iterations (the tile), that will be executed in parallel by the NPA. This transformation allows the constructor to not exceed the available global memory bandwidth, by performing register promotion in the inner loop nest, while minimizing the cost of the additional registers that this entails. Smaller tiles result in lower hardware cost but higher memory traffic. The constructor uses constrained combinatorial optimization to minimize the tile volume without exceeding the memory bandwidth allocated to the NPA. Next, the constructor transforms the inner loop nest into multiple, identical, synchronously parallel computations, one per PE. It does so by assigning a PE and a start time to each iteration in the tile while honoring data dependences between iterations. Furthermore, from the perspective of each processor, this iteration schedule is required to start an iteration every II cycles. This allows the constructor to express the computation on each PE as a single loop that performs all of the iterations assigned to that PE.

This loop is used to synthesize a single PE. Using user-specified pragmas regarding the requisite bit widths of variables, the constructor infers the minimum width requirements for all variables, temporaries, and operations. This allows the width of every register, FU and datapath to be minimized. A minimum-cost set of FUs are allocated and the operations of the loop body are assigned to these functional units and scheduled in time. Elcor is used to perform software-pipelining of the loop at the specified II using a variety of heuristics to minimize hardware cost. At this point the hardware for one PE is materialized, using the RTL components in the macrocell library. Register files, in the form of FIFOs with random read access are allocated to hold temporary values. Interconnections between the FIFOs and FUs are created only as needed, resulting in a sparse and irregular interconnect structure with connections to only certain of the FIFOs' elements.

An NPA consists of multiple instances of the PE, configured as an array. As many copies of the PE are created as specified by the design parameter, and interconnected in the specified geometry. The controller and the global memory interface are generated. The NPA, including its registers and array-level local memories, is accessed via a local memory interface by the VLIW processor. It may be initialized and examined using this interface. Finally, structural VHDL for the NPA is emitted.

The NPA constructor also performs some software synthesis. It takes the loop nest after tiling, with its additional outer loops, and removes the inner loop nest that has now been implemented as hardware. In its place, it generates the code that will invoke the NPA after making the appropriate initializations via the NPA's local memory interface. This new loop nest is inserted back into the application in place of the loop nest that was presented to PICO-N. Instead of executing the loop nest on the VLIW processor, the application will now trigger the computation on the NPA.

The cost evaluator estimates the chip area and gate count for the NPA, as described earlier for the VLIW processor. The performance evaluator for the NPA (which executes a predictable loop nest) is estimated using a formula instead of via simulation.

5.4   Cache Hierarchy Synthesis

The third major PICO subsystem automatically generates a parameterized set of Pareto sets for cache hierarchies that have been customized to the given application [2]. A design within the cache hierarchy framework consists of a first-level Dcache, a first-level Icache and a second-level Ucache. Just as at the system level, PICO decomposes the cache hierarchy design space into smaller design spaces for the Dcache, Icache and Ucache, respectively. Each of the three caches—Icache, Dcache and Ucache—is parameterized by the number of sets, the degree of associativity, the line size, and the number of ports. A valid cache hierarchy design must have parameters that are compatible as specified by the validity constraints, e.g. the porting on the Ucache must at least be equal to the Dcache porting (if data fetched from the Ucache is permitted to bypass the Dcache). A final parameter, dilation, is an attribute of the code size for each VLIW processor. This parameter determines not the design of the cache hierarchy, but rather the performance of the Icache and Ucache.

The spacewalker takes advantage of the fact that the cache hierarchy design space is decomposed into three smaller design spaces corresponding to the Icache, Dcache and Ucache, respectively. For each of these design spaces, a parameterized set of

Pareto sets is formed. Each design space's set contains member Pareto sets which are formed separately for each setting of the values of the parameters that participate in validity constraints The recomposition step uses these Paretos to form parameterized Pareto sets for the overall cache hierarchy, while enforcing validity constraints by only considering those combinations of Icache, Dcache and Ucache Pareto sets that are compatible.

The costs of the Icaches, Dcaches and Ucaches are evaluated using parameterized formulae. Trace-driven simulation is used to evaluate the miss rates, using the Cheetah cache simulator [13] which exploits inclusion properties between caches to simulate, in a single pass through the address trace, a range of cache designs with a common line size. This trace-driven cache simulation is done once for a reference VLIW processor. To a first order of approximation, the Dcache miss rate is assumed to be unaffected by the details of the VLIW processor. But this is not the case for the Icache and Ucache miss rates. Analytic techniques that use the dilation parameter, combined with interpolation of the reference processor's miss rate, are used to estimate the performance of these caches.

6 Conclusions

System-on-chip levels of VLSI integration are causing the center of gravity of the computer industry to move into embedded computing. The driving application for embedded computing will be a rich diversity of innovative smart products which depend, for their functionality, on the availability of extremely high-performance, low-cost embedded computer systems.

Successful computer architecture results from achieving a careful balance between the opportunities afforded by the latest technologies, on the one hand, and the requirements of the market, product and application, on the other. The opportunities, needs and constraints of embedded computing are quite distinct from those of general-purpose computing. This creates a new playing field and will lead to substantially different computer architectures, at both the system and the processor levels. Embedded architectures will be far more special-purpose, heterogeneous and irregular in their structure. There never was such a thing as the "one right architecture", even for general-purpose computing. This is even truer for embedded computing, which will trigger a renaissance in system and processor architecture. Custom and customizable architectures will assume a new importance.

Furthermore, we believe that the very large number of custom architectures required, due to the expected explosion in the number of smart products, will introduce a new theme into computer architecture: the automation of computer architecture. Our experience with PICO is that this is a perfectly practical and effective endeavor. The resulting designs are quite competitive with manual designs, but are obtained one or two orders of magnitude faster.

Acknowledgements

The ideas and opinions expressed in this paper have been greatly influenced by our colleagues, past and present, in the Compiler and Architecture Research group at HP Labs: Rob Schreiber, Shail Aditya, Vinod Kathail, Scott Mahlke, Santosh Abraham, Darren Cronquist, Mukund Sivaraman, Greg Snider, Sadun Anik and Richard Johnson. They, along with the authors, are responsible for the development of the PICO system.

References

1. Abraham, S., Rau, B.R. and Schreiber, R. Fast Design Space Exploration Through Validity and Quality Filtering of Subsystem Designs. Hewlett-Packard Laboratories Technical Report No. HPL-2000-98. Hewlett-Packard Laboratories, 2000.
2. Abraham, S.G. and Mahlke, S.A., Automatic and efficient evaluation of memory hierarchies for embedded systems. in *Proc. 32nd Annual International Symposium on Microarchitecture (MICRO '99)*, (1999), IEEE Computer Society, 114-125.
3. Aditya, S., Rau, B.R. and Kathail, V., Automatic architectural synthesis of VLIW and EPIC processors. in *Proc. International Symposium on System Synthesis, ISSS'99*, (San Jose, California, 1999), IEEE Computer Society and the ACM, 107-113.
4. Elliot, J.P. *Understanding Behaviorial Synthesis: A Practical Guide to High-Level Design*. Kluwer Academic, 1999.
5. Gonzalez, R.E. Xtensa: a configurable and extensible processor. *IEEE Micro, 20* (2). 60-70, 2000.
6. Kathail, V., Schlansker, M. and Rau, B.R. HPL-PD Architecture Specification: Version 1.1. Hewlett-Packard Laboratories Technical Report No. HPL-93-80 (R.1). Hewlett-Packard Laboratories, 2000.
7. Knapp, D.W. *Behaviorial Synthesis: Digital System Design Using The Synopsys Behaviorial Compiler*. Prentice Hall PTR, Upper Saddle River, New Jersey, 1996.
8. Lee, W.F. *VHDL: Coding and Logic Synthesis with Synopsys*. Academic Press, 2000.
9. Rau, B.R., Kathail, V. and Aditya, S. Machine-description driven compilers for EPIC and VLIW processors. *Design Automation for Embedded Systems, 4* (2/3). 71-118, 1999.
10. Schlansker, M.S. and Rau, B.R. EPIC: Explicitly Parallel Instruction Computing. *Computer, 33* (2). 37-45, 2000.
11. Schreiber, R., Aditya, S., Rau, B.R., Kathail, V., Mahlke, S., Abraham, S. and Snider, G., High-level synthesis of nonprogrammable hardware accelerators. in *Proc. International Conference on Application-Specific Systems, Architectures, and Processors (ASAP 2000)*, (Boston, Massachussetts, 2000), IEEE Computer Society, 113-124.
12. Sharma, A.K. *Programmable Logic Handbook: PLDs, CPLDs, and FPGAs*. McGraw Hill Companies, 1998.
13. Sugumar, R.A. and Abraham, S.G., Efficient Simulation of Caches under Optimal Replacement with Applications to Miss Characterization. in *Proc. ACM SIGMETRICS*, (1993), 24-35.

Instruction Level Distributed Processing

J.E. Smith

Dept. of Elect. and Comp. Engr.
1415 Johnson Drive
Univ. of Wisconsin
Madison, WI 53706

Abstract. Within two or three technology generations, processor architects will face a number of major challenges. Wire delays will become critical, and power considerations will temper the availability of billions of transistors. Many important applications will be object-oriented, multithreaded, and will consist of many separately compiled and dynamically linked parts. To accommodate these shifts in both technology and applications, microarchitectures will process instruction streams in a distributed fashion – instruction level distributed processing (ILDP). ILDP will be implemented in a variety of ways, including both homogeneous and heterogeneous elements. To help find run-time parallelism, orchestrate distributed hardware resources, implement power conservation strategies, and to provide fault-tolerant features, an additional layer of abstraction – the virtual machine layer – will likely become an essential ingredient. Finally, new instruction sets may be necessary to better focus on instruction level communication and dependence, rather than computation and independence as is commonly done today.

1 Introduction

Processor performance has been increasing at an exponential rate for decades; most computer users now take it for granted. In fact, this performance increase has come only through concerted effort involving the interplay of microarchitecture, underlying hardware technology, and software (compilers, languages, applications). Because of changes in underlying technology and software, future microarchitectures are likely to be very different from the complex heavyweight superscalar processors of today.

For nearly twenty years, microarchitecture research has emphasized instruction level parallelism (ILP) – improving performance by increasing the number of *instructions per cycle*. In striving for higher ILP, microarchitectures have evolved from pipelining to superscalar processing, with researchers pushing toward increasingly wide superscalar processors. Emphasis has been on wider instruction fetch, higher instruction issue rates, larger instruction windows, and increasing use of prediction and speculation. This trend has largely been based on *exploiting* technology improvements and has led to very complex, hardware-intensive processors.

Regarding applications, the focus for microarchitecture research has been SPEC-type applications, consisting of single threaded programs written in the conventional C and FORTRAN languages, following a *big static compile* model. That is, the entire binary is compiled at once, with very heavy optimization, sometimes with the assistance of profile feedback.

M. Valero, V.K. Prasanna, and S. Vajapeyam (Eds.): HiPC 2000, LNCS 1970, pp. 245–258, 2000.
© Springer-Verlag Berlin Heidelberg 2000

Starting with the conventional, big-compile view of software and ever-increasing transistor budgets, microarchitecture researchers made significant progress through the mid-90s. More recently, however, the problem has seemingly been reduced to one of finding ways of consuming transistors in some fashion. It is not surprising that the result is hardware-intensive and complex. Furthermore, the complexity is not just in critical path lengths and transistor counts; there is also high intellectual complexity resulting from increasingly intricate schemes for squeezing performance out of second and third order effects.

In the future, there will be substantial shifts in both software and hardware technology. In fact, these shifts are already underway, and the conventional ILP-based microarchitecture approach does not address them very well. In software the shift is toward object oriented programs that exploit thread level parallelism and dynamic linking. In hardware, long wire delays will dominate gate delays. These shifts will lead to general purpose microarchitectures composed of small, simple interconnected processing elements, running at a very high clock frequency. Multiple threads, some completely transparent to conventional software, will be managed by a thin layer of software, co-designed with the hardware and and hidden from the traditional software layers. This hidden software layer will manage the distributed hardware resources and dynamically optimize the executing threads, based on observed inter-instruction communication and memory access patterns.

In short, the microarchitecture focus will shift from Instruction Level Parallelism to Instruction Level Distributed Processing (ILDP) where emphasis will be on inter-instruction communication with dynamic optimization and a high level of interaction between hardware and software. In the following sections we expand on these ideas.

1.1   Where We Have Been

Fig. 1 is a simplified view of processor performance over the past three decades – beginning at the time when the CDC 7600 was the fastest available computer. At the left endpoint of the timeline, the 7600 was designed to 40 MHz, but shipped at a slightly slower rate, and was capable of sustaining about .25 instructions per cycle on real applications. Main memory was relatively fast compared with the processor speed, and there were no caches except for a small, loop-capturing instruction buffer. The compiler was capable of software pipelining (limited by the small register set).

The right endpoint of the timeline is today's superscalar processor that runs at about 1 GHz and is capable of sustaining perhaps 1.25 instructions per cycle on real applications (often less). Caches are heavily used because of the relatively slower main memory, and compilers are quite advanced.

All told, in the past 30 years, the performance improvement has been about 125 times, or 16-18 percent per year. Based on clock frequency, technology is responsible for a 25 times improvement, and the combination of microarchitecture and compilers are responsible for about a five times improvement in the past 30 years.

The MOS-based microprocessor performance curve has been much steeper, and is the one often cited when considering performance trends. However, as Fig. 1 shows, this is misleading when one considers the larger picture, and probably sets unattainable expectations. Recall that a microprocessor of the 1980 had a clock frequency of about 5

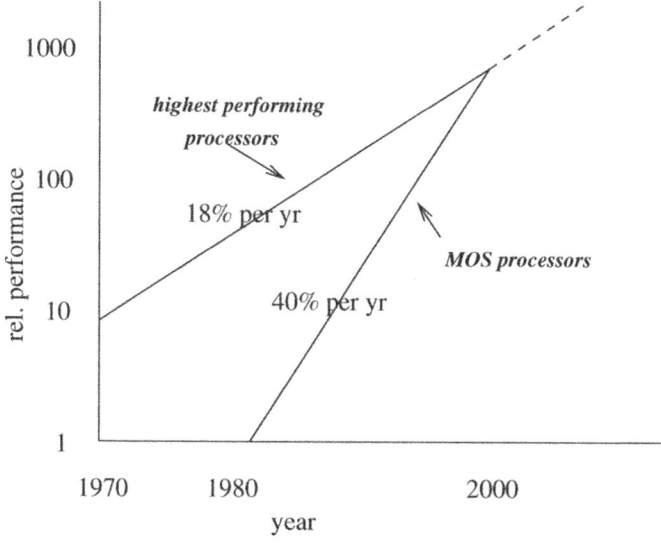

Fig. 1. High-end processor performance since 1970.

HHz with no pipelining, contrasted with the highest performance bipolar processors of the day which were deeply pipelined and ran in excess of 100 MHz.

Technology has provided faster transistors, as well as more gates available to the designer/architect. MOS technologies have added substantially to the number of available transistors and the rise of CMOS has gone hand-in-hand with the heavy emphasis on ILP for general purpose processing.

A second technology trend is that main memories have become quite large, but relatively slow when compared with processors. To accommodate this shift in technologies, microarchitects have developed very elaborate memory hierarchies, consisting of multi-level caches, and sophisticated prefetch and buffering schemes. These important developments do not explicitly appear in the five times microarchitecture/compiler contribution to overall performance, but they are crucial for mitigating major performance *losses* that would have resulted from the widening gap between processor and memory cycle times.

Turning now to applications and languages, in the early 1970s (and earlier), much of the emphasis in high performance computing was on numerical applications. Languages were FORTAN, and later, C. Individual users developed and compiled their programs (often with great care, to improve performance and to deal with small memories). A programs was often compiled for the individual machine on which the program was to be run, and forms of manual profiling were typically used as the code was developed. For large numerical problems containing high levels of inherent parallelism, vector instructions were often employed.

Over the years, the emphasis in general purpose computing has shifted from floating point applications to integer commercial applications, but the emphasis is still on conventional languages like C. And microarchitecture researchers still assume a heavy emphasis on compiler optimization and profiling, although one often hears that, in practice, profiling is seldom used, and unoptimized software is frequently shipped.

1.2 Where We Are Headed

Now consider where hardware technology and software are heading. With regard to technology, the two most important issues facing microarchitects are on-chip wire delays and power consumption.

In the case of local (short) wires, the problem is one of congestion. That is, for many complex, dense structures, transistor sizes do not determine area requirements, wiring does. With global (long) wires, delays will not scale as well as transistor delays, because wire aspect ratios will be more constrained than in the past and fringing capacitance becomes an important factor [1].

An interesting analysis of wire delay implications is given in [2] where reachable chip area, measured in terms of SRAM memory cells, is projected. Because of longer wire delays in the future, fewer bits of memory will be reachable in a single clock cycle than today. For example, in 35 nm technology, it is projected that the number or reachable bits will be about half of what is reachable today.

Of course, reachability is a problem with general logic as well. As logical structures become more complex (and take relatively larger area) global delays will increase simply because structures are farther apart, even if critical path delays are ignored. To put it another way, using simple logic will likely improve performance directly by reducing critical paths, but also indirectly by reducing area (and overall wire lengths). As is often the case, this is not a new observation – all the S. Cray designs benefited from this principle.

Finally, global multi-stop buses, i.e. buses with several receivers (and possibly multiple drivers) will have very long delays because of both wire capacitance and the loading on the bus. This type of on-chip, intra-processor bus is likely to disappear in the future because of its very poor delay characteristics.

With respect to power, dynamic power is related to voltage levels and transistor switching activity, i.e. dynamic power $\approx A V^2 f$, where A is a measure of switching activity, V is the supply voltage and f is the clock frequency. Higher clock frequencies and transistor counts have caused dynamic power to become a very big issue in recent years. Because of the dependence on voltage level, the trend is toward lower power supply voltage levels.

The power problem is likely to get much worse, however. To maintain high switching speeds with reduced supply voltages, transistor threshold voltages are also becoming lower [3]. This causes transistors to become increasingly "leaky"; i.e. current will pass from source to drain even when the transistor is not switching. In the future, the resulting static power consumption will likely become dominant. For example, consider Fig. 2 adapted from [3]. From this figure, trends indicate that static (standby) power will become more important than dynamic (active) power within the next few technology generations. There are relatively few solutions to the static power problem. One can selectively gate-off the power supply to unused parts of the processor, but this will be a higher-overhead process than clock-gating used for managing dynamic power and can lead to difficult-to-

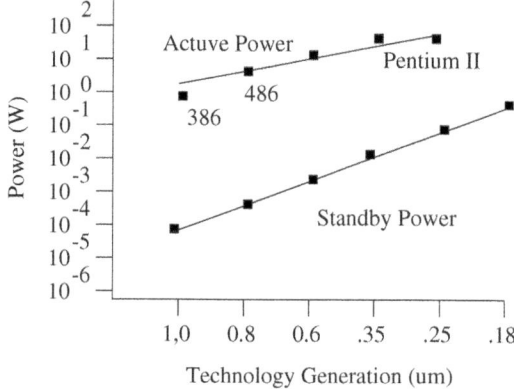

Fig. 2. Trends in dynamic and static power; from [3].

manage transient currents. Alternatively, one can use fewer transistors, or at least fewer fast (i.e. leaky) transistors.

In software, the trend is toward object oriented languages with dynamic linking. And as mentioned earlier, many application codes are not highly optimized anyway, and dynamic linking is not compatible with the big static compile model. Furthermore, with a variety of hardware platforms all supporting the same instruction set, it becomes difficult to optimize a single binary to be executed on all of them.

Finally, an important trend that brings together microarchitecture and applications is the use of on-chip multithreading. Multithreading has a very long tradition, but primarily in very high end systems where there has not been a broad software base. For example, multiprocessing became an integral component of large IBM mainframes and in Cray supercomputers in the early 1980s [4,5]. However, the *widespread* use of multithreading has been a chicken-and-egg problem that now appears to be near a solution.

As a way of continuing the aggressive MOS performance curve, single chips now support multiple threads or will soon [6,7]. This support is either in the form of multiple processors [8] or wide superscalar hardware capable of supporting multiple threads simultaneously [9,10]. In any case, the trend is toward widespread availability of hardware-supported on-chip multithreading, and general purpose software will no doubt take advantage of its availability. Furthermore, additional motivation for on-chip multithreading is the increasing number of important applications characterized by many parallel, independent transactions. For example, many web server applications are becoming throughput oriented; general purpose computing is shifting toward multiple latency-limited threads.

2   Instruction Level Distributed Processing

As was pointed out earlier, computer architecture innovation has done more than exploit technology; it has also been used to accommodate technology shifts. In the the case of cache memories, architecture innovation was used to avoid tremendous slow downs - by adapting to the shift in processor/RAM technologies. Considering the trends described above, we appear to be at a point where microarchitecture innovation will be driven by shifts in both technology and applications. The goal will be to maintain long term performance trends in the face of increasing on-chip wire delays, power consumption, and applications that are not compatible with big, static, highly optimized compilations.

A microarchitecture paradigm which deals effectively with technology and application trends is Instruction Level Distributed Processing (ILDP). A processor following the ILDP paradigm consist of a number of distributed functional units, each fairly simple with a very high frequency clock cycle (for example, Fig. 3).

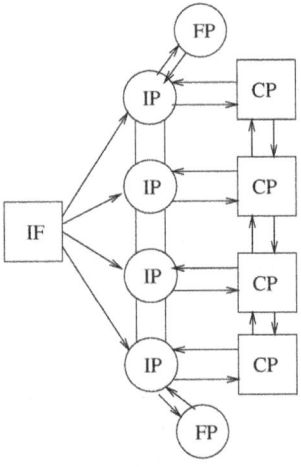

Fig. 3. An example ILDP microarchitecture consisting of an instruction fetch unit, integer processors, floating point processors and cache processors.

The presence of relatively long wire delays implies microarchitectures that explicitly account for inter-instruction and intra-instruction communication. As much as possible, communication should be localized to small units, while the overall structure should be organized for communication. Meanwhile, communication among units will be point-to-point (no busses), and will be measured in full clock cycles. Partitioning the system to accommodate these delays will be a significant part of the microarchitecture design effort. There may be relatively little low-level speculation (to keep the transistor counts low and the clock frequency high); determinism is inherently simpler than prediction and recovery.

With high inter-unit communication delays, the number of instructions executed per cycle may level off or decrease when compared with today, but overall performance can be gained by running the smaller distributed processing elements at a much higher clock rate. The structure of the system and clock speeds have implications for global clock distribution. There will likely be multiple clock domains, possibly asynchronous from one another.

Currently, there is an increasing awareness that clock speed holds the key to increased performance. For several years, the big push has been for ILP, and gains have been made, but they now appear to be diminishing, and it makes sense to push more in the direction of higher clock speeds. This idea is not new; the role of clock speed has long been the subject of debate among RISC proponents. This was certainly the Cray approach, and it is apparent in the evolution of in Intel processors [11,12,13] (see Fig. 4). In contrast to the Intel processors where the pipelines have been made extremely deep, the challenge will be to use simplicity to keep pipelines shallow, even with a very fast clock.

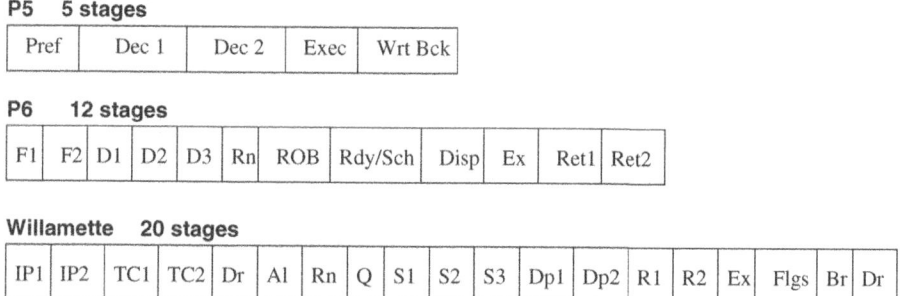

P5   5 stages

Pref	Dec 1	Dec 2	Exec	Wrt Bck

P6   12 stages

F1	F2	D1	D2	D3	Rn	ROB	Rdy/Sch	Disp	Ex	Ret1	Ret2

Willamette   20 stages

IP1	IP2	TC1	TC2	Dr	Al	Rn	Q	S1	S2	S3	Dp1	Dp2	R1	R2	Ex	Flgs	Br	Dr

Fig. 4. Evolution of Intel processor pipelines.

Turning to power considerations, the use of a very fast clock in an ILDP computer will by itself tend to increase dynamic power consumption. However, the very modular, distributed nature of the processor will permit better power management. With most units being replicated, resource usage and dynamic power consumption can be managed via clock gating. In particular, usage of computation resources can be monitored and subsets of replicated units can be used (or not) depending on computation requirements and priorities.

For static power, the high frequency clock will use fast leaky transistors more effectively. If transistors consume power even when they are idle, it is probably better to keep them busy with active work – which a fast clock will do. In addition, the replicated distributed units will make selective power gating easier to implement. Furthermore, some units may be just as effective if slower transistors are used, especially if multiple parallel copies of the unit are available to provide throughput.

For supporting on-chip multithreading, an ILDP provides the interesting possibility of a hybrid between chip multiprocessors and SMT. In particular, the computation units can be partitioned among threads. That is, with simple replicated units, different subsets of units can be assigned to individual threads. As a whole, the processor is shared as in SMT, but any individual unit services only one thread at a time, as in a multiprocessor. The challenge will be the management of threads and resources in such a fine-grain distributed system.

Following sections delve deeper into types of distributed microarchitectures.

2.1 Dependence-Based Microarchitecture

Clustered *dependence-based* architectures [14] are one important class of ILDP processors. The 21264 [15] is a commercial example; a much very earlier and little known example was an uncompleted Cray-2 design [16]. In these microarchitectures, processing units are organized into clusters and dependent instructions are *steered* to the same cluster for processing.

The 21264 microarchitecture there are two clusters, with different instructions routed to each at issue time. Results produced in one cluster require an additional clock cycle to be routed to the other. In the 21264, data dependences tend to steer communicating instructions to the same cluster. Although there is additional inter-cluster delay, a faster clock cycle compensates for the delay and leads to higher overall performance.

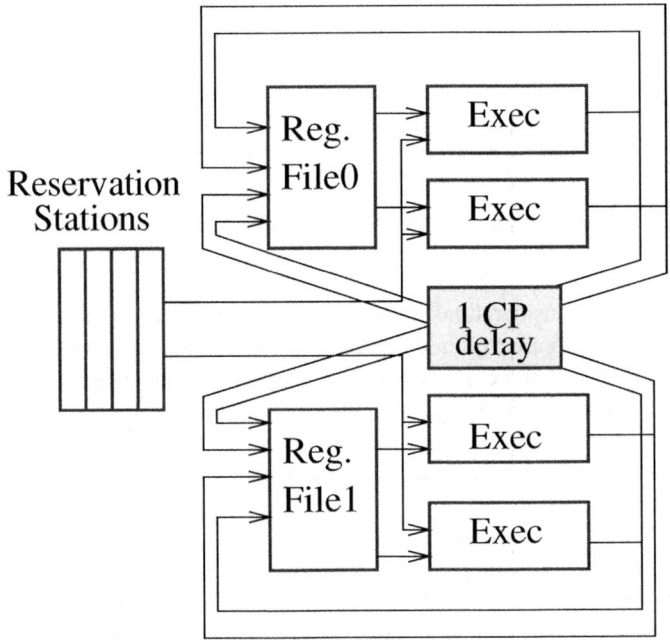

Fig. 5. Alpha 21264 clustered microarchitecture.

In general, a dependence based design may be divided into several clusters; cache processing can be separated from instruction processing; integer processing can be separated from floating point, etc. (see Fig. 3). In a dependence-based design, dependent instructions are collected together, so instruction control logic within a cluster is likely to be simplified, because there is no need to look for independence if it is known not to exist. For example, if all the instructions assigned to a cluster are known to form a dependence chain (or nearly so), they can be issued in order from a FIFO, greatly simplifying issue control logic.

The formation of dependent instructions can be done by the compiler, or at various stages in the pipeline, the dispatch stage being a good possibility. Waiting until the issue stage as in the 21264 may reduce inter-unit communication slightly, but at the expense of more complex issue logic.

2.2 Heterogeneous ILDP

Another model for an ILDP is a heterogeneous system where a simple core pipeline is surrounded by outlying *helper engines* (Fig. 6). These helper engines are not in the critical processing path, so they have non-critical communication delays with respect to the main pipeline, and may even use slower transistors to reduce static power consumption.

Examples of helper engines include the pre-load engine of Roth and Sohi [17] where pointer chasing can be performed by a special processing unit. Another is the branch engine of Reinman et al. [18]. An even more advanced helper engine is the instruction co-processor described by Chou and Shen [19]. Helper engines have also been proposed for garbage collection [20] and correctness checking [21].

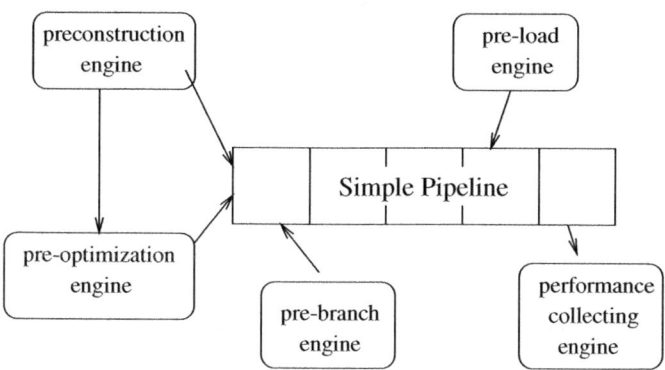

Fig. 6. A heterogeneous ILDP chip architecture.

3 Managing ILDP: Co-designed Virtual Machines

It seems clear that an ILDP computer will need some type of higher level management of the distributed resources used by executing instructions. This management involves the interactions among instructions making up a program, e.g. control and data dependences encoded in the instructions, and the interactions between the instructions and the computing/communication resources.

Determination of instruction interactions can be done by compiler-level software. At compile time, inter-instruction dependences and communication can be determined (or predicted), then this information can be encoded into machine level instructions. At runtime this information can be used to steer instruction control and data information through the distributed processing elements.

Alternatively, hardware can be used to determine the necessary inter-instruction attributes by using hardware tables to collect dynamic history information as programs are executed. Then, this history information can be accessed by later instructions for steering of control and data information.

Availability and usage of processor resources is another important consideration; for example, resource load balancing will likely be needed for good performance performance – at both the instruction level and thread level. For power efficiency, gating off unused or unneeded resources requires usage analysis and coordination, especially if power gating is widespread across a chip. Implementation of fault tolerance through replicated processing units also requires higher level management. This function can potentially be done via hardware or software, implemented as part of the OS.

While they are viable solutions, a big disadvantage of software approaches based on conventional OS and compilers is that they likely require re-compilation and OS changes to fit each particular ILDP hardware platform. Programs today must be recompiled to get maximum performance from the latest superscalar implementation, but they often get reasonable performance without recompilation (or even optimization in some cases). ILDP microarchitectures, however, may not be as forgiving as a homogeneous superscalar processor, although this remains to be seen. Disadvantages of the hardware-intensive solution are complex, power-consuming hardware and a rather limited scope for collecting information regarding the executing instruction stream.

An important alternative to using conventional software or hardware solutions is provided by currently evolving dynamic optimizing software and virtual machine technologies. A co-designed virtual machine is a combination of hardware and software that implements a virtual architecture (VA) [22,23,24]. Part of the implementation is in hardware – which supports an instruction set with implementation-specific features (the Implementation Architecture or IA). The other part of the implementation is in software, which translates the VA to the IA and which provides the capability of dynamically optimizing a program. A co-designed VM should be viewed as a way of giving hardware implementors a layer of software with which to work. This software layer provides flexibility in managing the resources that make up a ILDP microarchitecture. It also liberates the hardware designer from supporting a legacy VA purely in hardware.

With ILDP, the distributed processor resources must be managed with a global view. For example, instructions and data must be routed in such a way that resource usage is balanced and communication delays (among dependent instructions) are minimized, as with any distributed system. This would require high complexity hardware, if hardware alone were given responsibility. For example, the hardware would have to be aware of elements of program structure, such as data and control flow. However, conventional issue-window-based methods give hardware only a restricted view of the program, and the hardware would have to re-construct program structure information by viewing a small part of the instruction stream as it flows by.

Hence, in the co-designed VM paradigm (Fig. 7), software is responsible for determining program structure, dynamically re-optimizing code, and making complex decisions regarding management of ILDP. Hardware implements the lower level performance features that are managed by software. Hardware also can collect dynamic performance information and may trigger software when *unexpected* conditions occur.

Besides performance, the VM layer can also be used for managing resources to reduce power requirements [25] and to implement fault tolerance.

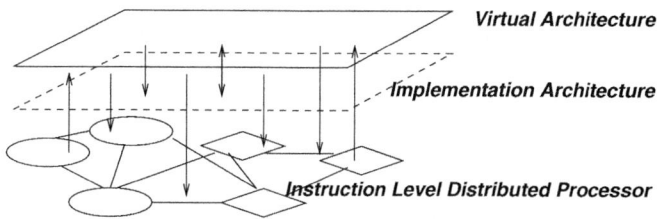

Fig. 7. Supporting an instruction level distributed processor with a co-designed Virtual Machine.

4   The Role of Instruction Sets

Historically, instruction sets have tended to evolve in discrete steps. After the original mainframe computers, instruction sets more-or-less stabilized until the early 1970s when minicomputers came on the scene. These machines used relatively inexpensive packaging and interconnections, and provided an opportunity to re-think instruction sets. Based on lessons learned from the relatively irregular mainframe instruction sets, regularity and "orthogonality" [26] became the goals. While incorporating these properties, mini computer ISAs typically supported relatively powerful, variable-length instructions; the PDP-11 and later VAX-11 instruction sets are good examples. As microprocessors evolved toward general purpose computing platforms, there was another re-thinking of instruction sets, this time with hardware simplicity as a goal. These RISC instruction sets, among other things, allowed a full pipelined processor implementation to fit on a single chip. As transistor densities have increased, we have reached the point where older, more complex microprocessor instruction sets can now be dynamically translated on-chip into RISC-like operations.

In retrospect, it seems that instruction set innovation has keyed off packaging/technology changes, from mainframes to minicomputers to microprocessors. And now may be a good time to have another serious investigation of instruction sets. This time, however, on-chip communication delays, high speed clocks, and advances in translation/virtual machine software provide the motivation.

Instruction sets can and should be optimized for ILDP. Features of new instruction sets should focus on communication and dependence, with emphasis on small, fast memory structures, including caches and registers. For example, variable length instructions lead to smaller instruction footprints and smaller caches. While legacy binaries have inhibited new instruction sets, virtual machine technology and binary translation enable new implementation-level instruction sets.

Most recent instruction sets, including RISC instructions sets, and especially VLIW instruction sets, have emphasized computation and independence. The view was that higher parallelism could be achieved by focusing on computational aspects of instruction sets and on placing independent instructions in proximity either at compile time or during execution time. For ILDP, however, instruction sets should be targeted at communication and dependence. That is, communication should be easily expressed and dependent instructions should be placed in proximity, to reduce communication delays.

There are at least three types of related information that are important for an ILDP instruction set to express: i) instruction dependence information; ii) instruction *steering* information (and possibly data steering information) to guide instructions and data to the proper distributed processing elements so that dependent instructions can be executed in proximity with each other, iii) value usage information; in particular, if a value is used only once, it does not have to be communicated beyond the consuming instruction.

One possibility for conveying this type of information is to add tag bits to conventional instruction sets. These tag bits can explicitly indicate instruction dependence information, instruction steering information, and value usage information. .ämbiguity in term value locality above .ïewrite the following This is similar to the independence bits added to long instruction words in the Intel IA-64 instruction set [27].

It is also possible to use implicit methods of specifying the above information. For example, a stack-based instruction set places the focus on communication and dependence. Dependent instructions communicate via the stack top; hence, communication is naturally expressed. Dependent instructions tend to be clustered together, and local values appear on the stack top and then are immediately consumed. Furthermore stack-based ISAs tend to have a small instruction footprints which will lead to smaller (faster) instruction cacheing structures.

As a more complete example, consider the following accumulator-based ISA. Assume 64 general purpose registers and a single accumulator are used for performing operations. All operations must involve the accumulator, so dependent operations are explicitly apparent as is local value communication. With such an ISA, there is need for only one general purpose register field per instruction and the ISA can be made quite compact, with instructions 1,2 or 4 bytes in length. For example, consider the following basic instruction types.

```
R  <- A                    1 byte
A  <- R                    1 byte
A  <- A op R               2 bytes
A  <- A op imm             4 bytes
A  <- M(R op imm)          4 bytes
M(R op imm) <- A           4 bytes
R  <- M(A op imm)          4 bytes
```

The first two instructions copy data to/from a register and the accumulator; A is the single accumulator and R is one of the general purpose registers. The next two instructions are examples of operations on data held in the accumulator (and general register file). The last three instructions are example loads and stores.

With an instruction set of this type, dependent instructions will naturally chain together via the accumulator and will be contiguous in the instruction stream. With a clustered ILDP implementation, all the instructions in a dependent chain can be steered simultaneously to the same cluster, with the next dependent chain being steered to another cluster. If the accumulator is re-named within each cluster [16], the parallelism among dependence chains can be exploited, with global communication taking place via the general registers. Because it contains only dependent instructions, the instruction issue queue in each cluster will be simplified as will local data communication through the accumulator.

The stack and single accumulator ISAs are simple examples of instruction sets that implicitly specify communication/dependence information; whether the implicit or explicit methods are better is not clear. The important point is that instruction sets deserved renewed study, and future technologies and ILDP microarchitectures provide fertile ground for innovation in this area.

5   Summary

Technology and application shifts are pointing toward instruction level distributed processing. These microarchitectures will contain distributed resources and will be explicitly structured for inter-unit communication. Helper processors may be distributed around the main processing elements to perform more complex optimizations and to perform highly parallel tasks. By constructing simple distributed processing elements, a very high clock rate can be achieved, probably with multiple clock domains. Replicated distribution processing elements will also allow better power management.

Virtual machines fit very nicely in this environment. In effect, hardware designers can be given a layer of software that can be used to coordinate the distributed hardware resources and perform dynamic optimization from a higher level perspective than is available to hardware alone. Finally, it is once again time that we reconsider instruction sets with the focus on communication and dependence. New instructions sets are needed to mesh with ILDP implementations, and they are enabled by the VM paradigm which makes legacy compatibility less important at the implementation architecture level.

Acknowledgements

This work was supported by National Science Foundation grant CCR-9900610, by IBM Corporation, Sun Microsystems, and Intel Corporation. This support is gratefully acknowledged.

References

1. Neil C. Wilhelm,"Why Wire Delays Will No Longer Scaler for VLSI Chips," SUN Microsystems Laboratories Technical Report TR-95-44, August 1995.
2. V. Agarwal, M. S. Hrishikesh, S. W. Keckler, D. Burger, "Clock Rate versus IPC: The End of the Road for Conventional Microarchitectures," *27th Int. Symp. on Computer Architecture*, pp. 248-259, June 2000.
3. S. Thompson, P. Packan, and M. Bohr,"MOS Scaling: Transistor Challenges for the 21st Century," *Intel Technology Journal*, Q3, 1998.
4. M. S. Pittler, D. Powers, D. L. Schnabel,"System Development and Technology Aspects of the IBM 3081 Processor Complex," *IBM Journal of Research and Development,*, pp. 2-11, Jan. 1982.

5. M. August, G. Brost, C. Hsiung, A. Schiffler, "Cray X-MP: The Birth of a Supercomputer," *IEEE Computer,* pp. 45-52, January 1989.
6. R. Eichemeyer, et al."Evaluation of Multithreaded Uniprocessors for Commercial Application Environments", *23rd Annual Int. Symp. on Computer Architecture,* pp. 203-212, June 1996.
7. K. Diefendorff,"Compaq Chooses SMT for Alpha," *Microprocessor Report*, pp. 1, 6-11, Dec. 6, 1999.
8. L. Barroso, et al.,"Prianha: A Scalable Architecture Based on Single-Chip Multiprocessing," *27th Int. Symp. on Computer Architecture,*, pp. 282-293, June 2000.
9. W. Yamamoto, M. Nemirovsky,"Increasing Superscalar Performance Through Multistreaming," 3rd Int. Symp. on Parallel Arch. and Compiler Techniques, June 1995.
10. D. Tullsen, S. Eggers and H. Levy, "Simultaneous Multithreading: Maximizing On-Chip Parallelism", *22nd Annual International Symposium on Computer Architecture,* June 1995.
11. B. Case,"Intel Reveals Pentium Implementation Details,", *Microprocessor Report*, pp. 9-17, March 29, 1993.
12. Linley Gwennap,"Intel's P6 Uses Decoupled Superscalar Design", *Microprocessor Report*, pp. 9-15, Feb. 16, 1995.
13. G. Hinton,"Willamette: Next Generation IA-32 Micro-architecture," *Intel Developer Forum Spring 2000*, Feb. 15, 2000. Intel Willamette
14. S. Palacharla, N. Jouppi, J. E. Smith, "Complexity-Effective Superscalar Processors," *24th Int. Symp. on Computer Architecture,*, pp. 206-218, June 1997.
15. L. Gwennap,"Digital 21264 Sets New Standard," *Microprocessor Report,* pp. 11-16, Oct. 1996.
16. Anonymous, *Cray-2 Central Processor*, unpublished document, 1979.
17. A. Roth and G. Sohi,"Effective Jump-Pointer Prefetching for Linked Data Structures," *26th Int. Symp. on Computer Architecture,*, pp. 111-121, May 1999.
18. 8. G. Reinman, T. Austin, B. Calder,"A Scalable Front-End Architecture for Fast Instruction Delivery," *26th Int. Symposium on Computer Architecture*, pp. 234-245, May 1999.
19. Yuan Chou and J. P. Shen,"Instruction Path Coprocessors," *27th Int. Symposium on Computer Architecture*, pp. 270-281, June 2000.
20. Timothy Heil and J. E. Smith,"Concurrent Garbage Collection Using Hardware-Assisted Profiling," *International Symposium on Memory Management (ISMM)*, October 2000.
21. T. Austin,"DIVA: A Reliable Substrate for Deep Submicron Microarchitecture Design," *32nd Int. Symposium on Microarchitecture*, pp. 196-297, Nov. 1999.
22. K. Ebcioglu and E. R. Altman,"DAISY: Dynamic Compilation for 100Compatibility," *24th Int. Symp. on Computer Architecture,*, June 1997.
23. A. Klaiber,"The Technology Behind Crusoe Processors," *Transmeta Technical Brief*, 2000.
24. J. E. Smith, T. Heil, S. Sastry, T. Bezenek,"Achieving High Performance via Co-Designed Virtual Machines," *Intl. Workshop on Innovative Architecture for Future Generation High-Performance Processors and Systems,*, pp. 77-84, Oct. 1998.
25. D. H. Albonesi,"The Inherent Energy Efficiency of Complexity-Adaptive Processors," *1998 Power-Driven Microarchitecture Workshop*, pp. 107-112, June 1998.
26. W. A. Wulf,"Compilers and Computer Architecture," *IEEE Computer,* pp. 41-48, July 1981.
27. L. Gwennap,"Intel, HP Make EPIC Disclosure," *Microprocessor Report,* pp. 1-9, Oct. 1997.

Speculative Multithreaded Processors

Gurindar S. Sohi and Amir Roth

Computer Sciences Department
University of Wisconsin-Madison
1210 W. Dayton St. Madison, WI 53706
sohi@cs.wisc.edu

Abstract. Architects of future generation processors will have hundreds of millions of transistors with which to build computing chips. At the same time, it is becoming clear that naive scaling of conventional (superscalar) designs will increase complexity and cost while not meeting performance goals. Consequently, many computer architects are advocating a shift in focus from high-performance to high-throughput with a corresponding shift to multithreaded architectures. Multithreaded architectures provide new opportunities for extracting parallelism from a single program via *thread level speculation*. We expect to see two major forms of thread-level speculation: *control-driven* and *data-driven*. We believe that future processors will not only be multithreaded, but will also support thread-level speculation, giving them the flexibility to operate in either multiple-program/high-throughput or single-program/high-performance capacities. Deployment of such processors will require innovations in means to convey multithreading information from software to hardware, algorithms for thread selection and management, as well as hardware structures to support the simultaneous execution of collections of speculative and non-speculative threads.

1 Introduction

The driving forces behind the tremendous improvement in processing speed have been semiconductor technology and innovative architectures and microarchitectures. Semiconductor technology has provided the "bricks and mortar" — increasingly greater numbers of increasingly faster on-chip devices. Innovations in computer architecture and microarchitecture (and accompanying software) have provided techniques to make good use of these building materials to yield high-performance computing systems. Computer designs are constantly changing as architects search for (and often find) innovations to match technology advances and important shifts in technology parameters; this is likely to continue well into the next decade.

The process of deciding how available semiconductor resources will be used can be decomposed into two. First, the architect must decide on the desired *functionality*: the techniques used to expose, extract and enhance performance. Then comes the problem of *implementation*: the techniques must translated to structures and signals which must themselves be designed, built and verified.

M. Valero, V.K. Prasanna, and S. Vajapeyam (Eds.): HiPC 2000, LNCS 1970, pp. 259–270, 2000.

Though described separately, these issues are, in practice, very tightly coupled in the overall design process. In the 1990s, novel functionality played the dominant role in microprocessor design. With a "reasonable" limit on the overall size of a design (e.g., fewer than tens of millions of transistors), the transistor budget could be divided by high-level performance metrics only. Verification was (relatively) simple and many of the problems encountered during implementation were manageable: wire delays were not significant as compared to logic delays, and power requirements were not exorbitant. In the future, however, implementation issues are likely to dominate even basic functionality. Monolithic designs occupying many tens or hundreds of millions of transistors will be very difficult to design, debug, and verify, and increasing wire delays will make intra-chip communication and clock distribution costly. These technology trends suggest designs that are made of replicated components, where each component may be as much as a complete processing element. Distributed, replicated organizations can "divide and conquer" the complexities of design, debug and verification, and can exploit localities of communication to deal with wire delays.

Fortunately, the twin goals of increasing single-program performance and easing implementation are not in conflict. In fact, with the right model for parallelism they can be synergistic. *Speculative multithreading* is such a model, making it a leading candidate for implementation in future-generation processors. In speculative multithreading, a processor is (logically) comprised of replicated processing elements that cooperate on the parallel execution of a conventional sequential program (also referred to as a conventional program thread) that has been divided into chunks called speculative threads. Speculation is a key element. Without speculation, programs can only be divided conservatively into threads whose mutual independence must be guaranteed. Speculation allows these guarantees to be bypassed, producing much more aggressive divisions into threads that are parallel with high probability.

2   Rationale for Speculative Multithreading

The motivation for using speculative multithreading comes from two directions. On one hand, the potential for further increasing single-program performance using known parallelism extraction techniques is diminishing. On the other, technology trends suggest processors that can execute multiple threads of code. These circumstances invite us to find those few innovations that will enable such multithreaded processors to support the parallel execution of a single program.

2.1   Limitations of Existing Techniques to Extract Parallelism

We begin by briefly reviewing the functionality and high-level operation of the incumbent model for achieving high single-program performance — the *superscalar* model. Imperative programs — programs written in imperative languages like Fortran, C, and Java — are defined by a static control flow in which individual instructions read and write named storage locations. At runtime, a superscalar processor unrolls the static control flow to produce a dynamic instruction stream. The positions of reader and writer instructions in this stream defines the way data flows from one operation to another, i.e., the algorithm itself. A super-scalar processor creates a dynamic *instruction window* (an unrolled contiguous segment of the dynamic instruction stream), repeatedly searches this window for un-executed, independent instructions, and attempts to execute these instructions in parallel. Sustained high-performance demands that any given window contain a sufficient number of independent instructions, i.e., a sufficient level of *instruction-level parallelism (ILP)*.

Unfortunately, the way in which imperative programs are written makes consistently high ILP a rarity. In order to preserve their sanity, programmers structure programs in certain ways, a basic technique being the static (and hence dynamic) grouping of dependent instructions. The spatial proximity of related statements helps programmers reason about programs in a hierarchical fashion but limits the amount of independent work that would available in a given window of dynamic instructions. Optimizing compilers attempt to improve the situation by transparently re-ordering instructions, mixing instructions from nearby program regions to improve the overall levels of window ILP. However, while very sophisticated, compiler scheduling is fundamentally limited by compilers' inability to perfectly determine the original intent of the programmer and their commitment to preserve the high-level structure of the original program.

The amount of parallel work being what it is, one option is to build a superscalar processor with an instruction window large enough to simultaneously contain code from different program regions (i.e., different functions or loop iterations). However, even if such a machine could be built — and there are many engineering obstacles to doing so — there is fundamental problem in keeping a large, contiguous instruction window full of *useful* instructions. Specifically, the decreasing accuracy of a series of branch predictions leads to an exponentially decreasing likelihood that instructions at the tail of the window will be useful.

Overcoming this problem requires a model that allows parallelism from different program regions to be exploited in a reasonably independent (i.e., non-contiguous, non-serial) manner. *Speculative multithreading* is such a model. In speculative multithreading, each program region is considered to be a *speculative thread*, i.e., a small program. By executing multiple speculative threads in parallel, additional parallelism can be extracted (especially if each thread is mostly sequential). The threads are subsequently merged to recreate the original program. Speculative multithreading allows a large instruction window to be created as an ensemble of smaller instruction windows, thereby facilitating implementation. In addition, a proper thread division can logically isolate branches in one thread from those in another [27], relieving the fundamental problem of diminishing instruction utility.

2.2   The Emergence of Multithreaded Architectures

Multithreaded processors — processors that support the concurrent execution of multiple threads on a single chip — are beginning to look as if they will dominate the landscape of the next decade. Two multithreaded processor models are currently being explored. *Simultaneous multithreading (SMT)* [5,7,14,32,33] uses a monolithic design with most resources shared amongst the threads. *Chip multiprocessing (CMP)* [12] proposes a distributed design (a collection of independent processing elements) with less resource sharing. The SMT model is motivated by the observation that support for multiple threads can be provided on top of a conventional ILP (i.e., superscalar) processor with little additional cost. The CMP model is more conventionally motivated by design simplicity and replication arguments. Both models target independent threads (multithreaded a multiprogrammed workloads) and use multithreading to improve processing *throughput.*

As technology changes, the distinction between the SMT and CMP microarchitectures is likely to blur. Increasing wire delays will require decentralization of most critical processor functionality, while flexible resource allocation policies will enhance the appearance of (perhaps asymmetric) resource sharing. Regardless of the specific implementation, multithreaded processors will logically appear to be collections of processing elements. The interesting question is whether this organization can be exploited to improve not only throughput but also the execution time of a single program. *Thread-level speculation* is the key to enabling this synergy. In addition to executing conventional parallel threads, the logical processors could execute *single programs that are divided into speculative threads.* Speculative multithreaded processors will provide not only high throughput but also high single-program performance when needed.

3   Dividing Programs into Multiple Threads

There are several ways in which to divide programs into threads. We categorize these divisions as *control-driven* and *data-driven* depending on whether threads are divided primarily along control-flow or data-flow boundaries. Each division strategy can be further sub-categorized as either *non-speculative* — the threads are completely independent from the point of view of the processor and any dependence is explicitly enforced using architectural synchronization constructs, or *speculative* — the threads may not be perfectly independent, or synchronized, and it is up to the hardware to detect and potentially recover from violations of the independence assumptions.

The threads obtained from a division of a program are expected to execute on different (logical) processing units. To achieve concurrency, *proximal* threads (i.e., threads that will simultaneously co-exist in the machine) need to be highly data-independent. If data-independence can be achieved, concurrency (and hence performance) can scale almost linearly with the number of threads even for small per-thread window sizes, and efficiency can be kept constant as bandwidth and (hopefully) performance are increased. We expect that speculation can allow data-independence criteria to be achieved more easily, giving speculative solutions distinct performance and applicability advantages over their more conventional non-speculative counterparts.

3.1   Control-Driven Threads

Although the object of multithreading a program is to divide it into data-independent (parallel) threads, the most natural division of an imperative program is along control-flow boundaries into control-driven threads. The architectural semantics of imperative programs are control-driven: instructions are totally ordered and architectural state is precisely defined only at instruction boundaries. Control-flow is explicit while data-flow is implicit in the total order. In control-driven multithreading, the dynamic instruction stream is divided into contiguous segments that can subsequently be "sewn" together end-to-end to reconstruct the sequential execution. The challenge of control-driven multithreading is finding division points that minimize inter-thread data dependences.

We should note here that control-driven multithreading is not the same as parallel programming. Parallel programs do execute multiple concurrent control-driven threads, but these threads exchange data in arbitrary ways. The semantics of a parallel program is rarely the semantics of the individual threads run in series. In contrast, control-driven multithreading is a way of imposing parallel execution on what is in essence a sequential program. Data flows between control-driven threads in one direction only, from sequentially "older" threads to "younger" ones.

Non-speculative Control-Driven Threads. Without support for detecting and recovering from data-dependence violations or to abort unnecessary threads and discard their effects, non-speculative control-driven multithreading requires strict guarantees about the *execution-certainty* and *data-integrity* of threads. Execution-certainty requirements spawn from the fact that thread execution cannot be *undone*, and mean that non-speculative control-driven threads can only be forked if their execution is known to be needed. In order to maximize concurrency, execution certainty is usually achieved by forking a thread at a previous control-equivalent point, e.g., forking of a loop iteration at the beginning of the previous iteration. Data-integrity refers to the requirement that access to thread shared data must occur (or appear to occur) in sequential order. When we speak of data-integrity, we are mainly concerned with memory-integrity. Support for direct inter-thread register communication is typically not available. We assume that if it is provided then appropriate synchronization is provided along with it. In contrast, inter-thread memory communication is naturally available, meaning that access to any memory location that could potentially be shared with other threads must be explicitly synchronized. Of course, data-sharing/synchronization should be kept to a minimum to allow for adequate concurrency among threads.

With such strict safety requirements, the division of a program into non-speculative threads has traditionally fallen into the realm of the programmer and compiler. The programmer has the deepest knowledge of the parallel dimensions of his algorithm and the potential for data-sharing among different divisions. However, performing thread division by hand is tedious, and manual attempts to minimize synchronization often lead to errors. In light of these difficulties, much effort has been placed into using the compiler to automatically multithread (parallelize) programs. Although (debugged) compilers don't make errors, and compiler tedium is less of an issue than programmer tedium, compiler multithreading has had success only in very limited domains.

Speculative Control-Driven Threads. Non-speculative control-driven multithreading suffers from two major problems. First, execution-certainty requirements limit thread division to control-independent program points, which may not satisfy the primary data-independence criteria. Second, even when proximal threads are data-independent, if this independence is unprovable, then conservative synchronization must be used to guard against the unlikely (but remotely possible) case of a re-ordered communication. Where synchronization is needlessly applied, concurrency and performance are unnecessarily lost.

Speculation can alleviate these problems. In speculative control-driven multithreading, memory does not need to be explicitly synchronized at all. The correct total order of memory operations can be reconstructed from the (explicit or implied) order of the threads. This ordering can be used as the basis for hardware support to detect and potentially recover from inter-thread memory-ordering violations [10,11]. With such support, access to thread shared data can proceed optimistically, with penalties incurred only in those cases when data is actually shared by proximal threads *and* the accesses occur in non-sequential order. Furthermore, since ordering violation scenarios are typically predictable, slight modifications to the basic mechanism allow it to learn to recognize these scenarios early and artificially synchronize the offending store/load pairs [4,17].

The execution-certainty constraints can be lifted using similar mechanisms. The ability to recover from inter-thread memory-ordering violations implies the presence of hardware that can buffer or undo changes to architected thread state. This support can be used to undo an entire thread, allowing threads to be spawned at points at which their final usefulness cannot be absolutely guaranteed, but where usefulness likelihood is high and the data-independence (parallelism) characteristics are more favorable.

Speculative control-driven multithreading has been the subject of academic research in the 1990's [1,6,9,13,16,27,29,34] and is slowly finding its way into commercial products. Sun's MAJC architecture [31] supports such threads, via its Space Time Computing (STC) model. More recently, NEC's Merlot chip [18] uses speculative control-driven multithreading to parallelize the execution of code that can't be parallelized by other known means. We expect that more processors will make use of speculative control-driven threads in the coming decade, as this technology moves from the research phase into commercial implementations.

3.2   Data-Driven Threads

Where control-driven multithreading divides programs along control-flow boundaries, data-driven multithreading uses data-flow boundaries as the major division criteria. Such a division naturally achieves the desired inter-thread data-independence and resulting parallelism [2,15,19,25,26].

Data-driven threads are almost ideal from a performance and efficiency standpoint. In its pure form, data-driven multithreading occurs at the granularity of a single instruction [2,19]. Data-driven instruction sequencing (i.e., fetch) is triggered by the availability of one of its input operands. Instructions enter the machine as soon as they may be able to execute but no sooner. This arrangement maximizes the amount of work that may be used to overlap with long latency instructions, while not wasting resources on instructions that are not ready to use them.

Instruction-level data-driven sequencing is not the only option. Data-driven sequencing may be used on a thread granularity with conventional, control-driven sequencing used at the instruction level [15,26,30]. In this organization, instructions from one or several related computations are packed into totally-ordered threads that implicitly specify data-flow relationships. Individual threads are assigned to processing elements and sequenced and executed in a control-driven manner. However, the data-flow relationships *between* threads are represented explicitly and thread creation is triggered in a data-driven manner (i.e., by the availability of its data inputs from the outcome of a previous thread). The data-driven threads we expect to see in future processors are likely to be of this form.

Non-speculative Data-Driven Threads. Non-speculative data-driven multithreading is difficult to implement for imperative languages. The main barrier is the incongruity of the requirement of an explicit data-flow program representation and the reality that for imperative programs, data-flow information is often impossible to explicitly specify *a priori* even as it applies to a few well-defined boundaries. A data-forwarding error, either of omission or false commission, changes the meaning of the program. The automatic conversion of imperative code to data-flow explicit form has been the subject of some research, but in general, data-driven program representations can only be constructed for code written in functional (data-driven) languages.

Speculative Data-Driven Threads. Non-speculative data-driven multithreading suffers from two major problems. First, programs can generally not be divided into data-driven threads. Second, even in cases where a division is possible, the resulting representation breaks the sequential semantics created by the programmer and the correlation between the executing program and the source code from which it was derived. Sequential semantics (or at least their appearance) is very important for program development, debugging, and the interaction with non-data-driven system components and tasks. The loss of sequential semantics is more serious than simply being a disturbance to the programmer.

Again, speculation is likely to be the key to solving these problems. However, a shift in approach regarding the role of multithreading may be needed first. Two complementary observations guide this new approach. First, program development and debugging will probably require the presence of a "main" or "architectural" thread whose execution will implement the sequential, control-driven semantics of the program. Second, programs inherently contain sufficient levels of ILP, but this ILP is hindered by long-latency microarchitectural events like cache misses and branch mis-predictions. The parallelism in the program can be extracted if these latencies — which are likely to get relatively longer — can be tolerated. These observations suggest a different role for multithreading, one which does not require dividing the program *per se*. Instead, the program is augmented with "helper" threads that run ahead and *pre-execute* or "solve" problem instructions before they have a chance to cause stalls in the "main" program thread. We believe that it is in this capacity, as high-powered "helper" threads, that speculative data-driven threads can best be used in an imperative context [3,8,20,21,23,24,28,35].

In the "helper" model, selected computations are copied from the program and packed into data-driven threads [8,23,24]. Now, the program is executed as a single control-driven thread, as usual. However, at certain points in the main program, data-driven threads are spawned in order to pre-execute the computation of some future problem instruction. When the main program thread catches up to the data-driven thread, it has the option of picking up the result directly [8,21,22,23] or simply repeating the work (albeit with a reduced latency) [20].

The role of speculation is intermingled with the reduced "helper" status of data-driven threads. The fact that the control-driven thread is present and ultimately responsible for the architectural interface, immediately relieves data-driven threads from any correctness obligations. Without these obligations, data-driven threads can be constructed using whatever data-flow information is available. In addition, they need not comprise a complete partitioning of the program; their use may be reserved only for those situations in which their parallelism-enhancing characteristics are most needed.

4 Practical Aspects

Whether future processors will also include support for speculative threads — either control-driven, data-driven or both — depends on the discovery of acceptable solutions to several practical problems. These problems range from the low-level (i.e., how threads should be implemented) to the high-level (i.e., how threads should be used) and cover all levels in between. We briefly touch upon some of these issues in this section.

4.1 System Architecture

The broadest decision that needs to be made and the one that will have the most impact on other decisions is the division of labor and responsibilities between the programmer, compiler, operating system and processor. It is obvious that the processor will execute the threads. However, the answer to the question of what entity should be responsible for other thread-related tasks — from selecting the threads themselves to spawning, scheduling, resource allocation and communication — is not clear. Placing all of the responsibility on the processor is one attractive option. With near-future processors having nearly one billion transistors, a few million can be dedicated to multithreading-specific management tasks (perhaps as a separate co-processor). A processor-only implementation has no forward or backward compatibility problems, it preserves the current system interface, while enhancing the performance of legacy software. Its drawbacks are added design complexity and the mandate rigidity and simplicity of the thread selection and management algorithms.

Since thread-selection is such an important and delicate problem, it seems logical to push at least that function to software or perhaps even the programmer. Thread-selection algorithms implemented in software can be more sophisticated and may produce better thread divisions. Thread divisions chosen by the programmer — who understands the program at its highest, algorithmic levels — and subsequently communicated to the compiler, may be better still. However, any path in which multithreading information flows from or through software to the hardware requires a change in the software/hardware interface. Such changes are typically met with some resistance, especially if they have architectural semantics that need to be implemented.

Our expectation is that speculative thread information is likely to be conveyed from software to hardware, but in an *advisory* form. An example of advisory information are prefetch instructions that are found in many recent architectures. The understanding is that the hardware may act upon this information either fully, partially or selectively, or even ignore it altogether, all without impacting correctness. The option to enhance or refine this information dynamically is left to the processor as well. Restricting speculative thread information to an advisory role relieves the architect from many functionality guarantees that would hamper future generation implementations.

4.2 Specific Hardware Support

For the full power of speculative multithreading to be realized, hardware support is required. Specifically, threads need to be made "lightweight" with mechanisms for fast thread startup and inter-thread communication and synchronization. Hardware support for speculation includes buffering for speculative actions and facilities for fast correct-speculation state commit and, likewise, mis-speculation recovery. The precise support required for control-driven and data-driven threads is somewhat different. An additional challenge is to provide this support, as well as support for conventional parallel threads, using a uniform set of simple mechanisms.

One apparent requirement for the implementation of lightweight threads (speculative and otherwise) is a mechanism for passing values from one thread to another via registers. Memory communication and synchronization is likely to be reasonably fast on a speculatively multithreaded processor, since the bulk of it will occur through the highest level of shared on-chip cache. However, a register path for communication and synchronization is likely to be faster still.

Inter-thread register communication will also allow thread register contexts to be initialized quickly, accelerating thread start-up.

We assume that the register-communication mechanism will implement inter-thread register synchronization. Another requirement is a mechanism for enforcing correct ordering of memory operations from different threads. At a high level, such a mechanism would buffer loads from young threads and compare them with colliding stores from older threads. Designs for inter-thread memory ordering mechanisms are known in both centralized [10] and distributed forms [11]. The distributed form uses a modified cache-coherence protocol that blends naturally with the protocol that implements general data-sharing for parallel threads. We expect this form to find widespread use in future processors.

5   Summary

Future processors will be comprised of a collection of logical processing elements that will collectively execute multiple program threads. To overcome the limitations in dividing a single program into multiple threads that can execute on these multiple logical processing elements, speculation will be used. A sequential program will be "speculatively parallelized" and divided into speculative threads. Speculative threads are not only a good match for the microarchitectures that are likely to result as technology advances, they have the potential to overcome the limitations of currently-known methods to extract instruction-level parallelism.

There are two main types of speculative threads that we expect to be used: control-driven and data-driven threads. Speculative control-driven threads have already begun to appear in commercial products (e.g., Sun's MAJC and NEC's Merlot), while speculative data-driven threads are still in the research phase.

Several technologies will have to be developed before speculative multithreading is commonplace in mainstream processors. These include means for conveying thread information from software to hardware, algorithms for thread selection and management, and hardware and software to support the simultaneous execution of a collection of speculative and non-speculative threads. Consequently we expect the next decade of processor development to be at least as exciting as previous decades.

Acknowledgements

This work was supported in part by National Science Foundation grants MIP-9505853 and CCR-9900584, donations from Intel and Sun Microsystems, the University of Wisconsin Graduate School and by an Intel Foundation Graduate Fellowship.

References

1. H. Akkary and M.A. Driscoll. A Dynamic Multithreading Processor. In *Proc. 31st International Symposium on Microarchitecture*, pages 226–236, Nov. 1998.

2. Arvind and R.S. Nikhil. Executing a Program on the MIT Tagged-Token Dataflow Architecture. *IEEE Transactions on Computers*, 39(3):300–318, Mar. 1990.

3. R.S. Chappell, J. Stark, S.P. Kim, S.K. Reinhardt, and Y.N. Patt. Simultaneous Subordinate Microthreading (SSMT). In *Proc. 26th International Symposium on Computer Architecture*, May 1999.

4. G.Z. Chrysos and J.S. Emer. Memory Dependence Prediction using Store Sets. In *Proc. 25th International Symposium on Computer Architecture*, pages 142–153, Jun. 1998.

5. G.E. Daddis and H.C. Torng. The concurrent execution of multiple instruction streams on superscalar processors. In *Proc. International Conference on Parallel Processing*, pages 76–83, May 1991.

6. P.K. Dubey, K. O'brien, K.A. O'brien, and C. Barton. Single-Program Speculative Multithreading (SPSM) Architecture: Compiler-Assisted Fine-Grained Multithreading. In *Proc. 1995 Conference on Parallel Architectures and Compilation Techniques*, pages 109–121, Jun. 1995.

7. J. Emer. Simultaneous Multithreading: Multiplying Alpha's Performance. Microprocessor Forum, Oct. 1999.

8. A. Farcy, O. Temam, R. Espasa, and T. Juan. Dataflow Analysis of Branch Mispredictions and Its Application to Early Resolution of Branch Outcomes. In *Proc. 31st International Symposium on Microarchitecture*, pages 59–68, Dec. 1998.

9. M. Franklin. *The Multiscalar Architecture*. PhD thesis, University of Wisconsin-Madison, Madison, WI 53706, Nov. 1993.

10. M. Franklin and G.S. Sohi. ARB: A Hardware Mechanism for Dynamic Reordering of Memory References. *IEEE Transactions on Computers*, May 1996.

11. S. Gopal, T.N. Vijaykumar, J.E. Smith, and G.S. Sohi. Speculative Versioning Cache. In *Proc. 4th International Symposium on High-Performance Computer Architecture*, pages 195–205, Feb. 1998.

12. L. Hammond, B.A. Nayfeh, and K. Olukotun. A Single-Chip Multiprocessor. *IEEE Computer*, 30(9):79–85, Sep. 1997.

13. L. Hammond, M. Willey, and K. Olukotun. Data speculation support for a chip multiprocessor. In *Proc. 8th International Conference on Architectural Support for Programming Languages and Operating Systems*, pages 58–69, Oct. 1998.

14. H. Hirata, K. Kimura, S. Nagamine, Y. Mochizuki, A. Nishimura, Y. Nakase, and T. Nishizawa. An Elementary Processor Architecture with Simultaneous Instruction Issuing from Multiple Threads. In *Proc. 19th Annual International Symposium on Computer Architecture*, pages 136–145, May 1992.

15. R.A. Iannucci. Toward a Dataflow/von Neumann Hybrid Architecture. In *Proc. 15 International Symposium on Computer Architecture*, pages 131–140, May 1988.

16. Z. Li, J.-Y. Tsai, X. Wang, P.-C. Yew, and B. Zheng. Compiler Techniques for Concurrent Multithreading with Hardware Speculation Support. In *Proc. 9th Workshop on Languages and Compilers for Parallel Computing*, Aug. 1996.

17. A. Moshovos, S.E. Breach, T.N. Vijaykumar, and G.S. Sohi. Dynamic Speculation and Synchronization of Data Dependences. In *Proc. 24th International Symposium on Computer Architecture*, pages 181–193, Jun. 1997.

18. N. Nishi, T. Inoue, M. Nomura, S. Matsushita, S. Toru, A. Shibayama, J. Sakai, T. Oshawa, Y. Nakamura, S. Shimada, Y. Ito, M. Edahiro, M. Mizuno, K. Minami, O. Matsuo, H. Inoue, T. Manabe, T. Yamazaki, Y. Nakazawa, Y. Hirota, and Y. Yamada. A 1 GIPS 1 W Single-Chip Tightly-Coupled Four-Way Multiprocessor with Architecture Support for Multiple Control-Flow Execution. In *Proc. 47th International IEEE Solid-State Circuits Conference*, Feb. 2000.

19. G. Papadopoulos and D. Culler. Monsoon: An Explict Token-Store Architecture. In *Proc. 17th International Symposium on Computer Architecture*, pages 82–91, Jul. 1990.
20. A. Roth, A. Moshovos, and G.S. Sohi. Dependence Based Prefetching for Linked Data Structures. In *Proc. 8th Conference on Architectural Support for Programming Languages and Operating Systems*, pages 115–126, Oct. 1998.
21. A. Roth, A. Moshovos, and G.S. Sohi. Improving Virtual Function Call Target Prediction via Dependence-Based Pre-Computation. In *Proc. 1999 Internation Conference on Supercomputing*, pages 356–364, Jun. 1999.
22. A. Roth and G.S. Sohi. Register Integration: A Simple and Efficent Implementation of Squash Re-Use. In *Proc. 33rd Annual International Symposium on Microarchitecture*, Dec. 2000.
23. A. Roth and G.S. Sohi. Speculative Data-Driven Multithreading. Technical Report CS-TR-00-1414, University of Wisconsin, Madison, Mar. 2000.
24. A. Roth, C.B. Zilles, and G.S. Sohi. Speculative Miss/Execute Decoupling. In *Proc. Workshop on Memory Access Decoupling in Superscalar and Multithreaded Architectures*, Oct. 2000.
25. S. Sakai, Y. Yamaguchi, K. Hiraki, Y. Kodama, and T. Yuba. An Architecture of a Dataflow Single Chip Processor. In *Proc. 16th Annual International Symposium on Computer Architecture*, pages 46–53, May 1989.
26. M. Sato, Y. Kodama, S. Sakai, Y. Yamaguchi, and Y. Koumura. Thread-based Programming for the EM-4 Hybrid Dataflow Machine. In *Proc. 19th Annual International Symposium on Computer Architecture*, pages 146–155, May 1992.
27. G.S. Sohi, S. Breach, and T.N. Vijaykumar. Multiscalar Processors. In *Proc. 22nd International Symposium on Computer Architecture*, pages 414–425, Jun. 1995.
28. Y.H. Song and M. Dubois. Assisted Execution. Technical Report #CENG 98-25, Department of EE-Systems, University of Southern California, Oct. 1998.
29. J.G. Steffan and T.C. Mowry. The Potential for Using Thread Level Data-Speculation to Facilitate Automatic Parallelization. In *Proc. 4th International Symposium on High Performance Computer Architecture*, Feb. 1998.
30. M. Takesue. A Unified Resource Management and Execution Control Mechanism for Data Flow Machines. In *Proc. 14th Annual International Symposium on Computer Architecture*, pages 90–97, Jun. 1987.
31. M. Tremblay. MAJC: An Architecture for the New Millenium. In *Proc. Hot Chips 11*, pages 275–288, Aug. 1999. http://www.sun.com/microelectronics/MAJC/documentation/docs/HC99sm.pdf.
32. D.M. Tullsen, S.J. Eggers, J.S. Emer, H.M. Levy, J.L. Lo, and R.L. Stamm. Exploiting Choice: Instruction Fetch and Issue on an Implementable Simultaneous Multithreading Processor. In *Proc. 23rd International Symposium on Computer Architecture*, pages 191–202, May 1996.
33. W. Yamamoto and M. Nemirovsky. Increasing Superscalar Performance Through Multistreaming. In *Proc. 1995 Conference on Parallel Architectures and Compilation Techniques*, Jun. 1995.
34. Y. Zhang, L. Rauchwerger, and J. Torrellas. Hardware for Speculative Run-Time Parallelization in Distributed Shared-Memory Multiprocessors. In *Proc. 4th International Symposium on High-Performance Computer Architecture*, Feb. 1998.
35. C.B. Zilles and G.S. Sohi. Understanding the Backward Slices of Performance Degrading Instructions. In *Proc. 27th International Symposium on Computer Architecture*, pages 172–181, Jun. 2000.

Session III-A

Cluster Computing and Its Applications
Chair: Hee Yong Youn
Information and Communications University, Korea

A Fast Tree-Based Barrier Synchronization on Switch-Based Irregular Networks

Sangman Moh[†], Chansu Yu[†], Hee Yong Youn[‡], Dongsoo Han[†],
Ben Lee[†§], and Dongman Lee[†]

[†]School of Engineering
Information and Communications University
58-4 Hwa-am, Yu-sung, Taejon, 305-348 KOREA
{smmoh,cyu,dshan,dlee}@icu.ac.kr
[‡]Department of Electrical and Computer Engineering
SungKyunKwan University, Suwon, KOREA
youn@ece.skku.ac.kr
[§]Department of Electrical and Computer Engineering
Oregon State University, Corvallis, OR 97331
benl@ece.orst.edu

Abstract. In this paper, we propose a *Barrier Tree for Irregular Networks* (*BTIN*) and a barrier synchronization scheme using BTIN for switch-based cluster systems. The synchronization latency of the proposed BTIN scheme is asymptotically $O(\log n)$ while that of the fastest scheme reported in the literature is bounded by $O(n)$, where n is the number of member nodes. Extensive simulation study shows that, for the group size of 256, the BTIN scheme improves the synchronization latency by a factor of $3.3 \sim 3.8$, and is more scalable than conventional schemes with less network traffic.

Index terms: Cluster systems, irregular networks, barrier synchronization, communication latency.

1   Introduction

Switch-based cluster systems have been widely accepted as cost-effective alternatives for high performance computers. Since computational nodes[1] or switches may be added to or detached from the network dynamically, it is generally assumed that the switches form an irregular topology [1,2]. Distributed reconfiguration algorithms identify the network topology before computation begins [3]. Here the irregular topology makes routing difficult to avoid *deadlock* among multiple packets traveling simultaneously. For example, *up/down routing* algorithm [1] prevents deadlock by restricting the sequences of *turns* in the routing paths. Collective operations need more attention than the operations with point-to-point communication since they often determine the execution time of the sequential part of a parallel program, which usually constitutes the bottleneck.

---

[1] Nodes, in this paper, actually mean PCs or workstations in a cluster system.

M. Valero, V.K. Prasanna, and S. Vajapeyam (Eds.): HiPC 2000, LNCS 1970, pp. 273–282, 2000.
© Springer-Verlag Berlin Heidelberg 2000

Such collective operations, however, become more complicated for switch-based cluster systems [4,5] due to the irregularity mentioned above.

A *barrier* is a synchronization point in a parallel program at which all processes participating in the synchronization must arrive before any of them can proceed beyond the synchronization point. In general, barrier synchronization is split into two phases – *reduction* and *distribution*. During the reduction phase, each participating process notifies the root process of its arrival at the barrier point. Upon the notification from all member processes, the distribution phase begins and the root process notifies them that they can proceed further. While *software barriers* inherently suffer from large communication latency, *hardware-supported barriers* are usually an order of magnitude faster than software barriers [6]. For switch-based irregular networks, even though hardware-supported multicast has been extensively studied [7,8,9], there have been, to the authors' knowledge, no works devoted to hardware-supported barrier synchronization.

In this paper, we propose a *Barrier Tree for Irregular Networks* (*BTIN*) for the switch-based cluster systems of irregular topology, which significantly reduces the synchronization latency and network traffic with no deadlock. It is a tree-based combining scheme which constructs a barrier tree and embeds it into the corresponding switches by putting special registers into the switches. The synchronization latency of the BTIN scheme is $O(\log n)$ while that of the fastest scheme reported in the literature is bounded by $O(n)$, where n is the number of member nodes. Extensive simulation study shows that, for the group size of 256, the BTIN scheme improves the synchronization latency by a factor of 3.3 ~ 3.8, and is more scalable than conventional schemes with less network traffic.

The rest of the paper is organized as follows. Switch-based cluster systems of irregular topology, the construction of a BTIN, and the corresponding switch operations are presented in the following section. Section 3 is devoted to analyzing the characteristics of the BTIN scheme including tree height and deadlock issue. The performance of the proposed scheme is evaluated and discussed in Section 4, and conclusions and future works are covered in Section 5.

2   Tree-Based Barrier Synchronization

We first introduce switch-based cluster systems of irregular topology. Then, the proposed BTIN and the corresponding operations are presented.

2.1   Switch-Based Cluster Systems

An example of switch-based cluster systems with irregular topology is drawn in Fig. 1. Each switch has a set of ports and each port is connected to a computational node or other switch. Some ports may be left open and can be used for further system expansion. In this example, the problem is to synchronize the 14 member nodes in a process group (the dark circles in Fig. 1) at a barrier point, which are selected for running a parallel application. Without loss of generality, in this paper, we consider *wormhole routing* [10] with input and output buffers of one flit wide each.

Hardware support for barrier synchronization is the barrier registers within switches. There will be a register assigned for each barrier, and similar concept has been assumed in the switch-based multicast approaches [11], where a processor can access a register within a switch. We assume that the barrier registers can hold an entire synchronization message. This can be justified by the fact that a synchronization message is very short and fixed in length since it needs not carry multiple destination addresses, and the size of synchronization data is quite small.

Collective operation primitives including barrier synchronization are included in most message-passing libraries. Among them, we target our discussion to *MPI* (*Message Passing Interface*) standard [12]. However, the algorithms presented here can be applied to other message-passing systems with little modifications.

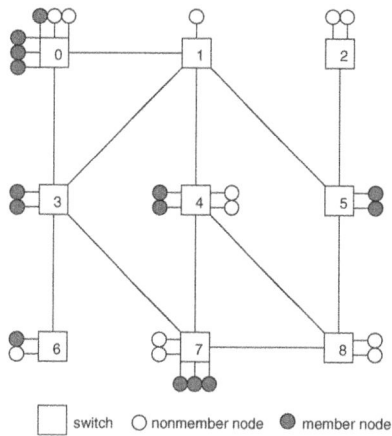

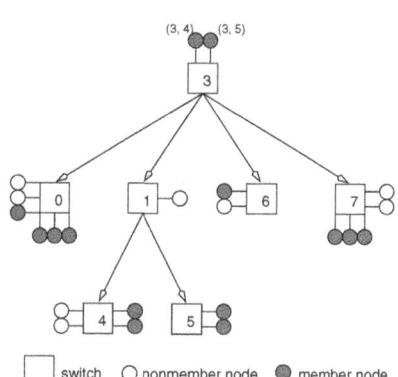

Fig. 1. A switch-based cluster system with irregular topology.

Fig. 2. A BTIN constructed from the cluster system in Fig. 1

2.2   Barrier Tree for Irregular Networks (BTIN)

In this paper, we define a *member switch* as a switch with at least one member node and a *nonmember switch* as one without any member node. Also note that a *representative member node* is defined as the member node attached to a member switch via the lowest numbered port. Given a root switch, a *breadth-first spanning (BFS)* tree is constructed by a distributed algorithm around the root switch. At the group creation time, the representative member node starts the algorithm to figure out the BFS topology, and then the nodes running the algorithm eventually agree on a unique BFS tree. Once a BFS tree is found, the algorithm checks whether there is any nonmember leaf switch in the tree. If such a leaf switch exists, it is removed from the BFS tree. Then, each representative member node sets up a barrier register in the corresponding switch properly to embed the resulting BTIN into the network.

Fig. 2 shows a BTIN at the distribution phase, which is constructed from the cluster system in Fig. 1 and contains the same 14 member nodes. The root switch is the switch labeled 3, and the root node is the node $(3, 4)$. Note here that the node notation $(3, 4)$ represents the node attached to port 4 of the switch 3. In Fig. 2, the distribution message follows the arrows in accordance with the BTIN routing algorithm. Note that the nonmember switches are not included in the tree unless they are intermediate switches in the tree.

Root Switch and Root Node

The root switch of a BTIN must be a member switch and is chosen so that the resulting BTIN has a minimum tree height among all possibilities. If two or more BFS trees have the same minimum height, one with the minimum number of edges is chosen. However, if two or more BFS trees have the same number of edges, one with the minimum number of leaves is selected. Finding the root switch is performed by every representative member node.

In the example of Fig. 2, the root switch is the switch labeled 3, and the node $(3, 4)$ is selected as the root node since it is connected via the lowest numbered port (port identifier of 4) between the two member nodes. Note here that, unlike the up/down routing tree, edges between siblings are not permitted in BTIN.

Switch Setup

Each barrier register contains a group identifier (GID), a parent port number (P), parent and children bits, arrival bits for children, and synchronization data as shown in Fig. 3. Unlike other fields, arrival bits and message fields are used when the synchronization message is processed rather than at the initial setup time. For example, A_0 identifies that a reduction message has arrived from the child switch connected to C_0 during the reduction phase.

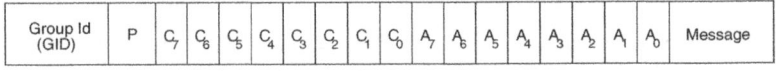

P: Port number for parent C_i : Parent or children for port i A_i : Arrived from C_i

Fig. 3. Structure of a barrier register.

Below we describe the distributed algorithm to setup the barrier register, which every representative member node runs at the BTIN construction time. A special operation is required to setup a barrier register in intermediate nonmember switches involved in BTIN as described in step 6.

Setup_Register$(S, M, s_r, m_r, s_l, m_l, \text{GID})$

1. Let $S = \{s_0, s_1, \cdots, s_{q-1}\}$ be all the switches, $M = \{m_0, m_1, \cdots, m_{n-1}\}$ be the addresses of member nodes, s_r be the root switch, m_r be the root node, s_l be the local switch, m_l be the local node, and GID be the group identifier.
2. If the local node m_l is not the representative member node of the local switch s_l, return.

3. Around the root switch s_r, establish the corresponding BFS tree, scanning switches in order of switch identifier (address).

4. Remove nonmember leaf switches from the found BFS tree until there is no such a nonmember leaf switch, making the tree become a BTIN.

5. Write the GID, the parent port number (P), and the parent and children bits into a barrier register in the local switch s_l.

6. If there exists any nonmember descendant switch which can be reached without passing through intermediate member switches, request a node of the nonmember descendant switch to setup a barrier register in the descendant switch by transmitting a point-to-point message. Then, the destinated node will write the GID, the parent port number (P), and the parent and children bits into a barrier register in the corresponding switch. Repeat this step 6 until there is no such a nonmember descendant switch.

2.3   Barrier Synchronization Using BTIN

Fig. 4 shows the format of synchronization message which contains message type, group identifier, and small synchronization data. For the example shown in Fig. 1, a synchronization message may comprise at most two bytes, *i.e.*, 2-bit message type, 8-bit group identifier for at most 256 different groups, and at most 6-bit synchronization data if any.

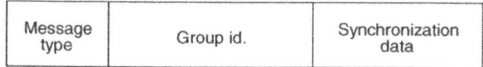

Fig. 4. Format of the barrier synchronization message.

The BTIN routing or *collective routing* performs message merging and replication at the reduction and distribution phase, respectively. Collective merging and replication are carried out at the switches. During the reduction phase, the reduction messages traverse in the up direction upward the root switch, being combined collectively at each branch switch. During the distribution phase, the distribution messages traverse in the down direction downward all the leaf switches, being replicated at each branch switch.

3   Characteristics of BTIN

Characteristics of BTIN are analyzed in this section. Then, we compare them with those of conventional approaches.

3.1   Tree Height of a BTIN

We define *connectivity*, or *connection ratio*, f of k-port switches as the ratio of the average number of connected ports over k [7]. Hence, fk is the average number of ports in a switch, which are connected to either other switches or computational nodes.

Synchronization latency is linearly proportional to the height of BTIN. We, thus, analyze the average tree height of a BTIN which is established on a randomly built irregular network. Without loss of generality, we assume that all the possibilities of network configuration with k-port switches of connectivity f are equally likely. The following Theorem 1 formally analyzes the average height of a BTIN under the assumption. (See [13] for a complete proof.)

Theorem 1: *For an irregular network with k-port switches of connectivity f, the average height, h, of a BTIN is asymptotically given by $h = \log_{(fk-p/q-1)} n$, where n is the number of member nodes, p is the number of nodes, and q is the number of switches.*

According to Theorem 1, the average height of a BTIN is $O(\log_{(fk-p/q-1)} n)$. It is simply rewritten by $O(\log n)$ because $fk - \frac{p}{q} - 1$ becomes a constant. Hence, the associated synchronization latency has a time complexity of $O(\log n)$.

3.2   Deadlock Freedom

In irregular networks as well as regular ones, a major issue with the wormhole routing is deadlock [14]. When the path of a message is blocked, the message head as well as the rest of the message are stopped where they are, holding the buffers and channels along the path. Deadlock could occur if these stoppages create a cyclic dependency. However, if the message size is small enough, deadlock could be easily avoided by holding the entire message in the switch. In our barrier synchronization scheme, the synchronization messages need not carry all the destination addresses, and thus the lengths are identical and very small.

The basic technique for proving that a network is deadlock-free is to articulate the dependences that can arise between channels as a result of message movement, and to demonstrate that there exists no cycle in the resulting channel dependence graph. This implies that no traffic patterns can lead to deadlock, where the traffic patterns include those incurred by three cases; a barrier synchronization, multiple concurrent synchronizations, and a mixture of synchronization messages and normal messages. See [13] for a complete discussion on how messages incurred by the above three cases do not create a deadlock situation.

3.3   Comparison of Characteristics

In this subsection, we compare the characteristics of BTIN with those of two conventional approaches, the method using point-to-point messages and the method using switch-based multicast at the distribution phase. For simplicity, in this paper, we call the two approaches the unicast scheme and the multicast scheme, respectively. The comparisons are summarized in Table 1. As can be seen from the table, the proposed BTIN possesses preferable characteristics for all the factors studied, which results in significantly better performance.

Table 1. Characteristics of barrier synchronization schemes on irregular networks.

	Unicast scheme	Multicast scheme	BTIN scheme
Initialization at group creation time			
Routing path/tree construction	Centralized at the root	Centralized at the root	Distributed at all members
Router setup	None	At member nodes (for distribution)	At member nodes
During a barrier operation			
Hardware support	None	Distribution phase	Reduction and distribution phase
Synchronization message size	Short (single destination address)	Long (multiple destination addresses)	Short (no destination address)
Number of startups	$2n$ (n for reduction and n for distribution)	$n+1$ (n for reduction and one for distribution)	2 (one for reduction and one for distribution)
Complexity of routing latency	$O(n)$	$O(n)$	$O(\log n)$
Primary weakness			
Primary weakness	Repetitive $2n$ unicast transfers (very slow)	Repetitive n unicast transfers and hardware for multicast	Hardware complexity at the router
	Hardware complexity at the router		

The unicast scheme requires repetitive n point-to-point message transfers for each of reduction and distribution phase, and thus it has the complexity of $O(2n)$ for routing latency, which is simply rewritten by $O(n)$. We assume that, in the multicast scheme, switch-based multicasting tree is used for the distribution phase in hardware level. In the multicast scheme, it is simple to see that the complexity of routing latency of the distribution phase is $O(\log n)$. However, for the reduction phase, repetitive n unicast message transfers are required, resulting in the complexity of $O(n)$. Hence, the complexity of the multicast scheme is $O(n + \log n)$, which is simply rewritten by $O(n)$.

4 Performance Evaluation

For different system configurations, the performance of the BTIN scheme is evaluated and compared to that of the multicast scheme, which is the most recent and efficient approach, using simulation.

4.1 Simulation Environment

We evaluate the performance of the proposed tree-based barrier synchronization scheme on two different system configurations; (i) 256 nodes and 75 switches and

(ii) 1024 nodes and 300 switches. We assume that the network is interconnected with 8-port switches having 75% connectivity. The member nodes are picked randomly and all the members are assumed to arrive at the barrier at the same time. Channel contention is not considered.

The *synchronization latency* is the most important performance metric of barrier synchronization, which is the interval from the time when the barrier synchronization is invoked until the time when all the member nodes finish the distribution phase. As another performance measure in our simulation, the *network traffic* incurred by the barrier synchronization is also investigated. This is measured by the number of links (hops) traversed by the synchronization messages during a barrier operation.

The default performance parameters have been assumed on the basis of overhead-minimized communication on advanced switches. We assume the following default performance parameters: communication startup time (t_s) of $2 \sim 10$ μsec, link propagation delay (t_p) of $20 \sim 40$ $nsec$, and switch (router) delay (t_r) of $300 \sim 500$ $nsec$. The startup time includes the software overheads for allocating buffers, copying messages, and initializing the router and DMA [15]. The router delay includes several steps of complicated operations and varies for various routing algorithms as Chien [16] analyzed. We also assume that the network interface delay is almost the same as the switch delay for our evaluation. Since synchronization messages do not need any data flits, the communication latency of a message transfer can be approximated to $t_s + d \cdot t_p + (d+1) \cdot t_r$, where d is the *distance* between the source and destination node in a communication.

4.2   Simulation Results and Discussion

Synchronization Latency

Fig. 5 shows the synchronization latency, where t_s, t_p, and t_r are assumed to be 2.0 μsec, 20.0 $nsec$, and 300.0 $nsec$, respectively. Here for each parameter set, 100 simulation runs are executed, and the results are averaged. In most of the cases, very small variance is observed. Both the number of nodes and the number of switches are shown in the parenthesis of labels in Fig. 5, where the connectivity f is 75%.

The synchronization latency of the BTIN scheme is significantly lower than that of the multicast scheme. Observe from the figure that the synchronization latency of the BTIN scheme is almost independent on the group size except for very small groups. This is mainly due to the fact that the tree height of BTIN is bounded by $\log_{(fk-p/q-1)} q$ on an irregular network with k-port switches of connectivity f, where p is the number of nodes and q is the number of switches. In a system of 1024 nodes and 300 switches, the group size can be more than 256. Even though it is not shown here, we increased the group size up to 1024. Then the synchronization latency converges to 11.5 μsec and 120.5 μsec for BTIN and multicast scheme, respectively. As shown in the figure, for the group size of 256, the BTIN scheme is faster than the multicast scheme by factors of 3.8 and 3.3 for the system of 256 nodes and 75 switches and the system of 1024 nodes and

300 switches, respectively. From the figure, it is obvious that the BTIN scheme is more scalable than the multicast scheme.

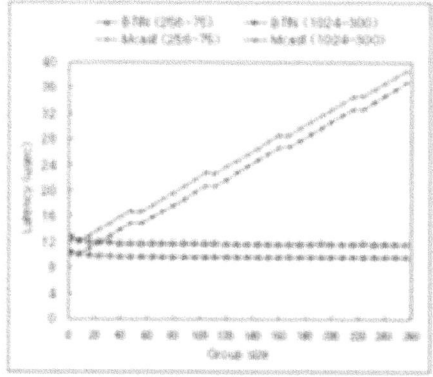

Fig. 5. Synchronization latency.

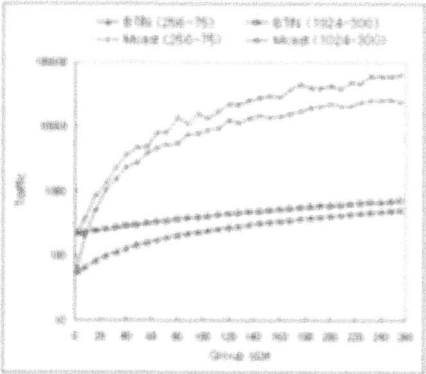

Fig. 6. Network traffic.

Network Traffic

As shown in Fig. 6, the network traffic of the BTIN scheme is significantly lighter than that of the multicast scheme. The performance of network traffic is more improved as the network size increases. For instance, for the group size of 256, the network traffic of the BTIN scheme is significantly lighter than the multicast scheme by factors of 46.8 and 88.6 for the system of 256 nodes and 75 switches and the system of 1024 nodes and 300 switches, respectively. As the group size is increased, the network traffic is also increased for both of the schemes because more nodes and switches of the network are involved in a barrier synchronization. The proposed BTIN scheme is clearly more scalable than the multicast scheme even in terms of network traffic. See [13] for more simulation results and discussion.

5   Conclusions

In this paper, we proposed a fast tree-based barrier synchronization scheme for switch-based irregular networks, which is, to the authors' knowledge, the first approach to hardware support for barrier synchronization on irregular networks. The BTIN tree is at most $(k - 1)$-ary, and the complexity of synchronization latency is $O(\log n)$ while that of the fastest scheme, which is the method using switch-based multicast at the distribution phase, is bounded by $O(n)$, where n is the number of member nodes. The proposed BTIN scheme has been analyzed and compared with the conventional schemes. Extensive simulation study shows that, for the group size of 256, the BTIN scheme improves the synchronization latency by a factor of $3.3 \sim 3.8$, and is more scalable than conventional schemes with less network traffic.

We are currently investigating the application of the BTIN scheme to other collective communications such as multicast or total exchange. It is also an interesting subject to consider the BTIN scheme for dynamic environment caused by load balancing and node/link failures.

References

1. M. D. Schroeder, *et.al.*, "Autonet: a High-speed, Self-configuring Local Area Network Using Point-to-point Links," *SRC Research Report*, No.59, Digital Equipment Corporation, April 1990.
2. A. M. Mainwaring, B. N. Chun, S. Schleimer, and D. S. Wilkerson, "System Area Network Mapping," *Proceedings of the Annual Symposium on Parallel Algorithms and Architectures*, 1997.
3. N. Boden, *et.al.*, "Myrinet: A Gigabit-per-Second Local Area Network," *IEEE Micro*, Vol. 15 No. 1, pp. 29-36, February 1995.
4. R. Buyya, *High Performance Cluster Computing: Architectures and Systems*, Prentice-Hall Inc., NJ, 1999.
5. G. F. Pfister, *In Search of Clusters*, 2nd Edition, Chapter 5, Prentice-Hall, Inc., NJ, 1998.
6. V. Ramakrishnan, I. D. Scherson, and R. Subramanian, "Efficient Techniques for Nested and Disjoint Barrier Synchronization," *Journal of Parallel and Distributed Computing*, Vol. 58, pp. 333-356, Aug, 1999.
7. R. Kesavan, K. Bondalapati, and D. K. Panda, "Multicast on Irregular Switch-Based Networks with Wormhole Routing," *Proceedings of the 3rd International Symposium on High-Performance Computer Architecture*, pp. 48-57, Feb. 1-5, 1997.
8. R. Libeskind-Hadas, D. Mazzoni, and R. Rajagopalan, "Tree-Based Multicasting in Wormhole-Routed Irregular Topologies," *Proceedings of the International Parallel Processing Symposium*, pp. 244-249, Mar. 30 - Apr. 3, 1998.
9. M. Gerla, P. Palnati, and S. Walton, "Multicasting Protocols for High-Speed, Wormhole-Routing Local Area Networks," *Proceedings of the International Conference on Applications, Techniques, Architectures, and Protocols for Computer Communication*, pp. 184-193, Aug. 28-30, 1996.
10. L. Ni and P. K. McKinley, "A Survey of Wormhole Routing Techniques in Direct Networks," *IEEE Computer*, Vol. 23, No. 2, pp. 62-76, Feb. 1993.
11. R. Sivaram, R. Kesavan, D. K. Panda, and C. B. Stunkel, "Where to Provide Support for Efficient Multicasting in Irregular Networks: Network Interface or Switch?," *Proceedings of the International Conference on Parallel Processing*, pp. 452-459, Aug. 10-14, 1998.
12. Message Passing Interface Forum, *MPI: A Message-Passing Interface Standard*, Version 1.1, June 12, 1995.
13. S. Moh, C. Yu, H. Y. Youn, D. Han, B. Lee, and D. Lee, "A Fast Tree-Based Barrier Synchronization on Switch-Based Irregular Networks," *Technical Report*, School of Engineering, Information and Communications University, July 2000.
14. S. Warnakulasuriya and T. M. Pinkston, "Characterization of Deadlocks in Irregular Networks," *Proceedings of the International Conference on Parallel Processing*, pp. 75-84, Sep. 21-24, 1999.
15. P. Pacheco, *Parallel Programming with MPI*, Morgan Kaufmann, San Francisco, CA, 1997.
16. A. A. Chien, "A Cost and Speed Model for k-ary n-Cube Wormhole Routers," *IEEE Transactions on Parallel and Distributed Systems*, Vol. 9, No. 2, pp. 150-162, Feb. 1998.

Experiments with the CHIME Parallel Processing System

Anjaneya R. Chagam[1], Partha Dasgupta[2], Rajkumar Khandelwal[1],
Shashi P. Reddy[1] and Shantanu Sardesai[3]

[1] Intel Corporation, Chandler, AZ, USA and Penang, Malaysia
[2] Arizona State University, Tempe, AZ, USA.   http://cactus.eas.asu.edu/partha
[3] Microsoft Corporation Redmond, WA, USA

Abstract: This paper presents the results from running five experiments with the Chime Parallel Processing System. The Chime System is an implementation of the CC++ programming language (parallel part) on a network of computers. Chime offers ease of programming, shared memory, fault tolerance, load balancing and the ability to nest parallel computations. The system has performance comparable with most parallel processing environments. The experiments include a performance experiment (to measure Chime overhead), a load balancing experiment (to show even balancing of work between slow and fast machines), a fault tolerance experiment (to show the effects of multiple machine failures), a recursion experiment (to show how programs can use nesting and recursion) and a fine-grain experiment (to show the viability of executions with fine grain computations.

1.   Introduction

This paper describes a series of experiments to test the implementation, features and performance of a parallel processing system called Chime. The experiments include runs of various scientific applications. Chime is a system developed at Arizona State University [1, 2] for running parallel processing applications on a Network Of Workstations (the NOW approach).

Chime is a full implementation of the parallel part of Compositional C++ (or CC++) [3], running on Windows NT. CC++ is a language developed at Caltech and is essentially two languages in one. It has two distinct subparts – a distributed programming language designed for NOW environments and a parallel programming language designed for shared memory multiprocessor environments. While shared memory multiprocessors are very good platforms for parallel processing, they are significantly costlier than systems composed of multiple separate computers. Parallel CC++ is an exceptionally good language, but it was not designed to run on NOWs. Chime solves this problem, by implementing parallel CC++ on the NOW architecture.

Chime uses a set of innovative techniques called "two-phase idempotent execution strategy" [4], "distributed cactus stacks" [1], "eager scheduling"[4], "dependency preserving execution" [2] and the well-known technique called "distributed shared

M. Valero, V.K. Prasanna, and S. Vajapeyam (Eds.): HiPC 2000, LNCS 1970, pp. 283–292, 2000
© Springer-Verlag Berlin Heidelberg 2000

memory" [5] to implement parallel C++. In addition, it has the extra features of load balancing, fault tolerance and high performance. Chime is the first (and of this writing, the only) parallel processing system that provides the above features, coupled with nested parallelism, recursive parallelism and synchronization (these are features of parallel CC++). This paper described the implementation of Chime, in brief and presents details on the experiments with Chime.

2.   Related Work

Shared memory parallel processing in distributed systems is limited to a handful of Distributed Shared Memory (DSM) systems that provide quite similar functions (Munin [6], Midway [7], Quarks [8], TreadMarks [9]). DSM systems care categorized by the type of memory consistency they provide. DSM systems do not provide a uniform view of memory i.e. some global memory is shared and some are not. In addition, the parallel tasks execute in an isolated context; i.e. they do not have access to variables defined in the parent's context. In addition, a parallel task cannot call a function that has an embedded parallel step (nesting of parallelism is not allowed).

The Calypso system [4, 17, 18] adds fault tolerance and load balancing to the DSM concept, but suffers from the lack of nesting and synchronization (except barrier synchronization). Chime is an extension to Calypso and absolves these shortcomings.

A plethora of programming systems for NOW based systems exist, that uses the message-passing technique. Two well-known systems are PVM [10] and MPI [11]. Fault tolerance and load balancing has been addressed, in the context of parallel processing by many researchers, a few examples are Persistent Linda [12], MPVM [13], Dynamic PVM [14] Piranha [15] and Dome [16]. The techniques used in most of these systems are quite different from ours and often add significant overhead for the facilities such as fault tolerance.

Most working implementations are built for the Unix platform (including Linux). Some have Windows NT implementations, but they are buggy at best. In our experience, we have not been able to make any non-trivial applications work correctly with these systems on the Windows platform. For this reason, we are unable to provide comparative performance tests.

3.   Chime Features

Shared memory multiprocessors are the best platform for writing parallel programs, from a programmer's point of view. These platforms support a variety of parallel processing languages (including CC++) which provide programmer-friendly constructs for expressing shared data, parallelism, synchronization and so on. However the cost and lack of scalability and upgradability of shared memory multiprocessor machines make them a less than perfect platform. Distributed Shared Memory (DSM) has been promoted as the solution that makes a network of computers look like a shared memory machine. This approach is supposedly more natural than the message passing method used in PVM and MPI. However, most programmers find this is not the case. The shared memory in DSM systems does not have the same access and sharing semantics as shared memory in shared memory multi-processors. For example, only a designated part of the process address space is shared, linguistic notions of global and local variables do not work intuitively, parallel functions cannot be nested and so on.

As stated before, Chime provides a multiprocessor-like shared memory programming model on network of workstations, along with automatic fault-tolerance and load balancing. Some of the salient features of the Chime system are:

1. Complete implementation of the shared memory part of the CC++ language. Hence programming with Chime is easy, elegant and highly readable.
2. Support for nested parallelism (i.e. nested barriers including recursion) and synchronization. For example, a parallel task can spawn more parallel tasks and tasks can synchronize amongst each other.
3. Consistent memory model, i.e. the global memory is shared and all descendants share the local memory of a parent task (the descendants execute in parallel).
4. Machines may join the computation at any point in time (speeding up the computation) or leave or crash at any point (slowdowns will occur).
5. Faster machines do more work than slower machines, and the load of the machines can be varied dynamically (load balancing).

In fact, there is very little overhead associated with these features, over the cost of providing DSM. This is a documented feature (see section 6.1) that Chime shares with its predecessor Calypso [4]. Chime runs on Windows NT and the released version can be downloaded from http://milan.eas.asu.edu.

3.1   Chime Programming Example

Chime provides a programming interface that is identical to the parallel part of Compositional C++ (or CC++) language Consider the following parallel CC++ program:

```
#include <iostream.h>
#include "chime.h"
#define N 1024

int GlobalArray[N];
void AssignArray(int from, to){
        if (from != to)
            par {
                    AssignArray(from, (from+to)/2);
                    AssignArray((from+to)/2 + 1, to);
            }
        else GlobalArray[from] = 0;
}
int main(int argc, char *argv[]) {
        AssignArray (0, N-1)
}
```

The above program defines a global (shared) array called GlobalArray, containing 1024 integers. Then it assigns the global array using a recursive parallel function called AssignArray. The AssignArray function uses a "par" statement. The par statement executes the list of statements within its scope in parallel, thus calling two instances of AssignArray in parallel. Each instance calls two more instances, and this recursion stops when 1024 leaf instances are running.

4.   Chime Technologies

The implementation of Chime borrows some techniques used in an earlier system called Calypso and adds a number of newer mechanisms. The primary mechanisms used in Chime are:

Eager Scheduling: In Chime, the number of parallel threads running a parallel application can change dynamically and is larger than the number of processors used. Each processor runs a "worker" process, and one designated processor runs the manager. A worker contacts the manager and picks up one thread and when it finishes, it requests the next thread. Threads are assigned from the pool of uncompleted jobs. This technology provides load balancing (faster workers do more work) and fault tolerance (failed workers do not tie up the system) using the same technique.

Two Phase Idempotent Execution Strategy (TIES): Since there is the possibility of multiple workers running the same thread, the execution of each thread must be idempotent. The idempotence is achieved by coupling eager scheduling with an atomic memory update facility implemented by Calypso DSM (see below).

Calypso DSM: This is a variant of the well-known RC-DSM (Release Consistent Distributed Shared Memory) technique. RC-DSM is modified so that the return of pages are postponed to the end of the thread, and the manager buffers all the returned pages and updates them in an atomic fashion and then marks the thread as completed. This ensures correct execution even when threads fail at arbitrary points [17].

Dependency Preserving Execution: Threads can create threads; threads can synchronize with other threads. This can cause unmanageable problems when multiple workers are executing the same thread or when threads fail after creating new threads (or fail after reaching a synch point). Dependency Preserving Execution solves this problem. Each time a thread created nested threads, it informs the manger of the new threads and the old thread mutates into a new thread itself. Similarly at synchronization points, the manger is informed, and a mutation step is performed. The complete description of this mechanism is beyond the scope of this paper and is described in [2].

Distributed Cactus Stack: A data structure that replicates the application stack amongst all machines to ensure correct nesting or parallel threads and scooping of variable local to functions [1].

5.   Experiments with Chime

We now describe a set of five experiments using various scientific applications to determine its performance and behaviors on a range of features. The five experiments shown below are the performance experiment, the load balancing experiment, the fault tolerance experiment, the recursion experiment and the fine grain execution experiment. Experiments were conducted at different points in time, at different locations by different people, hence all the equipment used are not the same (except that Intel machines with *Windows NT 4.0*, connected by a *100Mbps Ethernet* was used for all experiments). The systems used are stated along with the discussion of each experiment.

5.1   Performance Experiment

The performance experiments used several matrix multiply and ray-tracing programs, and both yielded similar results. We show the results of a ray-tracing program below using Pentium Pro 200 processors. The first step is to write the program in sequential C++ and measure its execution time. Then a parallel version is written in CC++ and run with Chime on a variable number of processors (from 1 to 5) and the speedups are calculated in respect to the sequential program. The results are shown in Figure 1.

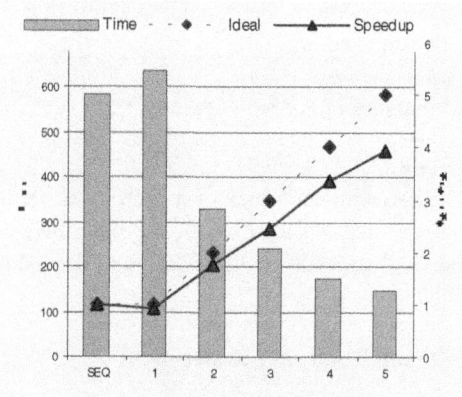

Fig. 1. Performance Experiment

Note that the single processor execution under chime is about 9% poorer than the sequential program and the execution speed scales with addition of processors – the degradation in performance is at most 21%. This makes Chime competitive with most parallel processing systems for NOWs even though Chime has significantly better features. This experiment shows that the overhead of Chime is quite small, in spite of its rich set of features. We have been unable to run (in spite of extensive attempts) any complicated programs with Windows NT implementations of systems such as PVM and MPI and hence cannot provide comparative performance numbers.

5.2   Load Balancing Experiment

The load balancing experiment involves the same ray-tracing program as above, but using machines of different speeds to run the parallel application. In many parallel-processing systems, the slowest machine(s) dictate performance; that is, fast machines are held up for the slow machines to finish the work allocated to them. Chime, does it differently. The application was executed on 4 slow machines (P-133, 64MB) and then a fast machine (P-200, 64MB) replaced one slow machine. This caused an increase in speed. Replacing more slow machines with fast machines kept the trend.

We calculate an "ideal" speedup of the program, as follows. Suppose a computation runs for T seconds on n machines, M_1, M_2, ..., M_n. Machine M_i has a performance index of p_i and is available for t_i seconds. (Performance index is the relative speed of a machine normalized to some reference.) Then the maximum theoretical speedup that may be achieved is:

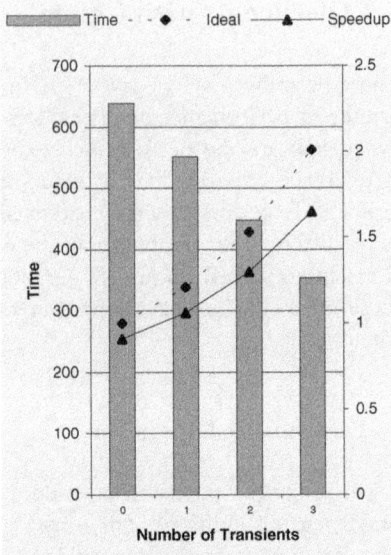

Fig. 2. Load Balancing Experiment

  Ideal speedup
  = number of equivalent machines

$$= \sum_{i=1:n}(p_i * t_i) / T$$

The performance index of a P-133 was set to 1 and the P-200 was measured to be 1.96. Note that the load balancing experiment shows that the actual speedup is close to the ideal speedup (within 22%) and the load is balanced well among slow and fast machines.

5.3   Fault Tolerance Experiment

The "fault tolerance" experiment, using the ray tracing program, shows the ability of Chime to dynamically handle failures as well as allowing new machines to join the computation. For this test up to four P-200 machines were used. Of these machines, one was a stable machine, and the rest were transient machines. The transient machines worked as follows:

  Transient machine: After 120 seconds into the computation, the transient machine joins the computation and then, after another 120 seconds, fails (without warning, i.e. crashes).

Figure 3 shows the effect of transient machines. The actual speedups and ideal speedups were computed according to the formula described earlier. Note that the ideal speedup measure takes into account the full power of the transient machines during the time they are up whether they are useful to the computation or not. The experimental results show that the transient machines do contribute to the computation. Note that the transient machines end their participation by crashing. Hence whatever they were running at that point is lost. Such crashes do not affect the correct execution of the program under Chime.

In fact this experiment shows the real power of Chime. The system handles load balancing and fault tolerance, with no additional overhead. The real speedups are close to the ideal speedups. In cases where machines come and go, the failure tolerance features of Chime actually provide more performance than an equivalent, non-fault-tolerant system.

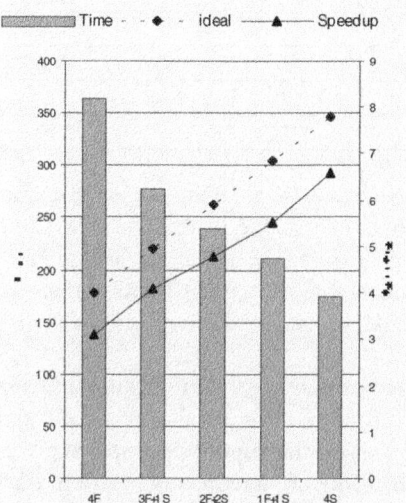

Fig. 3. Fault Tolerance Experiment

5.4   Nesting and Recursion Experiment

The nesting and recursion experiment is a test of Chime's ability to handle nested parallelism, especially in the case the program is recursive. We use a variant of the Fast Fourier Transform (FFT) algorithm for this experiment. This variant is called the Iterative FFT and has a significant computational complexity and hence scope for parallization. To calculate the Fourier Transform of an N-vector, we first compute the Fourier transform of two halves of the vector and then combine the results. The exact details of the FFT algorithm and its complexity analysis are omitted due to the space constraints.

This recursive program is written by writing a subroutine called ComputeFFT( ). This subroutine accepts the input vector size and the vector and then splits it into two parts and calls itself recursively and does it twice. Each invocation of the recursive call runs in parallel. Thus the program starts as one thread and then splits into two and then splits to 4 and so on, till the leaf nodes compute the FFT of 2 elements. For a data size of 32K (2^{15}) elements the recursion tree is 15 deep, the number of logical threads generated, is about 32,000.

The following pseudo code illustrates the core of the parallel algorithm.

```
ComputeFFT ( n, vector [size 2n ]){
    if (n==1) compute the FFT
    else  {
            Divide vector into odd and even halves;
            par {   // ** parallel step  **
                ComputeFFT(n-1, odd-half-of-vector);
                ComputeFFT(n-1, even-half-of-vector);
            }
        }
    assimilate results;
    return results to caller;
}
```

Input Data Size	Execution Time (seconds)		Percent speedup on three nodes
	1 node	3 nodes	
2^{15}	17	6	183 %
2^{16}	30	8	275 %
2^{17}	57	13	338 %
2^{18}	108	20	440 %

Note that the resulting parallel program is easy to write, readable and very elegant. This is one of the main appeal of Chime. The above program is run on data sets ranging from size 32K elements (215) to 265K (218) elements. The execution environment was three IBM Intellistation machines with Pentium-III 400 machines with 128MB of memory. The results are summarized below.

This experiment shows the ability of Chime to handle nesting and recursion, a feature that makes writing parallel programs simple and is not available on any NOW platform. Note the super linear speedup for the last two executions. This is due to the availability of large memory buffers when using three machines. While one-machine executions do page swapping, the buffering scheme built into Chime allows three machines to buffer the data and avoid having to use the paging disk. This causes a better than expected speedup, on some applications.

5.5 Fine Grain Execution Test

The fine grain test was run using two scientific applications, LU decomposition and computing Eigenvalues. We present the results from the LU Decomposition program. LU decomposition consists of transforming a matrix for a solution of a set of linear equations. The matrix transform yields a matrix whose lower triangle consists of zero. During forward elimination phase to reduce matrix A into L (Lower) and U (Upper) triangular matrices, each worker node works on subset of rows below the current pivot row to reduce them to row echelon form. The code that does this transformation is shown below:

```
for (i = 1 to N) {// N is the size of the array
    max = abs(a[i, i]);   pivot = i;
    for (j = (i+1) to N)   {
        if abs(a[j, i]) > max {
            max = abs(a[j,i];
            swap rows i and j;
        }
    }
    // at this point a[i, i] is the largest
    // element in the column I
    parfor (p = 1 to maxworker) { // ** parallel step **
        find start and end row from values of p and i;
```

```
for j = start to end {
    use the value in a[i,i] and a[j,i]
    to set the element a[j,i] to zero
}
// now all elements in column i from
// row i+1 down, is zero
}
```

As before, the code for the program is simple and readable and the structure of the parallelism is obvious. The program creates a set of maxworker threads on each iteration through the matrix. Depending on the value of N, these threads do varying amount of work, but the maximum work each thread does is about N simple arithmetic statements. Hence the program dynamically creates a lot of threads, in a sequential fashion and each thread is rather lightweight. Hence it is a test of Chime's ability to do fine-grain processing.

The above program was executed on three Pentium-133 machines with 64MB memory. As shown in the following table, the time to run on three machines, under Chime ranges between 30% and 150% faster. But the gap between single node and three nodes has been gradually reducing as the matrix sizes are increasing. This may be due to the fact that the communication overhead is more than the computational power desired on the virtual nodes.

Input Data Size	Execution Time (seconds)		Percent speedup on three nodes
	1 node	3 nodes	
100x100	46	19	142 %
200x200	94	45	109 %
300x300	156	89	75 %
400x400	242	172	41 %
500x500	349	272	28 %

6. Conclusions

Chime is a highly usable parallel processing platform for running computations of a set of non-dedicated computers on a network. The programming method used in Chime is the CC++ language and hence has all the desirable features of CC++. This papers shows, through a set of experiments how the Chime system can be used for a variety of different types of computations over a set of diverse scientific applications.

References

1. D. McLaughlin, S Sardesai and P. Dasgupta, Distributed Cactus Stacks: Runtime Stack-Sharing Support for Distributed Parallel Programs, 1998 Intl. Conf. on Parallel and Distributed Processing Technique and Applications (PDPTA'98), July 13-16, 1998,
2. Shantanu Sardesai, CHIME: A Versatile Distributed Parallel Processing System, Doctoral Dissertation, Arizona State University, Tempe, May 1997.
3. K. M. Chandy and C. Kesselman, CC++: A Declarative Concurrent, Object Oriented Programming Notation, Technical Report, CS-92-01, California Institute of Technology, 1992.
4. A. Baratloo, P. Dasgupta, and Z. M. Kedem. A Novel Software System for Fault Tolerant Parallel Processing on Distributed Platforms. In Proceedings of the 4th IEEE International Symposium on High Performance Distributed Computing, 1995.
5. K. Li and P. Hudak. Memory Coherence in Shared Virtual Memory Systems. ACM Transactions on Computer Systems, 7(4):321-359, November 1989.
6. J. B. Carter, Design of the Munin Distributed Shared Memory System, Journal of Parallel and Distributed Computing, 29(2), pp. 219-227, September 1995.
7. B. N. Bershad and M. J. Zekauskas and W. A. SawdonThe Midway Distributed Shared Memory System, Proc. of the 38th IEEE Int'l Computer Conf. (COMPCON Spring'93), pp. 528-537, February 1993.
8. Dilip R. Khandekar. Quarks: Distributed Shared Memory as a Basic Building Block for Complex Parallel and Distributed Systems. Master's Thesis. University of Utah. March 1996
9. C. Amza, A.L. Cox, S. Dwarkadas, P. Keleher, H. Lu, R. Rajamony, W. Yu, and W. Zwaenepoel. TreadMarks: Shared Memory Computing on Networks of Workstations, IEEE Computer, December 1995.
10. G. A. Geist and V. S. Sunderam. Network-Based Concurrent Computing on the PVM System. Concurrency: Practice and experience, 4(4):293-311, 1992.
11. W. Gropp, E. Lusk, A. Skjellum. Using MPI Portable Parallel Programming with the Message Passing Interface. MIT Press, 1994, ISBN 0-262-57104-8.
12. Brian Anderson and Dennis Shasha. Persistent Linda: Linda + Transactions + Query Processing. Workshop on Research Directions in High-Level Parallel Programming Languages, Mont Saint-Michel, France June 1991.
13. J. Casas, D. Clark, R. Konuru, S. Otto, R. Prouty, and J. Walpole. MPVM: A Migration Transparent Version of PVM, USENIX, 8(2): pages 171-616, Spring 1995.
14. L. Dikken, F. van der Linden, J. Vesseur, and P. Sloot, Dynamic PVM -- Dynamic Load Balancing on Parallel Systems, Proceedings Volume II: Networking and Tools, pages 273-277. Springer-Verlag, Munich, Germany, 1994.
15. David Gelernter, Marc Jourdenais, and David Kaminsky. Piranha Scheduling: Strategies and Their Implementation. Technical Report 983, Yale University Department of Computer Science, Sept. 1993.
16. E. Seligman and A. Beguelin, High-Level Fault Tolerance in Distributed Programs, Technical Report CMU-CS-94-223, School of Computer, Science, Carnegie Mellon University, December 1994.
17. A. Baratloo, Metacomputing on Commodity Computers, Ph.D. Thesis, Department of Computer Science, New York University, May 1999.
18. P. Dasgupta, Z. M. Kedem, and M. O. Rabin. Parallel Processing on Networks of Workstations: A Fault-Tolerant, High Performance Approach. In Proceedings of the 15th IEEE International Conference on Distributed Computing Systems, 1995.

Meta-data Management System for High-Performance Large-Scale Scientific Data Access

Wei-keng Liao, Xaiohui Shen, and Alok Choudhary

Department of Electrical and Computer Engineering
Northwestern University

Abstract. Many scientific applications manipulate large amount of data and, therefore, are parallelized on high-performance computing systems to take advantage of their computational power and memory space. The size of data processed by these large-scale applications can easily overwhelm the disk capacity of most systems. Thus, tertiary storage devices are used to store the data. The parallelization of this type of applications requires understanding of not only the data partition pattern among multiple processors but also the underlying storage architectures and the data storage pattern. In this paper, we present a meta-data management system which uses a database to record the information of datasets and manage these meta data to provide suitable I/O interface. As a result, users specify dataset names instead of data physical location to access data using optimal I/O calls without knowing the underlying storage structure. We use an astrophysics application to demonstrate that the management system can provide convenient programming environment with negligible database access overhead.

1 Introduction

In many scientific domains large volumes of data are often generated or accessed by large-scale simulation programs. Current techniques dealing with such I/O intensive problem use either high-performance parallel file systems or database management systems. Parallel file systems have been built to exploit the parallel I/O capabilities provided by modern architectures and achieve this goal by adopting smart I/O optimization techniques such as prefetching [1], caching [2], and parallel I/O [3]. However, there are serious obstacles preventing the file systems from becoming a real solution to the high-level data management problem. First of all, user interfaces of the file systems are low-level which forces the users to express details of access attributes for each I/O operation. Secondly, every file system comes with its own set of I/O interface, which renders ensuring program portability a very difficult task. The third problem is that the file system policies and related optimizations are in general hard-coded and are tuned to work well for a few commonly occurring cases only.

At the other end of using database management systems, a database provides a layer on top of file systems, which is portable, extensible, easy to use and maintain, and that allows a clear and natural interaction with the applications by abstracting out the file names and file offsets. However, their main target is to be general purpose and cannot provide high-performance data access. In addition, the data consistence and integrity semantics provided by almost all database management systems put an added obstacle

M. Valero, V.K. Prasanna, and S. Vajapeyam (Eds.): HiPC 2000, LNCS 1970, pp. 293–300, 2000.

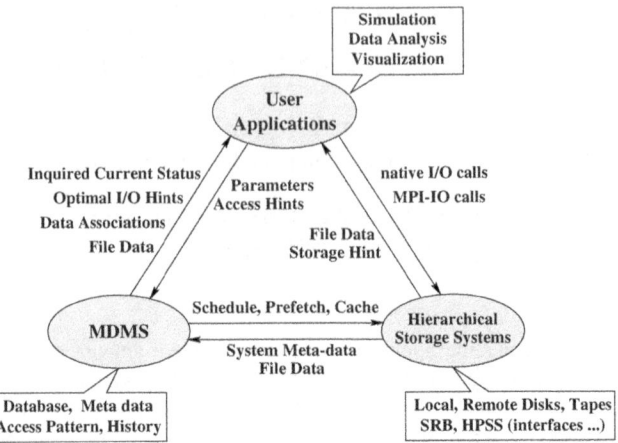

Fig. 1. The meta-data management system environment contains three key components. All three components can exist in the same site or can be located distributedly.

to high performance. Applications that process large amounts of *read-only* data suffer unnecessarily as a result of these integrity constraints [4].

This paper presents preliminary results for our ongoing implementation of a meta-data management system (MDMS) that manages meta data associated to the scientific applications in order to provide optimal I/O performance. Our approach tries to *combine* the advantages of file systems and databases and provides a user-friendly programming environment which allows easy application development, code reuse, and portability; at the same time, it extracts high performance from the underlying I/O architecture. It achieves these goals by using the management system that interacts with the parallel application in question as well as with the underlying hierarchical storage environment.

The remainder of this paper is organized as follows. In Section 2 we present the system architecture. The details of design and implementation is given in Section 3. Section 4 presents preliminary performance numbers using an astrophysics application. Section 5 concludes the paper.

2   System Architecture

Traditionally, the work of parallelization must deals with the problem of data structure used in the applications and the file storage configuration in the storage system. The fact that these two types of information are usually referred from the off-line documents increases the complexity and difficulty of application development. In this paper, we present a meta-data management system (MDMS) which is designed as a active middle-ware to connect users' applications and storage systems. The management system employs a database to store and manage all meta data associated to application's datasets and underlying storage devices. The programming environment of the data management system architecture is depicted in Figure 1. These three components can exist in the same site or can be fully distributed across distant sites.

MDMS provides a user-friendly programming environment which allows easy application development, code reuse, and portability; at the same time, it extracts high I/O performance from the underlying parallel I/O architecture by employing advanced I/O optimization techniques like data sieving and collective I/O. Since the meta-data management system stores information describing both application's I/O activity and the storage system, it can provide three easy-programming environments: data transparency through the use of data set names rather than file names; resource transparency through the use of the information about the abstract storage devices; and access function transparency through the automatic invocation of high-level I/O optimization.

3 Design and Implementation

The design of the MDMS is aimed to determine and organize the meta data; to provide a uniform programming interface for accessing data resources; to improve I/O performance by manipulating files within the hierarchical storage devices; to provide a graphic user interface to query the meta data.

3.1 Meta-data Management

There are four levels of meta data considered in this work that can provide enough information for designing better I/O strategies.

Application Level Two type of meta data exist in this level. The first describes users' applications which contains the algorithms, structure of datasets, compiling, and execution environments. The second type is the historical meta data, for instance, the time stamps, parameters, I/O activities, result summary, and performance numbers. The former is important for understanding the the applications and the management system can use it to provide browsing facility to help program development. The historical meta data can be used to determine the optimal I/O operations for the data access of the future runs.

Program Level This level of meta data mainly describes the attributes of datasets used in the applications. The attributes of datasets includes data type and structure. Since similar datasets may potentially perform the same operations and have the same access pattern, the dataset association provides an opportunity for performance improvement both on computation and I/O. The meta data with respect to I/O activity at this level includes file location, file name, I/O mode, and file structure.

Storage System Level For hierarchical storage system, the storage and file system configuration are considered as valuable meta data. In a distributed environment, since the storage device may not locate at the same site as the machine that runs the application, the meta data describing remote systems must be captured. The meta data at this level mainly deal with the file attributes among different physical devices and can be used to make a suitable I/O decision by moving files within the storage system aggressively.

Performance Level Besides the historical performance results, other valuable meta data includes I/O bandwidth of hierarchical storage system, bandwidth of remote connection, and performance of programming interfaces. The meta data that directly affects the I/O performance of parallel applications is the dataset processor partition pattern and

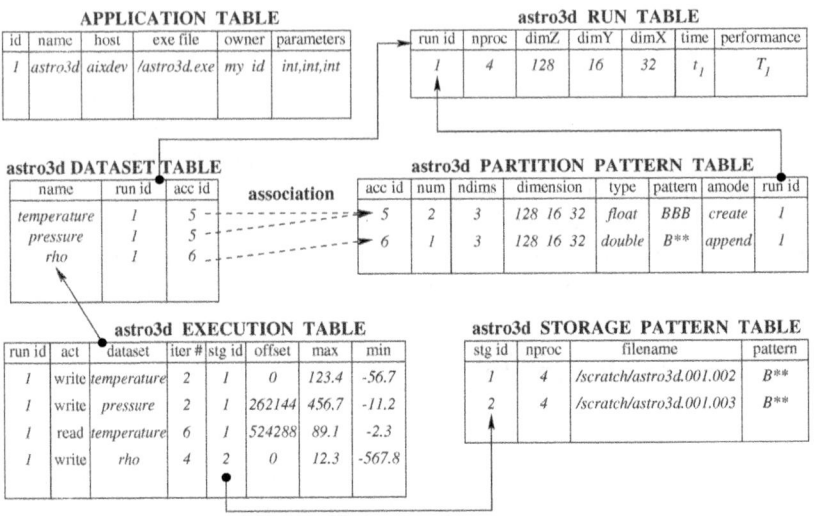

Fig. 2. The representation of meta-data in the database. The relationship of tables is depicted by the connected keys. Dataset objects with the same access pattern are associated together.

data storage pattern within the files. These two patterns can be used to determined the collective or non-collective I/O. In this work, we build the MDMS on top of MPI-IO while the proper meta data is passed to MPI-IO as file hints for further I/O performance improvement. For applications performing a sequence of file accesses, the historical access trail is typically useful for the access prediction.

We use a relational database to store the meta data and organize them into relation tables. Figure 2 shows several tables of our current implementation. For each registered application, five tables are created. Run table records the used run-time parameters for each specific run. The attributes of datasets used in the applications are stored in dataset and access pattern tables. Multiple datasets with the same structure and I/O behavior in terms of size, type, and partition pattern are associated together. In this example, two datasets, *temperature* and *pressure*, are associated together to the same row in the access pattern table. For the same I/O operations performed on the associated datasets, some resources can be re-used, eg. file view, derived data type, or even sharing the same file. The execution table stores all I/O activities for each runs. The storage pattern table contains the file locations and the storage patterns.

3.2 Application Programming Interface

The implementation of the meta-data management system uses a PostgreSQL database [5] and its C programming interface to store and manage the collected meta data. User applications communicate with MDMS through its application programming interface (API) which is built on top of PostgreSQL C programming interface. Since all meta-data queries to the database are carried out by using standard SQL, the overall implementation

Table 1. Some of the MDMS application programming interfaces.

Function name	Argument list	Description
initialization	(appName, argc, argNames, argValues)	Connect to database, register application, record a new run with a new run id
create_association	(num, datasetNames, dims, sizes pattern, numProcs, eType, handel)	Store dataset attributes into database, create dataset association, return a handle to be used in following I/O functions
get_association	(appName, datasetName, numProcs, handle)	Obtain the dataset metadata from the database, the return handle will be used in the following I/O calls
save_init	(handle, ghosts, status, extraInfo)	Determine file names, open files, decide optimal MPI-IO calls, define file type, set proper file view, calculate file offset
load_init	(handle, pattern, ghosts, status)	Find corresponding file names, open files, decide optimal calls, set proper file view, set file type, find file offsets
save	(handle, datasetName, buffer, count, dataType, iterNum)	Write datasets to files, update execution table in database
load	(handle, datasetName, buffer, count, dataType, iterNum)	Read datasets from files
save_final	(handle)	Close files
load_final	(handle)	Close files
finalization	()	Commit transaction, disconnect from database

of the MDMS is portable to all relational databases. Table 1 describes several APIs developed in this work.

MDMS APIs can be categorized into two groups: meta-data query APIs and I/O operation APIs. The first group of APIs, *initialization*, *create_association*, *get_association*, and *finalization*, is used to retrieve, store, and update meta data in the database system. Through this type of APIs, user application can convey information about its expected I/O activity to the MDMS and can request useful meta data from the MDMS including optimal I/O hints, data location, etc. Although user's application can use the inquired information to negotiate with the storage system directly, it may not be reasonable to require users to understand the details of the storage system to perform suitable I/O operations. Since the MDMS is designed to contain necessary information describing the storage system, its I/O operation APIs can act as an I/O broker and with the resourceful meta data inside the system this type of APIs can decide appropriate I/O optimizations.

Figure 3(a) shows a typical I/O application using the MDMS APIs. Through calling *create_association* or *get_association*, user applications can store or retrieve meta data. Functions *save_init* and *load_init* set up proper file view according the access patterns stored in the handle. Then, a sequence of I/O operations can be performed on the same group of associated datasets using *save* and *load*.

3.3 I/O Strategies

The design of I/O strategies focus on two levels of data access: I/O between memory and disk and I/O between disk and tape. The definition of data movement within a hierarchical storage system is given in Figure 3(b).

Data Access Between Memory and Disk For the parallel applications, the I/O costs is mainly determined by the partition pattern among processors and the file storage pattern in the storage system. When the two patterns are matched, the non-collective I/O performs best. Otherwise, non-collective I/O should be used. The MDMS I/O interface is built on top of MPI-IO [6]. The fact that MPI-IO features provide its I/O calls for

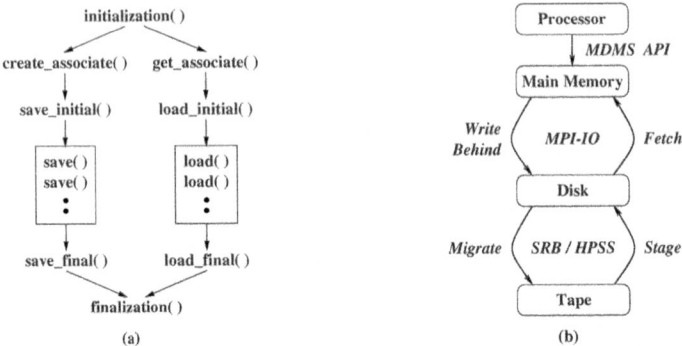

Fig. 3. (a) A typical execution flow of applications using the meta-data management system's APIs to perform I/O operations. (b) Data movement in the hierarchical storage system.

different storage platforms leads the implementation of MDMS I/O API to focus on I/O type determination. Other I/O strategies including data caching in the memory and pre-fetching from the disk can also be used to reduce the I/O costs.

Data Access Between Disk and Tape For data access of a single large file, the sub-filing strategy [7] has been designed which divides the file into a number of small chunks, called sub-files. These sub-files are maintained and transparent to the programmers. The main advantage of doing so is that the data requests for relatively small portions of the global array can be satisfied without transferring the entire global array from tape to disk. For accessing a large number of smaller files, we have investigated the techniques of native SRB container and proposed a strategy of super-filing [8].

3.4   Graphic User Interface

In order to provide users a convenient tool for understanding the meta data stored in the MDMS, we have developed a graphic interface for users to interact with the system [9]. The goal of developing this tool is to help users program their applications by examining the current status of underlying dataset configurations.

4   Experimental Results

We use the three-dimensional astrophysics application, *astro3d* [10], developed at University of Chicago as the testing program throughout the preliminary experiments. This application employs six float type datasets for data analysis and seven unsigned character type datasets for data visualization. The *astro3d* is performed in a simulation loop where for every six iterations, the contents of those six float type datasets are written into files for data analysis and checkpoint purposes and two of the seven unsigned character type datasets are dumped into files for every two iterations to represent the current visualization status. Since all datasets are block partitioned in every dimension among processors and have to be stored in files in row major, collective I/O is used. Let X, Y, and Z represent the size of datasets in dimension x, y and z, respectively, and N be the

Table 2. The amount of data written by *astro3d* with respective to the parameters N(number of iterations), X, Y, and Z(data sizes in three dimensions.)

		$X \times Y \times Z$		
N	No. I/O	$64 \times 64 \times 64$	$128 \times 128 \times 128$	$256 \times 256 \times 256$
6	33	20.97 Mbytes	167.77 Mbytes	1.34 Gbytes
12	56	35.13 Mbytes	281.02 Mbytes	2.25 Gbytes
24	102	63.44 Mbytes	507.51 Mbytes	4.06 Gbytes
36	148	91.75 Mbytes	734.00 Mbytes	5.87 Gbytes
48	194	120.06 Mbytes	960.50 Mbytes	7.68 Gbytes

number of iterations. Table 2 gives the amount of I/O performed with different values of parameters specified. The performance results were obtained on the IBM SP at Argonne National Laboratory (ANL) while the PostgreSQL database system is installed on a personal computer running Linux at Northwestern University. The parallel file system, PIOFS [11], on the SP is used to store the data written by *astro3d*. The experiments performed in this work employ 16 compute nodes.

Given different data size and iteration numbers, we compare the performance of original *astro3d* and its implementation using MDMS APIs. The original *astro3d* has already been implemented using optimal MPI I/O calls, that is, collective I/O calls. Therefore, we shall not see any major difference between the two implementations. However, our MDMS will outperform on other applications if they do not optimize their I/O. Figure 4 gives the performance results of overall execution time and the database access time for two data sizes with five iteration numbers. For the case of using $256 \times 256 \times 256$ data size, the total amount of I/O is from 1.34 to 7.68 Gbytes and the overall execution time ranges from 100s to 900s seconds. Since the connection between the IBM SP and the database is through the Internet, the database query times show variance but are all within 3 seconds. Comparing to relatively larger amount of I/O time, the overhead of database query time become negligible. Although using MDMS can result the overhead of negotiation with database, the advantage of dataset association can save the time of setting file views and defining buffer derived data types. For this particular application, *astro3d*, this advantage of using MDMS over the original program can be seen from the slight performance improvement shown in the Figure.

5   Conclusions

In this paper, we present a program development environment based on maintaining performance-related system-level meta data. This environment consists of user's applications, the meta-data management system, and a hierarchical storage system. The MDMS provides a data management and manipulation facility for use by large-scale scientific applications. Preliminary results obtained using an astrophysics application show negligible overhead of database access time comparing to the same application with I/O optimal implementation. The future work will extend the MDMS functionalities for hierarchical storage system including tape and remote file system.

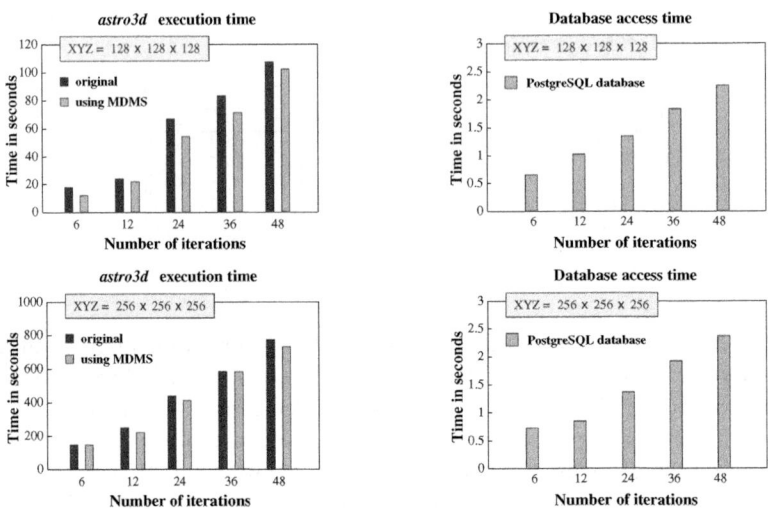

Fig. 4. Execution time of *astro3d* application and the database access time using MDMS. The timing results was obtained by running on 16 processors.

Acknowledgments

This work was supported by DOE under the ASCI ASAP Level 2, under subcontract No. W-7405-ENG-48. We acknowledge the use of the IBM SP at ANL.

References

1. C. Ellis and D. Kotz. Prefetching in File Systems for MIMD Multiprocessors. In *International Conference on Parallel Processing*, volume 1, pages 306–314, August 1989.
2. P. Cao, E. Felten, and K. Li. Application-Controlled File Caching Policies. In *the 1994 Summer USENIX Technical Conference*, pages 171–182, June 1994.
3. J. del Rosario and A. Choudhary. High Performance I/O for Parallel Computers: Problems and Prospects. *IEEE Computer*, March 1994.
4. J. Karpovich, A. Grimshaw, and J. French. Extensible File Systems (ELFS): An Object-Oriented Approach to High Performance File I/O. In *The Ninth Annual Conference on Object-Oriented Programming Systems*, pages 191–204, October 1994.
5. The PostgreSQL Development Team. *PostgreSQL User's Guide*, 1996.
6. W. Gropp, E. Lusk, and R. Thakur. *Using MPI-2: Advanced Features of the Message-Passing Interface*. The MIT Press, Cambridge, MA, 1999.
7. G. Memik et al. APRIL: A Run-Time Library for Tape Resident Data. In *NASA Goddard Conference on Mass Storage Systems and Technologies*, March 2000.
8. X. Shen and A. Choudhary. I/O Optimization and Evaluation for Tertiary Storage Systems. In *submitted to International Conference on Parallel Processing*, 2000.
9. X. Shen et al. A Novel Application Development Environment for Large-Scale Scientific Computations. In *International Conference on Supercomputing*, May 2000.
10. A. Malagoli et al. *A Portable and Efficient Parallel Code for Astrophysical Fluid Dynamics*. http://astro.uchicago.edu/Computing/On_Line/cfd95/camelse. html.
11. IBM. *RS/6000 SP Software: Parallel I/O File System*, 1996.

Parallel Sorting Algorithms
with Sampling Techniques on Clusters
with Processors Running at Different Speeds

Christophe Cérin[1] and Jean Luc Gaudiot[2]

[1] Université de Picardie Jules Verne, LaRIA, Bat CURI, 5 rue du moulin neuf, 80000 AMIENS - France
cerin@laria.u-picardie.fr,
http://www.laria.u-picardie.fr/~cerin/
[2] University of Southern California, Electrical Engineering-Systems, 3740 McClintock Ave., Los Angeles, CA 90089-2563
gaudiot@usc.edu,
* http://www-pdpc.usc.edu/

Abstract. In this paper we use the notion of quantile to implement Parallel Sorting by Regular Sampling (PSRS) on homogeneous clusters and we introduce a new algorithm for in-core parallel sorting integer keys which is based on the sampling technique. The algorithm is devoted to clusters with processors running at different speeds correlated by a multiplicative constant factor. This is a weak definition of non-homogeneous clusters but a first attempt (to our knowledge) in this direction.

1 Introduction and Motivations

Sorting in parallel records whose keys come from a linearly ordered set has been studied for many years[1]. The special case of *non homogeneous clusters*, we mean clusters with processors at different speeds is of particular interest for those who cannot replace instantaneously with a new processor generation whole the components of its cluster but shall compose with old and new processors. This paper deals with sorting on this particular class of clusters and it is innovative since all the papers (to our knowledge) about parallel sorting algorithms that work (we mean "implemented") always consider the special case of homogeneous computing platforms to ensure good load balancing properties.

The organization of the paper is as follows. In section 2 we review some techniques based on sampling for sorting in parallel. In section 3 we first introduce some notations then we describe a new parallel algorithm based on *the sampling technique* but we deal with the case of non homogeneous cluster. In section 4 we introduce the implementations conducted with BSP (Bulk Synchronous Parallel model) and we show experimental results. Sections 5 concludes the paper.

---

* The work reported in this paper is supported in part by NSF Grants #MIP 9707125 and #INT 9815742

[1] See our web link at http://www.laria.u-picardie.fr/~cerin/=paladin/ for a bibliography and a review of techniques about parallel sorting

M. Valero, V.K. Prasanna, and S. Vajapeyam (Eds.): HiPC 2000, LNCS 1970, pp. 301–309, 2000.
© Springer-Verlag Berlin Heidelberg 2000

2   Related Work on Sorting

In this section we explain why the PSRS (Parallel Sorting by Regular Sampling) technique is good for benchmarking clusters, what are the candidates for sorting on non homogeneous clusters and why we have chosen BSP as the programming target language.

2.1   Regular Sampling: An Efficient Technique for Sorting

In the Bulk Synchronous Programming model (BSP) the sorting problem has been approached with a technique known as "sorting by regular sampling" [1]. There is two available BSP distributions: one from the university of Oxford (UK): BSPlib; one from the university of Paderborn (Germany): PUB [2].

In this paper we only consider *one step merge-based algorithms*. Such algorithms have low communication cost because they move each element at most once (and at the "right place"). Their main drawback is that they have poor load balancing if we don't care about it: it is difficult to derive a bound to partition data into sublists of "equal sizes". Theoretically speaking, it can be shown that the computational cost of PSRS matches the optimal $\mathcal{O}(n/p \log n)$ bound. To obtain this result we must ensure that all the data are unique. In the case of no duplicates, PSRS *guarantees* to balance the work within a factor of two of optimal in theory, regardless of the value distribution. In practive we observe a few percent of optimal.

Until recently processors that sort in parallel according to the PSRS philosophy began by sequentially sorting their portions in the first phase of the algorithm and use regular sampling to select pivots. Even Shi and Schaeffer, the inventors of the PSRS technique in the original paper [1] said that "It appears to be a difficult problem to find pivots that partition the data to be sorted into ordered subsets of equal size without sorting the data first". In fact, an optimization can be done to bypass the first sorting phase if we concentrate (see [3]) on the notion of *quantiles*. Quantiles are a set of 'cut points' that divide a sample of data into groups containing (as far as possible) equal numbers of observations. The *p-quantiles* of an ordered set A of size n are the $p-1$ elements of rank $n/p, 2n/p, \cdots, (p-1)n/p$, splitting A into p parts of equal size. The *p-quantiles search problem* consists in determining those splitting elements of A. For instance, the problem is related to the problem of *finding the 'best'* (the maximum) as follows: the median - the $\lceil n/2 \rceil$th best - is given by a call to find(t,0,n-1,n/2) ($n = 2^k$) where find is the procedure that find the index of the ith largest element of vector t between indexes 0 and $n - 1$. To sum up, the randomized *find* algorithm that we have implemented is $\mathcal{O}(n)$ on average. Moreover, in practice we have that the number of required pivots is not of order n, but is related to the number of processors which is very inferior to the input size, we can expect near a 'linear time' (with a small constant) algorithm to select pivots which is better than a sequential sort in $\mathcal{O}(n \log n)$.

In other words, when implementing step 1 and step 4 of PSRS [1] we are faced to two choices if we want to be:

conform to the original paper . In this case we have to implement step 1 with a quicksort, then step 4 with a merge; This is done in the implementation of Sujithan for BSPLib;

not to be conform to the original paper , but conform to the spirit of the original paper (implementation choice). In this case, we implement step 1 with a quantile search (that do not offer ordered sequences after processing - the time complexity is linear in time for the worst case and for one quantile), then we implement step 4 with a quicksort (with time complexity $\mathcal{O}(n^2)$ in the worst case, but $\mathcal{O}(n \log n)$ for the average case).

We have implemented the two previous views. The experimental results are given in a forthcoming section to get the winner.

2.2   BSP as an Efficient Programming Language

Our programs are witten in BSP (bulk synchronous parallel). Thus, one advantage of BSP against competing approaches (CGM, LogP) is that some libraries are available for a large number of platforms on which they can effectively run BSP programs. We use the standard `BSPlib 1.4` library from Oxford university and also the `PUB` distribution from the university of Paderborn (Germany).

We have developped codes first for the Direct Remote Memory Access (DRMA) communication facilities implemented with `bsp_put`, `bsp_get`, `bsp_push_register`, `bsp_pop_register` primitives. The main disadvantage of using DRMA primitives is that the register step can be very time consuming, in particular with BSPLib (see [4] for an example). These two programming style allow us to compare the efficiency of communications. DRMA has the advantage that programmer deal with only one primitive to exchange data and there is no notion of 'sender' and 'receiver'. We have observed but it is not detailed in this paper that DRMA implementations for sorting provide a performance degradation of 10% for input sizes greater than 4Mb and with PUB, the Paderborn BSP implementation of BSP.

Let us now examine on Table 1 the execution times of our PSRS implementations that respect the original presentation of Shi and Schaeffer in [1] (the first step is implemented with a sort; the fourth step is implemented with a merge). Table 2 is about our implementation of PSRS with a quantile search in the first step and a quicksort algorithm in the fourth step. The experiments are conducted on our cluster of four alpha (EV56 21164, 500Mhz) processors interconnected with fast Ethernet. We conclude that there is no winner but our implementation with a quantile search requires less memory than with a merge (about n less on each processor). Note also that the cache size of each processor is 4Mb which is a big cache size for our experiment sizes. So we prefer using our modified implementation because at identical time performance, our implementation consumes less memory storage.

Tables 3 and 4 are related to the sublist expansion metric. We note the excellent coefficients very closed to 1, both on the two experimental platforms and for the two implementations.

Table 1. Configuration: 4 Alpha 21164, 500Mhz, 256Mb of RAM, 4Mo cache fast Ethernet

#PROCs	INPUT SIZE	MEAN (s)	VARIANCE	STANDARD DEVIATION
4	131072	0.13695885	0.00233757	0.04834850
4	262144	0.53295468	0.00650963	0.08068229
4	524288	1.06904605	0.02577709	0.16055246
4	1048576	1.64023283	0.07724168	0.27792388
4	2097152	2.69613435	0.03101140	0.17610054
4	4194304	5.44492585	0.09799507	0.31304164
4	8388608	11.23528133	0.28563781	0.53445095

Execution time metrics for benchmark 0
Algorithm used: original PSRS – code: `psrs_pub_merge_bench.c`

Table 2. Configuration: 4 Alpha 21164, 500Mhz, 256Mb of RAM, 4Mo cache, fast Ethernet

#PROCs	INPUT SIZE	MEAN (s)	VARIANCE	STANDARD DEVIATION
4	131072	0.13828583	0.00189696	0.04355412
4	262144	0.51180058	0.01442428	0.12010115
4	524288	1.01821333	0.03200934	0.17891154
4	1048576	1.61490733	0.06455982	0.25408625
4	2097152	2.55807533	0.14855640	0.38543015
4	4194304	5.42543410	0.12039131	0.34697451
4	8388608	11.35150307	0.29101192	0.53945521

Execution time metrics for benchmark 0
Algorithm used: modified PSRS with quantile search – code:
`psrs_pub_quantile_bench.c`

Table 3. Configuration: 5 Alpha 21164, 500Mhz, 256Mb of RAM, 4Mo cache, fast Ethernet

#PROCs	INPUT SIZE	Mean/Proc	MAX	MIN	SUBLIST (max)	SUBLIST (min)
4	131072	32768	33410	32291	1.01959228	.98544311
4	262144	65536	66392	64473	1.01306152	.98377990
4	524288	131072	132216	129707	1.00872802	.98958587
4	1048576	262144	264148	260381	1.00764465	.99327468
4	2097152	524288	527304	521123	1.00575256	.99396324
4	4194304	1048576	1051843	1044743	1.00311565	.99634456
4	8388608	2097152	2103125	2092089	1.00284814	.99758577

Sublist expansion metrics for benchmark 0
Algorithm used: modified PSRS with quantile search – code:
`psrs_pub_quantile_bench.c`

Concerning the Speedup, it can be shown that our sorting PSRS implementations are two times slower than the sequential case on a cluster of 8 Celerons interconnected by Ethernet but it is about 3 on the same cluster but interconnected with fast Ethernet (we divide by 6 approximatively the execution time

Table 4. Configuration: 8 Celerons 466Mhz, 96Mb of RAM, Ethernet

#PROCs	INPUT SIZE	Mean/Proc	MAX	MIN	SUBLIST (max)	SUBLIST (min)
8	131072	16384	17190	15696	1.04919433	.95800781
8	262144	32768	34099	31642	1.04061889	.9656372
8	524288	65536	67601	63880	1.03150939	.97473144
8	1048576	131072	133647	128841	1.01964569	.98297882
8	2097152	262144	265976	259652	1.01461791	.99049377
8	4194304	524288	528853	518734	1.00870704	.98940658
8	8388608	1048576	1057709	1039913	1.00870990	.99173831

Sublist expansion metrics for benchmark 0

Algorithm used: PSRS with quantile search – code: `psrs_pub_quantile_bench.c`

by the use of a network 10 times faster). This exemple demonstrates the gain of using a cluster with fast Ethernet instead of Ethernet as the communication layer.

3   An Efficient Load Balancing Algorithm with Two Kinds of Processors

The problem that we address in this paper is the following: n data (with no duplicate) that are physically distributed on p processors (disks) have to be sorted. The cluster is composed with p processors caracterized by their speeds that we denote by $s_i, (1 \leq i \leq p)$. The average sustained disk transfer rate and the interconnection network bandwidth is no matter here.

The *performance* P_i of a single machine is defined to be equal to s_i. Let $m = \max\{P_i\}, (1 \leq i \leq p)$. The speed ratio or *relative performance* denoted rp of each processor is defined as $rp_i = \dfrac{m}{P_i}, (1 \leq i \leq p)$. Our thesis is that a necessary condition to obtain good load balancing properties in the model is to fairly share the amount of data among processors: each processor recieves a ratio depending on rp_i, the constant of proportionality between processor performances.

For the sake of simplicity, we now consider only the case of two different sorts of processors. The number of processor is still equal to p.

3.1   Details of the Algorithm

We reuse mainly the notations of [1]. We adapt only pivot selection phases of the Parallel Regular Sampling Algorithm [1], the other phases stay identical. We assume p processors with $p = p_1 + p_2$ such that p_1 is the number of processors with the hihgest speed s_1 and p_2 is the number of processors with the slowest speeds s_2. Let k be $\max\{\frac{s_1}{s_2}, \frac{s_2}{s_1}\}$ and assume that these numbers are in the set of natural numbers. Initially, each p_1 processors has $n_1 = \frac{k.n}{(k+1)p_1}$ data and each p_2 processors has $n_2 = \frac{n}{(k+1)p_2}$ data that we assume to be natural numbers. Each processor sort its chunk with an appropriate sequential sort algorithm. Since each processor sort a number of data proportional to its speed we may assume that all of them finish at the same time.

From each of the p lists, $p-1$ samples are chosen, evenly spaced throughout the list. The p_1 processors keeps those values at indexes that are multiple of $\lfloor \frac{n_1}{p*k} \rfloor$, the p_2 processors keep values in their sorted lists at indexes that are multiple of $\lfloor \frac{n_2}{p*k} \rfloor$. The $p(p-1)$ pivots are gathered on a dedicated processor that sort them. Now, assume that $p-1$ values are chosen evenly spaced throughout the sorted list. Then, we can go on with step 4 of the "regular algorithm" we have presented erlier in the paper. In doing this, we can immediately apply the result of [1] to state that in phase 4, each processor merges less than $2.n_1$. This is a straight application of the main theorem of [1] since we have in our case that the number of data that one of our fastest processor is greater or equal to n/p, thus $2\frac{n}{p} \leq 2.n_1$. The following tighter bound can be obtained:

Theorem 1. *In phase 4 of the algorithm, under the assumption that $k \leq 2\frac{p_1}{p_2} \wedge$* $p_2 = p_1 \wedge n > p^3$, *each of the p_1 processor merges less than $2.n_1$ elements and each of the p_2 processor merges less than $2n_2$ elements.*

Proof. First of all we have to observe that the necessary condition $k \leq 2\frac{p_1}{p_2}$ is correct: if not we can have the possibility that a slow processor receive more than $2n_2$ elements. Consider for instance the case $k = 6, p_1 = p_2 = 2$. Assume that the first pivot in the list of ordered $p(p-1)$ pivots comes from the set of the fastest processors. In this case we have that $6 \leq 2\frac{2}{2}$ is false; thus a slow processor will receive more than $2n_2$ data. This condition is simply given by the relation:

$$\frac{k.n}{(k+1)p_1} \leq 2\frac{n}{(k+1)p_2} \iff k \leq 2\frac{p_1}{p_2}$$

At least, note that the assumption is reasonable in practice with current technology. The condition $p_2 = p_1$ will be explained later in the paper. It allow us to guaranty the bound on load balancing.

Second, the proof is concerned with the way we pick the $p-1$ pivots among the $p(p-1)$ ordered list $Y_1, Y_2, \cdots, Y_{p(p-1)}$. We have to distinguish pivots that come from the p_1 processors and those that come from the p_2 processors. We proceed as follows and we first select pivots for the slowest processors then for the fastest. Consider first Y_1. If it comes from p_1 we know that $\lfloor \frac{n_1}{p*k} \rfloor$ values on its left are lower or equal to it and reciprocally, if it comes from p_2 then $\lfloor \frac{n_2}{p*k} \rfloor$ values are lower or equal to it on its left. We sum all these numbers at each time we visit a new pivot (from left to right) until the sum becomes greater than $n_1 = \frac{k.n}{(k+1)p_1}$ or $n_2 = \frac{n}{(k+1)p_2}$ depending on the processor we deal with and we keep the pivot just before the pivot that stop the summation process.

The proof depends on the pivot rank i where $1 \leq i \leq (p-1)$ but we describe here only the case of the first selected pivot (the arguments for the other cases are similar):

Case $i = 1$: we have evinced at most $n_1 < 2.n_1 (n_2 < 2.n_2)$ elements according to the kind of processor when we made the choice of the first pivot. Note also that the index of the first pivot in the list of $p(p-1)$ pivot can not be greater than $(p-1)$ (it is equal when we pick pivots from the same processor). Moreover, the worst case (that produce the maximal number of elements that a processor have to sort in the last phase) is caraterized by the following state: $(p-1)$ pivots comes from a slow processor (thus $(p-1)(n_2/(p.k))$ elements are evinced). We have to show that the other $(p-1)$ processors have no more than $n_1, (n_2)$ elements that are lower than pivot i. Assume that the first pivot comes from a slow processor. Then there is no more than

$$S = (p_2 - 1)\left(\frac{n_2}{p.k} - 1\right) + p_1\left(\frac{n_1}{p.k} - 1\right)$$

values that can be add to the evinced values because otherwise the value at position $n_2/(p.k)$ or $n_1/(p.k)$ will be a selected pivot. The proof is to show that $S \leq n_2$. It is not difficult to observe that:

$$n_2 = \frac{n_1 p_1}{k.p_2}, n_1 = \frac{n_2.k.p_2}{p_1}$$

Since $p_1 = p_2$ we have that:

$$
\begin{aligned}
(p_2 - 1)\left(\frac{n_2}{p.k} - 1\right) + p_1\left(\frac{n_1}{p.k} - 1\right) &= (p_2 - 1)\left(\frac{n_2}{p.k} - 1\right) + p_2\left(\frac{n_2.k.p_2}{p_1.p.k} - 1\right)\\
&= (p_2 - 1)\left(\frac{n_2}{p.k} - 1\right) + p_2\left(\frac{n_2}{p} - 1\right)\\
&= n_2\left(\frac{p_2}{p.k} - \frac{1}{p.k} + \frac{p_2}{p}\right) - 2p_2
\end{aligned}
$$

which is lower than n_2 by the following argument: since $k \leq 2$, we have that

$$
\begin{aligned}
\frac{p_2}{p}\left(1 + \frac{1}{k}\right) - \frac{1}{p.k} &\leq \frac{p_2}{p}(1 + 1) - \frac{1}{p}\\
&\leq \frac{p_2.2}{2.p_2} - \frac{1}{p}\\
&\leq 1 - \frac{1}{p}\\
&< 1
\end{aligned}
$$

and thus $S \leq n_2$ as stated. Moreover, the proof for the special case with the first pivot that come from a fast processor is similar and it is not difficult to write now.

4   Experimental Setup, Implementation and Results

We have tested our ideas on many platforms (Pentium Pro interconnedted with Myrinet, Alpha processors with fast Ethernet, Celeron processors with Ethernet). The results we introduce now were obtained on a network of Celeron and a network of Alpha (alpha 21164 - EV56, Red Hat Linux 5.2, Kernel 2.2.11, 4Mb of cache) located in Amiens (http://www.laria.u-picardie.fr/PALADIN), France, interconnected by a fast ethernet network. The cluster is an homogeneous one but we are interesting in the sublist expansion metric here and not in the execution time.

The main experiment we conducted is about the computation of load balancing metrics: (i) the sublist expansion [5] and (ii) the load expansion [6]. We recall that the sublist expansion is defined as the ratio of the maximum sublist size to the mean sublist size. The load expansion metric is defined as the ratio of the time spent in sorting sublists on the most heavily loaded processor to the mean time spent over all processors.

Table 5 summarizes our results in the case of a 2×2 cluster with 2 processors running 2 times faster than the two others. In columns entitled "Opt.", we list the number of elements a slow (fast) processor is expected to merge in the last step of the algorithm. The "sublists" columns correspond to the sum of elements the slow (fast) processors have received in the last step, divided by the expected mean. We check that the sublist metric are below 2 as requested by our theorem. However, we note that the values are far from the optimal (which is a 1 value). Experimentally we obtain that the sublist expansion metrics are approximatively between the optimal and the theoretical upper bound. An open problem remains: " can we do better than this?".

Table 5. Configuration: 4 Celerons 466Mhz, 96Mb of RAM, Ethernet

#P	SIZE	Opt. Slow	Sublist (slow)	Opt. Fast	Sublist (fast)
4	131070	21845	0.7561	43690	1.1219
4	262140	43690	0.7540	87380	1.1229
4	524280	87380	1.4005	174760	0.7997
4	1048572	174762	1.6248	349524	0.6875
4	2097150	349525	1.5750	699050	0.7124
4	4194300	699050	1.5002	1398100	1.4995
4	8388606	1398101	1.5000	2796202	1.5001

Sublist expansion metrics for benchmark 0
Algorithm used: PSRS with a merge – Speed ratio = 2

5   Conclusion

In this paper we have implemented a modified version of PSRS with the notion of quantile which has good execution time and even a small memory advantage against the previous known implementation. Second we have introduced our approach to tackle the problem of sorting on non homogeneous clusters. The algorithms that we describe combine very good properties for load balancing that we can bound by a small constant. The algorithm is an adaptation of a known technique called PSRS (Parallel Sampling by Regular Sampling) with a particular choice of pivots. We exemplify the case of a cluster with processors running at two different speeds. Benchmark data sets used by Jájá in [7] are integrated in our codes and they serve as a reference.

To our knowledge, little work has been performed about sorting on non homogeneous network. If we try to reuse the main results [8,9,1,6] about sorting on homogeneous clusters as a general strategy, the main difficulty is to show that it will lead to good properties for load balancing, for execution time and also that the underlying algorithm is suitable for a concrete implementation. We are not yet convinced that other technique than the sampling technique will have so much properties.

We are also investigating the case where p processes are created on a $q < p$ processors machine. In this case we have to run several processes (threads) on the same processor so that the scheduling of threads becomes a challenge for our two PSRS implementation (modified and not modified). For the purpose we deal with the OVM (Out-of-order execution parallel Virtual Machine) tool which is developped in Orsay, France by Franck Cappello. OVM is based on RPC and it can be view as an execution model that integrate load balancing mechanisms.

References

1. H. Shi and J. Schaeffer, "Parallel sorting by regular sampling," *Journal of Parallel and Distributed Computing*, vol. 14, no. 4, pp. 361-372, 1992.
2. O. Bonorden, B. Juurlink, I. von Otte, and I. Rieping, "The paderborn university bsp (pub) library - design, implementation and performance," in *13th International Parallel Processing Symposium and 10th Symposium on Parallel and Distributed Processing*, 12 - 16 April, 1999, San Juan, Puerto Rico, available electronically through IEEE Computer Society, 1999.
3. N. S. Afonso Ferreira, "A randomized bsp/cgm algorithm for the maximal independent set problem," *Parallel Processing Letters*, vol. 9, no. 3, pp. 411-422, 1999.
4. C. Cérin and J.-L. Gaudiot, "Algorithms for stable sorting to minimize communications in networks of workstations and their implementations in bsp," in *IEEE Computer Society International Wokshop on Cluster Computing (IWCC'99)*, pp. 112-120, 1999.
5. G. Blelloch, C. Leiserson, and B. Maggs, "A Comparison of Sorting Algorithms for the Connection Machine CM-2," in *Proceedings of the ACM Symposium on Parallel Algorithms and Architectures*, July 1991.
6. H. Li and K. C. Sevcik, "Parallel sorting by overpartitioning," in *Proceedings of the 6th Annual Symposium on Parallel Algorithms and Architectures*, (New York, NY, USA), pp. 46-56, ACM Press, June 1994.
7. Helman and JáJá, "Sorting on clusters of SMPs," *Informatica: An International Journal of Computing and Informatics*, vol. 23, 1999.
8. M. Quinn, "Analysis and benchmarking of two parallel sorting algorithms: Hyperquicksort and quickmerge," *BIT*, vol. 29, no. 2, pp. 239-250, 1989.
9. X. Li, P. Lu, J. Schaeffer, J. Shillington, P. S.Wong, and H. Shi, "On the versatility of parallel sorting by regular sampling," *Parallel Computing*, vol. 19, pp. 1079-1103, Oct. 1993.

Evaluation of an Adaptive Scheduling Strategy for Master-Worker Applications on Clusters of Workstations

Elisa Heymann[1], Miquel A. Senar[1], Emilio Luque[1], and Miron Livny[2]

[1] Unitat d'Arquitectura d'Ordinadors i Sistemes Operatius
Universitat Autònoma de Barcelona
Barcelona, Spain
{e.heymann, m.a.senar, e.luque}@cc.uab.es

[2] Department of Computer Sciences
University of Wisconsin– Madison
Wisconsin, USA
miron@cs.wisc.edu

Abstract[*]. We investigate the problem arising in scheduling parallel applications that follow a master-worker paradigm in order to maximize both the resource efficiency and the application performance. We propose a simple scheduling strategy that dynamically measures application execution time and uses these measurements to automatically adjust the number of allocated processors to achieve the desirable efficiency, minimizing the impact in loss of speedup. The effectiveness of the proposed strategy has been assessed by means of simulation experiments in which several scheduling policies were compared. We have observed that our strategy obtains similar results to other strategies that use a priori information about the application, and we have derived a set of empirical rules that can be used to dynamically adjust the number of processors allocated to the application.

1. Introduction

The use of loosely coupled, powerful and low-cost commodity components (PCs or workstations, typically) connected by high-speed networks has resulted in the widespread usage of a technology popularly called cluster computing [1]. The availability of such clusters made them an appealing vehicle for developing parallel applications. However, not all parallel programs that run efficiently in a traditional parallel supercomputing environment can be moved to a cluster environment without significant loss of performance. In that sense, the Master-Worker paradigm is attractive because it can achieve similar performance in both environments as no high communication performance is usually required from the network infrastructure [2].

In this paradigm, a master process is responsible basically for distributing tasks among a farm of worker processes. Moreover, it is a good example of adaptive parallel computing because it can respond quite well to a scenario where applications are executed by stealing idle CPU cycles (we refer to these environments as non-dedicated clusters). The number of workers can be adapted dynamically to the number

---

[*] This work was supported by the CICYT (contract TIC98-0433) and by the Commission for Cultural, Educational and Scientific Exchange between the USA and Spain (project 99186).

M. Valero, V.K. Prasanna, and S. Vajapeyam (Eds.): HiPC 2000, LNCS 1970, pp. 310–319, 2000

of available resources in such an opportunistic environment so that, if new resources appear they are incorporated as new workers for the application.

However, the use of non-dedicated clusters introduces the need for complex mechanisms such as resource discovery, resource allocation, process migration and load balancing. In the case of master-worker applications, the overhead incurred in discovering new resources and allocating them can be significantly alleviated by not releasing the resource once the task has been completed. The worker will be kept alive at the resource waiting for a new task. However, by doing so, an undesirable scenario may arise in which some workers may be idle while other workers are busy. This situation will result in a poor utilization of the available resources in which all the allocated workers are not kept usefully busy and, therefore, the application efficiency will be low. In this case, the efficiency may be improved by restricting the number of allocated workers.

If we consider the execution time, a different criteria will guide the allocation of workers because the more workers allocated for the application the lower the total execution time of the application. Then, the speedup of the application directly depends on the allocation of as many workers as possible.

In general, the execution of a master-worker application implies a trade-off between the speedup and the efficiency achieved. On the one hand, our aim is to improve the speedup of the application as new workers are allocated. On the other hand, we want to also achieve a high efficiency by keeping all the allocated workers usefully busy.

Obviously, the performance of master-worker applications will depend on the temporal characteristics of the tasks as well as on the dynamic allocation and scheduling of processors to the application. So, in this work we consider the problem of maximizing the speedup and the efficiency of a master-worker application through both the allocation of the number of processors on which it runs and the scheduling of tasks to processors during runtime. We address this goal by first proposing a generalized master-worker framework which allows adaptive and reliable management and scheduling of master-worker applications running in a cluster composed of opportunistic computing resources. Secondly, we propose and evaluate by simulation a scheduling strategy that dynamically measures application efficiency and task execution times to control the assignment of tasks to workers.

The rest of the paper is organized as follows. Section 2 presents the model of the Master-Worker applications that we are considering in this paper. Section 3 gives a more precise definition of the scheduling problem, introduces our scheduling policy and reviews some related work. Section 4 presents some simulation results obtained in the evaluation of the proposed strategy, by comparing our policy with other scheduling policies. Section 5 summarizes the main results presented in this paper.

2. The Model for Master Worker Applications

In this work, we focus on the study of applications that follow a Master-Worker model that has been used to solve a significant number of problems such as Monte Carlo simulations [3] and material science simulations [4]. In this generalized master-worker model, the master process iteratively solves a batch of tasks. After completion of one task, the master process may perform some intermediate computations with the partial result obtained by the task. Subsequently, when the complete batch of tasks is finished the master may carry out some additional processing. After that, a new batch of tasks is assigned to the Master and this process is repeated several times until completion of the problem, that is, K cycles (which are later referred as *iterations*).

As can be seen in fig. 1, we are considering a group of master-worker applications with an iterative behavior. In these iterative parallel applications a batch of parallel tasks is executed K times (iterations). Workers execute *Function (task)* and *PartialResult* is collected by the master. The completion of a given batch induces a synchronization point in the iteration loop which facilitates also the collection of job's statistics in the Master process.

```
Initialization
Do
        For task = 1 to N
            PartialResult  =  +   Function (task)
        end
        act_on_batch_complete( )
while (end condition not met).
```

Fig. 1. A model for generalized Master-Worker applications.

In addition to these characteristics, empirical evidence has shown that, for a wide range of applications, the execution of each task in successive iterations tends to behave similarly, so that the measurements taken for a particular iteration are good predictors of near future behavior [4]. In the rest of the paper we will investigate to what extent an adaptive and dynamic scheduling mechanism may use historical data about the behavior of the master-worker application to improve its performance in an opportunistic environment.

3. Challenges for Scheduling of Master-Worker Applications

In this section we present the scheduling problem adopted in this work and we present also our proposed policy to solve it.

3. 1 Problem Statement and Related Work

Efficient scheduling of a master-worker application in a cluster of distributively owned resources should provide answers to the following questions:

- How many workers should be allocated to the application? A simple approach would consist of allocating as many workers as tasks are generated by the application at each iteration. However, this policy will result, in general, in poor resource utilization because some workers may be idle if they are assigned a short task while other workers may be busy if they are assigned long tasks.
- How should tasks be assigned to the workers? When the execution time incurred by the tasks of a single iteration is not the same, the total time incurred in

completing a batch of tasks strongly depends on the order in which tasks are assigned to workers.

We evaluate our scheduling strategy by measuring the efficiency and the total execution time of the application.

Resource efficiency [5] for n workers is defined as the ratio between the amount of time workers have actually spent doing useful work and the amount of time workers were able to perform work, i.e. the time elapsed since worker i is alive until it ends minus the amount of time that worker i is suspended.

Execution Time is defined as the time elapsed from when the application begins its execution until it finishes, using n workers.

The problem of scheduling master-worker applications on cluster environments has been investigated recently in the framework of middleware environments that allow the development of adaptive parallel applications running on distributed clusters. They include NetSolve [6], Nimrod [7] and AppLeS [5]. NetSolve and Nimrod provide APIs for creating task farms that can only be decomposed by a single bag of tasks. Therefore, no historical data can be used to allocate workers. The AppLeS (Application-Level Scheduling) system focuses on the development of scheduling agents for parallel applications but in a case-by-case basis, taking into account the requirements of the application and the predicted load and availability of the system resources at scheduling time.

There are other works in the literature that have studied the use of parallel application characteristics by processor schedulers of multiprogrammed multiprocessor systems, typically with the goal of minimizing average response time [8]. The results from these studies are not directly applicable in our case because they were focussed on the allocation of jobs in shared memory multiprocessors without considering the problem of task scheduling within a fixed number of processors. However, their experimental results also confirm that iterative parallel applications usually exhibit regular behaviors that can be used by an adaptive scheduler.

3. 2 Proposed Scheduling Policy

Our adaptive and dynamic scheduling strategy employs a heuristic-based method that uses historical data about the behavior of the application. It dynamically collects statistics about the average execution time of each task and uses this information to determine the order in which tasks are assigned to processors. Tasks are sorted in decreasing order of their average execution time. Then, they are assigned dynamically to workers in a list-scheme, according to that order. At the beginning of the application execution, as no data is available regarding the average execution time of tasks, tasks are assigned randomly. We call this adaptive strategy *Random & Average*, although the random assignment is done only once, simply as a way to obtain information about the tasks' execution time.

4. Experimental Study

In this section, we evaluate the performance of several scheduling strategies with respect to the efficiency and the execution time obtained when they are applied to schedule master-worker applications on homogeneous systems. As we have stated in previous sections, we focus our study on a set of applications that are supposed to exhibit a highly regular and predictable behavior. We will test different scheduling strategies that include both pure static strategies that do not take into account any runtime information and adaptive and dynamical strategies that try to learn from the application behavior.

As a main result from these simulation experiments, we are interested in obtaining information about how the proposed adaptive scheduling strategy performs on average, and some bounds for the worst case situations. Therefore, in our simulations we consider that the number of processors is available through the whole execution of the application (i.e. this would be the ideal case in which no suspensions occur).

4.1 Policies Description

The set of scheduling strategies used in the comparison were the following:

- **LPTF (Largest Processing Time First)**: For each iteration this policy first assigns the tasks with largest execution time. Before an iteration begins, tasks are sorted decreasingly by execution time. Then, each time a worker is ready to receive work, the master sends the next task of the list, that is, the task with largest execution time. It is well known that *LPTF* is at least ¾ of the optimum [9]. This policy needs to know the exact execution time of the tasks in advance, which is not generally possible in a real situation, therefore it is only used as a sort of upper bound in the performance achievable by the other strategies.

- **LPTF on Expectation**: It works in the same way as *LPTF*, but tasks are initially sorted decreasingly by the expected execution time. In each iteration tasks are assigned in that predefined order. If there is no variation of the execution time of the tasks, the behavior of this policy is the same as *LPTF*. This policy is static and non-adaptive, and represents the case in which the user has an approximately good knowledge of the behavior of the application and wants to control the execution of the tasks in the order that he specifies. Obviously, it is possible for a user to have an accurate estimation of the distribution of times between the tasks of the application, but in practice, small variations will affect the overall efficiency because the order of assignment is fixed by the user at the beginning.

- **Random**: For each iteration, each time a worker is ready to get work, a random task is assigned. This strategy represents the case of a pure dynamic method that does not know anything about the application. In principle, it would obtain the worst performance of all the presented strategies, therefore it will be used as a lower bound in the performance achievable by the other strategies.

4.2. Simulation Framework

All described scheduling policies have been simulated systematically, to obtain efficiency and execution time, with all the possible number of workers ranging from 1 to as many workers as numbers of tasks, considering the following factors:

- *Workload (W)*: This represents the work percentage done when executing the 20% largest tasks. We have considered 30%, 40%, 50%, 60%, 78% 80% and 90% workload values. A 30% workload would correspond to highly balanced applications in which near all the tasks exhibit a similar execution time. On the contrary, a 90% workload would correspond to applications in which a small number of tasks are responsible for the largest amount of work. Moreover, the 20% largest tasks can have similar or different execution times. They are similar if their execution time differences are not greater than 20%. The same happens to the other 80% of tasks. For each workload value we have undertaken simulations with the four possibilities (referred as *i-i* in figures of section 4.3).

- *Iterations (L)*: This represents the number of batches of tasks that are going to be executed. We have considered the following values: 10, 35, 50 and 100.
- *Variation (D)*: From the workload factor, we determine the base execution times for the tasks. Then, for each iteration a variation is applied to the base execution times of each task. Variations of 0%, 10%, 30%, 60% and 100% have been considered. When a 0% variation was used, the times of the task were constant along the different iterations. This case would correspond to very regular applications where the time of tasks is nearly the same in successive iterations. When a 100% variation was used, tasks exhibit significant changes in their execution time in successive iterations, corresponding to applications with highly irregular behavior.
- *Number of Tasks (T)*: We have considered applications with 30, 100 and 300 tasks. Thus we examine systems with a small, a medium or a large amount of tasks, respectively.

For each simulation scenario (fixing a certain value for *workload, iterations* and *variation*) the efficiency and execution time have been obtained using all the workers from 1 to *Number of Tasks*.

4.3. Simulation Results

Although we have conducted tests for all the commented values, in this section we present only those results that are the most interesting. We will illustrate with figures the results for 30 tasks since they prove to be representative enough for the results obtained with a larger number of tasks. Moreover, we emphasize those results with 30% and 100% deviation, representing low and high degrees of regularity. In real applications 100% deviation is not expected, but it allows us to evaluate the strategies under the worst case scenario.

In the rest of the section some relevant result figures for both efficiency and execution time are presented. The X-axis always contain the number of workers. The Y-axis contain the efficiency and the execution time values respectively. Five values *W, i-i, D, T* and *L* appear at the top of each figure. *W* stands for the workload, *i-i* describes the similarity of tasks, *D* stands for variation applied to task execution time at each iteration, *T* stands for the number of tasks and *L* for the number of iterations (loop). We now review the most relevant results obtained from our simulations.

Effect of the number of iterations (L): The number of iterations (L) that tasks are executed does not significantly affect efficiency for an adaptive strategy such as *Random & Average*. Figure 2 shows the effect of varying the number of iterations, considering 30% workload and 100% deviation. This is the case when the effect of the number of iterations is the most significant. As can be seen when the number of iterations varies from 10 to 35 the gain in efficiency is less than 5%. When the number of iterations was greater than 35, no significant gain in efficiency was observed. Therefore, our proposed strategy achieves a good efficiency without needing a long number of iterations to acquire a precise knowledge of the application.

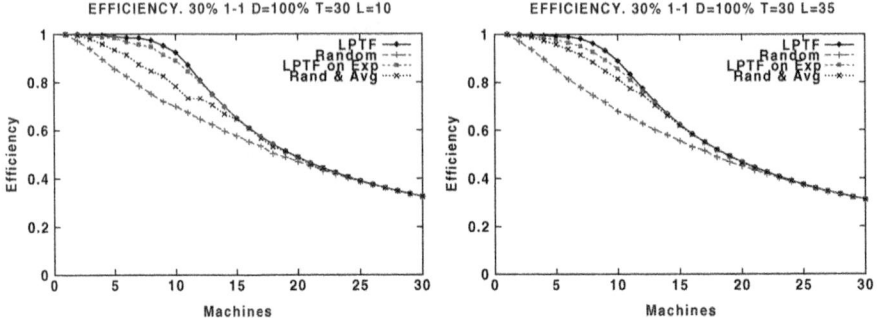

Fig. 2. Effect of varying the number of iterations. (a) L=10 (b) L=35

Effect of the workload (W): Figure 3 shows the effect of varying the workload, considering 30% and 60% workload, 0% deviation and the same execution time for all the largest tasks, and for all the smallest tasks. As expected, for large workloads the number of workers that can usefully be busy is smaller than for small workloads. Moreover, when the workload is higher, efficiency declines faster. A large workload also implies a smoother curve in efficiency. It is important to point out that in all cases there is a point from which efficiency continuously declines. Before that point, small changes in the number of workers may imply significant and contradictory changes in efficiency.

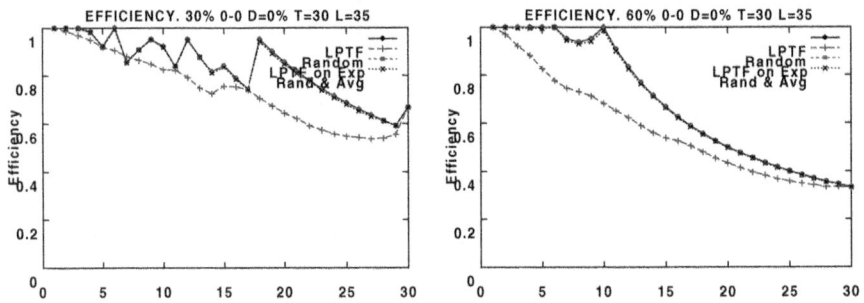

Fig. 3. Effect of varying the workload. (a) W=30% (b) W=60%

Effect of the tasks sizes (i-i): The 20% largest tasks determine when the drop of efficiency begins. If they have the same execution time the decay in efficiency is delayed. The 80% smallest tasks have less influence, they basically determine the smoothness of the efficiency curve. If the 80% smallest tasks have the same execution times the efficiency curve have more peaks.

Effect of the variation (D): When deviation is higher, efficiency declines more. But it is worth noting that it does not decline abruptly even when deviation is 100%. For all policies, even for high values of deviation (60% or 100%), efficiency was never worsen more than 10% of the efficiency obtained with 0% deviation.

Finally, Figure 4 illustrates the overall behavior that we have obtained for the execution time when using the different scheduling policies. The execution time is measured in terms of the relative differences with the execution time of *LPTF* policy. As can be see, the *Random* policy always exhibits the worst execution time, especially when an intermediate number of processors are used. *Random & Average* and *LPTF on Expectation* achieve an execution time comparable to the execution time of *LPTF* even in the presence of a high variation in the execution time of the tasks.

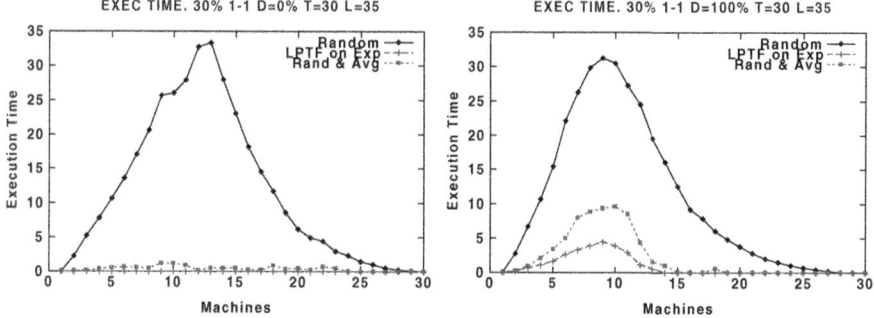

Figure 4. Execution time. (a) D=0% (b) D=100%

4.4. Discussion

We now summarize the main results that have been derived from all the simulations.

The number of iterations does not significantly affect either efficiency or execution time. The behavior of the policies was very similar for all the number of workers, but it was strongly affected by the variation of the execution times of the tasks in different iterations, by the workload and by having significant differences among the execution times of the 20% largest tasks.

Table 1 shows the efficiency bounds obtained for the previously described scheduling policies, always relative to *LPTF* policy. The first column contains the upper bound that is never surpassed in 95% of cases. The second column shows the upper bound for all the cases, which always corresponded to 30% 0-0 workload with D=100%, that is, tasks without significant execution time differences and with high variance. As can be seen, both *LPTF on Expectation* and *Random & Average* in most cases obtained an efficiency similar to the efficiency obtained by a policy such as *LPTF* that uses perfect information about the application. Even in the worst case (scenarios in which all tasks have a similar execution time but a high deviation (100%)) the loss of efficiency for both strategies was 17% approximately.

Slightly better results were obtained for execution time. *Random & Average* and *LPTF on Expectation* never performed worse than 4% in more than 95% of the cases. Only in the presence of high variations were the differences increased to 8%. In all cases, the execution time of the *Random* policy was always between 25% and 30% worse than *LPTF*.

Table 1. Worst efficiency bounds for scheduling policies.

	Eff. Bound in 95% of cases	Worst Efficiency Bound
Random	25,4 %	26,96 %
Random & Average	8,65 %	16,86 %
LPTF on Expectation	8,91 %	17,29 %

As a consequence of the simulations carried out, we can conclude that a simple adaptive strategy such as *Random & Average* will perform very well in terms of efficiency and execution time in most cases. Even in the presence of highly irregular applications the overall performance will not significantly worsen. Similar results have been obtained for the *LPTF on Expectation* policy, but the use of this policy implies that the user needs a good knowledge of the application. Therefore, *Random & Average* appears to be a promising strategy for solving the master-worker scheduling problem.

From our simulation we have also derived an empirical rule to determine the number of workers that must be allocated in order to get a good efficiency and a good execution time. The number of workers depends on the workload factor, on the differences among the execution times of the 20% largest tasks and on the variation of the execution times for different iterations. From our simulation results we have derived empirical table 2 which shows the number of processors that should be allocated, according to our simulations, for obtaining efficiency higher than 80% and execution time lower than 1.1 the time of executing the tasks with as many workers as tasks. This table gives an empirical value for the number of workers that ensures a smooth decrease in efficiency if more workers are added.

Table 2. Percentage of workers with respect to the number of tasks.

Workload	<30%	30%	40%	50%	60%	70%	80%	90%
%workers (largest tasks similar size)	Ntask	70%	55%	45%	40%	35%	30%	25%
%workers (largest tasks diff. size)	60%	45%	35%	30%	25%	20%	20%	20%

4.5. Implementation on a Condor Pool

The effectiveness of the *Random & Average* strategy has been tested in a real test bed, using a Condor [10] pool at the University of Wisconsin. Our applications consisted on a set of synthetic tasks that performed the computation of Fibonacci series. The execution of the application was carried out by using the services provided by MW [11]. In general, we have obtained efficiency values close to 0.8 and speedup values close to the maximum possible for the application [12].

5. Conclusions

In this paper we have discussed the problem of scheduling master-worker applications on clusters of homogeneous machines. We have proposed a scheduling policy that is both simple and adaptive, and takes into account the measurements taken during the execution of the tasks of the master-worker application. Our strategy tries to allocate and schedule the minimum number of processors that guarantee a good speedup by keeping the processors as busy as possible.

We have compared our strategy by simulation with several scheduling strategies using a large set of parameters to model different types of master-worker applications. And we also tested a preliminary version of the scheduling strategy on a cluster of machines, the resources of which were provided by Condor. The preliminary set of tests with synthetic applications allowed us to validate the results obtained in our simulations and the effectiveness of our scheduling strategy. In general, our adaptive scheduling strategy achieved an efficiency in the use of processors close to 80%, while the speedup of the applications was similar to the speedup achieved with a higher number of processors.

We will continue this work by first adapting the proposed scheduling strategy to handle an heterogeneous set of resources. Another extension will focus on the inclusion of additional mechanisms that can be used when the distance between resources is significant.

References

1. R. Buyya (ed.), "High Performance Cluster Computing: Architectures and Systems", Volume 1, Prentice Hall PTR, NJ, USA, 1999.
2. L. M. Silva and R. Buyya, "Parallel programming models and paradigms", in R. Buyya (ed.), "High Performance Cluster Computing: Architectures and Systems: Volume 2", Prentice Hall PTR, NJ, USA, 1999.
3. J. Basney, B. Raman and M. Livny, "High throughput Monte Carlo", Proc. of the Ninth SIAM Conf. on Parallel Processing for Scientific Computing, San Antonio Texas, 1999.
4. J. Pruyne and M. Livny, "Interfacing Condor and PVM to harness the cycles of workstation clusters", Journal on Future Generations of Computer Systems, Vol. 12, 1996.
5. G. Shao, R. Wolski and F. Berman, "Performance effects of scheduling strategies for Master/Slave distributed applications", TR-CS98-598, U. of California, San Diego, 1998.
6. H. Casanova, M. Kim, J. S. Plank and J. Dongarra, "Adaptive scheduling for task farming with Grid middleware", International Journal of Supercomputer Applications and High-Performance Computing, pp. 231-240, Volume 13, Number 3, Fall 1999.
7. D. Abramson, R. Sosic, J. Giddy and B. Hall, "Nimrod: a tool for performing parameterised simulations using distributed workstations", Symposium on High Performance Distributed Computing, Virginia, August, 1995.
8. T. B. Brecht and K. Guha, "Using parallel program characteristics in dynamic processor allocation policies", Performance Evaluation, Vol. 27 and 28, pp. 519-539, 1996.
9. L. A. Hall, "Aproximation algorithms for scheduling", in Dorit S. Hochbaum (ed.), "Approximation algorithms for NP-hard problems", PWS Publishing Company, 1997.
10. M. Livny, J. Basney, R. Raman and T. Tannenbaum, "Mechanisms for high throughput computing", SPEEDUP, 11, 1997.
11. J.-P. Goux, S. Kulkarni, J. Linderoth, M. Yoder, "An enabling framework for master-worker applications on the computational grid", Tech. Rep, U. of Wisc. – Madison, 2000.
12. E. Heymann, M. Senar, E. Luque, M. Livny. "Adaptive Scheduling for Master-Worker Applications on the Computational Grid". Proceedings of the First International Workshop on Grid Computing (GRID 2000). (to appear)

Session III-B

Architecture
Chair: Eduard Ayguade
Technical University of Catalonia

Multi-dimensional Selection Techniques for Minimizing Memory Bandwidth in High-Throughput Embedded Systems

Thierry J-F Omnès[1], Thierry Franzetti[2], and Francky Catthoor[1,3]

[1] Design Technology for Integrated Information and Communication Systems
(DESICS) division, IMEC VzW, Kapeldreef 75, **B**-3001 Leuven, Belgium
[2] Computer Science and Applied Mathematics Dept, INP-ENSEEIHT, Rue Charles
Camichel 2, **F**-31071 Toulouse Cédex, France
[3] EE Dept, KULeuven, Celestijnenlaan 200A, **B**-3001 Leuven, Belgium

Abstract. The idea of Force-Directed Scheduling (FDS) was first introduced by Paulin and Knight [1] to minimize the number of resources required in the high-level synthesis of high-throughput ASICs. In the frame
of our recent Data Transfer and Storage Exploration (DTSE) research
[7,15,18], we have extended FDS for low-cost[1] scheduling in real-time
embedded system synthesis. We have shown that FDS is in fact a projected solution to a more general *multi-dimensional* space/time scheduling problem [19]. By using this reformulation and by introducing (very)
low-complexity *dynamic* and *clustering* graph techniques [21], we have
shown that the interactive design of low-cost but still high-throughput
telecom networks, speech, image and video embedded systems is feasible using a runtime parameterizable Generalized Conflict-Directed Ordering (G-CDO(k)) algorithm. Because G-CDO(k) is based on a true
multi-dimensional design space exploration mechanism, it is in principle
able to analyze design bottlenecks with a much higher resolution than
any other technique. In this paper, we develop novel *multi-dimensional*
selection techniques to allow this powerful feature. Experiments of redesigning the large-scale and parallel Segment Protocol Processor (SPP)
from Alcatel give promising results.

1   Context

Many important software applications deal with large amounts of data, which in
turn have to be mapped onto a (partly) predefined hardware platform. This is
the case both in multi-dimensional signal processing applications such as video
and image processing, which handle indexed signals in the context of loops,
and in communication network protocols, which handle large sets of records
organized in tables. For these multimedia systems, typically a (very) large part

---

[1] By low-cost embedded system, we refer to an embedded system exhibiting a good
compromise between low-power, low-bandwidth, low-memory size (*i.e.* low-area) and
also low-runtime (*i.e.* fast). In our DTSE methodology, this compromise is parameterizable and is a pragmatic choice left to the designer.

M. Valero, V.K. Prasanna, and S. Vajapeyam (Eds.): HiPC 2000, LNCS 1970, pp. 323–334, 2000.
© Springer-Verlag Berlin Heidelberg 2000

of the area cost is due to memory units. The power is also heavily dominated by the storage and transfers. Hence, we believe that the dominating factor in the system-level design is provided by the organization of the global communication and data storage, and related algorithmic transformations. Therefore we have proposed a systematic design methodology in which the memory organization is optimized as a first step [2], **before** the detailed scheduling and mapping onto a (partly) predefined hardware platform is done. Support for this Data Transfer and Storage Exploration (DTSE) methodology is under development in our system-level exploration environment ACROPOLIS [4,15,18].

Storage bandwidth optimization [7] is one of the main steps in our DTSE script. It is executed prior to the Memory Allocation and Assignment (MAA) step [3,16]. The goal of the step is to minimize the required memory bandwidth within the given cycle budget, by adding ordering constraints in the memory access flow-Graph. This allows the memory allocation/assignment tasks to come up with a lower cost memory architecture with a small number of memories and memory ports. Our previous technique to solve the storage-bandwidth optimization step, namely Conflict Directed Ordering (CDO), has resulted in good quality of results for designing large-scale and/or parallel embedded systems under hard real-time constraints [19,20,21]. In this paper, we will enhance the quality of CDO by introducing *multi-dimensional* selection to enhance the quality of results further.

2   Related Work

Scheduling techniques optimizing the number of resources required to meet a deadline have been an active area of research for decades, involving a surprisingly large variety of research communities ranging from operations research for logistics, planning and other military applications; graph theory for high-performance computer-aided design; embedded synthesis and compiler research for more flexible and powerful computer technology to applied artificial intelligence research for next-generation programming environments. All this research relates to some extent to our Conflict-Directed Ordering (CDO) technique although, as will be developing in the coming sections, our particular storage bandwidth optimization formulation leads to a novel and yet still unsolved constrained scheduling problem.

First, the operations research community aims at seeking highly accurate solutions often at the price of complex modeling and large runtime [13]. When computer-aided engineering runtime becomes impractical, using exact solution schemes to a derived subproblem is generally proposed. Within this community, resource-minimal scheduling is typically referred to as a *complex* quadratic-assignment problem [11]. Hence, for practical runtime, *only* a quadratic-assignment problem is solved, which in turn optimizes the use of the bandwidth without analyzing *which* data is being accessed simultaneously.

In the embedded synthesis and compiler domain, most techniques operate on the scalar-level [1,6], which in turn lead to impractical runtime for real-life applications. Exceptions currently are the PHIDEO stream scheduler [6], the Notre-Dame rotation scheduler [5] and DRAM-aware optimizations from the University of California at Irvine [12]. Many of these techniques try to reduce the memory related cost by estimating the required number of registers for a given schedule. Only few of them try to reduce the memory bandwidth, which they do by minimizing the *number* of simultaneous accesses [6]. They do not take into account *which* data is being accessed simultaneously. Also no real effort is spent to minimize the memory access conflict graphs such that subsequent register/memory allocation can do a better job.

In the artificial intelligence domain, it was recently shown that next-generation programming environments [8,9] incorporate features that can solve problems such as the fast synthesis of digital signal processing systems [10]. Such constraint-logic programming environments exhibit powerful global optimization capabilities but, to our knowledge, were never applied for advanced memory bandwidth optimization. We believe this is mainly because a fast and effective problem implementation is required in a first step.

The main difference between our storage bandwidth optimization step and the related work discussed here is that we try to minimize the required memory bandwidth statically in advance by optimizing the access conflict graph for groups of scalars within a given cycle budget. We do this by putting ordering constraints on the flow graph, taking into account *which* memory accesses are being put in parallel (*i.e.* will show up as a conflict in the access conflict graph). Last but not least, our techniques operate on groups of scalars instead of on individual scalars [12], which we show in this paper can lead to highly accurate solutions when combined with *multi-dimensional* selection techniques.

3   Objectives and Contributions

The idea of Force-Directed Scheduling (FDS) was first introduced by Paulin and Knight [1] to minimize the number of resources required in the high-level synthesis of high-throughput Application-Specific Integrated Circuits (ASICs). FDS was later chosen as the basis of the Philips Research - PHIDEO silicon compiler, which resulted in an in-depth study of FDS implementation issues in [6]. State-of-the-art in FDS reports a runtime complexity of $\mathcal{O}(m^2 n^2)$, where n is the number of operations and m is the average time window of each operation, and a feasibility of solving problems of n equals 100 in one minute using the computers of the mid 90s.

Recently, based on the FDS algorithm, an extension called Conflict-Directed Ordering (CDO) was proposed [7] to minimize the required memory bandwidth of embedded multimedia systems. The CDO algorithm involves an important step in the IMEC - ATOMIUM/ACROPOLIS data transfer and storage exploration approaches [4,15]. Because the CDO solution process requires global system-level analysis, pre-existing optimization techniques do not fully hold and a straightforward implementation is left with a runtime complexity of $\mathcal{O}(n^4 m)$. This allows to solve problems for n equals 100 in ten hours, even using today's workstation series. Moreover, because the number of operations found in real-life embedded applications is today growing well beyond n equals 100 [14] and also because parallelism is often introduced which further increases the number of concurrent operations, a novel fast and effective algorithm for the CDO kernel is needed. To improve the designer interaction further [16], it is also desirable to have a controllable trade-off between user time and the quality of the solution (which directly influences the cost of the final solution).

In our recent research, we have revisited the original idea of Paulin and Knight [1] to show that FDS is in fact a projected solution to a more general *multi-dimensional* space/time scheduling problem. Based on this space/time formulation, we have introduced a Generalized Conflict-Directed Ordering (G-CDO) algorithm that instead of transforming data that statistically require too much bandwidth can go down to the precision of transforming $(data, ti - me)$ tuples directly. The high resolution of this technique allows a runtime complexity of $\mathcal{O}(n^2m^2)$ because a look-ahead is not mandatory any more for precision enhancement. Experiments of using the original CDO and GCDO(k) on redesigning Alcatel's Segment Protocol Processor (SPP) give promising results.

4   Generalized Conflict Directed Ordering

In [21], we present our Conflict Directed Ordering (CDO) solution technique. We prove that the original CDO algorithm exhibits a worst-case complexity (in Pairwise Probability of Conflict Computations, PPCCs) of $\mathcal{O}(n^4m)$, which can be reduced to $\mathcal{O}(n^3 ln(n)m)$ using *dynamic* graph techniques and further reduced to $\mathcal{O}(kn^2 ln(n))$ using *clustering* (also called *multilevel*) techniques. Based on those (very) significant quality and runtime improvements, we introduce the LCDO(k) algorithm.

In [19], we show that CDO is in fact a projected solution to a more general *multi-dimensional* space/time scheduling problem. In order to allow for multiple selection and multiple reductions (MSMR) in full, we introduce the multiple reductions (or *clustered* or *multilevel*) version of the Generalized Conflict Directed Ordering algorithm, G-CDO(k), **without** describing its *multi-dimensional* selection process, that is by implicitly reusing 1D selection techniques.

In the following, we apply *multi-dimensional* selection to G-CDO(k). We first show that a deterministic $\mathcal{O}(n^2m)$ implementation does not give satisfactory results. We therefore introduce several probabilistic selection techniques and couple them to the other selection improvements proposed earlier for LCDO in [21]. Thanks to its high-resolution and our selection tuning, allowing to remove space/time background memory conflict **exactly** where they are, we will see that G-CDO(k) outperforms LCDO(k) for designing large scale and/or parallel systems under stringent real-time constraints in runtime and in quality. The design of Alcatel's Segment Protocol Processor (SPP) will be used as a typical illustration throughout the presentation.

4.1   Deterministic Selection (DS)

G-CDO is a multi-dimensional space/time scheduler based on a Multi-Dimensional Conflict Graph (MDCG) [19]. As for the localized version of CDO (LCDO), the complexity of G-CDO can be largely reduced by applying *dynamic* graph techniques [21]. Let us first study the simplest way of minimizing bandwidth, that is by using the following cost function:

$$c_1(CG) = \max_t \sum_{(x,t)} \sum_{E(x,t)} w(e).1 \tag{1}$$

The effect of removing a node (x,t) from MDCG is first to decrease $\sum_{(x,t)}$ and then eventually to also decrement the maximum over t, $\max_t$. Because the selection process always picks-up the node (x,t) implying the highest cost decrease, it is easy to demonstrate that $\max_t$ is decremented only once in a while when all times exhibiting a high bandwidth cost have been decremented. This principle is general and could apply to any constantly decreasing cost function, even in the presence of global information like the chromatic number (number of memories). Hence, it is in general feasible to implement the above G-CDO selection process in $\mathcal{O}(ln(m))$ after a clever $\mathcal{O}(n^2m)$ initialization phase[2]. In turn, the reduction can involve a local ASAP/ALAP update such as the one of LCDO, leading to the following worst-case complexity:

$$w_2(FG) = \sum_{v \in EFG} \sum_{i=2}^{m(v)} \frac{1}{i} \times sf(v)$$
$$\equiv \mathcal{O}(nln(n)ln(m)) \tag{2}$$

$$\mathcal{O}(n^2m) + \mathcal{O}(nmln(m)) +$$
$$\sum_{i=1}^{n(m-1)} (1 + \frac{2}{n(m-1)}.w_2(FG)).d_i^t$$
$$\leq \mathcal{O}(n^2m + n^2ln(n)ln(m)) \tag{3}$$

By comparing formula (3) to formula (5) in [21], we prove that a multi-dimensional scheduler (G-CDO) can be less time consuming than its mono-dimensional counterpart LCDO.

Unfortunately, as shown in Figure 1, an optimally dynamic version of G-CDO does not give satisfactory results in practice. A tradeoff between quality and runtime is therefore needed to compete with the original LCDO.

---

[2] A max selection can be implemented in $\mathcal{O}(ln(m))$ by introducing a sorted list of $\sum_{(x,t)}$. The constantly decreasing property can guide the optimization of the sorted list accessor.

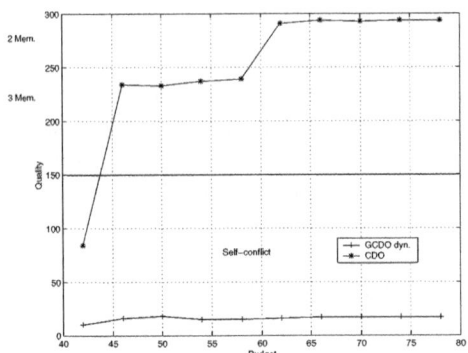

Fig. 1. Compared quality results for optimally dynamic G-CDO vs. LCDO using our global architecture-aware cost model. Only 2 concurrent memories are needed above 250, 3 concurrent memories above 200 and more below. Below 150, the introduction of expensive multi-port memories becomes unavoidable.

4.2   Time-Hashed Probabilistic Selection (THPS)

In order to improve the quality of G-CDO, we propose to also generalize the pairwise probability of conflict computation (PPCC) present in CDO. For pure bandwidth optimization, this is solved by using the following cost function:

$$c_2(CG) = \max_t \sum_{(x,t)} \sum_{E(x,t)} \omega(e).p(x) \tag{4}$$

$$N(x) = \sum_{(x,t)} 1 \tag{5}$$

$$p(x) = \frac{1}{N(x)} \tag{6}$$

Unfortunately, the effect of removing (x,t) from MDCG becomes complex. First, $\sum_{(x,t)}$ is decreased in $\mathcal{O}(d_i^t)$, where d_i^t is the degree of the nodes within the time projected conflict graph, $CG(t)$, at step i, $d_i^t \leq n$. Second, the number of time occurrences of access x, $N(x)$, is decreased by 1. Third, for all t where x is present, $p(x)$ is increased by $\frac{1}{N(x)} - \frac{1}{N(x)-1} = \frac{1}{N(N+1)(x)}$, involving $\mathcal{O}(d_i^t \times m)$ updates. Fourth, $\max_t$ has to be recomputed. In turn, maintaining feasibility of the solution is ideally achieved by local ASAP/ALAPs relying on $d_i^t \times m$ individual reductions in the loop:

$$\mathcal{O}(n^2 m) + \mathcal{O}(nm ln(m)) +$$

$$\sum_{i=1}^{n(m-1)} (1 + \frac{2}{n(m-1)}.c_2(FG)).d_i^t.m$$

$$\leq \mathcal{O}(n^2 m^2 + n^2 m ln(n) ln(m)) \tag{7}$$

By comparing formula (8) to formula (5) in [21], we prove that a fully generalized CDO (G-CDO) is slightly less time consuming than LCDO. This situation is improved when multi-level techniques are applied [19], thanks to a larger multi-dimensional search space.

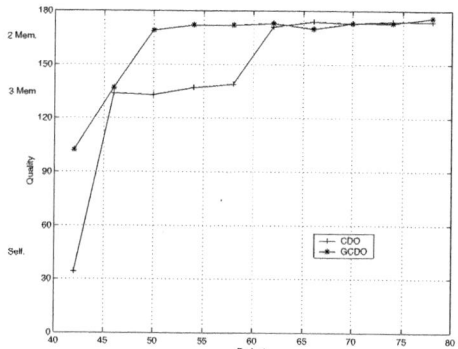

Fig. 2. Compared quality results for dynamic G-CDO vs. LCDO. using our global architecture-aware cost model. Only 2 concurrent memories are needed above 150, 3 concurrent memories above 120 and more below.

Fortunately, as shown in Figure 2, our time-hashed selection technique already gives better quality of results than CDO while still remaining dynamic.

4.3   Multi-dimensional Probabilistic Selection (MDPS)

In order to improve further the quality of G-CDO, it is possible to introduce a summation-based cost function, modeling a multi-dimensional selection process:

$$c_3(CG) = \sum_t \sum_{(x,t)} \sum_{E(x,t)} \omega(e).p(x) \tag{8}$$

The runtime complexity induced by c_3 is very similar to that of c_2, except that the selection becomes more complex. Indeed, in order to keep track of the node to be selected, one needs to first find the right t in $\mathcal{O}(ln(m))$ and then the right (x,t) in $\mathcal{O}(ln(n))$.

$$\mathcal{O}(n^2m) + \mathcal{O}(nmln(n)) +$$

$$\sum_{i=1}^{n(m-1)} (1 + \frac{2}{n(m-1)}.w_2(FG)).d_i^t.m$$

$$\leq \mathcal{O}(n^2m^2 + n^2mln(n)ln(m)) \tag{9}$$

By comparing formula (9) to formula (5) in [21], we prove that a multi-dimensional selection does not impact the runtime complexity of G-CDO.

Moreover, as shown in Figure 3, multi-dimensional selection can in some cases improve the quality of the original time-hashed solution.

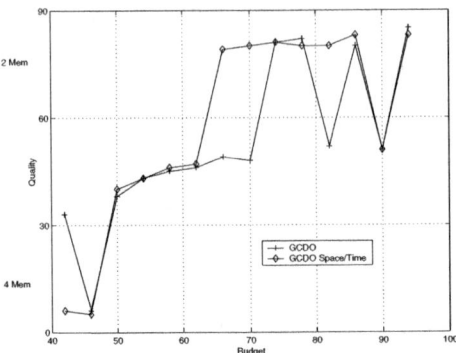

Fig. 3. Quality enhancement of multi-dimensional selection using our global architecture-aware cost model. Only 2 concurrent memories are needed above 60 and more below.

4.4   THPS with Probabilistic Constraint Propagation (CP)

When designing large systems, it is common practice to first divide the overall system into p smaller subsystems. When applying automatic design techniques such as G-CDO to those subsystems, it becomes much easier to design each subsystem independently in $\mathcal{O}(pn^2m^2)$ than to design and optimize the overall system in full in $\mathcal{O}((pn)^2m^2)$. This divide and conquer mechanism is especially interesting in the case of local optimization techniques, typically found in the heuristics used to make a system as fast as possible, because nothing more than just designing or optimizing p times is needed.

Unfortunately, optimizing for making a system as cost-effective as possible, requires **global** optimization techniques. In the case of CDO for example, deciding *which* data can be put in parallel requires (at least) a global track of *which* data have already been put in parallel on other parts of the system. Hence, it is still possible to apply divide and conquer techniques to G-CDO **only** if global constraints are propagated from one subsystem to another. This is for example the case in our data transfer and storage exploration (DTSE) methodology while optimizing over several nested loops [17]. Each nested loop is regarded as an individual system requiring an individual scheduling.

We propose to model constraint propagation by introducing a $\delta(e)$ in the cost function. This δ means that the knowledge of previous work realized on other parts of the system guides to either take into account a cost factor ($\delta(e) = 1$) because it is new or to cancel it ($\delta(e) = 0$) because it has already been taken into account previously.

$$c_4(CG) = \max_t \sum_{(x,t)} \omega . \delta(e) . p(x) \tag{10}$$

$$N(x) = \sum_{(x,t)} 1 \tag{11}$$

$$p(x) = \frac{1}{N(x)} \tag{12}$$

By default, all $\delta(e)$ are equal to 1. In a first step, introducing δ in the G-CDO algorithm impacts the cost of the individual MDCG edge cost computation (ECC), which remains a $\mathcal{O}(1)$ process anyway. In a second step, added computation is also introduced when deciding of keeping a conflict, hence necessitating to switch $\delta(e)$ to 0, again in $\mathcal{O}(1)$ provided the right data structure for ECC is put into practice[3]. Hence, constraint propagation exhibits the same upper-bound complexity than conventional G-CDO. Some overhead can however be expected in practice.

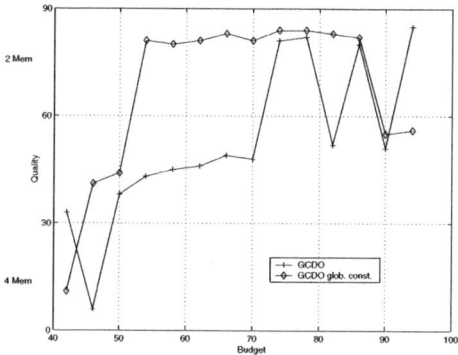

Fig. 4. Quality enhancement of constraint propagation using our global architecture-aware cost model. Only 2 concurrent memories are needed above 60, 3 concurrent memories above 30 and more below.

In Figure 4, we show that constraint propagation has a very positive impact on the quality of G-CDO. Indeed, it really makes sense to re-use the knowledge of problems encountered on other parts of the system to guide new optimizations for new subsystems.

4.5   THPS with Deterministic Constraint Propagation (CP)

It is also possible to implement constraint propagation on a deterministic conflict graph (CG), that is on a CG where edges are introduced only when a conflict is 100% certain. In that way, the G-CDO algorithm focuses on reducing nodes that are certain to generate a conflict. By introducing a deterministic CG, the G-CDO upper-bound complexity is not affected.

Surprisingly, as shown in Figure 5, a deterministic CG can contribute to further improve the quality of G-CDO. Indeed, it really makes sense to focus only on those design problems that have 100% chance to occur.

---

[3] A two-level pointing technique to compute $\omega(e).\delta(e)$ "on the fly".

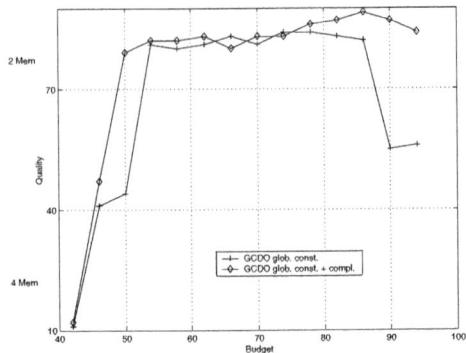

Fig. 5. Quality enhancement of deterministic constraint propagation reusing the axis definition of Figure 4.

5   Conclusion and Perspectives

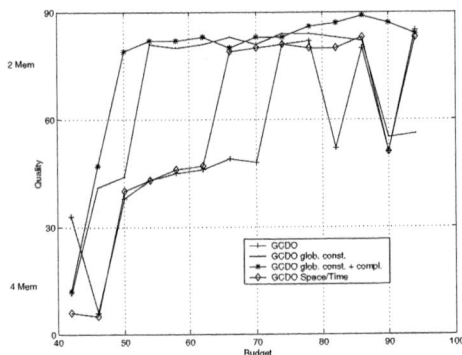

Fig. 6. Quality results for all presented G-CDO variants reusing the axis definition of Figure 4.

In Figure 6, we summarize the quality of our four G-CDO variants. By combining all quality enhancement techniques, our solution is very significantly improved for hard real-time designs. In particular, designing Alcatel's Segment Protocol Processor (SPP) for a real-time budget in the range of 50 to 70 requires two instead of initially three single-port memories.

Those experiments confirm that *multi-dimensional* selection can be a powerful quality-enhancing technique for designing large-scale and/or parallel low-cost but still high-throughput embedded systems under hard real-time constraints.

Acknowledgments

We thank Arnout Vandecappelle and Erik Brockmeyer for providing support to the use of their tools at IMEC. We wish to thank the European Commission for partly funding this work in the frame of the MEDEA System-Level Methods and Tools (SMT) project and G. de Jong from Alcatel for use of the Segment Protocol Processor (SPP) application.

References

1. P.G.Paulin and J.P.Knight, *"Force-Directed Scheduling in Automatic Data Path Synthesis"*, Proc. 24th ACM/IEEE Design Automation Conf., Miami Beach, FL, IEEE Computer Society Press, pp. 195–202, June 1987.
2. I.Verbauwhede, F.Catthoor, J.Vandewalle, H.De Man, *"Background memory management for the synthesis of algebraic algorithms on multi-processor DSP chips"*, Proc. VLSI'89, Int. Conf. on VLSI, Munich, Germany, pp. 2098–2118, Aug. 1989.
3. P.Lippens, J.van Meerbergen, W.Verhaegh, A.van der Werf, *"Allocation of multi-port memories for hierarchical data streams"*, Proc. IEEE Int. Conf. Comp. Aided Design, pp. 728–735, Santa Clara, Nov. 1993.
4. Lode Nachtergaele & al., *"Optimization of memory organization and partitioning for decreased size and power in video and image processing systems"*, IEEE Int. Workshop on Memory Technology, pp. 82–87, San Jose, CA, Aug. 1995.
5. N.Passos, E.Sha, *"Push-up scheduling : optimal polynomial-time resource constrained scheduling for multi-dimensional applications"*, Proc. IEEE Int. Conf. Comp. Aided Design, San Jose, CA, pp. 588–591, Nov. 1995.
6. W.Verhaegh, E.Aarts, J.Korst, P.Lippens, *"Improved Force-Directed Scheduling (IFDS) in high-throughput digital signal processing"*, IEEE Transactions on Computer-Aided Design, Vol. 14, p. 945–960, 1995.
7. S.Wuytack & al., *"Flow Graph Balancing for Minimizing the Required Memory Bandwidth"*, Proc. 9th ACM/IEEE Int. Symposium on System-Level Synthesis, La Jolla CA, pp.127–132, Nov. 1996.
8. Cosytec, *"Application development with the CHIP system"*, 1997. Available at: http://www.cosytec.fr/.
9. C.Schulte, *"Oz Explorer: A Visual Constraint Programming Tool"*, Proc. 14th Int. Conf. on Logic Programming, Lee Naish (eds), MIT Press, Cambridge, pp. 286–300, July 1997.
10. Jurgen Teich, E.Zitzler, S.Bhattacharyya, *"Optimized software synthesis for digital signal processing algorithms - an evolutionary approach"*, Proc. 1998 IEEE Workshop on Signal Processing Systems (SiPS), 1998.
11. Q.Zhao, S.Karisch, F.Rendl, H.Wolkowicz, *"Semidefinite programming relaxations for the quadratic assignment problem"*, University of Waterloo Tech. Report CORR-95-27, University of Copenhagen DIKU TR-96/32, Feb. 1998.
12. P.R.Panda, N.D.Dutt, A.Nicolau, *"Incorporating DRAM access modes into high-level synthesis"*, IEEE Trans. CAD Integrated Circuits and Systems, Vol. 17, No. 2, Feb. 1998.
13. C.Ancourt & al., *"Automatic data mapping of signal processing applications"*, ENSMP-CRI, UVSQ-PRiSM, Thomson CSF-LRC, 1998.
14. J.L.da Silva Jr. & al., *"Efficient System Exploration and Synthesis of Applications with Dynamic Data Storage and Intensive Data Transfer"*, Proc. 35th ACM/IEEE Design Automation Conf., pp. 76–81, 1998.

15. F.Catthoor & al., *"Custom Memory Management Methodology - Exploration of Memory Organization for Embedded Multimedia System Design"*, Kluwer Academic Publishers, ISBN 0-7923-8288-9, 1998.
16. A.Vandecappelle & al., *"Global Multimedia System Design Exploration using Accurate Memory Organization Feedback"*, Proc. 36th ACM/IEEE Design Automation Conf., pp. 327-332, June 1999.
17. E.Brockmeyer, S.Wuytack, A.Vandecappelle, F.Catthoor, *"Low power storage cycle budget distribution for hierarchical graphs"*, DATE Userforum, pp. 249–254, April 2000.
18. F.Catthoor, K.Danckaert, C.Kulkarni, T.Omnès, *"Data transfer and storage architecture issues and exploration in modern DSPs"*, Book chapter in Programmable Digital Signal Processors: Architecture, Programming, and Applications (ed. Y.H.Yu), Marcel Dekker, Inc., New York, 2000.
19. T.Omnès, T.Franzetti, F.Catthoor, *"Interactive algorithms for minimizing memory bandwidth in high throughput telecom and multimedia"*, Proc. 37th IEEE/ACM Design Automation Conf., pp. 328–331, June 2000.
20. Thierry Franzetti, *"Survey of Scheduling Algorithms for the Design of Embedded Multimedia Systems"*, Joint IMEC - Computer Science and Applied Mathematics Department of INP-ENSEEIHT internship report, June 2000.
21. T.Omnès, T.Franzetti, F.Catthoor, *"Dynamic algorithms for minimizing memory bandwidth in high throughput telecom networks, speech, image and video embedded systems"*, in special issue of Technique et Science Informatiques (TSI), Techniques de parallélisation automatique, C.Mongenet, S.Rajopadhye and Y.Robert (eds), Éditions Hermès, 2000.

Energy-Aware Instruction Scheduling

A. Parikh, M. Kandemir, N. Vijaykrishnan, and M.J. Irwin

Department of Computer Science and Engineering
The Pennsylvania State University
University Park, PA 16802-6106
{aparikh,kandemir,vijay,mji}@cse.psu.edu

Abstract. Energy consumption is increasingly becoming an important metric in designing computing systems. This paper focuses on instruction scheduling algorithms to reduce energy. Our experimentation shows that the best scheduling from the performance perspective is not necessarily the best scheduling from the energy perspective. Further, scheduling techniques that consider both energy and performance simultaneously are found to be desirable. We also validate the energy estimates of the instruction schedules obtained using the energy transition cost tables by executing the instructions on a cycle-accurate energy simulator. We conclude the paper with a discussion of alternate ways of building energy transition tables and estimating absolute energy consumption.

1 Introduction

Energy aware computing and communication are becoming increasingly important with the proliferation of battery-powered personal devices. The increasing software content in the mobile devices and the significant energy reduction that has been obtained using various compiler-based techniques motivate work in compilation area that has traditionally focused on optimizing performance.

This paper explores the influence of instruction scheduling on energy. In particular, we focus on the differences between various performance and energy oriented scheduling mechanisms. This work uses energy tables obtained using a cycle-accurate transition-sensitive energy simulator [11]. Some of the difficulties associated with current measurement-based models described in [5] are avoided using this approach, while providing the benefit of cycle-accurate energy measurements.

The cycle-accurate simulator used in this study, *SimplePower* [10], models energy consumption of an architecture, consisting of the processor data-path (five-stage pipeline), on-chip instruction and data caches, off-chip memory, and the interconnect buses between the core and the caches and between the caches and the off-chip memory. *SimplePower* outputs the energy consumed from one execution cycle to the next. It mines the transition sensitive energy models based on 0.8μ, 3.3V techno logy provided for each functional unit and sums them to estimate the energy consumed by each instruction cycle. The energy consumed by the instruction cache and data caches is evaluated using an analytical model

M. Valero, V.K. Prasanna, and S. Vajapeyam (Eds.): HiPC 2000, LNCS 1970, pp. 335–344, 2000.

that has been validated to be accurate (within 2.4% error) for conventional cache systems [2,6].

In the next section, we present different instruction scheduling algorithms. In Section 3, we evaluate the algorithms in terms of energy and performance. Section 4 presents some of the issues in developing energy models. Finally, we conclude in Section 5.

2   Instruction Scheduling Algorithms

2.1   Performance-Oriented Scheduling

Instruction scheduling is one of the most important optimizations performed by a compiler [3]. The scope of instruction scheduling might be a basic block, a number of basic blocks, a procedure, or an entire program. The most well-known approach to basic-block scheduling is *list scheduling* [3].

We can define *basic block,* in a given code, as a sequence of instructions where the thread of execution enters at the beginning (i.e., the first instruction) and exits at the end (i.e., the last instruction), without the possibility of branching except maybe at the end [1]. A common representation used in compilers for a basic block is a directed acyclic graph (DAG), in which each node represents an instruction and an edge between two nodes specifies a *data dependence* that needs to be preserved in any instruction reordering for optimization.

In the basic list scheduling algorithm,each node of the input DAG has a weight that denotes the execution time of the instruction and each edge between two nodes has a weight that gives the number of latency cycles between executions of these instructions. The approach starts by computing a *delay function* for each node that denotes the maximum possible delay from that node to the end of the basic block. Next, the DAG is traversed from the root towards the leaves, selecting the nodes to schedule. In doing so, the algorithm also keeps track of the current time and earliest time that each node can be scheduled.

At each step, the algorithm selects a node from a set of *candidate nodes* that have not been scheduled yet, but all of whose predecessors have been [3]. If there is only a single node in the candidate set, then it is the obvious choice; otherwise, the algorithm selects the node with the *maximum* potential delay as the next node to be scheduled.

2.2   Energy-Oriented Scheduling Algorithms

Top-down Scheduling: The algorithm for top-down scheduling is driven by a table called *energy transition cost* (ETC) *table.* The objective of this pure energy-oriented scheduling is to *minimize* the total transition costs for instructions in a given basic block by reordering them, taking into account the switching activity between successive instructions [4]. First, it selects one of the schedulable nodes (say i) and schedules it. In the next step, it attempts to select a node j such that the energy transition cost (obtained from ETC) between i and j is minimal among all possible alternatives; in the following step, a node k is selected such that the circuit-state cost between j and k is minimum, and so on. Note that this approach is also greedy. One noticeable difference between this approach and the classical performance-oriented list scheduling is that the latter considers the maximum delay among the candidate nodes whereas the former considers only the last scheduled node. The algorithm proposed by Tiwari et al. [7,8] uses a similar approach with one important exception. In addition to transition cost, they also define a *base cost* for each instruction; this corresponds to the energy spent in executing the instruction in isolation. We believe that in a pipelined architecture it is very difficult to obtain this cost exclusively; instead, we model the total energy cost due to an instruction sequence consisting of instructions i and j as the energy spent in running i and j one after another, which comprises all switching activity during this execution.

Look-ahead Approach: In this approach, which is a variant of the top-down scheduling scheme, instead of considering only the next immediate node to schedule, we consider the next two (or more) nodes at a time. In this way, the algorithm tries to obtain the minimum total transition cost (circuit-state effect) *not* up to (and including) i but up to (and including) the node that will immediately follow i in the final schedule.

Energy-with-Performance Scheduling: In this algorithm, we follow the top-down energy-oriented scheduling, but to select the first node to be scheduled and also to resolve the ties,when there exist multiple schedulable nodes with the same circuit-state cost, we take performance into consideration, i.e., we select and schedule the node with maximum delay. In case there are multiple nodes with the maximum delay, the choice is arbitrary. Since earlier, during energy-oriented scheduling, we were not considering performance at all, one would expect this algorithm to either improve performance or at least give the same performance. As far as energy is concerned, we do not expect a major deviation from the energy performance of the energy-oriented algorithm as we use the performance metric (delay function) only to break ties.

Performance-with-Energy Scheduling: In this technique, we follow performance-oriented list scheduling, but to resolve the ties we take energy criterion, i.e., the minimum inter-instruction energy transition cost (circuit state effect), into consideration. We expect this algorithm, as compared to the pure performance-oriented scheduling, to reduce energy consumption without a major variation in performance results (as we apply energy scheduling only when there is a tie). The magnitude of the reduction in energy consumption depends largely on how frequently the circuit state effects are used as tie-breakers.

Energy-Performance Scheduling: We also implemented another instruction scheduling algorithm which used a *product of delay* and *inter-instruction cost* as the main criterion to decide the next instruction to be scheduled. The idea is that such an energy×delay metric might be a good compromise between energy-oriented and performance-oriented scheduling techniques.

Table 1. The energy transition cost (ETC) table used in the experiments. [All values are in picoJoules (pJ)].

	lw	sw	mv	add	mult	sll
lw	574.0	302.7	874.9	904.9	660.6	480.4
sw	302.7	239.3	715.3	741.0	503.6	449.9
mv	877.9	718.6	244.7	272.3	406.9	319.7
add	907.9	744.2	272.3	244.2	405.4	350.6
mult	663.6	506.8	407.0	405.5	658.7	629.0
sll	496.2	465.6	331.1	361.9	638.8	248.1

3   Experiments

In this section, we first describe the different data structures used and then explain results of the experiments.

Building the Energy Transition Cost Table: We used *SimplePower* to build our ETC table. *SimplePower* simulates the energy consumption of a five-stage pipelined architecture that implements the integer subset of the SimpleScalar instruction set architecture and the associated memory hierarchy and bus interface. In this study, we focus on the impact of the scheduling algorithms on the datapath of the *SimplePower* architecture. This tool accepts either high-level C code or assembly instructions to estimate the energy consumed by the code. Figure 1 depicts a graphical overview of our system.

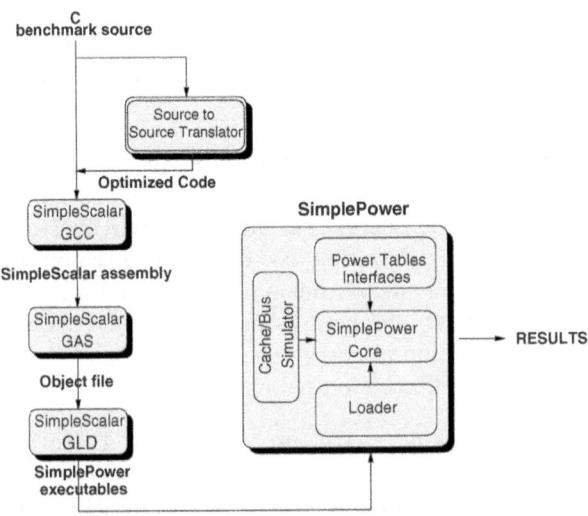

Fig. 1. *SimplePower* simulation flow overview.

In order to evaluate seven scheduling algorithms (including a *random-scheduling technique* which selects the next instruction to be scheduled randomly taking into account the dependence constraints), we selected five frequently used integer instructions from the SimpleScalar ISA: load word (`lw`), store word (`sw`), move (`mv`), add (`add`), multiply (`mult`), and shift-left logical (`sll`). To determine the transition cost between instructions i and j, we built an instruction sequence by repeating the pattern of instruction i followed by instruction j a number of times, thereby obtaining a large number of transitions (called *repetition count*).[1] Then, we measured the energy consumed by this sequence using our simulator, and after deducting the overhead costs, divided the remaining energy cost by the repetition count to obtain an entry in the ETC table. We found a repetition count value of 1600 to be acceptable. In fact, beyond 1600, increasing the repetition count did not change the computed transition costs. The final energy transition costs are given in Table 1.

DAG Generation: We wrote a program to generate random DAGs on which our scheduling algorithms are run. Our DAG generation algorithm takes the number of nodes, edges, and the types and frequencies of instructions as input and generates a random DAG . The latency upper and lower bounds can also be specified as inputs. We conducted experiments using DAGs of different sizes (in terms of nodes (N) and edges (E)). The nodes were assigned one of the five instructions of our target architecture randomly using a uniform distribution. Node sizes used are 4, 8, 16, 32, and 64, and for each node size, we used three different edge sizes, corresponding to three cases with small number of dependences, medium number of dependences, and large number of dependences, respectively. In our experiments, we used a latency value of 1 between each pair of dependent instructions.

Table 2. Execution cycles.

	Random Schdl.	Top-Down	Energy w/Perf.	Look-ahead	Perf. Schdl.	Perf. w/Energy	Energy -Perf.
(N=4,E=2)	14	14	14	14	14	14	14
(N=8,E=14)	20	19	19	19	19	19	19
(N=16,E=35)	31	29	29	29	30	30	29
(N=32,E=100)	51	50	50	50	49	50	51
(N=64,E=60)	94	93	94	93	91	90	90

Results: We have implemented all the scheduling algorithms discussed in this paper. We now present our experimental results in two parts. First, we present the execution cycles for the code sequences generated using different scheduling algorithms for five different DAG sizes. The representative results given in Table 2 are obtained through our simulator using the final code sequences generated

---

[1] The only difference in obtaining the inter-instruction transition cost between instructions j and i and instructions i and j is that the first instruction in the repeated pattern is instruction j in the former and instruction i in the latter.

by each scheduling algorithm. These results show that, as far as the execution cycles are concerned, all the algorithms exhibit similar behavior in most of the cases. So, *the instruction scheduling algorithms that take energy into account can generate similar performance results to those obtained using pure performance-oriented scheduling techniques.*[2]

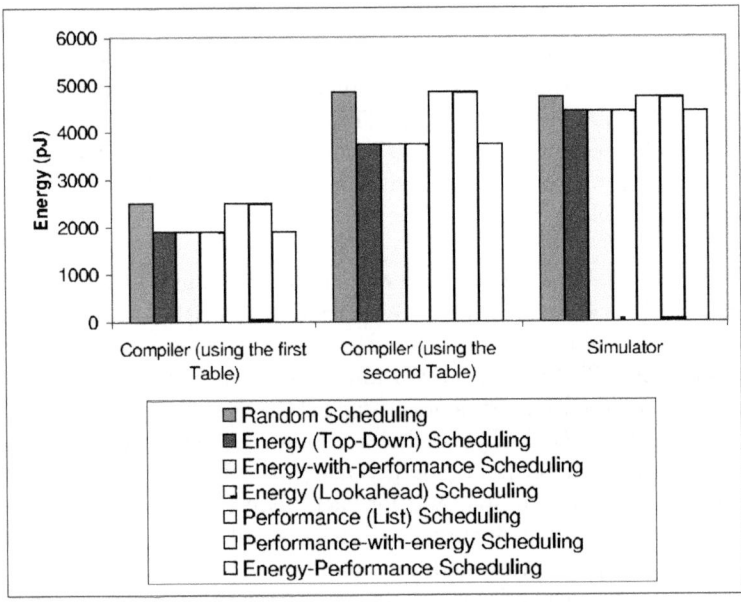

Fig. 2. Energy results obtained through simulator and compiler (with two different table-building methods).

We next present energy results for all DAG configurations we generated. First, we obtained the schedules for each of the proposed algorithms using the ETC table. Then, the sum of the transition costs of instructions in each resulting schedule gives us the compiler-estimated total energy consumption. We also fed the resulting instruction sequences obtained from the compiler to *Simple-Power*. The energy estimates obtained through the simulator are cycle-accurate and serve as a measure of the accuracy of the compiler estimates. In the graphs presented in Figure 3, for each DAG size, we compare the simulator results and compiler estimates. Let us first focus on compiler estimates (denoted `Compiler`). The first thing to note is that, with small DAGs, all scheduling algorithms generate similar energy results. With increasing DAG size, however, the difference between energy and performance scheduling becomes more pronounced. It is

---

[2] The results obtained using the compiler-based estimations instead were slightly different where the maximum difference between pure performance-oriented scheduling and top-down energy-oriented scheduling was 4 cycles in favor of the former.

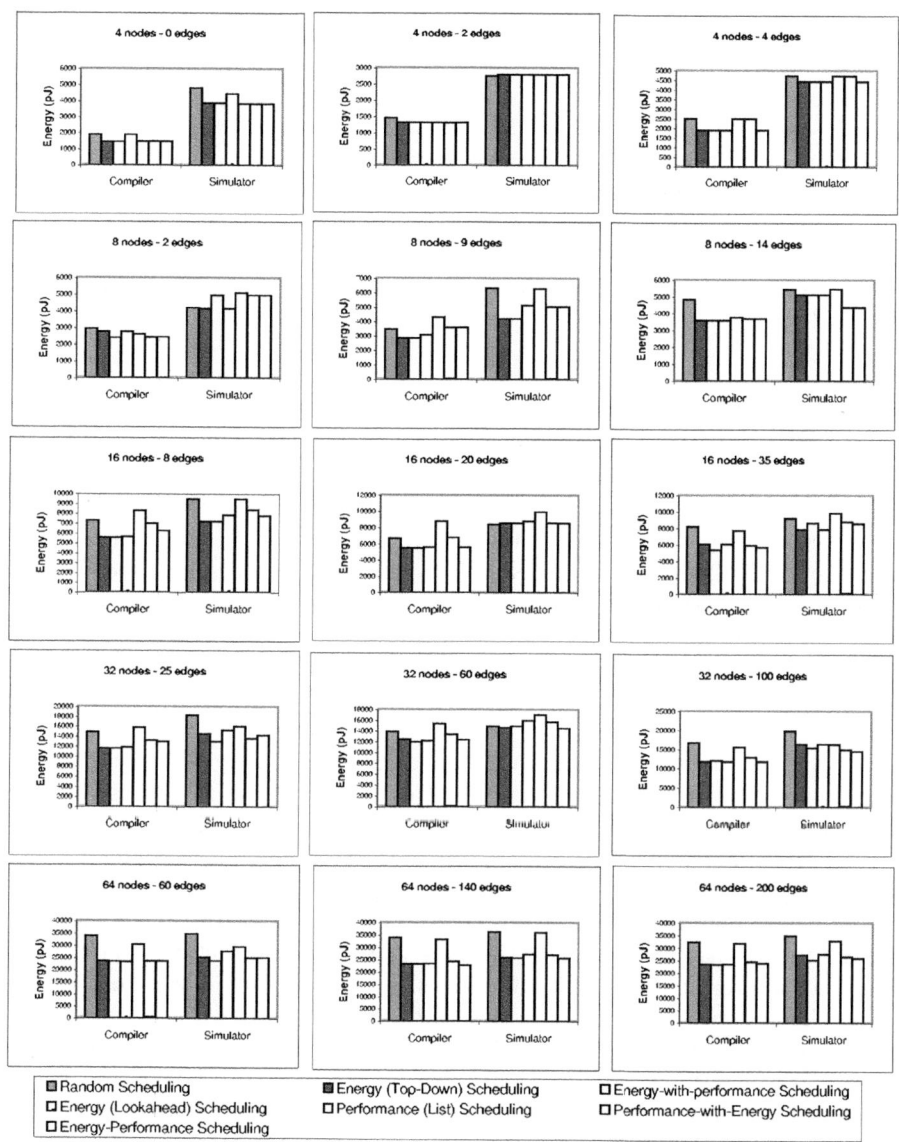

Fig. 3. Energy results.

interesting to note, however, that performance-with-energy scheduling achieves similar results to those obtained by top-down approach and in some cases the former even outperforms the latter. This is due to large number of schedulable nodes at each scheduling step for the DAGs we used. The larger the number of schedulable nodes, the higher the chances that the energy constraints will be taken into account during performance-with-energy scheduling. Another observation is that look-ahead scheduling does not provide any visible advantage over top-down approach. This is because,while the look-ahead scheme increases the greediness of the top-down approach, it also restricts the freedom of movement for the instructions yet to be scheduled. That is, the more greedy the algorithm is, the more restrictions we have for the remaining instructions. Finally, for many cases, top-down scheduling, energy-with-performance scheduling, performance-with-energy scheduling, and energy-performance scheduling generated very similar results. *Note that these experiments show that the best scheduling (or instruction sequence) from the performance point of view is not necessarily the best scheduling (or instruction sequence) from the energy point of view.*

4   Validation and Discussion

When we compare the scheduling algorithm (denoted `Compiler`) results with simulator results (denoted `Simulator`), we see that, as far as the general trend is concerned, the simulator results (i.e., those obtained by running the resulting codes through our simulator) follow the compiler(-estimated) results. Specifically, 11 out of 15 cases, the compiler and simulator found the same highest-energy code, and 10 out of 15 cases they found the same lowest-energy code. Thus, *if we are interested only relative performances (energy-wise) of different scheduling algorithms, we can use the compiler estimation as a first approximation in many cases.*

However, in many energy-aware environments, estimating absolute energy consumption of a given piece of code is of extreme importance. We see from Figure 3 that the compiler estimations are in general lower than the simulator results. To understand the reason, consider the architectural configuration simulated by *SimplePower*. In this model, a pipelined architecture is modeled with a pipeline depth of 5. So, two instructions that co-exist in the pipeline do not need to be consecutive; in fact, they might be three instructions apart. The table-based approach used in this paper does not capture the influence of factors such as the operands, instruction location, the interaction with instructions in other stages of the pipeline, and pipeline stalls while evaluating the energy table entries. Consequently, this results in an under-estimation of the total energy consumption for a given code sequence. *The conclusion therefore is that while the compiler estimation based on transitive costs between consecutive instructions can be used to some extent for measuring the relative performances of different scheduling algorithms, it is not a reliable method for estimating absolute energy consumption.*

We expect that a solution to this problem can be found by refining the table-driven approach used in the experiments. In particular, we can modify the way that the ETC table is constructed. Although one may think that defining base costs and transition costs separately (as done in Tiwari et al. [7]) would solve this problem, two observations refute this idea. First, determining base costs for a *pipelined machine* is very difficult. Second, as long as the set of instructions to be scheduled is *fixed*, a table-driven scheduling *cannot* change the total base cost and can only affect the total transition cost, which is defined only between two subsequent instructions. *Therefore, the same problem with table-driven approach exists even if we consider base costs separately from transition costs.*

We now explore another way of defining table entries . Instead of running replicated sequences of instruction pairs and dividing the total energy cost by the replication count, we calculate the energy consumed in six consecutive cycles starting with the first instruction in the fetch (IF) stage and ending with the second instruction in the write-back (WB) stage, assuming that there will be no pipeline stalls due to cache misses and that each of these two instructions will take five cycles. As a representative configuration, the entries in the new table for a DAG with N = 4 and E = 4,were found to be larger than our current table. When we estimate the total energy using the new table for the DAG configuration mentioned above, we found that the energy estimate was closer in magnitude to the simulator results (see Figure 2). However, we found that even this table-building method is *not* accurate as far as absolute energy value estimation is concerned. This is because, this method, still does not have all the information during static evaluation of energy table entries as opposed to the dynamic information (e.g., instruction location) available to the cycle-accurate energy simulator.

If we are interested only in estimation of energy consumed by a given code (e.g., for those found by scheduling techniques explained in this paper), we can achieve this using our cycle-accurate simulator and a more detailed ETC table. In this approach, for a pipeline of five stages (IF, ID, EXE, MEM, *and* WB), each table entry will be indexed by ($<i_1$,IF>, $<i_2$,ID>, $<i_3$,EXE>, $<i_4$,MEM>, $<i_5$,WB>) where each i_j is an instruction (which might also be a nop). Although such an approach increases the table size significantly, limiting the number of instructions considered may limit this table expansion. Our current research includes developing cycle-accurate, pipeline stage-sensitive scheduling algorithms based on such detailed energy tables.

5   Conclusions

In this paper, we evaluate several table-driven instruction scheduling algorithms from both energy and performance angles. Our results indicate that the best scheduling from the performance point of view is not necessarily the best scheduling from the energy point of view. Further, scheduling techniques that consider both energy and performance simultaneously obtain very good results in terms of both performance and energy. Our results also reveal that the compiler-estimated energy consumption of a given sequence of code is a reasonable indicator to compare energy consumption of different scheduling algorithms. However, the compiler-estimated absolute energy values are in general lower than the values obtained through a cycle-accurate simulator, which reflects on the underlying difficulty of determining accurate energy table entries. Our results motivate the need for further research on instruction scheduling algorithms that use more detailed energy transition cost tables.

References

1. A. V. Aho, R. Sethi, and J. Ullman. *Compilers: Principles, Techniques, and Tools.* Addison-Wesley, 1986.
2. M. Kamble and K. Ghose. Analytical energy dissipation models for low power caches. In Proc. *the International Symposium on Low Power Electronics and Design,* p. 143, August 1997.
3. S. S. Muchnick. *Advanced Compiler Design Implementation.* Morgan Kaufmann Publishers, San Francisco, California, 1997.
4. A. Parikh, M. Kandemir, N. Vijaykrishnan and M.J. Irwin. Instruction Scheduling Based on Energy and Performance Constraints. In Proc. *The IEEE CS Annual Workshop on VLSI,* April 2000.
5. J. Russell. Assembly code power analysis of a high performance embedded processor family. *Technical Report,* Department of Electrical and Computer Engineering, University of Texas at Austin.
6. W.-T. Shiue and C. Chakrabarti. Memory exploration for low power, embedded systems. *CLPE-TR-9-1999-20, Technical Report,* Center for Low Power Electronics, Arizona State University, 1999.
7. V. Tiwari, S. Malik, A. Wolfe, and T.C. Lee. Instruction level power analysis and optimization of software. *Journal of VLSI Signal Processing Systems,* Vol. 13, No. 2, August 1996.
8. V. Tiwari, S. Malik, and A. Wolfe. Power analysis of embedded software: A first step towards software power minimization. *IEEE Transactions on VLSI Systems,* December 1994.
9. M. C. Toburen, T. M. Conte, and M. Reilly. Instruction scheduling for low power dissipation in high performance microprocessors. In Proc. *the Power Driven Microarchitecture Workshop* (ISCA'98), Barcelona, Spain, June 1998.
10. N. Vijaykrishnan, M. Kandemir, M. J. Irwin, H. Y. Kim, and W. Ye. Energy-driven integrated hardware-software optimizations using SimplePower. In Proc. *the International Symposium on Computer Architecture (ISCA'00),* June 2000.
11. W. Ye. *Architectural Level Power Estimation and Experimentation.* Ph.D. Thesis, Comp. Sci. and Eng., The Pennsylvania State University, October 1999.

On Message-Dependent Deadlocks
in Multiprocessor/Multicomputer Systems*

Yong Ho Song and Timothy Mark Pinkston

SMART Interconnects Group, University of Southern California
Los Angeles, CA 90089-2562, USA
{yongho, tpink}@charity.usc.edu

Abstract. The existence of multiple message types and associated inter-message dependencies in multiprocessor/multicomputer systems may cause message-dependent deadlock in networks that are designed to be free of routing deadlock. In this paper, we characterize the frequency of message-dependent deadlocks using empirical methods. Results show that message-dependent deadlocks occur very infrequently under typical circumstances, thus rendering approaches based on avoiding them overly restrictive in the common case.

1   Introduction

Increased processing demands on parallel computing systems continue to impose greater performance demands on the interconnection network which provides communication paths for processing nodes in the system. Efficient and reliable communication between processing nodes is crucial for achieving high performance. Deadlock anomalies occurring as a result of cyclic hold-and-wait dependencies by messages on network resources reduce communication efficiency and reliability, consequently degrading network and system performance considerably. Recent research [1,2,3] has focused on the development of very efficient network routing techniques which assume that messages (or, alternatively, packets) in the network always sink upon arrival at their destinations. That is, it is assumed that the delivery of messages is not coupled in any way to the injection (generation) or reception (consumption) of any other message in the network or at network endpoints. This simplifying assumption is valid for networks with homogeneous message types, but it inaccurately represents network behavior when heterogeneous messages are routed in which dependencies between different message types exist. Deadlock-free routing algorithms designed using that assumption may provide efficient and deadlock-free communication paths between network endpoints (thus eliminating routing-dependent deadlocks), however they are still susceptible to deadlocks arising from the interactions and dependencies created at network endpoints between different message types.

---

* This research was supported in part by an NSF Career Award, grant ECS-9624251, and an NSF grant, CCR-9812137.

M. Valero, V.K. Prasanna, and S. Vajapeyam (Eds.): HiPC 2000, LNCS 1970, pp. 345–354, 2000.

The types of messages and the dependencies that can exist between them are defined by the communication protocol implemented by the system. For instance, in a typical shared memory multicomputer, a node receiving a *request* message for a data object generates a *reply* message in response to the request message in order to transfer the requested data object to the requesting node. Therefore, at any given end node in the system, there can be a coupling between the two message types: the generation of one message type, i.e., the reply generated by the destination, is directly coupled to the reception of another message type, i.e., the request received by the destination. As the coupling between message types is transferred to network resources, additional dependencies on network resources are created, referred to as *message dependencies.*

Resources along the message path inside each node (at network endpoints) and in the network (between network endpoints) are finite, subject to contention. Message dependencies occurring at network endpoints (i.e., on injection and reception resources) may prevent messages from sinking at their destinations. Furthermore, when message dependencies are added to the complete set of resource dependencies, knotted cycles [4] may form along escape resources [1], resulting in possible deadlock. We refer to deadlocks arising from this phenomenon as *message-dependent deadlocks.* As is the case for routing-dependent deadlock, approaches for addressing message-dependent deadlock can be based either on avoidance or on recovery. The primary distinction between these approaches is the trade-off made between routing freedom and deadlock formation. The advantages of techniques based on these approaches, therefore, depend on how frequently deadlocks occur and how efficient (in terms of resource cost and utilization) messages can be routed while guarding against deadlocks.

In this paper, we characterize the frequency of message-dependent deadlocks occurring in multiprocessor/multicomputer systems and evaluate various approaches for handling them. Empirical analysis is performed by simulating a CC-NUMA[1] system with a cache coherence protocol similar to the S-1 Multiprocessor [5]. However, restrictions that prevent the formation of message-dependent deadlocks are relaxed for the purposes of deadlock characterization. Synthetically generated traffic loads are used to stress the network in order to obtain broad characterization of message-dependent deadlock frequency. Critical network and network interface parameters are varied across the simulations to observe their effect on deadlock frequency.

The remainder of this paper is organized as follows. Section 2 provides background on deadlocks arising due to message dependencies. Section 3 reviews schemes for avoiding message-dependent deadlocks. Section 4 presents our deadlock frequency results. Related work and conclusions are given in Section 5.

2   Message-Dependent Deadlock

In shared memory systems, generic message types used to exchange information between communicating entities are *request* and *reply* messages. In addition to

---

[1] CC-NUMA stands for cache coherent non-uniform memory architecture.

these, many other message types—as defined by the communication protocol of the system—may be used to complete data transactions, resulting in many kinds (or classes) of message dependencies. A distinct class of message dependency is created for each pair of message types for which a direct coupling exists and is transferred to network resources.

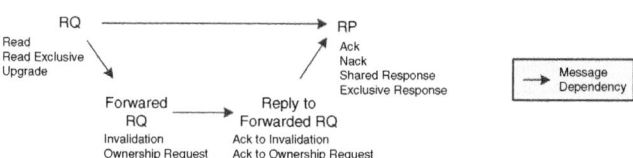

Fig. 1. Total ordering among message types in the Censier/Feautrier cache protocol.

For instance, the well-known cache coherence protocol designed by Censier and Feautrier [6] used in the S-1 Multiprocessor [5] permits data transactions to be composed of certain combinations of *request, forwarded-request, reply-to-forwarded-request*, and *reply* message types, as shown in Figure 1. Each combination may present different kinds of message dependencies and a corresponding message dependency chain on network resources. A *message dependency chain* represents a partially (or totally) ordered list of message dependencies allowed by the communication protocol. We define the partial order relation "$\prec$" between two message types m_1 and m_2 by the following: $m_1 \prec m_2$ if and only if m_2 can be generated by a node receiving m_1 for some data transaction. Message type m_2 is said to be *subordinate* to m_1 if $m_1 \prec m_2$, and the final message type at the end of the message dependency chain is said to be a *terminating* message type. The more kinds of message dependencies and chains allowed by a system, the more opportunity there is for the system to experience message-dependent deadlock. This is also influenced by the number of message types allowed within a message dependency chain, referred to as the chain *length*.

As previously mentioned, message-dependent deadlocks can form when dependency cycles on escape resources exist at least, in part, due to message dependencies. As a specific example, let us consider the case of resource sharing at network endpoints by messages of different types. Figure 2 illustrates this situation assuming only two message types exist in a system for which request-reply dependency exists (i.e., *request* $\prec$ *reply*). Depicted in the figure is a simple message-dependent deadlock represented by a resource dependency graph in which two nodes, Node A and Node B, are each sending request messages to one another and expecting to receive reply messages over a network free from routing-dependent deadlock. If the arrival rate of request messages exceeds the consumption rate, a backlog starts to form at the input message queue IQ_A at Node A. After a while, the backlog propagates backward in the direction of message injection at the output message queue OQ_B at Node B. The backlog eventually reaches the input message queue IQ_B at Node B and the output mes-

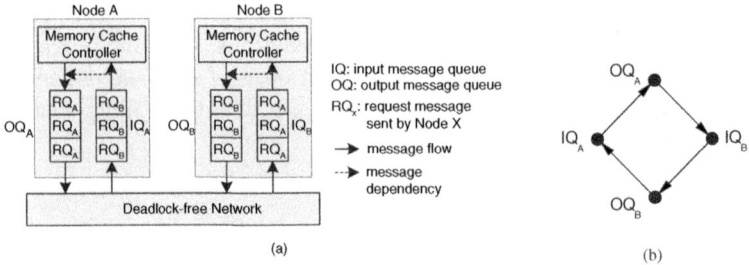

Fig. 2. (a) A simple example of message-dependent deadlock occurring between two nodes connected by a network free of routing deadlock. (b) The corresponding dependency graph for sharing of resources by message types at network endpoints.

sage queue OQ_A at Node A. At this point, a deadlock forms as no buffer space can be freed for reply messages needed by both nodes to continue execution.

This situation can be avoided by allocating separate queues to different message types in the network interfaces. However, decoupling the message types on resources in the network interfaces simply by separating messages into different queues based on type is necessary but not sufficient to completely avoid message-dependent deadlocks system-wide. The sharing of resources (channels) *between* network endpoints among different message types can also cause deadlocks due to message dependencies in a network that would otherwise be free from deadlock. This case of message-dependent deadlock occurring on resources across the network is considered in Figure 3 for the simple case of request and reply messages sharing network channels. When Node $R1$ sends a request message to Node $R3$ and Node $R3$ responds to the request by sending a reply message back to Node $R1$, a dependency from the high to low virtual channels in the network (shown as *dotted arcs* in the figure for Node $R3$ only) exists through the network interface due to the message dependency, represented by C_{R3} at Node $R3$. This completes the cycle in the dependency graph, making deadlocks possible.

3   Handling Message-Dependent Deadlocks

Message-dependent deadlock can be avoided by enforcing routing restrictions on network resources used to escape deadlock such that all dependencies on those resources, including message dependencies, are acyclic [1]. Alternatively, they can be avoided while allowing cyclic dependencies on escape resources by requiring some subset of escape resources (i.e., network interface queues) to be large enough such that they can never become fully occupied. Since sufficient resources and/or routing restrictions on a set of resources always prevent the formation of deadlock, these techniques for handling deadlock are said to be based on *deadlock avoidance*. The second technique can be implemented by providing enough buffer space in each node's network interface queues to hold at least as many messages as can be supplied, as in [7]. Although simple to implement, this technique is not very scalable since the size of the network interface queues

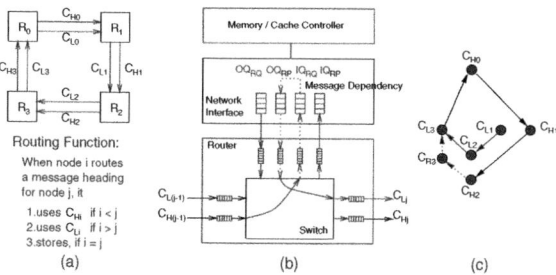

Fig. 3. (a) A four node system interconnected by a unidirectional ring network using two virtual channels to avoid routing deadlocks as described in [8]. (b) Message dependencies occurring inside network interfaces. (c) The corresponding resource dependency graph consisting of network channels and queue resources (i.e., C_{R3}) at Node $R3$.

grows as $O(P \times M)$ messages, where P is the number of processor nodes and M is the number of outstanding messages allowed by each node.

The first technique for avoiding message-dependent deadlock is more scalable and more commonly used. One way of guaranteeing acyclic dependencies on escape resources is to provide logically independent communication networks for each message type, implemented as either physical or virtual channels [9,10]. The partial ordering on message dependencies defined by the communication protocol is transferred to the logical networks so that the usage of network resources is acyclic. Figure 4 illustrates this for the previous example of a four node ring system which allows a message dependency chain length of two, i.e., request-reply dependency.

With this technique, the size of network resources does not influence deadlock properties, but at least as many logical networks are required as the length of the message dependency chain. For example, the Cavallino router/network-interface chip [10] (which can be used to build networks with thousands of nodes) has network interface queues of less than 1536 bytes each but requires two logically separated networks to handle request-reply dependency. Such partitioning of network resources decreases potential resource utilization and overall performance, particularly when message dependencies are abundant and resources (i.e., virtual channels) are scarce. For example, consider a system which supports a message dependency chain length of four such that $m_1 \prec m_2 \prec m_3 \prec m_4$ with Duato's adaptive routing algorithm. Two virtual channels are required for each message type m_i to escape from routing-dependent deadlocks in a torus network [8]. A total of eight virtual channels are required to escape from message-dependent deadlock, and only one of these is available to any message at any given time. If sixteen virtual channels were implemented, only three would be available to any message at any given time. Evidently, the main disadvantage of avoiding deadlock by disallowing cyclic dependencies on escape resources is the number of partitioned logical networks required.

It is believed that an appropriate combination of deadlock avoidance and recovery can reduce the amount of network resources (i.e., logical networks) required to handle deadlock, as compared to strictly avoiding deadlock. This would

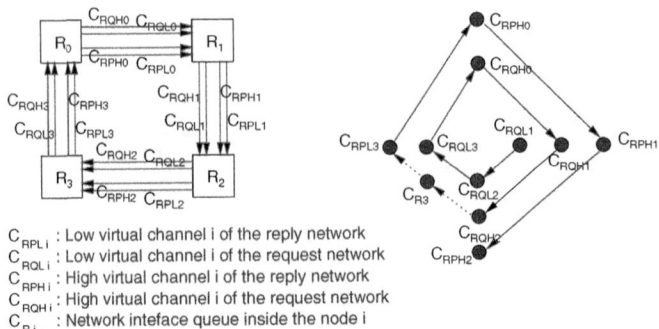

$C_{RPL\,i}$: Low virtual channel i of the reply network
$C_{RQL\,i}$: Low virtual channel i of the request network
$C_{RPH\,i}$: High virtual channel i of the reply network
$C_{RQH\,i}$: High virtual channel i of the request network
$C_{R\,i}$: Network inteface queue inside the node i

Fig. 4. Separation of request and reply networks avoids cyclic dependencies in the channel dependency graph, but it reduces channel utilization.

allow network resources to be used more efficiently and increase communication performance. However, to find the appropriate combination, it is necessary to characterize message-dependent deadlocks and to understand which design parameters predominantly affect their frequency and in what ways.

4   Evaluation

4.1   Simulation Methodology

Empirical analysis is performed using FlexSim 1.2 [11], a flit-level network simulator developed by the *SMART* Interconnects Group at USC. The simulator performs flit-level traffic flow within the interconnection network and maintains data structures that represent resource allocations and dependencies (resource wait-for relationships) occurring within the network.

Deadlock detection based on the channel wait-for graph(CWG) model implemented in FlexSim 1.2 is augmented to include message-level activities and dependencies in network interfaces. The CWG-based deadlock detection identifies all the cycles in CWG to examine the existence of knots [4] (deadlocks) every 50 cycles. However, this approach suffers from an explosive increase in the number of CWG cycles as network load increases. A newly added feature allows end nodes to detect potential deadlocks using locally available information: a deadlock is presumed to have occurred if both the input and output message queues at a node remain full for more than a threshold value, T network cycles, without making any progress. A threshold of 25 cycles is assumed since detection using the CWG method typically takes 25 cycles on average. Deadlock frequency is measured in relative terms by the normalized number of deadlocks—which is the ratio of the number of deadlocks to the number of messages delivered.

4.2   Characterization of Message-Dependent Deadlocks

Regular (k-ary n-cube toroidal) networks with the following default parameters have been simulated unless stated otherwise (see Table 1). Message lengths of 4 and 20 flits for request and reply, respectively, are assumed.

Table 1. Default simulation parameters.

Parameters	Values
Network Topology	8×8 torus
Link Transmission	Full-duplex
Switching Technique	Wormhole
Message Length	4 flits (Request), 20 flits (Reply)
Bristling Factor	1 processor/node
Virtual Channels per Link	2 virtual channels
Flit Buffers per Channel	2 flits
Routing Algorithm	Dimension Order Routing (DOR)
Message Types	2 (Request and Reply)
Message Service Time	40 clocks
Message Traffic Pattern	Random
Message Queue Size	16 messages
Handling Message-Dependent Deadlock	Minimal Regressive Recovery
Deadlock Detection	CWG-based and Local Detection

The simulator generates request messages at the rate specified as a simulation parameter. All subordinate message types are generated automatically–according to specified traffic statistics for the cache coherence protocol–upon completion of servicing messages at end nodes. Routing-dependent deadlocks are strictly avoided by using Dimension Order Routing [8]. When a deadlock (message-dependent) is detected, a minimum regressive recovery procedure is invoked which resolves the deadlock by killing one message out of an input message queue involved in the deadlock. Simulations are run for 30,000 simulation cycles starting from the point at which steady state is reached.

4.3   Simulation Results

Effect of Network Interface Message Queue Size on Deadlock. To observe the effects of increased queue size in the network interface on deadlock frequency, queues of size 4x, 8x, 16x, 32x and 64x message length were used with all other network parameters set to default values. These sizes are used in practical implementations (from several KBytes [10] to several hundred KBytes and more) and also reflect future trends.

As shown in Figure 5(a), no message-dependent deadlocks are observed even for small queues when the network operates below saturation, i.e., when the load rate is below 20% of network capacity. Beyond network saturation, all networks experience deadlocks irrespective of queue size. The network with a queue size of 4 has 6 to 8 deadlocks for every 100 messages delivered, but this number decreases to 1 to 2 for a queue size of 64. The network with the smaller queue starts experiencing message dependent deadlocks before the one with the larger queue. It also has a higher frequency of forming deadlocks, all of which follows intuitively.

However, it is also observed that once a deadlock forms and is minimally resolved (i.e., by killing only one message out of the deadlocked resources), another deadlock involving some of the messages in the previous deadlock is likely to form. It is possible to reduce the probability of this occurrence (and, therefore, deadlock frequency) by recovering from deadlock more aggressively, as shown in Figure 5(b). Here, every node involved in a message-dependent deadlock kills *all* request messages out of its input message queue, significantly reducing deadlock frequency (by 70-80%) and increasing network throughput (see Figure 5(c)). Although this method of aggressively recovering increases the degree of correlation required for subsequent deadlocks to form, it has the side-effect of victimizing 3–4 times more messages than the minimal approach. Thus, a more progressive recovery technique not based on killing potentially deadlocked messages would likely benefit most from this aggressive approach.

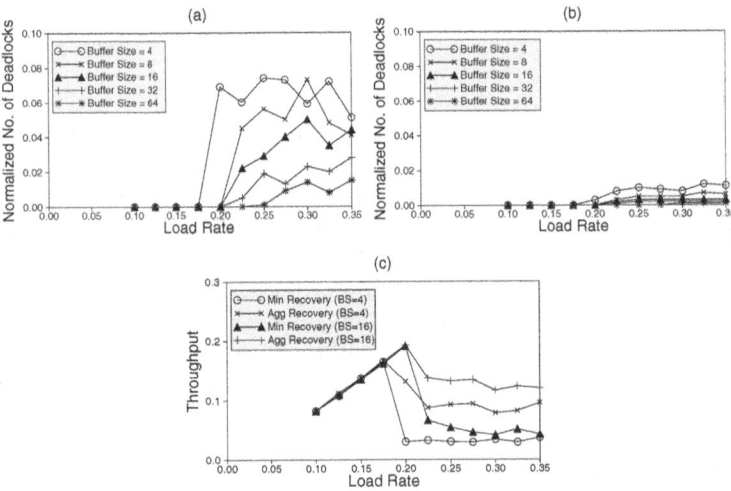

Fig. 5. Effect of message queue size at network interfaces. (a) Normalized number of deadlocks versus message generation rate under minimal recovery, (b) aggressive recovery, and (c) throughputs of both recovery schemes.

Effect of Message Service Latency on Deadlock Frequency. To observe the effects of message service latency at end-nodes on deadlock formation, 4x and 64x message-deep queues and 10, 40 and 160 network cycles of message service latency are assumed in each network interface, with all other network parameters set to default values. This covers the period from when a message is taken from the top of an input message queue to when the subordinate message is deposited into the output message queue. This is typically 50 cycles current machines (e.g., in the Origin 2000 [12]).

As shown in Figure 6(a), message service latency affects the load rate at which deadlocks start to form more significantly for the smaller network interface queue sizes than the larger ones. The network with a service latency of 160 cycles and the queue size of 4 messages experiences deadlock even at a load of 12.5% of the network capacity. As messages are serviced faster, the load rate at which deadlocks form increases but the frequency of deadlocks is unaffected. If the service latency is greater than the inter-message arrival interval at input queues from the network (which decreases with increasing load rate), a backlog forms which grows back across the network toward output message queues. This increases the likelihood of deadlocks forming.

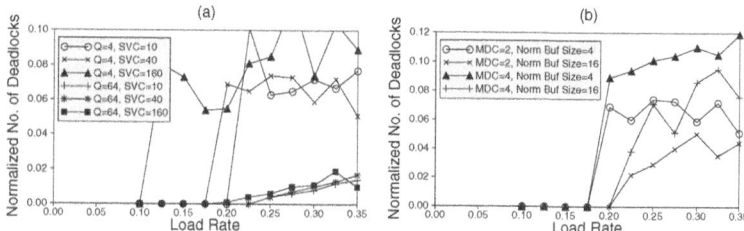

Fig. 6. (a) Effect of message service time at end-nodes, and (b) effect of message dependency chain length.

Effect of Dependency Chain Length on Deadlock Frequency. To observe the effects of message dependency chain length on deadlock formation, chain lengths of 2 and 4 message types are considered separately, with all other network parameters set to default values. This corresponds to the message dependency chain lengths allowed in the cache coherence protocol developed by Censier and Feautrier [6] (used in the S-1 Multiprocessor [5]) depicted in Figure 1.

As shown in Figure 6(b), the networks with a message dependency chain length of 4 experience roughly 20-50% more deadlocks than those with chain lengths of 2 message types. This follows intuitively considering the fact that as the message dependency chain length increases, network end-nodes have more chance of having coupled interactions—albeit, the degree of correlation needed for these interactions to lead to deadlock is substantially higher. The figure also indicates that the chain length has more of an effect on the frequency of deadlocks rather than the load rate at which deadlocks start to form, which also follows intuitively.

5   Conclusion

Previous work characterizes the effects of various network parameters on the frequency of blocked messages, resource dependency and deadlock [11]. The interrelationships between routing freedom, blocked messages, correlated resource dependencies and deadlock formation have been empirically quantified for a homogeneous set of messages. In contrast, this work characterizes deadlocks occurring due to the existence of heterogeneous messages sharing network resources.

Through empirical analysis, we observe that message-dependent deadlocks occur relatively infrequently. Key network interface and cache coherence protocol design parameters, such as message queue size, message service latency, and message dependency chain length, play a vital role in determining when and how frequent message-dependent deadlocks form. In the future, we will propose alternative techniques that relax restrictions, allowing the routing of packets and the handling of deadlocks to be done more efficiently.

References

1. J. Duato. A New Theory of Deadlock-free Adaptive Routing in Wormhole Networks. *IEEE Transactions on Parallel and Distributed Systems*, 4(12):1320, 1331 1993.
2. L. Schwiebert and D. N. Jayasimha. A Necessary and Sufficient Condition for Deadlock-free Wormhole Routing. *Journal of Parallel and Distributed Computing*, 32(1):103–117, January 1996.
3. Timothy Mark Pinkston. Flexible and Efficient Routing Based on Progressive Deadlock Recovery. *IEEE Transactions on Computers*, 48(7):649–669, July 1999.
4. Sugath Warnakulasuriya and Timothy Mark Pinkston. A Formal Model of Message Blocking and Deadlock Resolution in Interconnection Networks. *IEEE Transactions on Parallel and Distributed Systems*, 11(3):212–229, March 2000.
5. L. Widdoes Jr. and S. Correll. The S-1 Project: Developing High Performance Computers. In *Proc. COMPCON*, pages 282–291, Spring 1980.
6. L. M. Censier and P. Feautrier. A New Solution to Coherence Problems in Multicache Systems. *IEEE Transactions on Computers*, C-27:1112–1118, December 1978.
7. C.B. Stunkel et al. The SP2 high-performance switch. *IBM Systems Journal*, 34(2):185–204, 1995.
8. W. Dally and C. Seitz. Deadlock-free Message Routing in Multiprocessor Interconnection Net works. *IEEE Transactions on Computers*, 36(5):547–553, May 1987.
9. Charles E. Leiserson, Zahi S. Abuhamdeh, David C. Douglas, Carl R. Feynman, Mahesh N. Ganmukhi, Jeffrey V. Hill, W. Daniel Hillis, Bradley C. Kuszmaul, Margaret A. St. Pierre, David S. Wells, Monica C. Wong, Shaw-Wen Yang, and Robert Zak. The Network Architecture of the Connection Machine CM-5. In *Symposium on Parallel and Distributed Algorithms*, pages 272–285, 1992.
10. Joseph Carbonaro. Cavallino: The Teraflops Router and NIC. In *Proceedings of the Symposium on Hot Interconnects IV*, pages 157–160. IEEE Computer Society, August 1996.
11. Sugath Warnakulasuriya and Timothy Mark Pinkston. Characterization of Deadlocks in k-ary n-cube Networks. *IEEE Transactions on Parallel and Distributed Systems*, to appear, 1999.
12. James Laudon and Daniel Lenoski. The SGI Origin: A ccNUMA Highly Scalable Server. In *Proceedings of the 24th International Symposium on Computer Architecture*, pages 241–251. IEEE Computer Society, June 1997.

Memory Consistency and Process Coordination for SPARC Multiprocessors

Lisa Higham* and Jalal Kawash**

Department of Computer Science, The University of Calgary, Canada, T2N 1N4
{higham|kawash}@cpsc.ucalgary.ca

Abstract. Simple and unified non-operational specifications of the three memory consistency models Total Store Ordering (TSO), Partial Store Ordering (PSO), and Relaxed Memory Order (RMO) of SPARC multiprocessors are presented and proved correct. The specifications are intuitive partial order constraints on possible computations and are derived from natural successive weakening of Lamport's Sequential Consistency. The formalisms are then used to determine the capabilities of each model to support solutions to critical section coordination and both set and queue variants of producer/consumer coordination without resorting to expensive synchronization primitives. Our results show that none of RMO, PSO nor TSO is capable of supporting a read/write solution to the critical section problem, but each can support such a solution to some variants of the producer/consumer problem. These results contrast with the two previous attempts to specify these machines, one of which would incorrectly imply a read/write solution to the critical section problem for TSO, and the other of which is too complicated to be useful to programmers. Our general framework for defining and proving the correctness of the memory consistency models was key in uncovering the previous error and in achieving our simplification, and hence may be of independent interest.

1 Introduction

Sun Microsystems introduced the three memory memory consistency models Total Store Ordering (TSO), Partial Store Ordering (PSO), and Relaxed Memory Order (RMO). TSO and PSO were introduced in the version 8 architecture[16] and retained in version 9. RMO was introduced in the version 9 architecture[17]. These models are specified axiomatically — some partial orders are defined, relationships between them are specified, and an axiom given that determines the values returned by reads. These specifications are complicated and not intuitive. Hence it is challenging for a programmer to be certain of the possible program outcomes and to determine correct and efficient solutions to basic process coordination problems.

This dilemma is typical of current and proposed multiprocessor machines. Weakening the memory consistency model of a multiprocess system improves its performance and scalability. However, these models sacrifice programmability because they create

* Supported in part by the Natural Sciences and Engineering Research Council of Canada grant OGP0041900.
** Supported in part by a Natural Sciences and Engineering Research Council of Canada doctoral scholarship and an Izaak Walton Killam Memorial scholarship.

M. Valero, V.K. Prasanna, and S. Vajapeyam (Eds.): HiPC 2000, LNCS 1970, pp. 355–366, 2000.
© Springer-Verlag Berlin Heidelberg 2000

complex behaviors of shared memory. Without the use of expensive, built-in synchronization, these models exhibit poor capabilities to support solutions for fundamental process coordination problems [9]. This leads programmers to aggressively use these forms of synchronization, incurring additional performance burdens on the system.

Since use of synchronization primitives deteriorates performance, we are motivated to study the limitations and capabilities of weak memory consistency models without the use of these primitives. If the use of explicit synchronization is avoidable, then efficient libraries for certain classes of applications can be built. This would ease the job of distributed application programmers and make their applications more efficient.

In this paper, we derive a mathematical description of the behavior of the three SPARC variants in terms of partial order constraints on possible computations. Our specifications are surprisingly simple; each is a natural weakening of Sequential Consistency (SC) [12]. Because they are short and precise, we believe they give programmers a much improved tool for reasoning about the outcomes of their multiprocess programs. They also make it easier to construct verification tools that depend on automated reasoning. Finally, the simplicity of the descriptions facilitates the comparison of the SPARC models to each other and to several proposed consistency models including Processor Consistency[6], Causal Consistency[1], and Java Consistency[5,7].

Our definitions for TSO and PSO are proven to exactly capture the machine description of the version 8 architecture (and hence of version 9 since they are known to be equivalent). Our RMO definition is proven to exactly capture the version 9 specification. To achieve our proofs we establish a framework for describing operational and non-operational models and a technique for proving their equivalence. Because this framework is not restricted to any particular machine, it may be of independent interest. We have used it to prove correctness of other models [9].

We next use our partial order specifications to study the capabilities of the three SPARC memory consistency models TSO, PSO, and RMO to support solutions to fundamental process coordination problems without resorting to expensive synchronization primitives. The process coordination problems studied in this paper are critical section coordination and producer/consumer coordination. We distinguish two variants of producer/consumer coordination whose solution requirements differ: the set and queue variants. Our results show that the TSO (and hence PSO and RMO) model is incapable of supporting a read/write solution to the critical section problem, but even RMO (and hence PSO and TSO) can support such solutions to some variants of the producer/consumer problem.

Capturing the semantics of the three SPARC memory consistency models simply and precisely has proven to be surprisingly tricky. One earlier attempt to define TSO [11] resulted in a definition that is much stronger than what this architecture really provides and leads to erroneous conclusions about the coordination capabilities of TSO. In fact, we show that any program (with only read/write operations) that is correct for SC can be compiled into an equivalent program (with only read/write operations) that is correct for this erroneous version of Total Store Ordering [10]. To the contrary, this paper proves that read/write operations are insufficient to solve certain coordination problems for TSO. Another earlier definition for Total Store Ordering is one of our own that is also based on partial orders. Though it is proved equivalent to TSO [10], it is much more complicated than TSO as defined here and it is not completely non-operational. The original operational specifications of TSO and PSO [16] are also complex and are not particularly useful for studying the questions addressed in this paper. The subsequent update to version 9 definitions and the extension to RMO is renowned for the complexity of its many partial orders and especially of the value axiom. Park and Dill also study the Relaxed Memory Order [14]. The goal of their work is a verification tool for RMO constructed from the Murφ language, rather than a reformulation of the specifications. Their definition of RMO is taken unchanged from the SPARC Architecture Manual v9 .

Section 2 contains one of our main contributions — simple, unified, non-operational models of the SPARC TSO, PSO and RMO memory consistencies. The SPARC version 8 machine is overviewed in Section 3 in order to provide the basis of the proofs that our non-operational models for TSO and PSO are correct. Section 4 sketches this proof for TSO and overviews the corresponding proofs for PSO and RMO. Section 5 briefly examines the error with the earlier attempt at a simple partial order to capture the semantics of TSO. Section 6 contains our other main results. It establishes the possibilities and impossibilities for process coordination problems in SPARC TSO, PSO, and RMO machines. The SPARC models are compared with others in Section 7.

This presentation addresses only read and write operations to variables. SPARC machines support many other operations that affect shared memory such as swap-atomic and barrier (version 8) and the family of membar operations (version 9). It is straightforward to extend our models and our proofs to include these operations. Several proofs are omitted here and other only sketched. The full version of this paper [8] contains the complete models and all proofs.

2   The SPARC Memory Consistency Models

This section defines three memory consistency models, called TSO, PSO, and RMO based on progressive weakening of SC. The definitions refer only to the ordering of operations; they are very simple and natural; they do not depend on a machine description. We will see in Section 4, however, that these definitions do indeed capture exactly the behavior of the corresponding SPARC architectures.

We first summarize the way that we formalize any memory consistency model. A multiprocess system can be modeled as a collection of processes operating on a collection of shared data objects. The only shared data objects considered here are variables supporting read and write operations. The full version of this paper [8] provides the extension to other SPARC operations. The notation $r(x)v$ and $w(x)u$ denotes, respectively, a read operation of variable x returning v and a write operation to x of value u. The *invocation of operation* o, is just the operation without its output value determined and is denoted $in(o)$. A read $r(x)v$ and a read invocation with v as yet undetermined, are distinguished by writing $in(r(x)v) = r(x)v?$. Since a write invocation has no response value, $in(w(x)u) = w(x)u$.

It suffices to model a *process* as a sequence of read and write invocations, and a *multiprocess system* as a collection of processes together with the shared variables. Henceforth, we denote a multiprocess system by the pair (P, J) where P is a set of processes and J is a set of variables. A *process computation* is the sequence of operations obtained from the process by augmenting each read operation invocation with its matching response. A *system computation* is a collection of process computations, one for each process. Let O be all the (read and write) operations in a computation of a system (P, J). Then, $O|p$ denotes all the operations that are in the process computation of process $p \in P$; $O|x$ denotes all the operations that are applied to variable $x \in J$; $O|w$ denotes all the write operations; and $O|r$ denotes all the read operations. These notations are also combined to select the combined restriction of operations. For example, $O|w|x|p$ is the set of all write operations by process p to variable x.

A sequence of read and write operations to variable x is *valid* if and only if each read operation in the sequence returns the value written by the most recently preceding write operation. Given any collection of read and write operations O on a set of variables J, a *linearization of* O is a (strict) linear order $(O, \xrightarrow{L})$ such that for each variable x in J, the subsequence $(O|x, \xrightarrow{L})$ of $(O, \xrightarrow{L})$ is valid. For any relation R, the notation $s_1 R s_2$ means $(s_1, s_2) \in R$. A read operation is *foreign* if the value returned by the read was written by a process different from the one invoking the read. Otherwise, it is called *domestic*.

Let O be a set of operations in a computation of a system (P, J). Define the *program order*, denoted $(O, \xrightarrow{prog})$, by $o_1 \xrightarrow{prog} o_2$ if and only if o_2 follows o_1 in the computation of p. Consider the following conditions on o_1 and o_2 where $o_1 \xrightarrow{prog} o_2$:

same-variable: $o_1, o_2 \in O|x$, for some $x \in J$.
preceding-read: $o_1 \in O|r$ and o_1 is foreign.
following-write: $o_2 \in O|w$.

The *RMO partial program order*, denoted $(O, \xrightarrow{rmo})$, is the transitive closure of the relation: $(O, \xrightarrow{prog})$ intersect same-variable.

The *PSO partial program order*, denoted $(O, \xrightarrow{pso})$, is the transitive closure of the relation: $(O, \xrightarrow{prog})$ intersect (either same-variable or preceding-read).

The *TSO partial program order*, denoted $(O, \xrightarrow{tso})$, is the transitive closure of the relation: $(O, \xrightarrow{prog})$ intersect (either same-variable or preceding-read or following-write).

Definition 1. *Let O be all the operations of a computation C of a multiprocess system* (P, J). *Then*
C is TSO if there exists a linearization $(O, \xrightarrow{L})$ such that $(O, \xrightarrow{tso}) \subseteq (O, \xrightarrow{L})$,
C is PSO if there exists a linearization $(O, \xrightarrow{L})$ such that $(O, \xrightarrow{pso}) \subseteq (O, \xrightarrow{L})$,
C is RMO if there exists a linearization $(O, \xrightarrow{L})$ such that $(O, \xrightarrow{rmo}) \subseteq (O, \xrightarrow{L})$.

Notice that RMO, when restricted to reads and writes, is just the well-known consistency model known as Coherence [3].

3   The SPARC v8 Multiprocessor Machines

In SPARC v8 (a two process version is in Figure 1) there is one write-buffer associated with each process in the system. The main memory is single ported with a nondeterministic switch providing one memory access at a time. When a process performs a write, it is sent to its write-buffer, which is responsible for committing the pending writes to main memory. The process, in the meantime, can continue executing. When a read is issued by a process, the associated write-buffer is checked for any pending writes to the same variable. If there is any such write, the value "to be written" by the last such write is returned. If there is no such pending write in the buffer, the read accesses main memory in the normal manner. The order in which the write-buffer commits the writes to main memory differentiates between two machine variants of the SUN SPARC v8. If the buffer is FIFO, the resulting machine is called Total Store Ordering and is denoted M_{TSO}. If it is FIFO only on a per-variable basis, the resulting machine is Partial Store Ordering, denoted M_{PSO}. SPARC v8 does not define a Relaxed Memory Order machine.

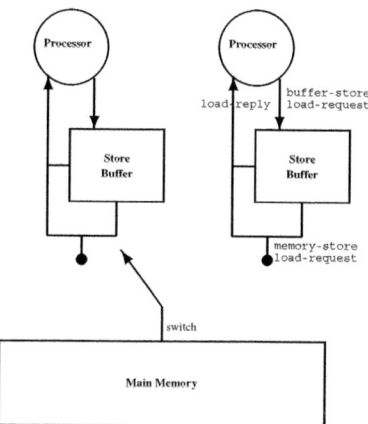

Fig. 1. A two-process SPARC architecture

To describe the precise operational behavior of the SPARC machine, we specify the sequence of events that is triggered by each operation invocation and the rules that constrain how these events interleave and instantiate variables. This description follows in a straightforward way from the description provided by SUN Microsystems [16]. Because of page limitations, we only provide the details for M_{TSO}.

The M_{TSO} is a machine that accepts read and write operation invocations, and for each executes a sequence of events. M_{TSO} is specified by the triple $(\mathcal{E}, \mathcal{I}, \mathcal{R})$ as follows.

$\mathcal{E}$, the set of *event types*, contains load-request, load-reply, buffer-store, and memory-store. A load-request (p, x) is a request by process p to load the value stored in variable x. The event load-reply (p, x, v) returns a value, v, of x to p. The event buffer-store (p, x, u) is a store request by p to x of a value u; the request is placed in the store-buffer associated with p. A memory-store (p, x, u) commits p's request, buffer-store (p, x, u), by removing the request from the buffer and applying it to main memory.

$\mathcal{I}$, the *implementation function*, defines the sequence of two events that occurs for each operation invocation.

- A write in M_{TSO} is implemented by the ordered pair that represents placing the write in the buffer and later committing it to main memory.

$$\mathcal{I}(w(x)u, p) = \langle \texttt{buffer-store}\,(p, x, u)\,, \texttt{memory-store}\,(p, x, u)\,\rangle$$

- A read is implemented by an ordered pair that represent a request for a value followed by return of the requested value.

$$\mathcal{I}(r(x)v?, p) = \langle \texttt{load-request}\,(p, x)\,, \texttt{load-reply}\,(p, x, v?)\,\rangle$$

The events that implement an operation invocation *correspond* to that operation and two such events are *matching*.

$\mathcal{R}$, the set of *machine rules*, $\{\rho_i | 1 \le i \le 4\}$, further restricts the machine behavior as follows. The implementation function, $\mathcal{I}$, produces an ordered pair of two events for each read or write in (P, J). The set E of all such events occur in some total order $(E, \xrightarrow{\Xi})$ that respects the ordering of these pairs and the additional constraints ρ_1, ρ_2 and ρ_3 and has variables instantiated according to ρ_4. $(E, \xrightarrow{\Xi})$ is also denoted by Ξ. Recall that $\xrightarrow{prog}$ denotes program order.

- ρ_1 (buffers are FIFO): Let e_1 and e_2 be buffer-store events by the same process. Also, let e_1' and e_2' be the matching memory-store events of e_1 and e_2, respectively. If $e_1 \xrightarrow{\Xi} e_2$ then $e_1' \xrightarrow{\Xi} e_2'$.
- ρ_2 (loads are blocking): Let e represent a load-request and e' represent its matching load-reply event. If there is an event $\hat{e}$ by the same process such that $e \xrightarrow{\Xi} \hat{e} \xrightarrow{\Xi} e'$, then $\hat{e}$ is necessarily a memory-store event.
- ρ_3 (request order matches program order): Let $\mathcal{I}(in(o_1), p) = \langle e_1, e_1' \rangle$ and $\mathcal{I}(in(o_2), p) = \langle e_2, e_2' \rangle$. If $o_1 \xrightarrow{prog} o_2$ then $e_1 \xrightarrow{\Xi} e_2$.
- ρ_4 (variable instantiation): A $\texttt{load-reply}\,(p, x, v?)$ event e instantiates $v?$ as follows. If the most recent buffer-store that precedes e in $\Xi|p|x$ is $\texttt{buffer-store}$ (p, x, u) such that its matching $\texttt{memory-store}\,(p, x, u)$ follows e in $\Xi|x$, then we have $\texttt{load-reply}\,(p, x, v? \leftarrow u)$. Otherwise, $\texttt{load-reply}\,(p, x, v? \leftarrow u)$ where the most recent memory-store event that precedes e in $\Xi|x$ is $\texttt{memory-store}\,(q, x, u)$.

Call any such sequence of events $(E, \xrightarrow{\Xi})$ that is generated by $\mathcal{I}$ acting on (P, J) and satisfying $\mathcal{R}$, an *execution* of (P, J) on M_{TSO}. An execution, Ξ, of the system (P, J) induces a computation C_Ξ of (P, J) — simply the one that attaches to each read invocation of the system the response value of the corresponding load-reply event.

4   Correctness of the SPARC Definitions

This section illustrates one of our proof techniques by sketching the proof that the machine M_{TSO} of Section 3 and the memory consistency model TSO of Section 2 are equivalent in the sense of the commuting diagram in Figure 2. Or goal is to prove that a computation of a system (P, J) could arise from an execution on M_{TSO} if and only if that computation satisfies TSO. This goal is achieved through two lemmas. Let (P, J) be a multiprocess system with read and write operation invocations. If a read r returns the value written by a write w, r and w are said to be *causally-related*.

Lemma 1. *Any computation that is induced by an execution of the system (P, J) on M_{TSO} is TSO.*

Proof Sketch: Let O be all the operations of a computation C_Ξ that is induced by an execution Ξ of the system (P, J) on M_{TSO}. First, construct a sequence of operations X from Ξ as follows. For each load-reply $(-, x, v)$ (respectively, memory-store $(-, x, u)$) event in Ξ place a $r(x)v$ (respectively, $w(x)u$) operation in X, with read and write operations ordered in X as the corresponding load-replies and memory-stores are ordered in Ξ. For a read that completed at main memory, the corresponding load-reply instantiates its parameter variable according to a preceding memory event. So, this construction of X guarantees that such a read (domestic or foreign) follows its causally-related write. This does not hold for a read that corresponds to a load-reply that completed at the buffer level. Such a read precedes its causally-related write in X, violating both validity and $\xrightarrow{tso}$. Because its causally-related write is (necessarily) applied to the same variable as the read, the program order of such a write followed by that read must be maintained in $\xrightarrow{tso}$ (same-variable condition). Now we adjust X by moving the domestic reads that violate validity as follows. Initially, mark every domestic read operation in X as *unvisited*. Iterate through X examining each unvisited domestic read operation o in turn. Let e be o's corresponding load-reply event in Ξ. Let e' be the memory event in Ξ such that e returns the value written by e'. By construction of X, e' has a corresponding write operation o' in X. If o' precedes o in X, then mark o as visited and continue with the first unvisited read in X. If o' follows o in X, then move o in X such that it immediately follows the latest of o' or the last moved read operation. Mark o as visited, and continue with the first unvisited read in X. Finally, define $(O, \xrightarrow{L})$ to be this adjusted X. The proof proceeds by showing with careful case analysis that $(O, \xrightarrow{L})$ is a linearization that extends $(O, \xrightarrow{tso})$. ∎

Lemma 2. *Any TSO computation of the system (P, J) is a computation induced by some execution of the system (P, J) on M_{TSO}.*

Proof Sketch: Let C be a TSO computation of system (P, J). We construct an execution Ξ of (P, J) on M_{TSO} that induces the computation C. Let O be all the operations resulting from C.

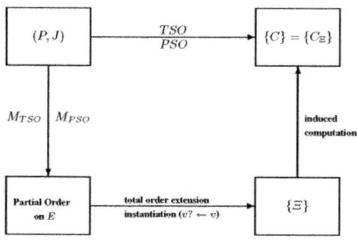

Fig. 2. Proving the correctness of TSO and PSO

Given the linearization $(O, \xrightarrow{L})$ guaranteed by Definition 1 (for TSO), first construct a sequence X of operations as follows. Initially X is empty. For each process p computation of C, maintain a pointer $\downarrow_p$ that initially points at the first operation in p's computation. If $\downarrow_p$ points at o, then we say $\downarrow_p = o$. Similarly, maintain the pointer $\downarrow_L$ to operations in $(O, \xrightarrow{L})$. Initially, $\downarrow_L$ points at the first operation in $(O, \xrightarrow{L})$. When there are no more operations to consider in a sequence s, we say $\downarrow_s = \perp$. Also, advancing $\downarrow_s$ means the pointer is incremented to point at the next operation in s. Initially, all operations are *unmarked*. Let $\hat{o}$ denote the marked copy of operation o.

Repeat until $\downarrow_L = \perp$:
> **If** $\downarrow_L = o \in O|p$ for some p, and $\downarrow_p = o$ **then**
>> Append o to X. Advance both $\downarrow_L$ and $\downarrow_p$.
>
> **Else-if** $\downarrow_L = o \in O|p$ and o is unmarked but $\downarrow_p = o' \neq o$ **then**
>> Append o' to X, and record o' as marked (denoted $\hat{o}'$). Advance $\downarrow_p$ only.
>
> **Else** ($\downarrow_L = o$ and o is marked)
>> **If** o is a write, **then** append $\hat{o}$ to X.
>> Advance $\downarrow_L$.

In each iteration of the repeat loop, either $\downarrow_L$ is advanced or an operation in $(O, \xrightarrow{L})$ is marked, and no operation is marked more than once. Furthermore, $\downarrow_L$ always advances over marked operations. Therefore this procedure terminates for finite computations. Notice that some writes are duplicated in X with the unmarked copy preceding the marked one. At the end (when $\downarrow_L = \perp$), if there are write operations that have not been duplicated in X, then for each such write o insert a marked copy $\hat{o}$ immediately after o.

Now, construct Ξ from X as follows. Iterating through X, consider each operation o. If o is a $r(x)v$ by p, then append to Ξ `load-request` (p, x) immediately followed by `load-reply` (p, x, v). If o is a $w(x)u$ by p, then if o is unmarked, append `buffer-store` (p, x, u) to Ξ and if o is marked, append `memory-store` (p, x, u) to Ξ.

To complete the proof, we need to show that execution Ξ complies with $\mathcal{I}$ and $\mathcal{R}$ of M_{TSO}. This is achieved with a series of sublemmas and case studies as given in the full version of this paper [8]. ∎

Lemmas 1 and 2 together imply:

Theorem 1. *A computation is induced by an execution of the system (P, J) on M_{TSO} if and only if it is TSO.*

The correctness of our definition of PSO is proved similarly to that for TSO [10].

To show that our definition of RMO is correct requires a different strategy because RMO is not defined as a machine. Rather, RMO is introduced in SPARC v9 [17] and specified by a complicated series of definitions. Again, let O be all the operations of a computation of system (P, J). First a *dependence order* is defined on O as a three-part restriction of program order. Then *memory order* is defined as a total order on O that extends the union of a restriction of dependence order and a different restriction of program order. Finally, a *value axiom* determines from the memory order how the value returned by a read is related to the value written by a write. In the full version of the paper [8], we show that such a memory order exists if and only there is a different order on

O that is both a linearization and that maintains program order on a per-variable basis. Hence, the formal specification of Relaxed Memory Order given by SUN Microsystems [17] is equivalent to that of Definition 1.

5   A Small Change with a Huge Consequence

Kohli et al. attempted to provide a non-operational definition for Total Store Ordering [11]. We refer to their definition by TSO-K. TSO-K differs from TSO only in the preceding-read condition, which applies to all reads rather than just foreign reads. The K-partial program order, denoted $(O, \xrightarrow{kpo})$, is the transitive closure of the relation: $o_1 \xrightarrow{prog} o_2$ intersect $(o_1, o_2 \in O|x$, for some x, or $o_1 \in O|r$ or $o_2 \in O|w)$. Kohli et al. defined the Total Store Ordering model from the point of view of processes but a definition equivalent to theirs is as follows (see [9]):

Definition 2. *Let O be all the operations of a computation C of a multiprocess system (P, J). Then C is TSO-K if there exists a linearization $(O, \xrightarrow{L})$ such that $(O, \xrightarrow{kpo}) \subseteq (O, \xrightarrow{L})$.*

Definition 2 does not capture TSO semantics. In fact, TSO-K would have substantial implications on the process coordination capabilities of the SPARC architecture (see [10] for proof).

Theorem 2. *Let $\mathcal{P}$ denote a multiprocess program that uses just reads and writes to variables in J, and let $\mathcal{P}'$ denote the program obtained from $\mathcal{P}$ by adding a read invocation of variable x immediately after every write invocation to x, for every variable x in $\mathcal{P}$. Then any computation of $(\mathcal{P}', J)$ on a TSO-K system has exactly the same outcome as some computation of $(\mathcal{P}, J)$ on a SC system.*

There are several solutions to the critical section problem for SC machines using only read and writes of shared variables. These can be automatically compiled into solutions for machines satisfying TSO-K, using the technique of Theorem 2 [10,8]. This compilation will not work if the target machine satisfies only TSO because the added reads could be domestic and thus impose no additional ordering constraints. In fact, as will be confirmed in Section 6, there is no solution to the critical section problem using only reads and writes for machines satisfying only TSO.

6   Process Coordination for SPARC multiprocessors

6.1   Critical Sections Coordination

In the Critical Section Problem (CSP) [15], (also called the Mutual Exclusion Problem) a set of processes coordinate to share a resource. Each process repeatedly cycles through the four procedures <remainder>, <entry>, <critical section>, <exit> such that (1) *Mutual Exclusion:* At any time there is at most one process in its <critical section> and (2) *Progress:* If at least one process is in <entry>, then eventually one will be in <critical section>.

We denote a CSP problem by CSP(n) where $n \geq 2$ is the number of processes in the system.

Theorem 3. *There does not exist an algorithm that solves CSP(n) for TSO, for any $n \geq 2$.*

Proof Sketch: Assume that there is an algorithm A that solves CSP(n) for TSO for some $n \geq 2$. If A runs with processor p in <entry> and processor q in <remainder> and the other processors not participating, then by the Progress property, p must enter its <critical section> producing a partial computation of the form of Computation 1, where λ denotes the empty sequence and o_i^p denotes p's i^{th} operation.

Computation 1 $\begin{cases} p : o_1^p, o_2^p, ..., o_k^p \ \ (p \text{ is in its } < critical\ section >) \\ q : \lambda \end{cases}$

Similarly, if A runs with q's participation only, Progress guarantees that Computation 2 exists.

Computation 2 $\begin{cases} p : \lambda \\ q : o_1^q, o_2^q, ..., o_l^q \ \ (q \text{ is in its } < critical\ section >) \end{cases}$

Now, computations 1 and 2 are used to construct Computation 3 where both p and q are participating, and both are in their <critical section>. By showing that Computation 3 satisfies TSO, we reach a contradiction because Mutual Exclusion is violated.

Computation 3 $\begin{cases} p : o_1^p, o_2^p, ..., o_k^p \ \ (p \text{ is in its } < critical\ section >) \\ q : o_1^q, o_2^q, ..., o_l^q \ \ (q \text{ is in its } < critical\ section >) \end{cases}$

To see that Computation 3 satisfies TSO, imagine a situation where p and q enter their critical sections before the contents of their store-buffers are committed to main memory. So, the domestic reads by q complete at the buffer level, while foreign reads necessarily return initial values.

More formally, construct a sequence, S, of operations as follows. Initially, $S = \lambda$. Set sequence Q to be q's computation $\langle o_1^q, o_2^q \cdots, o_l^q \rangle$. Examine each o_i^q in Q in order from $i = 1$ to l. If o_i^q is a foreign read, append o_i^q to S and remove it from Q. When there are no foreign reads left in Q, append to S p's computation $\langle o_1^p, o_2^p, \cdots, o_k^p \rangle$. Finally, append Q (with foreign reads removed) to S.

Define $(O, \xrightarrow{L})$ to be S. $(O, \xrightarrow{L})$ consists of three segments. The first consists entirely of foreign reads by q, the second consists entirely of p's computation, and the third consists entirely of q's computation minus operations in the first segment.

It is straightforward to check that $(O, \xrightarrow{L})$ is a linearization. To see that $(O, \xrightarrow{L})$ satisfies TSO, note first that $(O|p, \xrightarrow{L}) = (O|p, \xrightarrow{prog})$. Consequently, $(O|p, \xrightarrow{tso}) \subseteq (O|p, \xrightarrow{L})$. Second, program order is maintained in the first segment by construction and also in the third segment. Finally, each read moved to the first segment is foreign and returns the initial value. Therefore, such a read is not preceded in q's computation by any writes to the same variable. So the moved reads do not violate $(O, \xrightarrow{tso})$. Therefore, $(O, \xrightarrow{tso}) \subseteq (O, \xrightarrow{L})$. Thus, Computation 3 satisfies TSO.   ∎

Since the preceding impossibility result made no assumption about fairness or about size of variables, it implies impossibility even of unfair solutions or solutions using unbounded variables.

6.2   Producer/Consumer Coordination

Producer/Consumer [2] objects are frequently used for process coordination. The producer (respectively, consumer) is a process that repeatedly cycles through the procedures <entry>, <producing> (respectively, <consuming>), <exit>. Let m and n be, respectively, the number of producer and consumer processes. We distinguish two problems the solution requirements of which vary: the *producer/consumer set* problem, denoted $P_m C_n$-set, and the *producer/consumer queue* problem, denoted $P_m C_n$-queue. A solution to $P_m C_n$-set must satisfy (1) *Safety:* Every produced item is consumed exactly once, and every consumed item is produced exactly once, and (2) *Progress:* If a producer (respectively consumer) is in <entry>, then it will eventually be in <producing> ((respectively <consuming>) and subsequently in <exit>. A solution to $P_m C_n$-queue must satisfy Safety, Progress, and (3) *Order:* Consumption order must respect the production order, where these orders are defined as follows. Item x *precedes in production order* item y if the production of x completes before the production of y begins. Consumption order is defined similarly.

Impossibilities: The general queue problems $P_m C_n$-queue, $P_1 C_n$-queue, or $P_m C_1$-queue are unsolvable for TSO, PSO, and RMO without using explicit synchronization. The proof of the following theorem is very similar to that of Theorem 3.

Theorem 4.   *There does not exist an algorithm that solves $P_m C_n$-queue or $P_m C_n$-queue for TSO when $m + n \geq 3$.*

Possibilities: Proofs of the following theorems are in the full version of the paper [8].

Theorem 5.   *There is a single-writer solution for $P_1 C_1$-queue for TSO (which fails for PSO and hence RMO).*

Theorem 6.   *There is a multi-writer solution for $P_1 C_1$-queue for RMO (and hence for PSO and TSO).*

Theorem 7.   *$P_m C_n$-set is solvable for RMO (and hence for TSO or PSO).*

7   SPARC Models versus Other Consistency Models

In the full paper [8] we compare TSO, PSO and RMO with other known memory models including Weak Ordering (WO) [3], Java [5,7], Causal Consistency (CC) [1], Processor Consistency (PC-G) [6], and Pipelined-Random Access Machine (P-RAM) [13], Coherence [3], and Sequential Consistency (SC) [12]. A summary of these comparisons is given in Figure 3 In this figure, an arrow from model A to model B means that the constraints of A imply those of B.

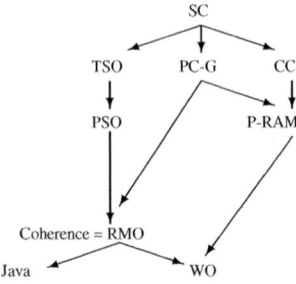

Fig. 3. Relationships between SPARC models and other models

References

1. M. Ahamad, G. Neiger, J. E. Burns, P. Kohli, and P. W. Hutto. Causal memory: Definitions, implementations, and programming. *Distributed Computing*, 9:37–49, 1995.
2. E. W. Dijkstra. Cooperating sequential processes. Technical Report EWD-123, Technological University, Eindhoven, the Netherlands, 1965. Reprinted in [4].
3. M. Dubois, C. Scheurich, and F. Briggs. Memory access buffering in multiprocessors. In *Proc. 13th Int'l Symp. on Computer Architecture*, pages 434–442, June 1986.
4. F. Genuys, editor. *Programming Languages*. Academic Press, 1968.
5. A. Gontmakher and A. Schuster. Characterizations of Java memory behavior. In *Proc. 12th Int'l Parallel Processing Symp.*, April 1998.
6. J. Goodman. Cache consistency and sequential consistency. Technical Report 61, IEEE Scalable Coherent Interface Working Group, March 1989.
7. L. Higham and J. Kawash. Java: Memory consistency and process coordination (extended abstract). In *Proc. 12th Int'l Symp. on Distributed Computing*, pages 201–215, September 1998.
8. L. Higham and J. Kawash. Specifications for the SPARC version 9 memory consistency models. Technical report, Department of Computer Science, The University of Calgary, 2000.
9. J. Kawash. *Limitations and Capabilities of Weak Memory Consistency Systems*. Ph.D. dissertation, Department of Computer Science, The University of Calgary, January 2000.
10. J. Kawash and L. Higham. Memory consistency and process coordination for SPARC v8 multiprocessors. Technical Report 99/646/09, Department of Computer Science, The University of Calgary, December 1999. A brief announcement appeared in *Proc. 19th ACM Symp. on Principles of Distributed Computing*, page 335, July 2000.
11. P. Kohli, G. Neiger, and M. Ahamad. A characterization of scalable shared memories. In *Proc. 1993 Int'l Conf. on Parallel Processing*, August 1993.
12. L. Lamport. How to make a multiprocessor computer that correctly executes multiprocess programs. *IEEE Trans. on Computers*, C-28(9):690–691, September 1979.
13. R. J. Lipton and J. S. Sandberg. PRAM: A scalable shared memory. Technical Report 180-88, Department of Computer Science, Princeton University, September 1988.
14. S. Park and D. L. Dill. An executable specification and verifier for relaxed memory order. *IEEE Trans. on Computers*, 48(2):227–235, February 1999.
15. M. Raynal. *Algorithms for Mutual Exclusion*. The MIT Press, 1986.
16. SPARC International, Inc. *The SPARC Architecture Manual version 8*. Prentice-Hall, 1992.
17. D. L. Weaver and T. Germond, editors. *The SPARC Architecture Manual version 9*. Prentice-Hall, 1994.

Improving Offset Assignment on Embedded Processors Using Transformations

Sunil Atri[1], J. Ramanujam[1*], and Mahmut Kandemir[2**]

[1] ECE Dept., Louisiana State University, Baton Rouge LA 70803, USA
[2] CSE Dept., The Pennsylvania State University, University Park, PA 16802, USA

Abstract. Embedded systems consisting of the application program ROM, RAM, the embedded processor core and any custom hardware on a single wafer are becoming increasingly common in areas such as signal processing. In this paper, we address new code optimization techniques for embedded fixed point DSP processors which have limited on-chip program ROM and include indirect addressing modes using post increment and decrement operations. These addressing modes allow for efficient sequential access but the addressing instructions increase code size. Most of the previous approaches to the problem aim to find a placement or layout of variables in the memory so that it is possible to subsume explicit address pointer manipulation instructions into other instructions as a post-increment or post-decrement operation. Our solution is aimed at transforming the access pattern by using properties of operators such as commutativity so that current algorithms for variable placement are more effective.

1 Introduction

Embedded processors (e.g., fixed-point digital signal processors, micro-controllers) are found increasingly in audio, video and communications equipment, cars, etc. thanks to the falling cost of processors [6]. These processors have limited code and data storage. Therefore, making efficient use of available memory is very important. On these processors, the program resides in the on-chip ROM; therefore, the size of the code directly impacts the required silicon area and hence the cost. Current compiler technology for these processors typically targets code speed and not code size; the generated code is inefficient as far code size is concerned. An unfortunate consequence of this is that programmers are forced to hand optimize their programs. Compiler optimizations specifically aimed at improving code size will therefore have a significant impact on programmer productivity [4,5].

DSP processors such as the TI TMS320C5 and embedded micro-controllers provide addressing modes with auto-increment and auto-decrement. This feature allows address arithmetic instructions to be part of other instructions. Thus, it eliminates the need for

---

* Department of Electrical and Computer Engineering. Louisiana State University. Baton Rouge, Louisiana 70803-5901. {jxr,sunil}@ee.lsu.edu
** Department of Computer Science and Engineering. Pennsylvania State University. University Park, PA 16802-6106. kandemir@cse.psu.edu

M. Valero, V.K. Prasanna, and S. Vajapeyam (Eds.): HiPC 2000, LNCS 1970, pp. 367–374, 2000.

explicit address arithmetic instructions wherever possible, leading to decreased code size. The memory access pattern and the placement of variables has a significant impact on code size. The auto-increment and auto-decrement modes can be better utilized if the placement of variables is performed after code selection. This delayed placement of variables is referred to as *offset assignment.*

This paper considers the *simple offset assignment* (SOA) problem where there is just one address register. A solution to the problem assigns optimal frame-relative offsets to variables of a procedure, assuming that the target machine has a single indexing register with only the indirect, auto-increment and auto-decrement addressing modes. The problem is modeled as follows. A basic block is represented by an *access sequence*, which is a sequence of variables written out in the order in which they are accessed in the high level code. This sequence is in turn further condensed into a graph called the *access graph* whose nodes represent variables and with weighted undirected edges. The weight of of an edge (a, b) is the number of times variables a and b are adjacent in the access sequence. The SOA problem is equivalent to a graph covering problem, called the *Maximum Weight Path Cover* (MWPC) problem. A solution to the MWPC problem gives a solution to the SOA problem. This paper presents a technique that modifies the access pattern using algebraic properties of operators such as commutativity. The goal is to reduce the number of edges of non-zero weight in the access graph. Rao and Pande have proposed some optimizations for the access sequence based on the laws of commutativity and associativity [7]. Their algorithm is exponential. In this paper, we present an efficient polynomial time heuristic.

2   Commutative Transformations

The access sequence has a great impact on the cost of offset assignment. We explore opportunities for better assignments that can be derived by exploiting the commutativity and other properties of operators in expressions. We propose a heuristic that attempts to modify the access sequence so as to achieve savings in the cost of the final offset assignment. We base our heuristic on the assumption that reducing the number of edges in an access graph will lead to a low cost assignment (see [1] for a justification). Towards this end, we identify edges that can be possibly be eliminated by a reordering transformation of the access sequence, which is in turn (because of a transformation in a statement in the C code basic block) based on the rule of commutativity. The edges we target for this kind of transformation are those with weight one. We consider the case where there are only two or less variables on the RHS of a statement. The statements considered are those that are amenable to commutative reordering. Let us focus now on the set of ordered statements as in Figure 1(a); they have an access sequence as shown in Figure 1(b). The variables shown in the boxes are those whose relative position *cannot* be changed. A change in l_2 for example will require statement reordering as shown in Figure 1(c). The weight change will be:

$$(1)\ \ w(l_2, f_3)-- \qquad (2)\ \ w(l_1, f_2)-- \qquad (3)\ \ w(l_0, f_1)--$$
$$(4)\ \ w(l_1, f_3)++ \qquad (5)\ \ w(l_2, f_1)++ \qquad (6)\ \ w(l_0, f_2)++$$

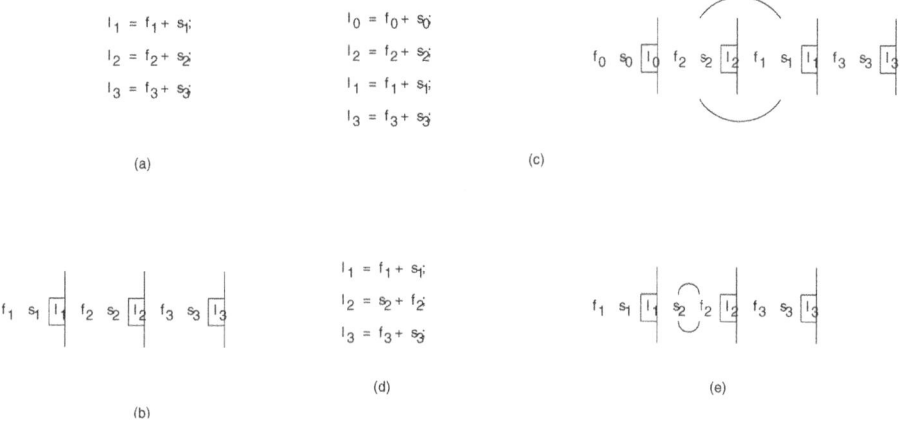

$$l_1 = f_1 + s_1;$$
$$l_2 = f_2 + s_2;$$
$$l_3 = f_3 + s_3;$$

(a)

$$l_0 = f_0 + s_0;$$
$$l_2 = f_2 + s_2;$$
$$l_1 = f_1 + s_1;$$
$$l_3 = f_3 + s_3;$$

(c)

(b)

$$l_1 = f_1 + s_1;$$
$$l_2 = s_2 + f_2;$$
$$l_3 = f_3 + s_3;$$

(d)

(e)

Fig. 1. Commutative transformation concept.

where the trailing ++ or -- indicate a increase in weight by one or a decrease in weight by one, respectively. This can be seen in Figure 1(c). We will explore variable reordering due to commutativity rather than statement reordering; the latter requires data dependence analysis.

There are different ways to evaluate the cost or benefit of a transformation. The one we propose is computationally much less expensive than Rao and Pande's procedure [7,1]. Consider again the access sequence shown in Figure 1(b). Without loss of generality, we can assume that the weight of the edge $w(s_2, l_2) = 1$. Then if we try the sequence shown in Figure 1(b), we have a modification in the access sequence, which, corresponds to the commutative transformation shown in Figure 1(d). This transformation can now be evaluated in two ways. The first, is to do a local evaluation of the cost, and the second one, is to run Liao's algorithm (or Incremental-Solve-SOA) on the changed access sequence. We propose two procedures based on these two methods of evaluating benefit.

We discuss the local evaluation first. As before, consider Figure 1(b) which is the access sequence, with the variables that cannot be moved marked by a box. As we had assumed the weight of the edge (s_2, l_2) was one and also that reducing the number of edges is possibly beneficial, reducing the weight of (s_2, l_2) to zero will effectively remove the edge from the access graph. If we wish to reduce $w(s_2, l_2)$ from one to zero then the following four edge weight reductions will occur.

(1) $w(l_1, s_2)$++
(2) $w(l_1, f_2)$- -
(3) $w(s_2, l_2)$- -; note that $w(s_2, l_2) = 0$ is possible here
(4) $w(f_2, l_2)$++

We define the *primary benefit* as the following value: (the number of non-zero edges turning zero) − (number of zero edges turning non-zero). In addition, we define a *secondary benefit* = (the sum of the increases in the weights of already present edges) + (the sum of the increases in the weights of self-edges).

If there are two edges with the same primary benefit, then the tie-break between them is done using the secondary benefit measure. The second method of evaluating benefit is to run Liao's heuristic, compute the cost, and select the edge giving the least cost. Of course, this option has a higher complexity.

2.1   Detailed Explanation of the Commutative Transformation Heuristic

We now discuss in detail our commutative transformation heuristic, called Commutative-Transformation-SOA(AS), given in Figure 3. This heuristic is greedy, in that, it performs the transformation that appears to be the best at that point in time. Lines 1 and 2 indicate that the input is the unoptimized access sequence AS and the output is the optimized access sequence AS°. We build the access graph by a call to the *AccessGraph*(AS) function. We also initialize the optimized access sequence AS° with the starting access sequence AS. Lines 8 to 14 constitute the first loop. Here we assign the Primary Benefit and Secondary Benefit measure to the edges of weight one. Edges with negative primary benefit are not considered and so we compute the secondary benefit only if there is a non-negative primary benefit. The data structure E_{sort}^1 is used to hold the various edges in descending order of the primary benefit, and if the the primary benefit is the same, then the secondary benefit is used as a tie-breaker in finding the ordering of the edges. T hold the set of edges that are considered as a set in the **while** loop. The compatibility of an edge is required as it is possible that a transformation made to AS° could be undone by the transformation motivated by another incompatible edge. As an example, consider Figure 2(a). The primary benefit of removing edge (a, b) from the access graph is 2, as edges (a, b) and (c, d) become zero weight edges. Now, if we try to remove edge (d, a) from the access graph, whose primary benefit is 1, we reintroduce edges (a, b) and (c, d), while removing edge (d, a). This is not the desired change, so when the transformation to remove edge (a, b) is made then the edge (d, a) should not be considered anymore.

We have included the **if** statement since, if there is no primary benefit, i.e., Primary Benefit is zero, it need not be included for consideration in the subsequent **while** loop. In the **while** loop from line 19 to line 27 for the same highest primary and secondary benefit measure, the maximal set of compatible edges are extracted from E_{sort}^1 and assigned to H. Transformation to AS° is performed in line 25. Once the transformations are done, The set of chosen edges is updated to contain the edges in C. Finally, when E_{sort}^1 is empty, we return the optimized access sequence AS°.

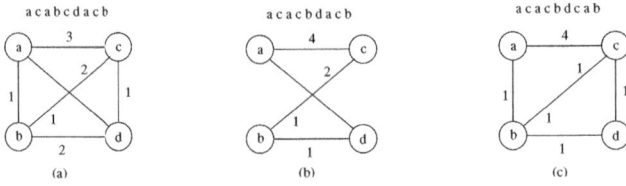

Fig. 2. Incompatible edges (a, b), and (d, a).

```
1   // INPUT :   Access Sequence AS
2   // OUTPUT :   Optimized Access Sequence AS°
3   Procedure Commutative-Transformation-SOA(AS)
4   (G = (V, E)) ← AccessGraph(AS)
5   AS° ← AS
6   S ← the set of all edges with weight one
7   E¹_sort ← φ
8   for (each edge (u, v) ∈ S) do
9      Compute_Primary_Benefit((u, v))
10     if (Primary Benefit (u, v) is positive)
11        Compute_Secondary_Benefit((u, v))
12        Add (u, v) to E¹_sort
13     endif
14   enddo
15   // In the next sorting step, the Secondary Benefit is used to break ties
16   Sort the entries in E¹_sort in descending order of Primary Benefit
17   // T holds the set of edges of weight one that are chosen
18   T ← φ
19   while (E¹_sort ≠ φ)
20      // Extract_Edges extracts (removes) the set of all edges with the
21      // highest primary benefit and the same largest secondary benefit.
22      // Let H = {e₀, e₁, ..., eᵣ} be these edges
23      H ← Extract_Edges(E¹_sort)
24      C ← the maximum compatible subset of H ∪ T
25      Perform_Transformation(AS°, C)
26      T ← T ∪ C
27   endwhile
28   return(AS°)
```

Fig. 3. Commutative-Transformation-SOA.

2.2   Commutative Transformation Heuristic Example

We now show the working of the heuristic through an example. Consider the access sequence Figure 4(b), with the resulting access graph in Figure 4(c). The edges of weight 1 are: (a, f), (a, b), (b, c), (c, d), (e, f), (d, e), (b, f) and (d, f). Table 1 summarizes the different computation of the benefits. The first column lists the edges and the second column shows the transition which would need to occur in the access sequence in Figure 4(a). The third column shows the number of non-zero edges turning zero and the fourth column gives the number of zero edges turning non-zero. The fifth and the sixth columns give the increase in the weight of the non zero edge and the decrease in the weight of the non zero edge, respectively. Also shown are the increase and decrease in the weight of the self edge. The second last column (P.B.) show the primary benefit, i.e., difference between column three and four. The last column shows the secondary benefit which is the sum of column five and seven. The function $Assign\text{-}Benefit$ formulates Table 1 and assigns the Primary and Secondary Benefit value. The $Assign\text{-}$

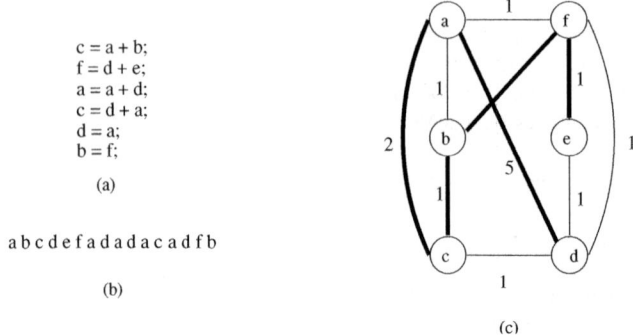

```
c = a + b;
f = d + e;
a = a + d;
c = d + a;
d = a;
b = f;
```

(a)

abcdefadadacadfb

(b)

(c)

Fig. 4. Example code and the associated access graph.

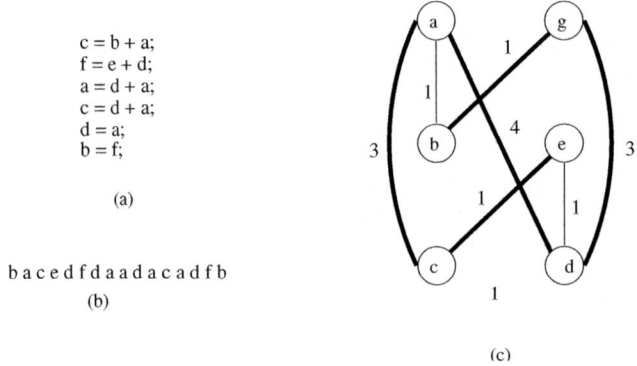

```
c = b + a;
f = e + d;
a = d + a;
c = d + a;
d = a;
b = f;
```

(a)

bacedfdaadacadfb

(b)

(c)

Fig. 5. Optimized code, the associated access graph, and the offset sequence.

Compatibility function call checks for the compatibility of each weight 1 edge with other weight 1 edge. As all the edges listed, except the last two ones, have a positive primary benefit, they are stored in E^1_{sort} in the order same as in the table. Only the last two entries will not be in E^1_{sort}. Here the primary benefit is all 1, and the secondary benefit for edge(a, f) is higher than the rest. The Extract_Edges function call will return all the edges in the table as they are all compatible with each other. The following three transformation will be performed in line 18: (1) $fada \rightarrow fdaa$; (2) $abcd \rightarrow bacd$; and (3) $cdef \rightarrow cedf$. This transformation will result in the access sequence shown in Figure 5(b). This access sequence AS^o will be returned in line 21. The Liao cost of the offset assignment obtained now has fallen from 5 to 2 as shown in Figure 4(c) and Figure 5(c).

Let us concentrate now on the example shown in Figure 6(b). We follow the same procedure as explained above. The primary and secondary benefit values that will be generated are shown in Table 2. The transformations that would be performed are: (1) $aefd \rightarrow afed$, and (2) $fbaa \rightarrow faba$. The input access sequence is shown in Figure 6(b). and the output access sequence is shown in Figure 6(c). The Liao's cost for this example has fallen from 4 to 2. In the case of edge(e, d) the primary benefit is -1, so it would

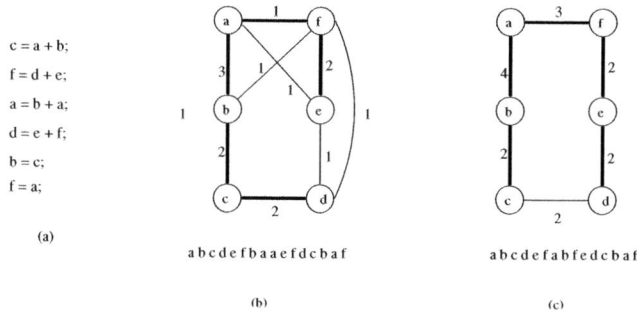

(a)

c = a + b;
f = d + e;
a = b + a;
d = e + f;
b = c;
f = a;

abcdefbaaefdcbaf

(b)

abcdefabfedcbaf

(c)

Fig. 6. Optimal solution of the SOA.

Table 1. Primary and secondary benefit measures for the example in Figure 4(b).

Edge	Trans.	NZ → 0	0 → NZ	NZ ↑	NZ ↓	self ↑	self ↓	P.B.	S.B.
(a, f)	$fada \rightarrow fdaa$	1 (a, f)	0	1 (f, d)	0	1 (a, a)	0	1	2
(a, b)	$abcd \rightarrow bacd$	1 (b, c)	0	1 (a, c)	0	0	0	1	1
(b, c)	$abcd \rightarrow bacd$	1 (b, c)	0	1 (a, c)	0	0	0	1	1
(c, d)	$cdef \rightarrow cedf$	2 (c, d) (e, f)	1 (c, e)	0	0	0	0	1	0
(e, f)	$cdef \rightarrow cedf$	2 (c, d) (e, f)	1 (c, e)	0	0	0	0	1	0
(d, e)	$cdef \rightarrow cedf$	2 (c, d) (e, f)	1 (c, e)	0	0	0	0	1	0
(b, f)	$fixed$								
(d, f)	$fixed$								

not be included in E^1_{sort} because of the **if** statement in line 9. Experimental results and additional details have been omitted for lack of space; see [2] for this.

3   Conclusions

Optimal code generation is important for embedded systems in view of the limited area available for ROM and RAM. Small reductions in code size could lead to significant changes in chip area and hence reduction in cost. The offset assignment problem is useful in reducing code size on embedded processors. In this paper, we explored the use of commutative transformations in order to reduce the number of edges in the access graph so that the probability of finding a low cost cover is increased. We have considered commutative transformations, but it is also possible to look at the others transformations, like associative and distributive transformations and even statement reordering. Rao and Pande's solution [7] computes all the possible transformations, which is exponential. The heuristic presented in this paper identifies specific edges and selects the corresponding transformation to perform. As the number of transformation which we perform is

Table 2. Primary and secondary benefit measures for the example in Figure 6(b).

Edge	Trans.	NZ → 0	0 → NZ	NZ ↑	NZ ↓	self ↑	self ↓	P.B.	S.B.
(a,e)	$aefd \to afed$	2 (a,e) (f,d)	0	2 (a,f) (e,d)	0	0	0	2	0
(f,d)	$aefd \to afed$	2 (a,e) (f,d)	0	2 (a,f) (e,d)	0	0	0	2	0
(b,f)	$fbaa \to faba$	1 (b,f)	0	1 (a,b)	0	0	1	1	1
(e,d)	$cdef \to cedf$	0	1 (c,e)	2 (e,d) (d,f)	1	0	0	-1	2
(f,d)	*fixed*								

bounded by the number of edges, this algorithm is much faster. We are currently exploring several issues. First, we are looking at the effect of statement reordering on code density. Second, we are evaluating the effect of variable life times and static single assignment on code density. In addition, reducing code density for programs with array accesses is an important problem.

Acknowledgments

The work of J. Ramanujam is supported in part by NSF Young Investigator Award CCR–9457768 and NSF grant CCR–0073800.

References

1. S. Atri. *Improved Code Optimization Techniques for Embedded Processors.* M.S. Thesis, Dept. Electrical and Computer Engineering, Louisiana State University, Dec. 1999.
2. S. Atri, J. Ramanujam, and M. Kandemir. *The effect of transformations on offset assignment for embedded processors.* Technial Report, Louisiana State University, May 1999.
3. R. Leupers and P. Marwedel. Algorithms for address assignment in DSP code generation. In *Proc. International Conference on Computer Aided Design*, pages 109–112, Nov. 1996.
4. S. Y. Liao, *Code Generation and Optimization for Embedded Digital Signal Processors*, Ph.D. Thesis. MIT, June 1996.
5. S. Y. Liao, S. Devadas, K. Keutzer and S. Tjiang, and A. Wang. Storage Assignment to Decrease code Size Optimization. In *Proc. 1995 ACM SIGPLAN Conference on Programming Language Design and Implementation.* pages 186-195, June 1995.
6. P. Marwedel and G. Goossens, editors. *Code Generation for Embedded Processors*, Kluwer Acad. Pub., 1995.
7. Amit Rao and Santosh Pande. Storage assignment optimizations to generate compact and efficient code on embedded DSPs. *Proc. 1999 ACM SIGPLAN Conference on Programming Language Design and Implementation.* pages 128–138, June 1999.

Session IV-A

Applied Parallel Processing
Chair: Partha Dasgupta,
Arizona State University

Improving Parallelism in Asynchronous Reading of an Entire Database

Subhash Bhalla

Database Systems Laboratory
University of Aizu
Aizu-Wakamatsu, Fukushima 965-8580, JAPAN
bhalla@u-aizu.ac.jp

Abstract. To recover from media failures, a database is 'restored' from an earlier backup copy. A recovery log of transactions is used to roll forward from the backup version to the desired time (the current time). High availability requires - backup activity to be fast, and on-line with ongoing update activity. Such concurrent generation of a database copy, interferes with system activity. It introduces blocking and delays for many update transactions. We study the performance of revised algorithms, to highlight the level of concurrent activity permitted by these algorithms, in parallel. Subsequently, the interference between global database copy activity and transaction updates is minimized based on a new algorithm for asynchronous generation of a copy of the database.

1   Introduction

Taking frequent backup is an essential part of database operations. Many applications require reading an entire database. Existing algorithms [1,8,9] introduce delays for other executing update transactions. Consider a few examples [1], an inventory official may need to read the entire inventory to compile a stock report, or a computer operator may wish to take a database backup without suspension of services. If conventional two-phase locking protocol is used to maintain consistency, then the global-read transaction renders a considerable part of the database inaccessible for the update transactions [2]. Other algorithms have similar drawbacks. These algorithms, either do not guarantee serializability [8,9], or restrict certain type of transactions [1,8,9], and allow few concurrent updates.

A transaction is an update transaction, if it updates any data within the database. A read-only transaction is a transaction that does not update data but reads a part of the database. A transaction is *two-phase,* if it acquires locks during a growing phase, and releases locks during a shrinking phase. An update transaction acquires exclusive locks (write locks). A read-only transaction acquires the read (shared) locks, that can be shared among executing read-only transactions. A global-read transaction is an *incremental* transaction that acquires read locks for reading, a few at a time. It releases locks and acquires more locks in small steps.

M. Valero, V.K. Prasanna, and S. Vajapeyam (Eds.): HiPC 2000, LNCS 1970, pp. 377–384, 2000.
© Springer-Verlag Berlin Heidelberg 2000

A global-read produces an inconsistent database version, in the presence of update transactions. For example [1,8], suppose, it is desired to calculate the total deposits in a bank, with a global-read that incrementally sums the checking and savings accounts. If a client, executes an update transaction to move a thousand dollars from a savings to a checking account during a global-read, the summation may be a thousand dollars short, if checking account is read by global-read, before it is updated (given, the savings account is read after it is has been updated).

Proposals for exploiting histories for database backup has been studied in [5,6]. We have considered transaction level backup and recovery. The rest of the paper is organized as follows. In section 2, an early global-read algorithm has been studied for making a system model. Section 3 studies a revised proposal. In section 4, and 5 new proposals for global-reading have been put forward. Section 6 considers proof of correctness. Section 7 outlines the summary.

2   The Original Global-Read Algorithm

In the original proposal of Pu [8], a global-read transaction divides the database in two parts (Figure 1.). The items read by global-read are treated as black. Others are considered as white items. An ordinary transaction can update the database within one of these two portions. All other transactions are not permitted. It was shown by the later study [1], that the algorithm does not guarantee serializability. A cyclic precedence could be setup, if a white transaction, read items updated by a black transaction.

```
BEGIN
  STEP 1 :  Set color of each entity to white
  STEP 2 :  WHILE    there are 'white' entities
            DO  IF   all white entities are exclusively locked,
                     Request shared lock on a white entity  -
                wait until lock is granted
                ELSE  -  Lock any sharable white entity
                      -  Read entity -  Color entity black
                      -  Release entity lock
            END WHILE
END.
```

Fig. 1. The incremental global-read algorithm.

The original global-read algorithm changes the color of the entities within the database, along with the progress of the global-read transaction (T_{gr}). Initially, all items are colored as white. The early transactions tend to be white as these update white entities. With the progress of T_{gr}, the number of grey transactions increases. During the final phase, most transactions tend to be black, as these update database items that have already been read by T_{gr}. From the point of view of serializability, all black transactions are considered to have occurred after the global-read. Similarly, the white transactions are considered to occur before the global-read transaction, because their updates are incorporated before the global-read transaction ($T_w \longrightarrow T_{gr} \longrightarrow T_b$).

For sake of simplicity, we assume that the update transactions access the entities in the database uniformly. Let us consider a database with n entities, of which r have been painted as black. An update transaction that writes on k items, will not conflict with a T_{gr}, if all its write-set entities are either, black or white. The probability of this event is -

$$P_{n,k}(k) = \frac{C_k^r}{C_k^n} + \frac{C_k^{n-r}}{C_k^n} \tag{1}$$

The probability of an abort in the case of grey transactions, after algebraic transformations [9] is given by -

$$P(k) = \frac{2}{n+1} \sum_{r=k}^{n} \prod_{i=0}^{k-1} \frac{r-i}{n-i} \tag{2}$$

These probabilities can be computed [9]. The proportion of aborted transactions increases with k. For k = 2, 3, 4, 5, we obtain - 0.333, 0.500, 0.600, and 0.667, respectively [8]. For a normal range of values of k (between 5 - 10), more than 65 - 80 % aborts are likely to be encountered.

2.1   Simulation Model

The simple model uses the input and output parameters listed in Table 1. The validity of the model was verfied by adopting the above results in (2) for the original global-read algorithm [8].

Table 1. Input parameters and Output measurement variables.

Input Parameters		
Number of entities in the database	:	1000
Number of entities read by a transaction (r)	:	1 - 6
Number of entities updated by a transaction (k)	:	1 - 6
Distribution of the values of ' r '	: uniform	
Distribution of the values of ' k '	: uniform	
Output Parameters		
Proportion of white transactions (T_w)	:	Pw
Proportion of black transactions (T_b)	:	Pb
Proportion of aborted (grey) transactions (T_g)	:	Pg
Proportion of aborted white transactions (T_{wa})	:	Pwa

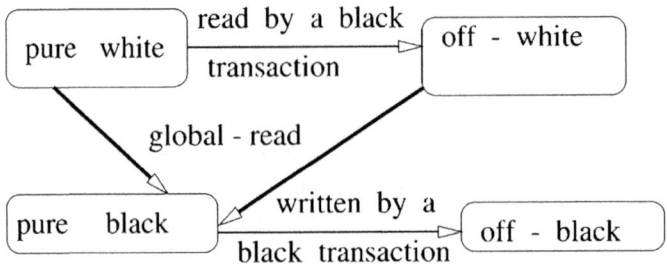

Fig. 2. Revised state transitions for data entities.

3   Revised Global-Read Algorithm

The revised global-read algorithm [1] divides the database entities into two colors : white and black. Also, a shade is associated with data entities that have been read or written by a transaction (Figure 2.). The revised algorithm does not permit the following data operations, and transactions (Figure 3.) (please also see [1]).

– All grey transactions that write on both black and white items;
– A white transaction can not write on an off-white entity; and
– A white transaction can not read an off-black entity.

As shown in Figure 3 and 4., the revised global-read algorithm further increases the number of transactions to be aborted [7].

Other Parallel Transaction	Definition	Notion of Serialization with global-read Transaction (T_{gr})
Read-only Transaction (T_r)	(Write-Set) $WS = \phi$	Concurrent
White update Transaction (T_w)	WS = white entities	Before global-read
Prohibited white trans-actions (T_{wa})	WS = off white; (Read-Set) RS = off black	Rejected
Black update Transaction (T_b)	WS = pure black entities	After global-read
Grey Transaction (T_g)	WS = white and black entities	Rejected

Fig. 3. Serialization of Transactions in the revised algorithm [$\mathbf{T_w} \longrightarrow \mathbf{T_{gr}} \longrightarrow \mathbf{T_b}$; **Reject** T_g, **and** T_{wa}]

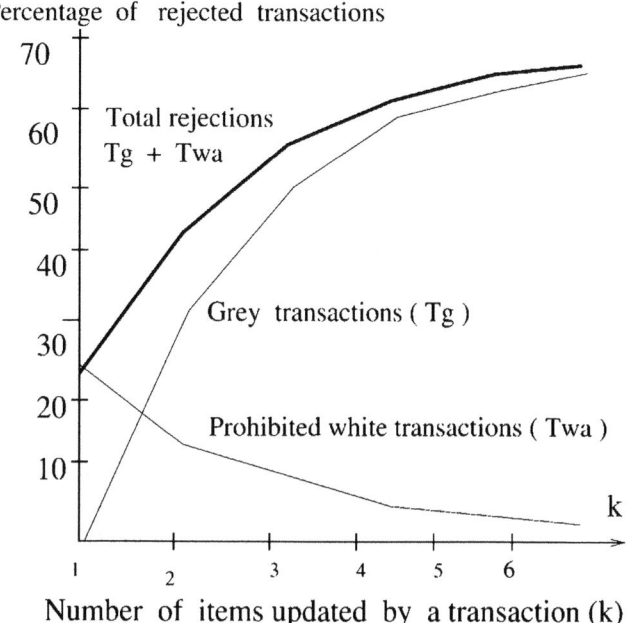

Fig. 4. Total percentage of rejected transactions, (k=r)

4   Save Some Strategy

As a way, for executing grey transactions, Pu [8] propose to isolate their updates
that concern white entities. In this case, the database system needs to make a
copy of the older version of data, so that the global-read transaction T_{gr} can read
an earlier (consistent) version. This strategy allows T_g to be treated as black
transactions, that occurs after the global-read. Similarly, the white transactions,
that are rejected by the revised algorithm, can be permitted by allowing the
updated white entities to keep an earlier version of the data. Thus, their updates,
do not interfere with the T_{gr}. This proposal, requires additional storage of earlier
versions of data. In the worst case, an additional 100 % disk space capacity may
be needed.

5   Save All Strategy

Based on the idea of 'State Transformation' by the committing transactions [4],
a list of committed transactions is generated by the system as a log, during the
execution of a global-read transaction. This pool of information is later used to
perform generation of a consistent database copy.

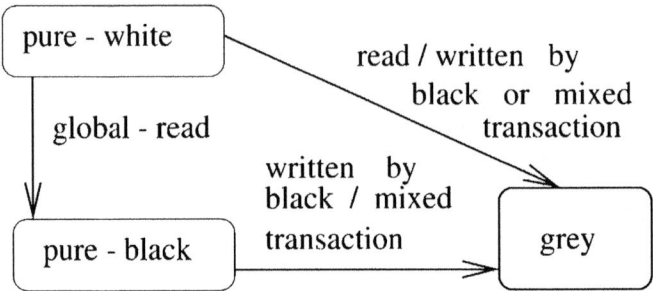

Fig. 5. Proposed state transition diagram for data entities.

5.1 Global-Read Processing

The revised algorithm executes in two phases (Figure 6.). In phase 1, it generates a log during the execution of a global-read transaction. This log (called the **color log**) contains a color marking for each update transaction. On completion of the global-read, in phase 2, modifications are applied to make the copied data become consistent.

At the beginning, all database entities are white. Gradually, entities read by a T_{gr} are colored black. In addition, entities that are written by black or mixed transactions are colored as grey. White data entitity that are read by black or mixed transaction are also colored as grey (Figure 5.). Normal update transactions are colored white or black depending on the color of data entities being updated by them. A transaction is termed as mixed (or grey), if either -

- it updates a mixture of black and white data entities; or
- it reads a data entity that is colored as grey; or
- it writes on a data entity that is colored grey.

The color of a transaction can be determined at the time of its commit by examining the color of data items in its, read-set (RS) and write-set (WS). Normal read-only transactions are not colored and proceed normally subject to the constraints of the two-phase locking protocol.

5.2 Data Management

The color of all data entities is set as white, at the beginning of a (T_{gr}). It changes to black immediately after a read access by a T_{gr} transaction. Of the update transactions, only black or grey transactions can change the color of a data entity to grey. The change of color must occur before the black or grey transaction releases the shared and exclusive locks after data updation.

Before the start of a T_i, a Transaction-id is assigned to it on its arrival. It contains information related to site-of-origin, site-status, transaction type (read-only class/ duration class/ priority class; (if known)), time of the day, and local sequence number. Site-status value indicates the Commit Sequence Number (CSN) of the most recent transaction committed.

```
/*  Begin phase 1 --  Generate Color Log  */
BEGIN
 csn = 0 ;
 WHILE        global-read
     FOR  ( each transaction commit ) csn = csn + 1;
          WRITE  in color log -
          (csn, Transaction-id, Read-set, Write-set, color)
 END WHILE
 last-csn  =  csn
END


/*  Begin phase 2  -- Over write from the Log */
BEGIN
     FOR   csn  1, last-csn
     IF color  =  'grey'
        WRITE  transaction Write-set into DATABASE copy
END
```

Fig. 6. Proposed algorithms for Generation of Database Copy

In phase 1, the algorithm carries out an on-the-fly reading (no consistency) of the contents of the database. A color log of mixed (or grey) transactions T_g, is created (Figure 6.). In phase 2, the algorithm generates a consistent copy of the database by overwriting the updates from its color log. In the color log, it maintains a table of entries of completed grey transactions as a Committed Transaction List (CTL). This list contains entries for:
{ Transaction-id; Read-set; Write sets and data values; and Allotted Commit Sequence Number (CSN) },   Where, next CSN value = Previous CSN value + 1.

6   Proof of Correctness

The proposed algorithm considers only one global-read at a time. Its extension to multiple global-reads and for considering a distributed system has been considered in [8]. A formal proof in the case of the proposed algorithm is similar to the proof in [8]. Informally, a white transaction writes on (only) pure white entities. These transactions do not affect the contents of the copied database. Each black and mixed transaction is selected and its updates are over-written. This eliminates the chances of data inconsistency.

6.1   Proof of Consistency

An informal proof is being presented on similar lines as [1,3,8]. A transaction is a sequence of n read/write action steps :
T = ($(T, a_1, e_1), \ldots, (T, a_i, e_i), \ldots, (T, a_n, e_n)$), where $1 =< i =< n$. Where T is the transaction, and a_i is the action at step i, and e_i is the entity acted upon at

the step. A schedule (S) of transaction dependencies, can be written. A schedule S is consistent, if there exists an equivalent serial schedule. All schedules formed by well formed transactions following 2-phase locking are consistent [3].

Assertion 1 : Given that, T_1 is a read-only transaction. A schedule (S) of transaction dependencies (T_1, e, T_2), can be transformed to schedule (S'), as (T_2, e, T_1), if all writes of T_2, are later overwritten on data read by T_1. Consider,

- In the given setup, white transactions precede, the black transactions (please see [1] for the proof).
- (by definition) T_w precedes, the global-read transaction;
- For various items accessed by a grey transaction, two kind of dependency relations can exist. (T_{gr}, e_i, T_{grey}), and (T_{grey}, e_j, T_{gr}). The dependencies, (T_{gr}, e_i, T_{grey}), are transformed to (T_{grey}, e_j, T_{gr}) during phase 2, by **Assertion 1** .
- (by definition) T_{gr} precedes black transactions

7   Summary and Conclusion

The paper presents algorithms for for processing global-read of an entire database copy. In this scheme, the other update transactions that execute in parallel face no blocking of update activity. The proposed algorithm is simple to implement. It preserves the consistency of the backup version, as per the notion of serializability.

References

1. P. Amann, Sushil Jajodia, and Padmaja Mavuluri, "On-The-Fly Reading of Entire Databases", IEEE Transactions of Knowledge and Data Engineering, Vol. 7, No. 5, October 1995, pp. 834-838.
2. P.A. Bernstein, V. Hadzilacos, and N. Goodman, "Concurrency Control and Recovery in Database Systems", Addison-Wesley, 1987.
3. Eswaran, K.P., J.N. Gray, R.A. Lorie, and I.L. Traiger, "The Notion of consistency and predicate locks in a Database System, " Communications of ACM, Vol. 19, Nov. 1976.
4. J. Gray, "Notes on Database Operating System", IBM Technical report RJ2188 (Feb 1978), also LNCS Vol. 60, published by Springer-verlag.
5. D.B. Lomet, "High Speed On-line Backup When Using Logical Log Operations", Proceedings of ACM SIGMOD Annual Conference in SIGMOD record, Vol. 29, No., 2, June 2000.
6. D.B. Lomet and and B. Salzberg, "Exploiting a History Database for Backup", VLDB Conference, Dublin (Sept. 1993), 380-390.
7. Y. Ohtsuka, "Algorithms for Online Generation of Global Database Checkpoint", Master's Thesis, March 2000, University of Aizu, Fukushima, Japan, pp. 1-34.
8. C. Pu, "On-the-fly, incremental, consistent, reading of entire databases", Algoritmica, Vol. 1, No. 3, pp. 271-287, Oct. 1986.
9. C. Pu, C.H. Hong, and J.M. Wha, "Performance Evaluation of global-reading of entire databases", Proc. of Intl. Symposium on Databases in Parallel and Distributed Systems, pp. 167-176, Austin, Texas, Dec. 1988.

A Parallel Framework for Explicit FEM*

Milind A. Bhandarkar and Laxmikant V. Kalé

Parallel Programming Laboratory,
Department of Computer Science,
University of Illinois at Urbana-Champaign, USA
{milind,kale}@cs.uiuc.edu,
http://charm.cs.uiuc.edu/

Abstract. As a part of an ongoing effort to develop a "standard library" for scientific and engineering parallel applications, we have developed a preliminary finite element framework. This framework allows an application scientist interested in modeling structural properties of materials, including dynamic behavior such as crack propagation, to develop codes that embody their modeling techniques without having to pay attention to the parallelization process. The resultant code modularly separates parallel implementation techniques from numerical algorithms. As the framework builds upon an object-based load balancing framework, it allows the resultant applications to automatically adapt to load imbalances resulting from the application or the environment (e.g. timeshared clusters). This paper presents results from the first version of the framework, and demonstrates results on a crack propagation application.

1 Introduction

When a parallel computation encounters dynamic behavior, severe performance problems often result. The dynamic behavior may arise from intrinsic or external causes: intrinsic causes include dynamic evolution of the physical system being simulated over time, and algorithmic factors such as adaptive mesh refinement. External causes include interference from other users' jobs on a time-shared cluster, for example. Significant programming efforts are required to make an application adapt to such dynamic behavior.

We are developing a broad approach that facilitates development of applications that automatically adapt to dynamic behavior. One of the key strategies used in our approach is multi-partition decomposition: the application program is decomposed into a very large number of relatively small components, implemented as objects. The underlying object-based parallel system (Charm++) supports multiple partitions on individual processors, and their execution in a data-driven manner. The runtime system controls the mapping and re-mapping of these objects to processors, and consequent automatic forwarding of messages. Automatic instrumentation is used to collect detailed performance data, which

---

* Research funded by the U.S. Department of Energy through the University of California under Subcontract number B341494.

M. Valero, V.K. Prasanna, and S. Vajapeyam (Eds.): HiPC 2000, LNCS 1970, pp. 385–394, 2000.

is used by the load balancing strategies to adaptively respond to dynamic behavior by migrating objects between processors. This load balancing framework has been demonstrated to lead to high-performance even on large parallel machines, and challenging applications, without significant effort on part of the application programmer [4].

In order to promote programmer productivity in parallel programming, it is desirable to automate as many parallel programming tasks as possible, as long as the automation is as effective as humans. The load balancing framework accomplishes such automation for the broad categories of parallel applications. However, to use this framework, one must write a program in a relatively new, and unfamiliar parallel programming language (in this case Charm++). In contrast, an application scientist/engineer is typically interested in modeling physical behavior with mathematical techniques; from their point of view, parallel implementation is a necessary evil. So, the effort required to develop a parallel application is clearly detrimental to their productivity, even when adaptive strategies are not called for.

We are continuing this drive towards automation by considering special classes of applications that can be further automated. In this paper, we describe the beginnings of our effort to make such a special class: computations based on the finite element method (FEM). The framework whose design and implementation we describe in this paper allows an application programmer to write sequential code fragments specifying their mathematical model, while leaving the details of the finite element computation, including the details of parallelization, to the "system" — the finite element framework, along with the underlying load balancing framework. The code written in this fashion is succinct and clear, because of the clear and modular separation between its mathematical and parallelization components. Further, any application developed using this framework is automatically able to adapt to dynamic variations in extrinsic or intrinsic load imbalances.

Other FEM frameworks such as Sierra [6] (Sandia National Laboratories) exist. However, a detailed review of related work is not possible here due to space limitations. We note that most other frameworks are not able to deal with dynamic behavior as ours does.

In the next section, we describe the overall approach to dynamic computations that we're pursuing. Section 3 provides a background for finite element computations, while section 4 describes our parallel FEM framework. Our experiences in porting an existing crack propagation application (developed by P. Guebelle et al at the University of Illinois), and the relevant performance data are presented in section 5. This framework is only a first step in our effort to facilitate development of complex FEM codes. Some directions for future research are identified in the last section.

2   Parallelizing Dynamic Computations

In *dynamic* applications, the components of a program change their behavior over time. Such changes tend to reduce performance drastically, especially on a large number of processors. Not only does the performance of such applications tend to be poor, the amount of effort required in developing them is also inordinate.

We are developing a broad approach for parallelization of such problems, based on *multi-partition decomposition* and data-driven execution. The application programmer decomposes the problem into a large number of "objects." The number of objects is chosen independently of the number of processors, and is typically much larger than the number of processors. From the programmer's point of view, all the communication is between the objects, and not the processors. The runtime system is then free to map and re-map these objects across processors. The system may do so in response to internal or external load imbalances, or due to commands from a time shared parallel system.

As multiple chunks are mapped to each processor, some form of local scheduling is necessary. A data-driven system such as Charm++ can provide this effectively. On each processor, there is a collection of objects waiting for data. Method invocations (messages) are sent from object to object. All method invocations for objects on a processor are maintained in a prioritized scheduler's queue. The scheduler repeatedly picks the next available message, and invokes the indicated method on the indicated object with the message parameters.

Scientific and engineering computations tend to be iterative in nature, and so the computation times and the communication patterns exhibited by its objects tend to persist over time. We call this "the principle of persistence", on par with "the principle of locality" exhibited by sequential programs. Even the dynamic CSE applications tend to either change their pattern abruptly but infrequently (as in adaptive refinement) or continuously but slowly. The relatively rare case of continuous and large changes can still be handled by our paradigm, using more dynamic and localized re-mapping strategies. However, for the common case, a runtime system can employ a "measurement based" approach: it can measure the object computation and communication patterns over a period of time, and base its object re-mapping decisions on these measurements. We have shown [4] that such measurement-based load balancing leads to accurate load predictions, and coupled with good object re-mapping strategies, to high-performance for such applications.

Based on these principles, we have developed a load balancing framework within the Charm++ parallel programming system [2]. The framework automatically instruments all Charm++ objects, collects their timing and communication data at runtime, and provides a standard interface to different load balancing strategies (the job of a *strategy* is to decide on a new mapping of objects to processors). The framework is sufficiently general to apply beyond the Charm++ context, and it has been implemented in the Converse [3] portability layer. As a result, several other languages on top of Converse (including threaded MPI) can also use the load balancing functionality.

Overall, our approach tries to automate what "runtime systems" can do best, while leaving those tasks best done by humans (deciding what to do in parallel), to them. A further step in this direction is therefore to "automate" commonly used algorithms in reusable libraries and components. The parallel FEM framework is an attempt to automate the task of writing parallel FEM codes that also adapt to dynamic load conditions automatically, while allowing the application developer to focus solely on mathematical modeling techniques, as they would if they were writing sequential programs.

3   Finite Element Computations

The Finite Element Method is a commonly used modeling and analysis tool for predicting behavior of real-world objects with respect to mechanical stresses, vibrations, heat conduction etc. It works by modeling a structure with a complex geometry into a number of small "elements" connected at "nodes". This process is called "meshing". Meshes can be structured or unstructured. Structured meshes have equal connectivity for each node, while unstructured meshes have different connectivity. Unstructured meshes are more commonly used for FEM.

The FEM solver solves a set of matrix equations that approximate the physical phenomena under study. For example, in stress analysis, the system of equations is:

$$F = kx$$

where F is the stress, k is the stiffness matrix, and x is the displacement. The stiffness matrix is formed by superposing such equations for all the elements. When using FEM for dynamic analysis, these equations are solved in every iteration, advancing time by a small "timestep". In explicit FEM, forces on elements are calculated from displacements of nodes (strain) in the previous iteration, and they in turn cause displacements of nodes in the next iteration.

In order to reduce errors of these approximations, the element size is chosen to be very small, resulting in a large number of elements (typically hundreds of thousands). The sheer amount of computations lends this problem to parallel computing. Also, since connectivity of elements is small, this results in good computation to communication ratios. However, the arbitrary connectivity of unstructured meshes, alongwith dynamic physical phenomena introduce irregularity and dynamic behavior, which are difficult to handle using traditional parallel computing methods.

4   The Parallel FEM Framework

Our FEM framework treats the FEM mesh as a bipartite graph between nodes and elements. In every iteration of the framework, elements compute attributes (say, stresses) based on the attributes (displacements) of surrounding nodes, and the nodes then update their attributes based on elements they surround. Nodes are shared between elements, and each element is surrounded by a (typically fixed) number of nodes.

We partition the FEM mesh into several "chunks". In order to reduce communication, we take locality into account by using spatial partitioning. We divide the elements into chunks. Nodes shared by multiple elements within the same chunk belong to that chunk. Nodes shared between elements that belong to different chunks are duplicated in those chunks. In order to identify association of a node with its mirrors, we represent each node with a global index. However, for the purpose of iterating over the nodes in a chunk, we also have an internal local index for each node. Each chunk also maintains a table that maps its shared nodes to a (chunk-index, local-index) pair.

The FEM Framework implements all parallelization features, such as a chare array of chunks, a data-driven driver for performing simulation timesteps, messages for communicating nodal attributes etc. It makes use of generic programming features of C++, and consists of several Charm++ class "templates", with the user-defined datatypes as template parameters. Another approach for our framework could have been to provide base classes for each of these asking the user to provide derived classes that implement the application-specific methods. However, this approach adds overheads associated with virtual method invocation. It also misses on several optimization at compile time as can be done with our template-based approach.

Application code and the FEM framework have interfaces at various levels. The FEM framework provides parallelization, communication data structures, as well as a library for reading the file formats it recognizes. The application-specific code provides the data structures, function callbacks, and data that controls the execution of the FEM framework.

Figure 1 shows the algorithm used by this framework.

```
Chunk 0:
   Initialize Configuration Data [FEM_Init_Conf_Data]
   Broadcast Configuration data to all chunks.

All Chunks:
   Initialize Nodes [FEM_Init_Nodes]
   Initialize Elements [FEM_Init_Elements]
   for i = 0 to FEM_Get_Total_Iter
     for all elements
       FEM_Update_Element(Element, Node[], ChunkData)
     for all nodes
       FEM_Update_Node(Node, ChunkData)
     FEM_Update_Chunk(ChunkData)
     Communicate shared NodeData to neighboring chunks
   end
```

Fig. 1. Pseudo-code for the Data-Driven Driver of FEM Framework

4.1   Application-Specific Datatypes

All FEM applications need to maintain connectivity information to represent the bipartite graph among nodes and elements. This information is maintained by the FEM framework in its data structures associated with each chunk. In addition, each application will have different attributes for nodes, elements etc. Only these need to be provided by the user code.

Node: This user-defined datatype contains nodal attributes. It also provides instance methods called getData and update. getData returns the NodeData (see below), and update combines the NodeData received from other chunks with the node instance.

NodeData: NodeData is a portion of Node, which is transmitted to the mirror of that node on another partition. It consists of nodal attributes that are needed to compute element attributes in the next iteration. In stress analysis, these typically include force vectors. The Node type may contain other nodal attributes that may be fixed at initialization, so one does not need to communicate them every iteration, and thus can be excluded from NodeData.

Element: This datatype should contain attributes of each element, such as material, density, various moduli etc. These properties usually do not change, and maybe put in the ChunkData(see below). However, separating them into a different datatype makes the code much cleaner.

ChunkData: Most of the information about a chunk such as number of nodes, elements, and communication data, is maintained by the framework itself. Application data that is not an attribute of a node or an element, but has to be updated across iterations belongs to ChunkData. A typical usage is to store the error estimate of a chunk that needs to be used for testing convergence.

Config: This read-only data structure is initialized only on one processor and broadcast by the framework after the initialization phase. This may include various material properties, filenames for output files, number of timesteps, and other parameters.

4.2   Application-Specific Callbacks

In addition to above datatypes, user needs to provide only the following function callbacks. These functions will be called at various stages during a simulation.

At the initialization stage, chunk 0 calls FEM_Init_Conf_Data function to initialize the application-specific configuration data, and broadcasts it to every other chunk. Each chunk then initializes the application-specific node and element data structures by calling FEM_Init_Nodes and FEM_Init_Elements callbacks. FEM_Get_Total_Iter callback returns the number of iterations to perform.

The core computations are performed by the functions FEM_Update_Element, FEM_Update_Node, and FEM_Update_Chunk. These functions are called by the framework on all elements and nodes on every chunk once per iteration. Typ-ically, an element is updated by calculating the stresses using the nodal at-tributes of surrounding nodes. In more complicated applications, such as the crack-propagation, it calculates the progress of the crack. A node is updated by calculating its velocity and acceleration. This is where the simulation spends most of its time, so care should be taken to optimize these functions.

5   Crack Propagation

We have ported an initial prototype of an application that simulates pressure-driven crack propagation in structures (such as the casing of a solid rocket booster.) This is an inherently dynamic application. The structure to be simu-lated is discretized into an FEM mesh with triangular elements[1]. The simulation starts out with initial forces on boundary elements, calculating displacements of the nodes, and stresses on the elements. Depending on the geometry of the struc-ture and pressure on boundary elements, a crack begins to develop. As simulated time progresses, the crack begins to propagate through the structure. The rate of propagation and the direction depends on several factors such as the geometry of the computational domain, material properties, and direction of initial forces. In order to detect the presence of a crack, several "zero-volume" elements are added near the tip(s) of the crack. These elements are called "cohesive" elements. In the current discretization, these cohesive elements are rectangular, and share nodes with regular (triangular) elements.

In order to parallelize this application, we may start out with a load-balanced mesh partitioning, produced with Metis [5]. However, as the crack propagates, cohesive elements are added to partitions that contain tip(s) of the propagat-ing crack, increasing computational load of those partitions, and thus causing severe load imbalance. A partial solution to this problem is to insert cohesive elements everywhere in the mesh initially. This tries to reduce load imbalance by introducing redundant computations and memory requirements (increasing the number of elements). Since our framework supports dynamic load balancing for medium-grained partitions, one can eliminate such redundancy.

5.1   Programming Effort

We started out with a sequential program of 1900 lines that simulated crack propagation on a 2-D structure, and converted it to our framework. Since the parallelization aspects were handled by the framework, all that was required to convert this program was to locate node-specific and element-specific compu-tations, separating them out into user-defined data types, and supplying them to the framework. The resultant code was only 1200 lines of C++ (since the connectivity information between nodes and elements is also handled by our

---

[1] This initial prototype simulates a 2-D structure. Final versions will simulate 3-D structures with tetrahedral meshes. No changes will have to be made to the frame-work code and interfaces between framework and application code.

framework, code for reading it in, and forming a bipartite graph, and iterating on this graph through indirection, was completely eliminated.) The framework itself has about 500 lines of C++ code, including dynamic load balancing mechanisms. Even when we add facilities for adaptive refinements, and oct-tree based meshes, reduction in application-specific code is expected to be similar.

5.2   Performance

Since many FEM codes are written using Fortran (with better optimization techniques for scientific programs), we compared the performance of our C++ code for crack-propagation with similarly partitioned Fortran code. C++ code is about 10% slower. We expect this difference to become smaller with advances in C++ compiler technology.

We divide the original mesh into a number of small chunks; many more than the available number of processors, mapping multiple chunks on the same processor. Therefore, one may be concerned about the overhead of scheduling these chunks. However, as Figure 2 shows, mapping multiple chunks to each processor does not add significant overhead. In fact, as number of chunks per processor increases, performance may improve, as the data of a smaller chunk is more likely to fit in cache! This also eliminates the difference between C++ and Fortran versions of our codes because the advantages of Fortran to perform better cache-specific optimizations are no longer relevant.

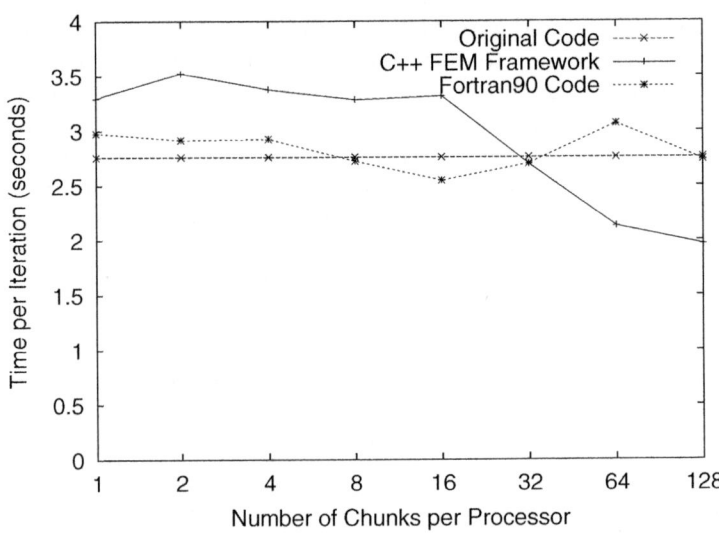

Fig. 2. Effects of partitioning the mesh into chunks more than number of processors.

5.3   Adaptive Response

As described earlier, the pressure-driven crack propagation application exhibits dynamic behavior that affects load balance as simulation progresses. In order to judge the effectiveness of our framework in handling such imbalance, we ran this application on 8 processors of NCSA Origin2000, splitting the input mesh of 183K nodes into 32 chunks, i.e. 4 chunks per processor. Figure 3 shows the results where performance (indicated by number of iterations per second) is sampled at various times. As the simulation finishes 35 timesteps, a crack begins to appear[2] in one of the partitions, increasing the computational load of that partition. Although the load of only one partition (and therefore of the processor that partition is on) increases, the overall throughput of the application decreases dramatically, because of dependences among partitions. That is, other processors wait for the overloaded processor to process its partitions. This demonstrates the impact of load imbalance in such an application. At timestep 25, 50 and 75, the load balancer is activated. On steps 25 and 75, it does not do any significant remapping since there are no changes in load conditions. At step 50, however, it instructs the FEM framework of the new mapping, which it carries out by migrating the requested chunks. Improvement in performance is evident immediately.

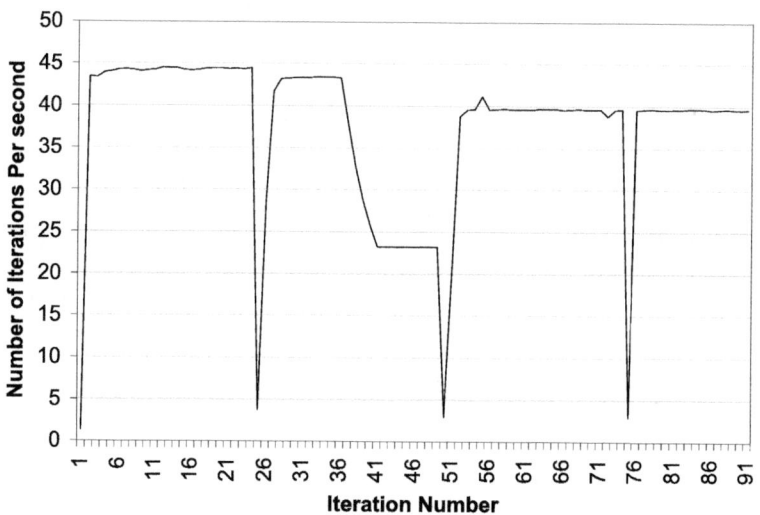

Fig. 3. Crack Propagation with Automatic Load Balancing

---

[2] In actual simulations, the crack may appear much later in time. However, in order to expedite the process, we emulated the effects of crack propagation.

6   Conclusion and Future Work

We described a C++ based framework for developing explicit finite element applications. Application programmers do not need to write any parallel code, yet the application developed will run portably on any parallel machine supported by the underlying Charm++ parallel programming system (which currently includes almost all available parallel machines, including clusters of workstations). Further, this framework handles dynamic application-induced imbalances, variations in processor speeds, interference from external jobs on timeshared clusters, evacuation of machines when demanded by their owners, and so on.

We are currently engaged in porting several dynamic FEM applications to our framework, such as 3-D simulations of dendritic growth in materials during casting and extrusion processes.

In our future work, we aim at even more complex FEM applications. Specifically, we first plan to support addition and deletion of elements, adaptive refinement of FEM meshes, oct-tree and quad-tree based meshes. Further extension to this work can be in the form of support for implicit solvers, multi-grid solvers and interface to matrix-based solvers.

References

1. Neil Hurley Darach Golden and Sean McGrath. Parallel adaptive mesh refinement for large eddy simulation using the finite element methods. In *PARA*, pages 172-181, 1998.
2. L. V. Kale, Milind Bhandarkar, and Robert Brunner. Run-time Support for Adaptive Load Balancing. In *Proceedings of 4th Workshop on Runtime Systems for Parallel Programming (RTSPP) Cancun – Mexico*, March 2000.
3. L. V. Kale, Milind Bhandarkar, Robert Brunner, and Joshua Yelon. Multiparadigm, Multilingual Interoperability: Experience with Converse. In *Proceedings of 2nd Workshop on Runtime Systems for Parallel Programming (RTSPP) Orlando, Florida – USA*, Lecture Notes in Computer Science, March 1998.
4. Laxmikant Kalé, Robert Skeel, Milind Bhandarkar, Robert Brunner, Attila Gursoy, Neal Krawetz, James Phillips, Aritomo Shinozaki, Krishnan Varadarajan, and Klaus Schulten. NAMD2: Greater scalability for parallel molecular dynamics. *Journal of Computational Physics*, 151:283-312, 1999.
5. George Karypis and Vipin Kumar. A fast and high quality multilevel scheme for partitioning irregular graphs. TR 95-035, Computer Science Department, University of Minnesota, Minneapolis, MN 55414, May 1995.
6. L. M. Taylor. Sierra: A software framework for developing massively parallel, adaptive, multi-physics, finite element codes. In *Presentation at the International Conference on Parallel and Distributed Processing Techniques (PDPTA 99)*, Las vegas, Nevada, USA, June 1999.
7. J.B. Weissman, A.S. Grimshaw, and R. Ferraro. Parallel Object-Oriented Computation Applied to a Finite Element Problem. *Scientific Computing*, 2(4):133-144, February 1994.

Analyzing the Parallel Scalability
of an Implicit Unstructured Mesh CFD Code

W.D. Gropp[1], D.K. Kaushik[1], D.E. Keyes[2], and B.F. Smith[1]

[1] Math. & Comp. Sci. Division, Argonne National Laboratory, Argonne, IL 60439,
{gropp,kaushik,bsmith}@mcs.anl.gov
[2] Math. & Stat. Department, Old Dominion University, Norfolk, VA 23529,
ISCR, Lawrence Livermore National Laboratory, Livermore, CA 94551,
and ICASE, NASA Langley Research Center, Hampton, VA 23681,
keyes@icase.edu

Abstract. In this paper, we identify the scalability bottlenecks of an unstructured grid CFD code (PETSc-FUN3D) by studying the impact of several algorithmic and architectural parameters and by examining different programming models. We discuss the basic performance characteristics of this PDE code with the help of simple performance models developed in our earlier work, presenting primarily experimental results. In addition to achieving good per-processor performance (which has been addressed in our cited work and without which scalability claims are suspect) we strive to improve the implementation and convergence scalability of PETSc-FUN3D on thousands of processors.

1 Introduction

We have ported the NASA code FUN3D [1,3] into the PETSc [4] framework using the single program multiple data (SPMD) message-passing programming model, supplemented by multithreading at the physically shared memory level. FUN3D is a tetrahedral vertex-centered unstructured mesh code originally developed by W. K. Anderson of the NASA Langley Research Center for compressible and incompressible Euler and Navier-Stokes equations. FUN3D uses a control volume discretization with a variable-order Roe scheme for approximating the convective fluxes and a Galerkin discretization for the viscous terms. In reimplementing FUN3D in the PETSc framework, our effort has focused on achieving small time to convergence without compromising scalability, by means of appropriate algorithms and architecturally efficient data structures [2].

The solution algorithm employed in PETSc-FUN3D is pseudo-transient Newton-Krylov-Schwarz (ψNKS) [8] with block-incomplete factorization on each subdomain of the Schwarz preconditioner and with varying degrees of overlap. This code spends almost all of its time in two phases: flux computations (to evaluate conservation law residuals), where one aims to have such codes spent almost *all* their time, and sparse linear algebraic kernels, which are a fact of life in implicit methods. Altogether, four basic groups of tasks can be identified based on the criteria of arithmetic concurrency, communication patterns, and the ratio of operation complexity to data size within the task. These four distinct phases, present in most implicit codes, are: vertex-based loops, edge-based loops, recurrences, and global reductions. Each of these groups of tasks

M. Valero, V.K. Prasanna, and S. Vajapeyam (Eds.): HiPC 2000, LNCS 1970, pp. 395–404, 2000.

stresses a different subsystem of contemporary high-performance computers. Analysis of PETSc-FUN3D shows that, after tuning, the linear algebraic kernels run at close to the aggregate memory-bandwidth limit on performance, the flux computations are bounded either by memory bandwidth or instruction scheduling (depending upon the ratio of load/store units to floating point units in the CPU), and parallel efficiency is bounded primarily by slight load imbalances at synchronization points [6,7].

Achieving high sustained performance, in terms of solutions per second, requires attention to three factors. The first is a scalable implementation, in the sense that time per iteration is reduced in inverse proportion to the the number of processors, or that time per iteration is constant as problem size and processor number are scaled proportionally. The second is algorithmic scalability, in the sense that the number of iterations to convergence does not grow with increased numbers of processors. The third factor arises since the requirement of a scalable implementation generally forces parameterized changes in the algorithm as the number of processors grows. However, if the convergence is allowed to degrade the overall execution is not scalable, and this must be countered algorithmically. The third is good per-processor performance on contemporary cache-based micropro-cessors. In this paper, we only consider the first two factors in the overall performance in Sections 3 and 4, respectively. For a good discussion of the optimizations done to improve the per processor performance, we refer to our earlier work [7,6].

2   Large Scale Demonstration Runs

We use PETSc's profiling and logging features to measure the parallel performance. PETSc logs many different types of events and provides valuable information about time spent, communications, load balance, and so forth, for each logged event. PETSc uses manual counting of flops, which are afterwards aggregated over all the processors for parallel performance statistics. In our rate computations, we exclude the initialization time devoted to I/O and data partitioning. Since we are solving large fixed-size problems on distributed memory machines, it is not reasonable to base parallel scalability on a uniprocessor run, which would thrash the paging system. Our base processor number is such that the problem has just fit into the local memory.

Fig. 1 shows aggregate flop/s performance and a log-log plot showing execution time for the 2.8 million vertex grid on the three most capable machines to which we have thus far had access. In both plots of this figure, the dashed lines indicate ideal behavior. Note that although the ASCI Red flop/s rate scales nearly linearly, a higher fraction of the work is redundant at higher parallel granularities, so the execution time does not drop in exact proportion to the increase in flop/s.

3   Implementation Scalability

Domain-decomposed parallelism for PDEs is a natural means of overcoming Amdahl's law in the limit of fixed problem size per processor. Computational work on each evalua-tion of the conservation residuals scales as the volume of the (equal-sized) subdomains, whereas communication overhead scales only as the surface. This ratio is fixed when

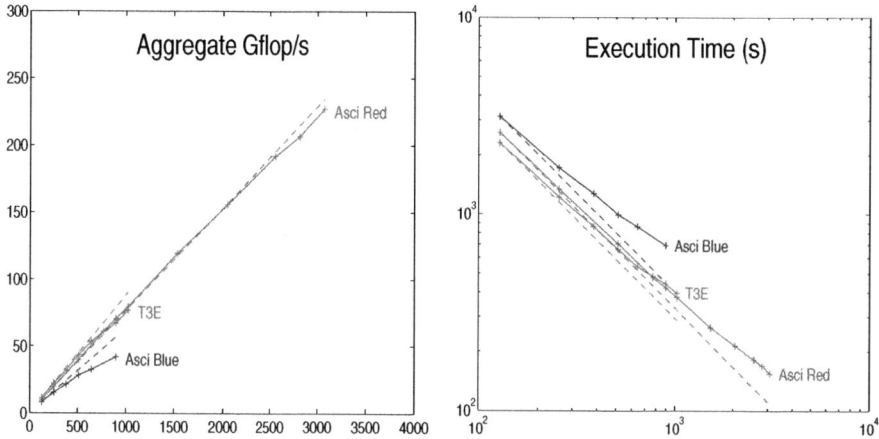

Fig. 1. Gigaflop/s ratings and execution times on ASCI Red (up to 3072 2-processor nodes), ASCI Pacific Blue (up to 768 processors), and a Cray T3E (up to 1024 processors) for a 2.8M-vertex case, along with dashed lines indicating "perfect" scalings.

problem size and processors are scaled in proportion, leaving only global reduction operations over all processors as an impediment to perfect performance scaling.

When the load is perfectly balanced (which is easy to achieve for static meshes) and local communication is not an issue because the network is scalable, the optimal number of processors is related to the network diameter. For logarithmic networks, like a hypercube, the optimal number of processors, P, grows directly in proportion to the problem size, N. For a d-dimensional torus network, $P \propto N^{d/d+1}$. The proportionality constant is a ratio of work per subdomain to the product of synchronization frequency and internode communication latency.

3.1 Scalability Bottlenecks

In Table 1, we present a closer look at the relative cost of computation for PETSc-FUN3D for a fixed-size problem of 2.8 million vertices on the ASCI Red machine, from 128 to 3072 nodes. The intent here is to identify the factors that retard the scalability. The overall parallel efficiency (denoted by $\eta_{overall}$) is broken into two components: η_{alg} measures the degradation in the parallel efficiency due to the increased iteration count (Section 4) of this (non-coarse-grid-enhanced) NKS algorithm as the number of subdomains increases, while η_{impl} measures the degradation coming from all other nonscalable factors such as global reductions, load imbalance (implicit synchronizations), and hardware limitations.

From Table 1, we observe that the buffer-to-buffer time for global reductions for these runs is relatively small and does not grow on this excellent network. The primary factors responsible for the increased overhead of communication are the implicit synchronizations and the ghost point updates (interprocessor data scatters).

Interestingly, the increase in the percentage of time (3% to 10%) for the scatters results more from algorithmic issues than from hardware/software limitations. With an

Table 1. Scalability bottlenecks on ASCI Red for a fixed-size 2.8M vertex mesh. The preconditioner used in these results is block Jacobi with ILU(1) in each subdomain. We observe that the principle nonscaling factor is the implicit synchronization.

Number of Processors	Its	Time	Speedup	Efficiency		
				$\eta_{overall}$	η_{alg}	η_{impl}
128	22	2,039s	1.00	1.00	1.00	1.00
256	24	1,144s	1.78	0.89	0.92	0.97
512	26	638s	3.20	0.80	0.85	0.94
1024	29	362s	5.63	0.70	0.76	0.93
2048	32	208s	9.78	0.61	0.69	0.89
3072	34	159s	12.81	0.53	0.65	0.82

Number of Processors	Percent Times for			Scatter Scalability	
	Global Reductions	Implicit Synchronizations	Ghost Point Scatters	Total Data Sent per Iteration (GB)	Application Level Effective Bandwidth per Node (MB/s)
128	5	4	3	3.6	6.9
256	3	6	4	5.0	7.5
512	3	7	5	7.1	6.0
1024	3	10	6	9.4	7.5
2048	3	11	8	11.7	5.7
3072	5	14	10	14.2	4.6

increase in the number of subdomains, the percentage of grid point data that must be communicated also rises. For example, the total amount of nearest neighbor data that must be communicated per iteration for 128 subdomains is 3.6 gigabytes, while for 3072 subdomains it is 14.2 gigabytes. Although more network wires are available when more processors are employed, scatter time increases. If problem size and processor count are scaled together, we would expect scatter times to occupy a fixed percentage of the total and load imbalance to be reduced at high granularity.

The final column in Table 1 shows the scalability of the "application level effective bandwidth" that is computed by dividing the total amount of data transferred by the time spent in scatter operation. It includes the message packing and unpacking times plus any contention in the communication. That is why it is far lower than the achievable bandwidth (as measured by the "Ping-Pong" test from the message passing performance (MPP) [10] tests) of the networking hardware. The Ping-Pong test measures the point to point unidirectional bandwidth between any two processors in a communicator group. It is clear that the Ping-Pong test results in Table 2 are not representative of the actual communication pattern encountered in the scatter operation. To better understand this issue, we have carried out the "Halo" test (from the MPP test suite) on 64 nodes of the ASCI Red machine. In this test, a processor exchanges messages with a fixed number of neighbors, moving data from/to contiguous memory buffers. For the Halo test results

Table 2. MPP test results on 64 nodes of ASCI Red. The Ping-Pong results measure the unidirectional bandwidth. The Halo test results (measuring the bidirectional bandwidth) is more representative of the communication pattern encountered in the scatter operation.

Message	Bandwidth, MB/s	
Length, KB	Ping-Pong	Halo
2	93	70
4	145	94
8	183	92
16	235	106
32	274	114

in Table 2, each node communicated with 8 other nodes (which is a good estimate of the neighbors a processor in PETSc-FUN3D will need to communicate). The message lengths for both these tests (Ping-Pong and Halo) have been varied between 2KB to 32 KB since the average length of a message in the runs for Table 1 varies from 23 KB to 3 KB as the number of processor goes up from 128 to 3072. We observe that the bandwidth obtained in the Halo test is significantly less than that obtained in the Ping-Pong test. This loss in performance perhaps can be attributed to the fact that a processor communicates with more than one neighbor at the same time in the Halo test. In addition, as stated earlier, the scatter operation involves the overhead of packing and unpacking of messages at the rate limited by the achievable memory bandwidth (about 145 MB/s as measured by the STREAM benchmark [14]). We are currently investigating the impact of these two factors (the number of pending communication operations with more than one neighbor and the memory bandwidth) on the performance of scatter operation further.

3.2 Effect of Partitioning Strategy

Mesh partitioning has a dominant effect on parallel scalability for problems characterized by (almost) constant work per point. As shown above, poor load balance causes idleness at synchronization points, which are frequent in implicit methods (e.g., at every conjugation step in a Krylov solver). With NKS methods, then, it is natural to strive for a very well balanced load. The p-MeTiS algorithm in the MeTiS package [12], for example, provides almost perfect balancing of the number of mesh points per processor. However, balancing work alone is not sufficient. Communication must be balanced as well, and these objectives are not entirely compatible. Figure 2 shows the effect of data partitioning using p-MeTiS, which tries to balance the number of nodes and edges on each partition, and k-MeTiS, which tries to reduce the number of noncontiguous subdomains and connectivity of the subdomains. Better overall scalability is observed with k-MeTiS, despite the better load balance for the p-MeTiS partitions. This is due to the slightly poorer numerical convergence rate of the iterative NKS algorithm with the p-MeTiS partitions. The poorer convergence rate can be explained by the fact that the p-MeTiS partitioner generates disconnected pieces within a single "subdomain," effectively increasing the number of blocks in the block Jacobi or additive Schwarz algorithm

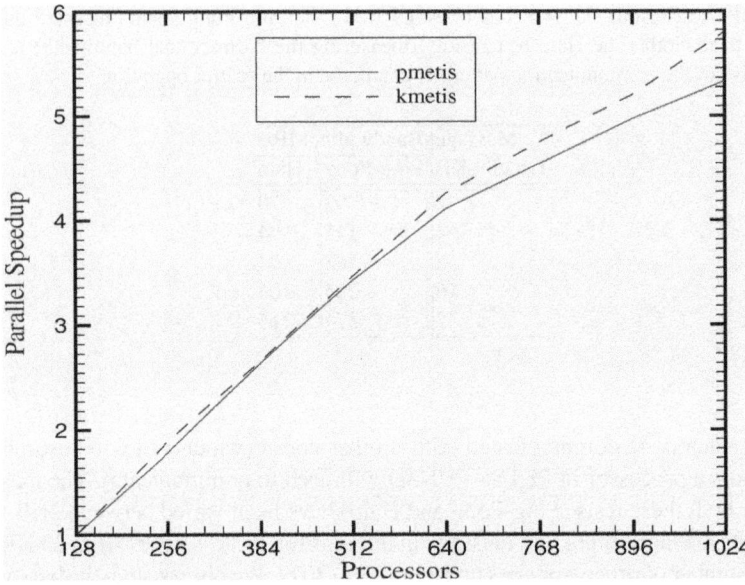

Fig. 2. Parallel speedup relative to 128 processors on a 600 MHz Cray T3E for a 2.8M vertex case, showing the effect of partitioning algorithms k-MeTiS, and p-MeTiS.

and increasing the size of the interface. The convergence rates for block iterative methods degrade with increasing number of blocks, as discussed in Section 4.

3.3 Domain-Based and/or Instruction-Level Parallelism

The performance results above are based on subdomain parallelism using the Message Passing Interface (MPI) [9]. With the availability of large scale SMP clusters, different software models for parallel programming require a fresh assessment. For machines with physically distributed memory, MPI has been a natural and successful software model. For machines with distributed shared memory and nonuniform memory access, both MPI and OpenMP have been used with respectable parallel scalability. For clusters with two or more SMPs on a single node, the mixed software model of threads within a node (OpenMP being a special case of threads because of the potential for highly efficient handling of the threads and memory by the compiler) and MPI between the nodes appears natural. Several researchers (e.g., [5,13]) have used this mixed model with reasonable success.

We investigate the mixed model by employing OpenMP only in the flux calculation phase. This phase takes over 60% of the execution time on ASCI Red and is an ideal candidate for shared-memory parallelism because it does not suffer from the memory bandwidth bottleneck (see next section). In Table 3, we compare the performance of this phase when the work is divided by using two OpenMP threads per node with the performance when the work is divided using two independent MPI processes per node. There is no communication in this phase. Both processors work with the same amount of

Table 3. Execution time on the 333 MHz Pentium Pro ASCI Red machine for function evaluations only for a 2.8M vertex case, comparing the performance of the hybrid (MPI/OpenMP) and the distributed memory (MPI alone) programming models.

	MPI/OpenMP Threads per Node		MPI Processes per Node	
Nodes	1	2	1	2
256	483s	261s	456s	258s
2560	76s	39s	72s	45s
3072	66s	33s	62s	40s

memory available on a node; in the OpenMP case, it is shared between the two threads, while in the case of MPI it is divided into two address spaces.

The hybrid MPI/OpenMP programming model appears to be a more efficient way to employ shared memory than are the heavyweight subdomain-based processes (MPI alone), especially when the number of nodes is large. The MPI model works with larger number of subdomains (equal to the number of MPI processors), resulting in slower rate of convergence. The hybrid model works with fewer chunkier subdomains (equal to the number of nodes) that result in faster convergence rate and shorter execution time, despite the fact that there is some redundant work when the data from the two threads is combined due to the lack of a vector-reduce operation in the OpenMP standard (version 1) itself. Specifically, some redundant work arrays must be allocated that are not present in the MPI code. The subsequent gather operations (which tend to be memory bandwidth bound) can easily offset the advantages accruing from the low latency shared memory communication. One way to get around this problem is to use some coloring strategies to create the disjoint work sets, but this takes away the ease and simplicity of the parallelization step promised by the OpenMP model.

4   Convergence Scalability

The convergence rates and, therefore, the overall parallel efficiencies of additive Schwarz methods are often dependent on subdomain granularity. Except when effective coarse-grid operators and intergrid transfer operators are known, so that optimal multilevel preconditioners can be constructed, the number of iterations to convergence tends to increase with granularity for elliptically controlled problems, for either fixed or memory-scaled problem sizes. In practical large-scale applications, however, the convergence rate degradation of single-level additive Schwarz is sometimes not as serious as the scalar, linear elliptic theory would suggest (e.g. see the 2nd column in Table 1). Its effects are mitigated by several factors, including pseudo-transient nonlinear continuation and dominant intercomponent coupling. The former parabolizes the operator, endowing diagonal dominance. The latter renders the off-diagonal coupling less critical and, therefore, less painful to sever by domain decomposition. The block diagonal coupling can be captured fully in a point-block ILU preconditioner.

4.1 Additive Schwarz Preconditioner

Table 4 explores two quality parameters for the additive Schwarz preconditioner: sub-domain overlap and quality of the subdomain solve using incomplete factorization. We exhibit execution time and iteration count data from runs of PETSc-FUN3D on the ASCI Red machine for a fixed-size problem with 357,900 grid points and 1,789,500 degrees of freedom. These calculations were performed using GMRES(20), one subdomain per processor (without overlap for block Jacobi and with overlap for ASM), and ILU(k) where k varies from 0 to 2, and with the natural ordering in each subdomain block. The use of ILU(0) with natural ordering on the first-order Jacobian, while applying a second-order operator, allows the factorization to be done in place, with or without over-lap. However, the overlap case does require forming an additional data structure on each processor to store matrix elements corresponding to the overlapped regions.

From Table 4 we see that the larger overlap and more fill helps in reducing the total number of linear iterations as the number of processors increases, as theory and intuition predict. However, both increases consume more memory, and both result in more work per iteration, ultimately driving up execution times in spite of faster convergence. Best execution times are obtained for any given number of processors for ILU(1), as the number of processors becomes large (subdomain size small), for zero overlap.

The execution times reported in Table 4 are highly dependent on the machine used, since each of the additional computation/communication costs listed above may shift the computation past a knee in the performance curve for memory bandwidth, commu-nication network, and so on.

4.2 Other Algorithmic Tuning Parameters

In [7] we highlight some additional tunings (for ψNKS Solver) that have yielded good results in our context. Some subsets of these parameters are not orthogonal, but interact strongly with each other. In addition, optimal values of some of these parameters depend on the grid resolution. We are currently using derivative-free asynchronous parallel direct search algorithms [11] to more systematically explore this large parameter space.

5 Conclusions

Unstructured implicit CFD solvers are amenable to scalable implementation, but careful tuning is needed to obtain the best product of per-processor efficiency and parallel effi-ciency. In fact, the cache blocking techniques (addressed in our earlier work) employed to boost the per-processor performance helps to improve the parallel scalability as well by making the message sizes longer. The principle nonscaling factor is the implicit syn-chronizations and not the communication. Another important phase is the scatter/gather operation that seems limited more by the achievable memory bandwidth than the net-work bandwidth (at least on ASCI Red). Given contemporary high-end architecture, critical research directions for solution algorithms for systems modeled by PDEs are (1) multivector algorithms and less synchronous algorithms, and (2) hybrid programming models. To influence future architectures while adapting to current ones, we recommend adoption of new benchmarks featuring implicit methods on unstructured grids, such as the application featured here.

Table 4. Execution times and linear iteration counts on the 333 MHz Pentium Pro ASCI Red machine for a 357,900-vertex case, showing the effect of subdomain overlap and incomplete factorization fill level in the additive Schwarz preconditioner. **The best execution times for each ILU fill level and number of processors are in boldface in each row.**

ILU(0) in Each Subdomain					
Number	Overlap				
of	0		1		2
Processors	Time	Linear Its	Time	Linear Its	Time	Linear Its
32	**688s**	930	**661s**	816	696s	813
64	**371s**	993	374s	876	418s	887
128	**210s**	1052	230s	988	222s	872

ILU(1) in Each Subdomain					
Number	Overlap				
of	0		1		2
Processors	Time	Linear Its	Time	Linear Its	Time	Linear Its
32	598s	674	**564s**	549	617s	532
64	**334s**	746	335s	617	359s	551
128	**177s**	807	178s	630	200s	555

ILU(2) in Each Subdomain					
Number	Overlap				
of	0		1		2
Processors	Time	Linear Its	Time	Linear Its	Time	Linear Its
32	**688s**	527	786s	441	—	—
64	**386s**	608	441s	488	531s	448
128	**193s**	631	272s	540	313s	472

Acknowledgments

We are indebted to Lois C. McInnes and Satish Balay of Argonne National Laboratory, to W. Kyle Anderson, formerly of the NASA Langley Research Center, for collaborations leading up to the work presented here. Gropp and Smith were supported by the Mathematical, Information, and Computational Sciences Division subprogram of the Office of Advanced Scientific Computing Research, U.S. Department of Energy, under Contract W-31-109-Eng-38. Kaushik's support was provided by a GAANN Fellowship from the U.S. Department of Education and by Argonne National Laboratory under contract 983572401. Keyes was supported by the National Science Foundation under grant ECS-9527169, by NASA under contracts NAS1-19480 and NAS1-97046, by Argonne National Laboratory under contract 982232402, and by Lawrence Livermore National Laboratory under subcontract B347882. Debbie Swider of Argonne National Laboratory was of considerable assistance in performing ASCI platform runs. Computer time was supplied by Argonne National Laboratory, Lawrence Livermore National Laboratory, NERSC, Sandia National Laboratories, and SGI-Cray.

References

1. W. K. Anderson and D. L. Bonhaus. An implicit upwind algorithm for computing turbulent flows on unstructured grids. *Computers and Fluids*, 23:1–21, 1994.
2. W. K. Anderson, W. D. Gropp, D. K. Kaushik D. E. Keyes, and B. F. Smith. Achieving high sustained performance in an unstructured mesh CFD application. In *Proceedings of SC'99*. IEEE Computer Society, 1999. Gordon Bell Prize Award Paper in Special Category.
3. W. K. Anderson, R. D. Rausch, and D. L. Bonhaus. Implicit/multigrid algorithms for incompressible turbulent flows on unstructured grids. *J. Computational Physics*, 128:391–408, 1996.
4. S. Balay, W. D. Gropp, L. C. McInnes, and B. F. Smith. The Portable Extensible Toolkit for Scientific Computing (PETSc) version 28.
   http://www.mcs.anl.gov/petsc/petsc.html, 2000.
5. S. W. Bova, C. P. Breshears, C. E. Cuicchi, Z. Demirbilek, and H. A. Gabb. Dual-level parallel analysis of harbor wave response using MPI and OpenMP. *Int. J. High Performance Computing Applications*, 14:49–64, 2000.
6. W. D. Gropp, D. K. Kaushik, D. E. Keyes, and B. F. Smith. Toward realistic performance bounds for implicit CFD codes. In D. Keyes, A. Ecer, J. Periaux, N. Satofuka, and P. Fox, editors, *Proceedings of Parallel CFD'99*, pages 233–240. Elsevier, 1999.
7. W. D. Gropp, D. K. Kaushik, D. E. Keyes, and B. F. Smith. Performance modeling and tuning of an unstructured mesh CFD application. In *Proceedings of SC2000*. IEEE Computer Society, 2000.
8. W. D. Gropp, L. C. McInnes, M. D. Tidriri, and D. E. Keyes. Globalized Newton-Krylov-Schwarz algorithms and software for parallel implicit CFD. *Int. J. High Performance Computing Applications*, 14:102–136, 2000.
9. William Gropp, Ewing Lusk, and Anthony Skjellum. *Using MPI: Portable Parallel Programming with the Message Passing Interface,* 2nd edition. MIT Press, Cambridge, MA, 1999.
10. William D. Gropp and Ewing Lusk. Reproducible measurements of MPI performance characteristics. In Jack Dongarra, Emilio Luque, and Tomàs Margalef, editors, *Recent Advances in Parallel Virtual Machine and Message Passing Interface*, volume 1697 of *Lecture Notes in Computer Science*, pages 11–18. Springer Verlag, 1999. 6th European PVM/MPI Users' Group Meeting, Barcelona, Spain, September 1999.
11. P. D. Hough, T. G. Kolda, and V. J. Torczon. Asynchronous parallel pattern search for nonlinear optimization. Technical Report SAND2000-8213, Sandia National Laboratories, Livermore, January 2000. Submitted to SIAM J. Scientific Computation.
12. G. Karypis and V. Kumar. A fast and high quality scheme for partitioning irregular graphs. *SIAM J. Scientific Computing*, 20:359–392, 1999.
13. D. J. Mavriplis. Parallel unstructured mesh analysis of high-lift configurations. Technical Report 2000-0923, AIAA, 2000.
14. J. D. McCalpin. STREAM: Sustainable memory bandwidth in high performance computers. Technical report, University of Virginia, 1995.
   http://www.cs.virginia.edu/stream.

Process Interconnection Structures in Dynamically Changing Topologies

Eugene Gendelman, Lubomir F. Bic, and Michael B. Dillencourt

Department of Information and Computer Science
University of California, Irvine, CA, USA
{egendelm, bic, dillenco}@ics.uci.edu

Abstract. This paper presents a mechanism that organizes processes in the hierarchy and efficiently maintains it in the presence of addition/removal of nodes to the system, and in the presence of node failures. This mechanism can support total order of broadcasts and does not rely on any specific system features or special hardware. In addition, it can concurrently support multiple logical structures, such as a ring, a hypercube, a mesh, and a tree.

1 Introduction

Distributed systems consisting of a network of workstations or personal computers are an attractive way to speed up large computations. There are several coordination protocols that must be used during system execution. Among these protocols are those responsible for checkpointing [1], stability detection [3], and maintaining a global virtual time [4]. Usually, protocols that involve a central coordinator are more efficient, as they require fewer communication messages than distributed protocols, and are simpler to construct. However, in large systems the bottleneck of the central coordinator makes these protocols unusable. A hierarchical coordination can be used instead [3,5]. Hierarchical protocols have the efficiency and simplicity of the centralized protocols and scalability of the distributed protocols. This paper presents a mechanism, called *Process Order*, for building and maintaining a hierarchy of processes in a dynamically changing distributed system.

2 Performance of Centralized and Distributed Coordination

We measured the time it takes to broadcast a message and receive an acknowledgement for centralized and hierarchical schemes. Such broadcast is a part of many communication protocols [2-5,8].

Inter-process communication was implemented with TCP/IP, as it provides guaranteed message delivery, which is essential for many coordination protocols. The application program was implemented in Java. Broadcasts were running in parallel with distributed application. For more details about this study please refer to [7].

M. Valero, V.K. Prasanna, and S. Vajapeyam (Eds.): HiPC 2000, LNCS 1970, pp. 415–414, 2000
© Springer-Verlag Berlin Heidelberg 2000

There is a limit on the number of socket connections that could be open at the same time. Solaris 5.1 allows only 64 file descriptors open at any time. Considering this, a new connection is set up for every send and receive by the coordinator in the centralized broadcast. This is not necessary in the case of the hierarchical broadcast, which was measured with and without reconnections.

Table 1. Time (in seconds) spent by different schemes on 100 broadcasts.

	63 PEs		16 PEs		15 PEs
Centralized (C)		463.2s		39.6s		39.6s
Hierarchical 1 (H1)	(6 levels)	24.2s	(5 levels)	14.1s	(4 levels)	7.5s
Hierarchical 2 (H2)	(6 levels)	4.8s	(5 levels)	3.5s	(4 levels)	1.3s

Table 1 presents the results of running 100 broadcasts for each protocol on the sets of 15, 16, and 63 processing elements (PEs). A new cycle of broadcasting is not started until the previous one is completed. With 63 PEs the hierarchical broadcast with reconnection (H1) takes 19 times less time than centralized broadcast (C), and hierarchical broadcast without reconnection (H2) is 5 times faster than H1.

For the centralized broadcast in the first two columns of table 1 the number of processors decreased by a factor of four, and the running time decreased by a factor of almost twelve. This reflects the difference in load on the central coordinator induced by 63 and 16 PEs.

For hierarchical broadcast the difference in performance between columns one and two is similar to that of columns two and three. The reason for this is that in both cases, i.e., when moving from 63 PEs to 16 PEs and when moving from 16 PEs to 15 PEs, the depth of the tree is decreased by one level. As all the nodes at the same depth of the tree broadcast to their children in parallel, the broadcasting time grows with the depth of the tree, and not with the number of PEs.

In summary, our experiments confirm the intuition that as the system gets large, hierarchical broadcasting scales much better than the centralized one.

3   Process Order

Each process in the computation has a unique system id, such as the combination of the process id and IP address. If there is only one process per physical node, the IP address is sufficient to provide a unique system id.

The processes in the system exchange their system ids during the initialization phase. Then each process sorts all processes by their system ids. After the processes are sorted, each process is given a Process Order Id (POID) according to its place in the sorted list, starting from 0. This information is stored in a *Process Order Table*. An example of a Process Order Table for a system with 6 PEs is shown in figure 1(a).

The processes can be arranged in a hierarchy implicitly with the Process Order using formula (1).

$$POID_{coord} = (POID_{self} - 1) \text{ DIV } K,$$

Where K is the maximum number of processes coordinated by any single   **(1)**
process.

POID	0	1	2	3	4	5
IP	21	32	33	43	54	60
Name	a	c	D	E	g	H

(a) Process Order Table (POTable)

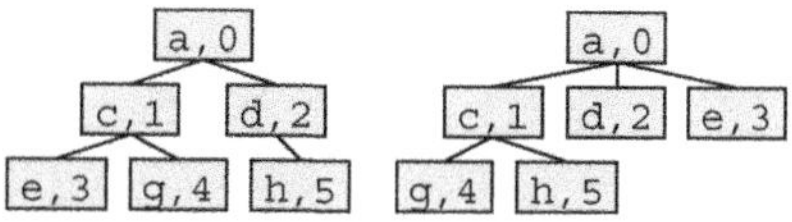

(b) Hierarchy K=2   (c) Hierarchy K=3

Fig. 1. Constructing hierarchy with POTable and formula (1).

Using this scheme each process can identify its coordinator without exchanging any messages with other processes. By changing only the branching coefficient K in formula (1), processes automatically organize in different hierarchies (figure 1 (b), (c)).

4   Process Order in a Dynamically Changing Failure-Free System

This section describes how the consistency of the Process Order mechanism is preserved when processes can join or leave the system at run time. We assume that all deletions and insertions of processes are initiated by special modification messages sent to the current coordinator process.

Definition 1:   The system modification is successful, iff
- *All broadcasts initiated before a modification message are delivered to the old PE set.*
- *All live processes receive the modification message.* This includes any process being deleted or inserted.
- *All broadcasts initiated after the modification message are delivered to the new PE set.*
- *All broadcasts are delivered in total order.*

When a process joins the system it sends a *SystemJoin* message to the root coordinator containing its IP address. This is illustrated in figure 2(a), as node *B* joins the system. The root updates its Process Order Table (POTable), and sends a *POModify* message to its newly calculated children, as shown in figure 2(b), where *POModify* messages are shown by thick arrows. The *POModify* message includes the information about the new process. After receiving a *POModify* message a process modifies its POTable. If the children list changed, the process establishes connections with its new children and forwards the *POModify* to them (fig 2(c)).

Similarly, when a process is exiting the system, it sends a *SystemLeave* message to the coordinator, as does node *X* in figure 2(d). The coordinator modifies its POTable and broadcasts *POModify* message to its new children (fig 2(e)). This is repeated down the tree, as shown in figure 2(f). The old parent of the leaving process sends a

POModify message to its new children, as well as to the leaving process. In figure 2(f) *B* signals *X* to leave.

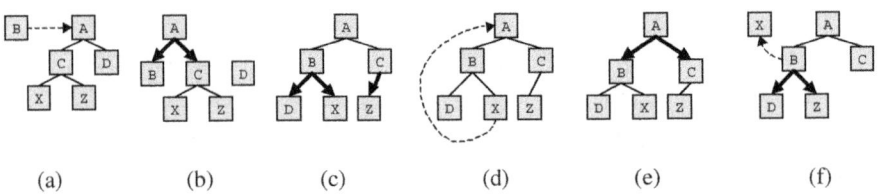

(a)   (b)   (c)   (d)   (e)   (f)

Fig. 2. Hierarchy in dynamically changing system. (a)-(c): add node B, (d)-(f): remove node X

Each process keeps a modification log (ModLog), which is a collection of the received *POModify* messages. Each process also keeps the log of the broadcasts (BLog). The modification algorithms will be clarified, as their correctness is shown.

Proof of Correctness

*   *All broadcasts initiated before a modification message are delivered to the old PE set.*

As the topology of the system changes, messages that would normally arrive in the FIFO order, can be interchanged, as shown in Figure 3. In this example (fig.3(a)) broadcast b1 is started before adding node B, which is initiated by a message m (fig.3(b)). This is followed by broadcast b2 (fig.3(c)). Node D receives b1 from A, and m and b2 from B. Theoretically b1 could reach D after m and even after b2. In this case D would broadcast b1 only to Y, and b1 would never be delivered to node X (fig. 3(d)).

This problem is fixed by appending the ModLog of the root coordinator to the broadcasted message. This allows processes to reconstruct the hierarchy as it was at the time of the broadcast. As ModLog of A did not contain the addition of node B, D will also send b1 to X.

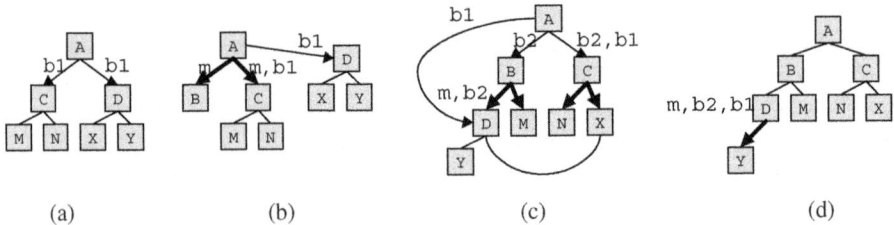

(a)   (b)   (c)   (d)

Fig. 3. Broadcasts while adding node B. b1, b2 - broadcasts, m - modification message.

*   *All live processes receive the modification message.*

Each process processes the *POModify* message, modifies its POTable, and then broadcasts *POModify* message to its new children. As this is done consistently by every process, every process in the new hierarchy receives the modification message.

If a process is being deleted then it receives the *POModify* message from its original parent.

- *All broadcasts initiated after the modification message are delivered.*

When a new broadcast is initiated, it traverses a new hierarchy built by *POModify* messages. As TCP/IP provides FIFO channels, a broadcasting message cannot reach its destination before the *POModify* message, and therefore the broadcast is delivered to the whole new hierarchy.

- *All broadcasts are delivered in total order.*

In a modification-free execution all broadcasts are delivered in total order. Therefore, all broadcasts that happened before a modification, and all broadcasts sent after a modification are delivered in the correct order. However, broadcasts sent before the modification (point 1 above) could mix with broadcasts initiated after the modification, as was shown on figure 3.

To solve this problem, the root coordinator assigns all the broadcasts a unique, monotonically increasing broadcast index (BI). Each process knows the BI of the last message it received, and therefore can postpone processing out-of-order messages. Reliable communication channels guarantee the delivery of all messages, or failure will be detected otherwise. Therefore, deadlock is not possible.

5   Process Order in the Presence of Failures

5.1 Failure Detection

TCP/IP generates an interrupt on the sender side if the receiver socket is not responding. This feature is used to detect process failures. The participating processes are arranged in a logical ring using Process Order. Each process sends a heartbeat message to the next process in the ring. Processes are arranged into a ring with Process Order using the formula

$$POID_{receiver} = (POID_{sender} + 1) \bmod n, \text{ where n is a number of nodes } (n > 1) \tag{2}$$

5.2 Failure Recovery

Consistency of the Process Order mechanism in the presence of failures depends on the recovery algorithm used in the system. If the rollback-recovery algorithm presented in [6] is used, the consistency of the Process Order is preserved automatically. In this scheme, once failure is detected, the whole system rolls back and recovers from the previously saved consistent system state. When the recovery algorithm is finished, all the processes are aware of all other live processes in the system, and therefore Process Order tables are consistent. This recovery scheme might be good for applications requiring barrier synchronization, individual-based simulations relying on the virtual time, and grid applications.

For other applications this recovery scheme might produce large overhead, especially in large systems. In such cases recovery protocols that restart only the failed process are used. One way to provide such a recovery is by using a causal logging protocol [9, 13]. The rest of this section describes how Process Order consistency is preserved in case of such a recovery.

Non-root node failures. The node failures can be divided into three types: 1. Failures occurring before processing a broadcast message, which was received, but not propagated down the hierarchy. 2. Failures occurring in the middle of the broadcast, when the process sends a message to some of its children, and then fails. 3. Failures occurring after the broadcast, when the node is not participating in the broadcast. The *POModify* messages are treated in the same way as other broadcast messages.

The idea is to merge the first two types of failures with the simpler third type. This is done by comparing the BI index of the node's parent to the BIs of the node's children. If the failed node's child misses some messages received by the failed node's parent, these messages are sent to that node. After this is done, the failure could be considered belonging to type 3. Multiple failures are handled in a similar way. In figure 3(d) if node B fails, the BIs of D and M are compared to BIs of A. If D also fails, the BIs of Y and M are compared to A.

Type 3 failures are handled in the following way. After failure is detected, the coordinator is notified, and it acts in the same way as if the failed process willingly leaves the system: the coordinator broadcasts *POModify* message to its children.

If a process during a broadcast discovers that its child in the hierarchy is down, it sends a notification message to the root coordinator, and propagates the broadcast to the children of the failed process. For example, if node D in figure 3(d) failed and B tried to send it a broadcast, B would notify A of D's failure and propagate the broadcast to Y and to M. As a result all the processes, except the failed one, receive the broadcasted message. The same rule works in the presence of multiple failures.

Coordinator failures. The failures of the coordinator can be divided into the same three types as the failure of any other node. Type 3 failure is the same in both cases but the recovery from type 1 and 2 failures is different, since the coordinator has no parent.

Type 1 failure occurs when some node sends a message to the coordinator in order to initiate its broadcast. If the coordinator fails before broadcasting the message, then the message will be lost. To prevent this, the initiator of the message is responsible for making sure the message is broadcast. The initiator appends a unique monotonically increasing broadcast number to the message. When the initiator is notified of the coordinator failure, it checks if it received all the messages that it initiated. If not, then it sends a message to the new coordinator asking to repeat the broadcast.

Type 2 failures occur when the coordinator fails during the broadcast. To circumvent this, a new coordinator compares its BIs with the BIs of all the children of the failed coordinator. If some of the children miss a certain message, it is rebroadcast to that particular part of the old tree. Each of the rebroadcast messages contains the history of changes to the topology that happened by the time the original broadcast was made. This way each node in the system can figure out its descendants in the tree before the failure, and the whole old hierarchy is reconstructed.

The type 3 failures of the coordinator are handled in the same way as the type 3 failures of other nodes, except that the failure notification message is sent to the new coordinator. The new coordinator is always the one with the lowest POID.

Node recovery. To preserve the total order of broadcasts, a restarted process must also receive all the broadcast messages it missed while being down. Comparing the BIs of the restarted process with BIs of the coordinator determines the messages missed by the failed process. When the failed process restarts, these missed broadcast messages are supplied to the failed process along with other logged messages needed for the recovery. When the failed process is ready to rejoin the system, the total order of broadcasts in the system is preserved.

The ModLog and BLog are periodically cleared along with traditional causal logging protocol logs during the capture of a consistent system snapshot, or by a specialized message.

6   Extensions to the Basic Scheme

6.1 Broadcasting with Total Order

The use of the hierarchy can easily be extended to a more general broadcasting mechanism, where all the processes in the system can broadcast messages. To do this, the initiator of the broadcast sends its message to the root coordinator, which then propagates the message down the hierarchy. As TCP/IP provides reliable FIFO channels, this broadcasting mechanism supports a total order of broadcasts.

6.2 Broadcasting without Total Order

If the total order of broadcasts is not a necessity, the initial message does not need to go through the coordinator. Instead, the sender process becomes a root of another broadcast tree. The POID of the coordinator is sufficient for all other processes to construct the desired tree on the fly. Figure 4 shows how this tree is constructed with node D as a coordinator. The indices 0-5 are simply rotated so that D is assigned the POID zero. This results in the tree where D is the root. Both trees, with coordinators A and D, respectively, can coexist. All topology modifications are still going through the tree with the coordinator A. Modifications to the topology do not cause inconsistencies in D's tree as the sender's ModLog is attached to the message, so that every node can calculate its children, as was shown in section 4. This scheme reduces message traffic and, most importantly, it removes the bottleneck resulting from a central coordinator.

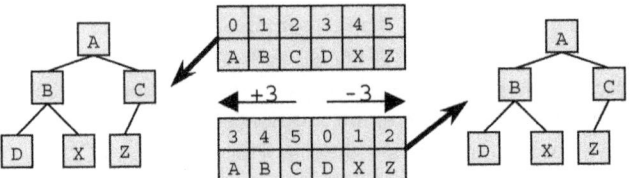

Fig. 4. Constructing tree with D as a coordinator

6.3 Possible Optimizations

Modifying the system topology also requires restructuring of the tree, which might involve breaking old and establishing new TCP/IP connections. The penalty of restructuring the tree is discussed in section 2. As was shown in section 4, appending the list of changes to the broadcast allows delivery of the message to intended recipients. This fact can be used to optimize restructuring. Instead of updating the tree with every change, the tree is updated only occasionally (lazy update). Figure 5 illustrates this approach. When a node joins the system, it is added at the leaf level of the tree (fig 5(b) and fig 5(e)). When a node exits the system, the parent of the exited node takes responsibility for broadcasting the message to the exiting node's children (fig 5(c) and fig 5(d)). These modifications are recorded in a special log. When this log is cleared, together with ModLog, the tree is recomputed, taking into account all the changes in the logs (fig 5(f)).

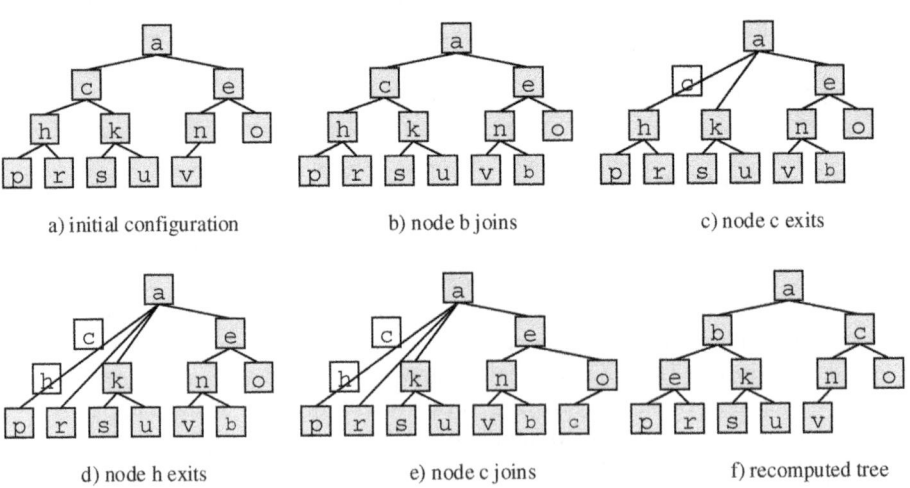

Fig. 5. Lazy update.

For the scheme presented in section 6.1, where the broadcasting tree changes for each broadcasting process, the cost of reconnecting the processes could be high. A possible solution is to swap the broadcasting node with the original root node, as shown on figure 6.

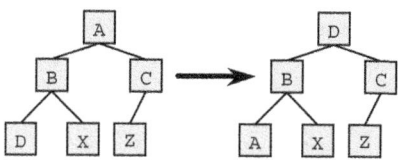

Fig. 6. Constructing tree with D as a coordinator.

Another optimization can be implemented when processes are distributed through the WAN. The hierarchy could be built in two steps to minimize the communications between different parts of the network. First a hierarchy is built between hosts in different subnets, and then within the subnets. This is possible, as the location of the machine is specified in its IP address.

7   Conclusion

The Process Order mechanism is a flexible and simple tool for creation and maintenance of logical structures in distributed system. Its main advantage over other approaches is that nodes only need to have a list of currently active processes. Each node can then determine the complete topology of the system locally, using a formula. Thus the topology of the system never needs to be broadcasted to other nodes. The approach can be implemented using the TCP/IP protocol, which allows its use on practically any machine. This is an important factor for the systems using non-dedicated workstations.

Aside from the hierarchy, it is often desirable to arrange processes in other logical structures, such as a ring for fault detection, a hypercube [10], to implement stability detection algorithms, a mesh [11], for grid computations, or a group [12], for load balancing. Process Order can be used to construct and maintain these structures in a similar way as the tree. The usability of the Process Order mechanism depends on how well the desired logical structure can be expressed with a formula.

The cost of maintaining the Process Order is a total of n messages when adding or removing a new process to the system, where n is the number of processes. The cost of maintaining the Process Order Table consists of inserting and deleting elements in a sorted list. Even with the most straightforward algorithm this operation is of the order of $O(n)$. Maintaining the PO Table also requires minimal extra space: to maintain the unique ids of the participating processes. The cost of maintaining Process Order is independent of how many structures are being concurrently supported by the mechanism.

References

1. E. N. Elnozahy, D. B. Johnson, Y. M. Wang. "A Survey of Rollback-Recovery Protocols in Message Passing Systems.", *T.R. CMU-CS-96-181*, School of Computer Science, Carnegie Mellon University, Oct. 1996

2. J. Leon, A. L. Fisher, and P. Steenkiste. "Fail-Safe PVM: A portable package for distributed programming with transparent recovery." *Tech. Rep. CMU-CS-93-124*, Carnegie Mellon Univ., February 1993

3. K. Guo. "Scalable Message Stability Detection Protocols." PhD thesis, Department of Computer Science, Cornell University, 1998

4. M. Fukuda. "MESSENGERS: A Distributed Computing System Based on Autonomous Objects." PhD thesis, Department of Information and Computer Science, University of California, Irvine, 1997

5. E. Gendelman, L. F. Bic, M. Dillencourt. "An Efficient Checkpointing Algorithm for Distributed Systems Implementing Reliable Communication Channels." *18th Symposium on Reliable Distributed Systems*, Lausanne, Switzerland 1999

6. E. Gendelman, L. F. Bic, M. Dillencourt. "An Application-Transparent, Platform-Independent Approach to Rollback-Recovery for Mobile Agent Systems" *20th IEEE International Conference on Distributed Computing Systems*. Taipei, Taiwan 2000

7. Eugene Gendelman, Lubomir F Bic, Michael B.Dillencourt. " Process Interconnection Structures in Dynamically Changing Topologies", Univ. of California, Irvine. TR #00-27 http://www.ics.uci.edu/~egendelm/prof/processOrder.ps

8. E. L. Elnozahy, D. B. Johnson, and W. Zwaenepoel. The performance of consistent checkpointing. In *Proc. of the 11th Symposium on Reliable Distributed Systems*, pages 39-47, October 1992

9. K. Kim, J. G. Shon, S. Y. Jung, C. S. Hwang. Causal Message Logging Protocol Considering In-Transit Messages. In *Proc. of the ICDCS 2000 workshop on Distributed Real-Time Systems*. Taipei, Taiwan 2000

10. R. Friedman, S. Manor, and K. Guo. "Scalable Stability Detection Using Logical Hypercube." *18th Symposium on Reliable Distributed Systems, Lausanne*, Switzerland 1999

11. K. Solchenbach and U. Trottenberg. "SUPRENUM: System essentials and grid applications." *Parallel Computing* 7 (1988) pp. 265-281

12. Corradi, L. Leonardi, F. Zambonelli. "Diffusive Load-Balancing Policies for Dynamic Applications". *Concurrency*. January-March 1999.

13. L. Alvisi, B. Hoppe and K. Marzullo, "Nonblocking and Orphan-Free Message Logging Protocols," *Proceedings of the 23rd Fault-Tolerant Computing Symposium*, pp.145-154, June 1993.

Conservative Circuit Simulation on Multiprocessor Machines *

Azzedine Boukerche

Parallel Simulations and Distributed Systems (PARADISE) Research Lab
Department of Computer Sciences, University of North Texas, Denton, TX. USA
boukerche@cs.unt.edu

Abstract. Distributed computation among multiple processors is one approach to reducing simulation time for large VLSI circuits designs. In this paper, we investigate conservative parallel discrete event simulation for logic circuits on multiprocessor machines. The synchronization protocol makes use of the Semi-Global Time-of-Next- Event (SGTNE/TNE) suite of algorithms. Extensive simulation experiments were conducted to validate our model using several digital logic circuits on an Intel Paragon machine.

1 Introduction

As the size of VLSI circuit designs becomes very large, involving thousands of gates, simulation of the design becomes a critical part of the validation and design analysis process prior to fabrication, since probing and repairing VLSI systems is impractical. Rather than dealing with voltages and currents at signal nodes, discrete logic states are defined, and simple operations are used to determine the new logic value at each node. This gives rise to the event driven approach for efficiently simulating VLSI systems.

Logic simulators are straightforward to implement if a centralized system is used. Unfortunately, logic simulation algorithms are well known for their computational requirements, and their execution time which can far exceed even the fastest available uniprocessor machines. It is therefore natural to attempt to use multiple processors to speed up the simulation process by exploiting the features of parallel architectures, and taking advantage of the concurrency inherit in VLSI systems [14].

Parallel simulation has proven to be an important application for parallel processing in general. Accordingly, several approaches [2,5,10] have been proposed to perform simulation on multiprocessor machines. These methods can be classified into 2 groups, the *conservative* algorithms and *optimistic* algorithms. Both approaches view the simulated system as a collection of logical processes (LPs) that communicate via timestamped messages. While conservative synchronization techniques [8, 10, 11, 12, 13] rely on *blocking* to avoid violation of

---

* This work was supported by the UNT Faculty Research Grant

M. Valero, V.K. Prasanna, and S. Vajapeyam (Eds.): HiPC 2000, LNCS 1970, pp. 415–424, 2000.

dependence constraints, *optimistic* methods [2, 5, 7, 9] rely on detecting synchronization errors at run-time and on recovery using a *rollback* mechanisms.

Nevertheless, the use of logic simulation and simulators for designing large and complex systems VLSI systems have brought several challenges to the PDES[1] community. The challenges require not only extension and advances in current parallel logic simulation methodologies, but also the discovery of innovative approaches and techniques to deal with the rapidly expanding expectations of the VLSI system designers. In this paper, we focus upon conservative methodology for parallel simulation. Specifically, we propose a generalized hierarchical scheduling of the Time-of-Next-Event Algorithm in the context of digital logic circuit simulation on an Intel Paragon Machine.

2   Overview of Time Next Event (TNE) Algorithm

In this section, we describe the intuition behind the TNE algorithm and present pseudo-code for TNE. The reader familiar with TNE may wish to skeep this section.

We employ the example presented in Fig. 1 to informally describe TNE.

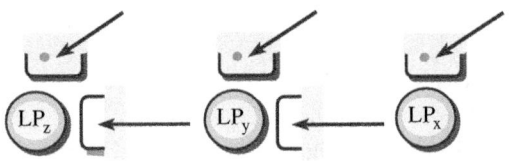

Fig. 1. LPs connected by empty links

Figure 1 shows three logical processes LP_x, LP_y and LP_z connected by directed empty links. Let LST_x, LST_y and LST_z be the local simulation times[2] and T_{min_x}, T_{min_y} and T_{min_z} be the smallest timestamp at LP_x, LP_y and LP_z respectively. $T_{smin_{i,j}}$ denotes the smallest time-stamp increment an event sent from LP_i to its neighbor LP_j. We define T_i to be the smallest timestamp which can be sent by LP_i and $T_{i,j}$ the TNE (i.e., the time of the next the event) of the link (i, j).

Consider the empty link from LP_x to LP_y. LP_x cannot send an event message with a smaller timestamp than T_x, where $T_x = Max(LST_x, T_{min_x})$. On its way[3], the message has to pass through LP_y as well. A service time at LP_x has to be added to each output sent by LP_x. Therefore LP_y cannot expect an event message from LP_x with smaller timestamp than $T_{x,y}$, where $T_{x,y} = T_x + T_{smin_{x,y}}$.

---

[1] Parallel Discrete Event Simulation

[2] Local Simulation Time (LST) is the time of the last event that has been processed an LP.

[3] We assume that it is safe for LP_x to process the event with the smallest time stamp T_x.

Thus a new T_y is computed as : $T_y = Min(T_y, T_{x,y})$. LP_z can not expect a message with a smaller timestamp than $T_{y,z}$ from LP_y for the same reason.

Based on these observations, a shortest path algorithm may be employed to compute the TNEs. The TNE algorithm explores the directed graph of LPs connected by empty links. It finds estimates of the next timestamps on all empty links that belongs to the same processor. If the estimate(s) of the future timestamps at all of an LP's empty links are higher than the LP's smallest timestamps, then the event with the smallest timestamp can be processed, i.e., the LP is unblocked. Provided that the TNE algorithm will be executed independently and frequently enough in all of the processors, estimates on future timestamps can unblock several LPs.

We could improve the lower bound produced by the TNE by precomputing a portion of the computation for future events [11]. The priority queue (PQ) might be implemented as a splay tree. The TNE algorithm is not computationally very expensive. TNE can be mapped to the shortest path problem, and therefore has the same complexity, i.e., $O(nlog(n))$ where n is the number of nodes in the graph. The TNE algorithm helps to unblock LPs and breaks all local deadlocks-see [2, 3]. However, it doesn't have access to any global information, and this unfortunately makes *inter-processor deadlocks* possible. Thus another algorithm is needed to take care of this kind of deadlock see [2]).

3   Hierarchical Scheduling Schemes

While the TNE algorithm is executed over a cluster of processes, SGTNE, a Semi-Global TNE algorithm is executed over a cluster of processors. TNE helps to breaks local deadlocks, while SGTNE help to break inter-processor deadlocks by accessing data pertaining to processes assigned to different processors.

The SGTNE algorithm is executed by a single processor (the *manager*[4]) while the other processors are still running. Hence, a simulation can progress while SGTNE is running. Making use of the SGTNE algorithm, we find estimates for the next time stamps on all empty links belonging to the same cluster of processors. If the estimates of the future timestamps at all of an LP's empty input links are higher than the LP's smallest time stamp, then the event with the smallest timestamp can be processed (i.e., the LP is unblocked). Provided that the SGTNE algorithm is executed frequently enough, estimates on future timestamps can unblock several LPs and processors, therefore speeding up the simulation.

SGTNE, like the TNE, is basically a variation of the shortest path algorithm. The algorithm computes the greatest lower bound for the timestamp of each next event to arrive at each input link within a given cluster of processors. Although, TNE and SGTNE are not computationally expensive, simulation efficiency must still be considered.

Several hierachical schemes were investigated. In this section, we present a two-level scheduling scheme and a generalized hierarchical scheme.

---

[4]   The manager is dedicated to executing SGTNE

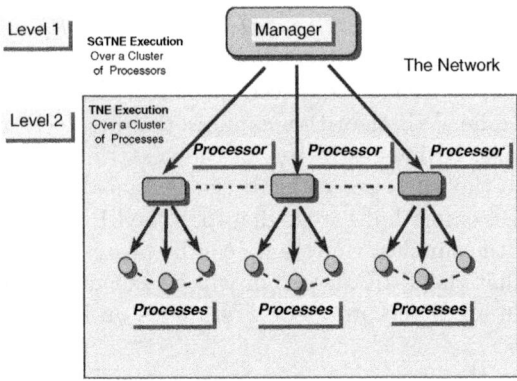

Fig. 2. The Two-Level Hierarchical Scheduling Scheme

3.1   Two-Level Scheduling

In this approach, we wish to prevent the formation of deadlocks. At level 2, TNE is called when a processor becomes idle. In the event that it does not unblock the processor, SGTNE is called. Executing SGTNE will help to break inter-processor deadlocks involving a cluster of processors.

To reduce the overhead of this approach, we make use of the following strategy: when a processor (Pr_i) sends a *request* message to the manager asking for SGTNE to be executed over a cluster of processors, Pr_i sends an updated version of the sub process-graph with the *request* message to the manager. When the manager receives a request message from processor Pr_i, it updates its sub process-graph and sends a request process-graph message to each processor Q in the same cluster. Once the manager receives these graphs in response to all of its request messages, it updates its process-graph. If there are any changes to the graph, it executes the SGTNE algorithm. If we use the process graph and execute SGTNE without making sure that the portion of the process-graph associated with *each* processor within the cluster is updated, we obtain worse results.

Figure 2 illustrate a conceptual model for this two-level scheme. A manager keeps a global data consisting of the process graph. If SGTNE helps to unblock some LPs, then the manager informs the processors Pr_i and Q, where $Q \in Sphere_{Pr_i}$ (defined below), of the changes. Once Pr_i receives a *reply* message from the manager, it updates its sub-graph and unblocks all possible LPs if there is any change in the graph.

We define a neighborhood of a processor to be a sphere centered at a processor p with radius r as:

$$Sphere_p = \{q|Dist(p,q) = r\} \cup \{p\};$$

where $Dist(p,q) = min_{ij}(\delta_{ij})$, such that $LP_i \in p$ and $LP_j \in q$, δ_{ij} is the (shortest) number of hops between LP_i and LP_j. In our experiments $r = 1$, as we made use of 16 processors. We also determine statically the neighborhood

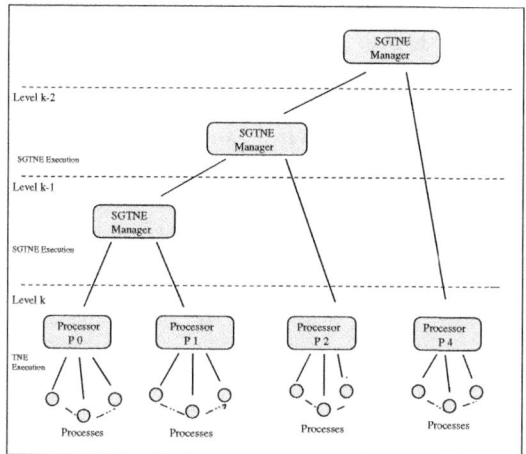

Fig. 3. Generalized Hierarchical Scheduling Scheme

of each processor at compile time before the simulation starts. Further work will be directed at determining these neighborhoods at run time, i.e., during the execution of the simulation.

3.2   Generalized Hierarchical Scheduling

Figure 3 illustrates a conceptual model for this multi-level scheme with 4 processors. A manager keeps a global data consisting of the process graph.

In this approach, we wish to reduce the overhead of the bottleneck that might arises with the two-level scheme; prevent the formation of deadlocks; and increase the efficiency of the SGTNE/TNE suite of algorithms. In order to do this, we make use of the connectivity matrix M= $[m_{ij}]$. An entry m_{ij} in the matrix represents the number of interconnections between processor P_i and procesor P_j. The diagonal entries m_{ii} represent internal connections and will be set to zero.

The basic idea of the multi-level hierarchical scheme is based on the connectivity between the processors. We choose to arrange the processors, according to their connectivity with each others, as a chain of processors $P_0, P_1, ..., P_n$, where n is the total number of processors. In order t to determine the chain, we choose to follow this simple strategy:

We pick 2 processors that have the highest connectivities, and we identify them as P_0 and P_1. Then, we choose the next processors that has the highest connectivity to processor P_1; and so on, until, we have identified all processors. As show in in Figure 3, At the (lowest) level k[5] TNE is called when a processor P_i (for $i \neq 0$) becomes idle. In the event that it does not unblock the processor, a request to execute SGTNE over processors P_i and P_{i-1} is sent to the SGTNE

---

[5] k is basically equal to the depth of the tree.

Manager. In the event that it does not unblock LPs within either P_i or P_{i-1}, a request to execute SGTNE over processors $\{P_i, P_{i-1}\}$ and P_{i+1} is sent to the SGTNE Manager (at a higher level), and so on.

In the event that P_0 becomes idle, a request to execute SGTNE over P_0 and P_1 is sent to the SGTNE Manager. When the manager receives a request message from processor Pr_i, it updates its sub process-graph and sends a request process-graph message to processors P_i and P_{i-1}. Once the manager receives these graphs in response to both of its request messages, it updates its process-graph. If there are any changes to the graph, it executes the SGTNE algorithm. Note that the process graph might be updated by the SGTNE Manager at a lower level (see Fig. 3.

Our first implementation of the SGTNE manager is to dedicate one single processor as a manager. Future work will be directed towards a distributed (local) SGTNE manager. A manager keeps a global data consisting of the process graph. It is easy to see that with this approach any eventual inter-processor deadlock will be broken.

4   Simulation Model and Experiments

The main goals of these tests, were to determine the effectiveness of the generalized hierarchical scheduling scheme of TNE in achieving good speed up of VLSI logic simulation. In this approach, the VLSI system is viewed as a collection of subsystems that are simulated by a set of processes that communicate by sending/receiving timestamped events. A subsystem may represent a collection of logical elements at varying level of abstraction; e.g., transistors, NOR gates, Adders, flip-flops, etc... State changes such as the change in output value of individual gate, are represented by event (messages) in the simulation.

Our logic simulation model uses three discrete logic values: 1, 0, and undefined. To model the propagation delay, each gate has a constant service time. All of the common logical gates were implemented: AND, NAND, NOR, Flip-Flops, etc.

The circuits used in our study are digital sequential circuits selected from the ISCAS'89 benchmarks. many circuits of different sizes have been tested (see 1 below), since they are both representative of the results we obtained with the other circuits. A program was written to read the description file of the ISCAS benchmarks and to partitionthem into clusters. We used a string partitioning algorithm, because of its simplicity and especially because the results have shown that it favors concurrency over cone partitioning.

The implementation of the Logic Simulator was performed on an an Intel Paragon[6] A4, a distributed memory multiprocessor consisting of 56 compute nodes, 3 service nodes and 1 HIPPI node. The compute nodes are arranged in a two-dimensional mesh comprised of 14 rows and 4 columns [6]. The compute nodes use the Intel i860 XP microprocessor and have 32 MBytes of memory each. The A4 has a peak performance of 75 MFlops (64-bit), yielding a system peak speed of 4.2 GFlops, which is the reason for the "4" in the name "A4". The

---

[6] Paragon is a registered trademark of the Intel Corporation.

Table 1. ISCAS Circuits: C1 and C2

Circuits	Inputs	Outputs	Flip-Flops	Total
C1: s38417	28	106	1,636	23,949
C2: s3858	12	278	1,452	20,995

total memory capacity of the compute partition is approximately 1.8 GBytes. The service nodes also use the Intel i860 XP microprocessor. Two of the service nodes have 16 MBytes of memory, and one service node has 32 MBytes. The HIPPI node also uses the Intel i860 XP microprocessor and has 16 MBytes of memory. A single HIPPI node controls both HIPPI "send" and HIPPI "receive".

4.1   Experiments

We conducted 2 categories of experiments: one to assess the speedup obtained by our scheduling schemes relative to the fastest sequential logic simulator. The sequential circuit simulation make use of a splay tree data structure. Experimental studies showed that splay tree is among the fastest techniques for implementing an event list [3]. The second test of experiments to determine the overhead implicit in the use of our generalized hierarchical scheme, pointing out the the importance of the inter-processor communication overhead.

The usual measurement of the effectiveness of a parallel simulation is *Speedup*. We define speedup $(SP(n))$ to be the time T_1 it takes a single processor to perform a simulation divided by the time T_n it takes the parallel simulation to perform the same simulation when n processors are used, i.e., $SP(n) = T_1/T_n$. The speedup can be thought of as the effective number of processors used for the simulation.

The speedup curves of the parallel simulation algorithms relative to a sequential circuit simulation for different circuits for both Two-Level and Multi-Level schedulings, and for both circuits are shown in Figure 4. As we can see, both scheduling yield to a good speedup. This is largely due to the the TNE's efficiency in unblocking LPs and breaking local deadlocks, and the hierarchical scheduling which provide enough information and LPs update of LP located in the neighboring processors. We also observe that the Multi-Level scheduling outperforms the Two-Level scheduling when we use more than 16 processors. The factor of improvement increases as we increase the number of processors. This is due to the the fact that the Multi-Level scheduling executes the SGTNE first on a small cluster of processors and try to break inter-processors deadlocks among neighboring processors. Indeed, there is no need to access data pertaining to a larger pool of processes which are assigned to different processors to break the same inter-processor deadlocks. (there is no need to execute the SGTNE on

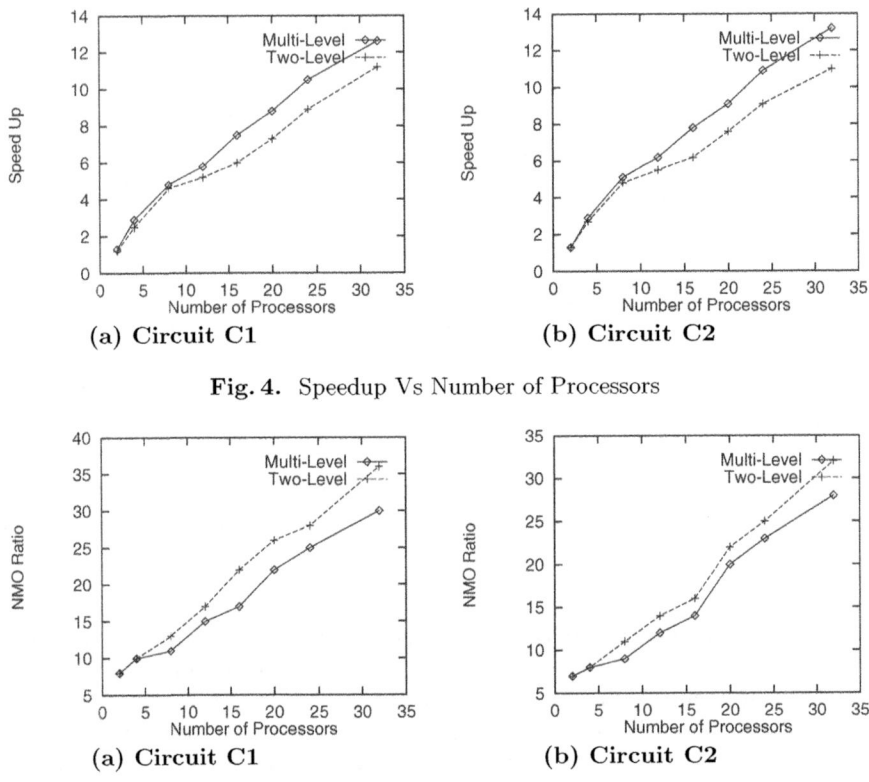

Fig. 4. Speedup Vs Number of Processors

Fig. 5. Synchronization Overhead Vs Number of Processors

a larger cluster to break the same inter-processor deadlock.). This is the main purpose of the Multi-Level scheduling. Thus, less time will be spent executing SGTNE over a cluster of processors.

Our results indicate clearly the following observations: (1) The difference in speedup between the two- scheduling and the multi-level scheduling increases with the number of processors; and (2) careful scheduling is an important factor in the SGTNE/TNE suite of algorithms.

The Message Synchronization Overhead (MSO) is defined as the total number of request messages (NM_{req}) to execute SGTNE divided by the total number of messages involved in the simulation (NM_{tot}), i.e.,

$$MSO = \frac{NM_req}{NM_{tot}};$$

MSO can be seen as the efficiency of the scheduling strategy used to schedule the TNE/SGTNE suite of algorithms.

Synchronization overhead (MSO) curves for both synchronization schedulings and for both circuits are shown in Figure 5. The results show that MSO increases as we increase the number of processors. We also observe that the

Table 2. Communication/Execution Percentage.

Number of Processors	Circuit C1 s (%)	Circuit C2 (%)
2	10.80	11.8
4	17.32	18.9
8	22.3	23.9
16	25.9	27.8
32	31.2	33.5

(a) Processor $< -- >$ Processor

Number of Processors	Circuit C1 %	Circuit C2 %
2	5.1	5.1
4	7.2	7.5
8	10.6	11.03
16	13.9	14.4
32	15.9	16.3

(b) Processor $< -- >$ SGTNE-Manager

synchronization overhead of the Multi-Level scheme is slightly higher than the two-level scheme. Hence, the overhead of the multi-level scheme is modest when we compare it to the two-level scheme.

Note that this did not affect the speedup of our simulation. indeed, idle processors receive updated information about their blocked LPs much faster when we use the multi-level scheme.

Communication/execution time percentages are reported in Table 2. Our first way of computing the overhead of the communication overhead engendered by SGTNE with the two-level scheme is by dividing the average time which an LP spends in sending messages to or receiving messages from neighboring processors divided by the execution time of the simulation. Our second measure is directed toward the communication overhead engendered by either of the 2 hierarchical schedulings, i.e., by dividing the average time which a processor spends in sending messages to or receiving messages from the (processor) manager divided by the execution time of the simulation.

As can be seen from Table 2 the inter-processor communication plays an important role in determining the speed of the simulation when we employ a large number of processors. The results also miror the synchronization overhead obtained earlier. They show that the communication overhead, due to the message passing between the idle processor and the the SGTNE-manager that receives the request to execute the SGTNE algorithm over the cluster of processors, is slightly higher when we use the multi-level scheme as opposed to the two-level scheme.

5   Conclusion

We have described in this paper, our experience with several hierarchical scheduling of the Time-of-Next-Event (TNE) algorithm in the context of digital circuit simulation on multiprocessor machine. We have proposed a generalized hierarchical scheduling strategy, which we refer to as a multi level schemes, and implemented both schemes (two/multi levels) on an Intel Paragon machine. The objective of mult-level schems was to make use of information in neighboring processors to unblock LPs rather than only relying upon the information about LPs present in the processor. In doing so, it helps to avoid the formation of inter-processor deadlocks, and improve the efficiency of TNE. Extensive simulation experiments on several digital logic circuits were reported on an Intel Paragon machine. Our results indicate that the multi-level scheduling outperformed the two-level scheduling, and that careful scheduling is an important factor in the SGTNE's success for digital logic simulations.

Acknowledgment

This research was performed in part using the Intel Paragon System operated by the California Institute of Technology on behalf of the Concurrent Supercomputing Consortium. Thanks are in order to Paul Messina, director of the CCSF at Caltech for providing us with super-computer access.

References

1. Ayani, R., "Parallel Simulation Using Conservative Time Windows", *Proc. of the 1992 Winter Simulation Conference*, Dec. 1992, pp. 709-717.
2. Boukerche, A., "Time Management in Parallel Simulation", Chapter 18, pp. 375-395, in High Performance Cluster Computing, Ed. R. Buyya, Prentice Hall (1999).
3. Boukerche A., and Tropper C., " Parallel Simulation on the Hypercube Multiprocessor", Distributed Computing, Spring Verlag 1995, pp.
4. Chandy,K.M, and Misra, J., "Distributed Simulation: A Case Study in Design and Verification of Distributed Programs", *IEEE Trans. on Software Engineering*, SE-5, Sept.1979, 440-452.
5. Fujimoto, R. M., "Parallel Discrete Event Simulation", in *CACM*, 33(10), Oct. 1990, pp. 30-53.
6. Hockney, R. W., "The Communication Challenge for MPP: Intel Paragon and Meiko CS-2", *Parallel Computing*, Vol. 20, 1994. pp. 383-398.
7. Jefferson, D. R., "Virtual Time", *ACM Trans. Prog. Lang. Syst.* 77, (3), July 1985, pp. 405-425.
8. Lin, Y. B., and Lazowska, E. D., "Conservative Parallel Simulation for Systems with no Lookahead Prediction", *TR 89-07-07*, Dept. of Comp. Sc. and Eng., Univ. of Washington, 1989.
9. Lomow, G., Das, S. R., and Fujimoto R.,"User Cancellation of Events in Time Warp", *Proc. 1991, SCS Multiconf. on Advances in Dist. Simul.*, Jan. 1991, pp. 55-62.
10. Misra, J., "Distributed Discrete-event Simulation", ACM *Computing Surveys*, 18(1), 1986, pp. 39-65.
11. Nicol, D. M, "Parallel Discrete-Event Simulation of FCFS Stochastic Queuing Networks", in *Proc. of the ACM SIGPLAN Symp. on Parallel Programming, Environment, Applications, and Languages*, 1988.
12. Wagner, D. B., Lazowska, E.D, and Bershad, B., "Techniques for Efficient Shared Memory Parallel Simulation", *Proc. 1989 SCS Multiconf. on Distributed Simulation*, 1989, pp. 29-37.
13. Wood K., and Turner, S. J., "A Generalized Simulation- The *Carrier Null Message* Approach", *Proc. of the SCS Multiconf. on Distributed Simulation*, Vol. 22, No. 1, Jan. 1994.
14. J. Keller, T. Rauber, and B. Rederlechner, "Conservative Circuit Simulation on Shared-memory Multiprocessors"; PADS'96, pp. 126-134.

Session IV-B

Networks
Chair: C.S. Raghavendra
University of Southern California

Exact Evaluation of Multi-traffic for Wireless PCS Networks with Multi-channel*

Wuyi Yue[1] and Yutaka Matsumoto[2]

[1] Dept. of Applied Mathematics, Konan University
8-9-1 Okamoto, Higashinada-ku, Kobe 658-8501 JAPAN
yue@konan-u.ac.jp
[2] I.T.S., Inc.
2-2 Hiranoya Shinmachi, Daito-City, Osaka 574-0021 JAPAN
its@osk.3web.ne.jp

Abstract. In this paper we propose two different procedures of multi-channel multiple access schemes with the slotted ALOHA protocol for both data and voice traffic and present an exact analysis to numerically evaluate the performance of the systems. In scheme I, there is no limitation on access between data transmissions and voice transmissions, i.e., all channels can be accessed by all transmissions. In scheme II, a channel reservation policy is applied, where a number of channels are used exclusively for voice packets while the remaining channels are used for both data and voice packets. We call the system using scheme I "Non-reservation system" and call the system using scheme II "Reservation system". Performance characteristics include the loss probability for voice traffic, average packet delay for data traffic and channel utilization for both traffic. The performance of the scheme I and II, and the effects of the design parameters are numerically evaluated and compared to a wide-bandwidth conventional single-channel slotted ALOHA system.

1 Introduction

Recently, wireless communication has become an important field of activity in telecommunications. The demand for services is increasing dramatically and the required service is no longer restricted to the telephone call. Wireless data communication services, such as portable computing, paging, personal email, etc., play an increasingly important role in current wireless systems. It is recognized that these systems will be required to support a wider range of telecommunication applications, including packet data, voice, image, and full-motion video. By implementation of a truly multimedia and Personal Communication System (PCS), the user will be released from the bondage of the telephone line and will enjoy greater freedom of telecommunications. To this end, the system must be able to provide diverse quality of service to each type of traffic.

---

* This work was supported in part by GRANT-IN-AID FOR SCIENTIFIC RESEARCH (No.11650404).

M. Valero, V.K. Prasanna, and S. Vajapeyam (Eds.): HiPC 2000, LNCS 1970, pp. 427–438, 2000.

In any communication system it is desirable to efficiently utilize the system bandwidth. In the recent literature, random channel access protocols in wireless communications have been actively studied to utilize the limited spectrum among all the users efficiently. In particular, the slotted ALOHA protocol has been considered in a number of papers, for example [1]-[4]. Slotted ALOHA allows a number of relatively-uncoordinated users to share a common channel in a flexible way. It is therefore interesting to investigate it in more depth and to exploit some characteristics in the mobile radio environments to obtain some gain in performance.

There are some studies on multi-channel radio communication systems or on packet reservation for a number of random access protocols to achieve higher channel utilization [2]-[5]. These studies showed that the multi-channel scheme can improve the channel throughput rate, packet delay and other important performance measures by reducing the number of users who make simultaneous access on the same channel. In addition, it has many favorable characteristics such as easy expansion, easy implementation by frequency division multiplexing technology, high reliability and fault tolerance.

However, we notice that in most of these studies, multi-channel systems have been analyzed to obtain performance measures for data traffic only. To the authors' knowledge, there is no work on the performance analysis of multi-channel multi-traffic wireless networks supporting data traffic and real-time traffic simultaneously. Increasing demands on communication networks require the design and development of high capacity transmission systems for simultaneous transportation of different kinds of traffic, such as video, voice and digital data.

To satisfy the huge service demand and the multi-traffic requirement with limited bandwidth, this paper proposes two different procedures of multi-channel multiple access schemes with the slotted ALOHA operation for both data and voice traffic and presents exact analysis to numerically evaluate the performance of the systems. In scheme I, there is no limitation on access between data transmissions and voice transmissions, i.e., all channels can be accessed by all transmissions. In scheme II, a channel reservation policy is applied, where a number of channels are used exclusively for voice packets while the remaining channels are used for both data packets and voice packets. This paper is organized as follows. In Section 2, we present the multi-channel multi-traffic system models with two specific schemes. Performance of schemes I and II is analyzed in Section 3 and 4. The performance measures are given in Section 5, they include loss probability for voice traffic, average delay for data packets, average number of successful channels, and channel utilization for both traffic. The performance of two schemes, and the effects of the design parameters are numerically evaluated and compared in Section 6 and conclusions are drawn in Section 7.

2   System Model and Its Assumptions

The system consists of a finite population of N users accessing a set of parallel M channels that have equal bandwidth in Hz. The time axis is slotted into segments of equal length τ seconds corresponding to the transmission time of a packet. All users are synchronized and all packet transmissions over the chosen channel are started only at the beginning of a time slot. If two or more packets are simultaneously transmitted on the same channel at the same slot, a collision of the packets occurs.

Traffic is classified into two types: data traffic, which is strict with regard to bit errors but relatively tolerant of instantaneous delivery, and real-time bursty traffic (such as voice or video), which is strict with regard to instantaneous delivery but relatively tolerant of bit errors. In this paper we assume that each user can generate both data and voice packets. Namely, the user can be in one of three modes: idle mode, data transmission mode, or speech mode. At the beginning of a slot, each user in the idle mode enters the data transmission mode by generating a data packet with probability λ_d, or enters the speech mode with probability λ_v, or remains in the idle mode with probability $1 - \lambda_d - \lambda_v$.

Users in the data transmission mode transmit a data packet in the slot when they enter the data transmission mode with probability one (immediate-first-transmission). If this first transmission attempt is successful, they return to the idle mode at the end of the slot. Otherwise, they schedule its retransmission after a time interval that obeys a geometric distribution with mean $1/\mu_d$ slots.

Users in the speech mode transmit a voice packet in the slot when they enter the speech mode with probability one (immediate-first-transmission) and in the succeeding slots with probability μ_v until they quit the speech mode to return to the idle mode. The length of a user's speech mode obeys a geometric distribution with mean $1/q_v$ slots. We assume that if voice packets are involved in collision, they are just discarded and no retransmission is necessary.

"Non-reservation system" and "Reservation system" differ in channel access method. In the non-reservation system, all M channels can be accessed by both voice and data packets which choose one of the channels for transmission with equal probability $1/M$. In the reservation system, voice packets can access all M channels as in the non-reservation system, while access of data packets is restricted to $M - R_v$ channels and they choose one of the channels for transmission with equal probability $1/(M - R_v)$.

We define the system state as a pair of the number of users in the data transmission mode and the number of users in the speech mode. We observe the system state at the end of slots and denote it by $\{x(t), t = 0, 1, 2, ...\}$. Then we can model the present system as a finite Markov chain. Let $I = [i_d, i_v]$ denote a system state that there are i_d users in the data transmission mode and i_v users in the speech mode at the end of a slot. We note that as $0 \le i_d + i_v \le N$, the total number A of system states is given by $A = (N + 2)(N + 1)/2$.

3 Analysis of Non-reservation System

In this section, we first analyze the non-reservation system. In the non-reservation system, all M channels can be accessed by both voice and data packets which choose one of the channels for transmission with equal probability $1/M$.

Let n_d and n_v represent the numbers of users in the data transmission mode and in the speech mode, respectively, who transmit their packets at the beginning of a slot. Among $n_d + n_v$ packets, let c_d and c_v represent the numbers of data and voice packets, respectively, which are transmitted successfully in the slot. Because one of the M channels is chosen randomly with equal probability and the transmission is successful if no other users transmit packets over the same channel at the same time slot, the probability $Q(c_d, c_v | n_d, n_v, M)$ that c_d data and c_v voice packets are successfully transmitted, given that n_d data and n_v voice packets access one of M channels randomly, is given by

$$Q(c_d, c_v | n_d, n_v, M) = (-1)^{c_d + c_v} M! \binom{n_d}{c_d} \binom{n_v}{c_v} (n_d + n_v - c_d - c_v)! M^{-(n_d + n_v)}$$

$$\cdot \sum_{j = c_d + c_v}^{\min(n_d + n_v, M)} \frac{(-1)^j (M - j)^{n_d + n_v - j}}{(j - c_d - c_v)! (M - j)! (n_d + n_v - j)!}$$

$$(0 \le c_d + c_v \le \min(n_d + n_v, M), 0 \le n_d + n_v \le N). \tag{1}$$

The derivation of this equation is possible by extending the proof for the single-traffic [2], [3]. See [6].

Let us consider the transition of the system state in equilibrium. Given that there are i_d and i_v users in the data transmission mode and in the speech mode, respectively, at the beginning of the t-th slot (notationally, $x(t) = I$), the following five events must occur to be j_d and j_v users in the data transmission mode and in the speech mode, respectively, at the beginning of the $t + 1$-st slot (notationally, $x(t + 1) = J$), where $t \to \infty$.

(1) At the beginning of the slot, l_d and l_v users in the idle mode enter the data transmission mode and the speech mode by generating new data and voice packets, respectively, where l_d and l_v are integer random variables with scope $0 \le l_d \le N - i_d - i_v, 0 \le l_v \le N - i_d - i_v - l_d$.

(2) At the beginning of the slot, among i_d users in the data transmission mode, $n_d - l_d$ users retransmit data packets, where n_d is an integer random variable with scope $l_d \le n_d \le i_d + l_d$.

(3) At the beginning of the slot, among i_v users in the speech mode, $n_v - l_v$ users transmit voice packets, where n_v is an integer random variable with scope $l_v \le n_v \le i_v + l_v$.

(4) Given n_d data and n_v voice packets are transmitted over M channels, $i_d + l_d - j_d$ data and c_v voice packets succeed in transmission, where c_v is an integer random variable with scope $0 \le c_v \le \min(n_v, M - i_d - l_d + j_d)$.

(5) At the end of the slot, among $i_v + l_v$ users in the speech mode, $i_v + l_v - j_v$ users quit the speech mode and return to the idle mode.

We note that the probabilities for events (1) to (5) are given by $F(N - i_d - i_v, l_d, l_v)$, $H(i_d, n_d - l_d, \mu_d)$, $H(i_v, n_v - l_v, \mu_v)$, $Q(i_d + l_d - j_d, c_v \mid n_d, n_v, M)$, and $H(i_v + l_v, i_v + l_v - j_v, q_v)$, respectively, where we define

$$H(a, b, \alpha) \overset{\triangle}{=} \binom{a}{b} \alpha^b (1 - \alpha)^{a-b},$$

$$F(N - i_d - i_v, l_d, l_v) \triangleq \binom{N - i_d - i_v}{l_d} \binom{N - i_d - i_v - l_d}{l_v} \lambda_d^{l_d} \lambda_v^{l_v}$$
$$\cdot (1 - \lambda_d - \lambda_v)^{N - i_d - i_v - l_d - l_v}.$$

To obtain a one-step transition probability, we firstly define a conditional probability $P(l_d, l_v, n_d, n_v, c_v)$ for the non-reservation system as follows:

$$P(l_d, l_v, n_d, n_v, c_v)$$
$$= F(N - i_d - i_v, l_d, l_v) H(i_d, n_d - l_d, \mu_d) H(i_v, n_v - l_v, \mu_v)$$
$$\cdot H(i_v + l_v, i_v + l_v - j_v, q_v) Q(i_d + l_d - j_d, c_v \mid n_d, n_v, M)$$
$$(0 \le l_d \le N - i_d - i_v, 0 \le l_v \le N - i_d - i_v - l_d, l_d \le n_d \le i_d + l_d,$$
$$l_v \le n_v \le i_v + l_v, 0 \le c_v \le \min(n_v, M - i_d - l_d + j_d)). \tag{2}$$

Let $P_{IJ}^{(I)}$ to be the one-step transition probability for the non-reservation system as $P_{IJ}^{(I)} \triangleq \lim_{t \to \infty} \Pr\{x(t+1) = J | x(t) = I\}$. Then $P_{IJ}^{(I)}$ can be obtained as follows:

$$P_{IJ}^{(I)} = \sum_{l_d} \sum_{l_v} \sum_{n_d} \sum_{n_v} \sum_{c_v} P(l_d, l_v, n_d, n_v, c_v). \tag{3}$$

4   Analysis of Reservation System

In the reservation system, a channel reservation policy is applied, where a number of channels can be accessed exclusively by voice packets, while the remaining channels can be accessed by both data and voice packets. Let R_v $(0 \le R_v < M)$ denote the number of reserved channels that can be exclusively accessed by voice packets.

Let $P(c_{v_1} | n, R_v)$ denote the conditional probability that c_{v_1} voice packets are successfully transmitted given that n voice packets are transmitted over R_v channels. $P(c_{v_1} | n, R_v)$ is given as follows:

$$P(c_{v_1} | n, R_v) = (-1)^{c_{v_1}} R_v! n! (c_{v_1}! R_v^n)^{-1} \sum_{j = c_{v_1}}^{\min(n, R_v)} \frac{(-1)^j (R_v - j)^{n-j}}{(j - c_{v_1})! (R_v - j)! (n - j)!}$$
$$(0 \le c_{v_1} \le \min(n, R_v),\ 0 \le n \le n_v). \tag{4}$$

As in the case of the reservation system, let us consider the transition of the system state in equilibrium. Given that there are i_d and i_v users in the data transmission mode and in the speech mode, respectively, at the beginning of the t-th slot, the seven events must occur to be j_d and j_v users in the data transmission mode and in the speech mode, respectively, at the beginning of the $t + 1$-st slot, where $t \to \infty$. Among these seven events, the first three events are the same as in the case of the non-reservation system, so we describe the adding four events as follows:

(4) Among n_v voice packets which are transmitted over M channels, n voice packets belong to R_v reserved channels, where n is an integer random variable with scope $0 \leq n \leq n_v$.

(5) Given n voice packets are transmitted over R_v channels, c_{v_1} voice packets succeed in transmission, where c_{v_1} is an integer random variable with scope $0 \leq c_{v_1} \leq \min(n, R_v)$.

(6) Given n_d data and $n_v - n$ voice packets are transmitted over $M - R_v$ free channels, $i_d + l_d - j_d$ data and c_{v_2} voice packets succeed in transmission, where c_{v_2} is an integer random variable with scope $0 \leq c_{v_2} \leq \min(n_v - n, M - R_v - i_d - l_d + j_d)$.

(7) At the end of the slot, among $i_v + l_v$ users in the speech mode, $i_v + l_v - j_v$ users quit the speech mode and return to the idle mode.

We note that the probabilities for events (4) to (7) are given by $H(n_v, n, \frac{R_v}{M})$, $P(c_{v_1}|n, R_v)$, $Q(i_d + l_d - j_d, c_{v_2} \mid n_d, n_v - n, M - R_v)$, and $H(i_v + l_v, i_v + l_v - j_v, q_v)$, respectively. Therefore we can write the conditional probability $P(l_d, l_v, n_d, n_v, n, c_{v_1}, c_{v_2})$ and the one-step transition probability $P_{IJ}^{(II)}$ in the stationary state for the reservation system as follows:

$$P(l_d, l_v, n_d, n_v, n, c_{v_1}, c_{v_2})$$
$$= F(N - i_d - i_v, l_d, l_v)H(i_d, n_d - l_d, \mu_d)H(i_v, n_v - l_v, \mu_v)$$
$$\cdot H(i_v + l_v, i_v + l_v - j_v, q_v)H\left(n_v, n, \frac{R_v}{M}\right)P(c_{v_1} \mid n, R_v)$$
$$\cdot Q(i_d + l_d - j_d, c_{v_2} \mid n_d, n_v - n, M - R_v)$$
$$(0 \leq l_d \leq N - i_d - i_v, 0 \leq l_v \leq N - i_d - i_v - l_d, l_d \leq n_d \leq i_d + l_d,$$
$$l_v \leq n_v \leq i_v + l_v, 0 \leq n \leq n_v, 0 \leq c_{v_1} \leq \min(n, R_v),$$
$$0 \leq c_{v_2} \leq \min(n_v - n, M - R_v - i_d - l_d + j_d)), \qquad (5)$$

$$P_{IJ}^{(II)} = \sum_{l_d}\sum_{l_v}\sum_{n_d}\sum_{n_v}\sum_{n}\sum_{c_{v_1}}\sum_{c_{v_2}} P(l_d, l_v, n_d, n_v, n, c_{v_1}, c_{v_2}). \qquad (6)$$

5 Performance Measures

Let $\boldsymbol{P}^{(I)} = [P_{IJ}^{(I)}]$ and $\boldsymbol{P}^{(II)} = [P_{IJ}^{(II)}]$ denote the state transition probability matrixes for the non-reservation system and the reservation system, respectively, obtained in the preceding sections, where $\boldsymbol{I} = [i_d, i_v]$ $(0 \leq i_d \leq N, 0 \leq i_v \leq N - i_d)$ and $\boldsymbol{J} = [j_d, j_v]$ $(\max(0, i_d - M) \leq j_d \leq N - i_v, 0 \leq j_v \leq \min(N - j_d, N - i_d))$. $\boldsymbol{P}^{(I)}$ and $\boldsymbol{P}^{(II)}$ are $A \times A$ matrixes, respectively, where A is the total number of the system states. Let $\boldsymbol{\Pi}^{(I)} = \left[\pi_0^{(I)}, \pi_1^{(I)}, ..., \pi_{A-1}^{(I)}\right]$ and $\boldsymbol{\Pi}^{(II)} = \left[\pi_0^{(II)}, \pi_1^{(II)}, ..., \pi_{A-1}^{(II)}\right]$ denote the A-dimensional row vector of the stationary distributions for the non-reservation system and the reservation system, respectively. $\boldsymbol{\Pi}^{(I)}$ and $\boldsymbol{\Pi}^{(II)}$ can be determined by solving $\boldsymbol{P}^{(I)}$ and $\boldsymbol{P}^{(II)}$ with the same method as follows:

$$[\Pi]_J = \sum_I [\Pi]_I P_{IJ} \qquad \text{or} \qquad \Pi = \Pi P, \qquad \sum_J [\Pi]_J = 1. \quad (7)$$

By using the state transition probabilities and the stationary distribution, we can evaluate the performance of the systems. We define the channel utilization as the average number of packets successfully transmitted on an arbitrary channel per slot. Let ρ_d and ρ_v denote the channel utilization for data traffic and voice traffic, respectively. They are given as follows:
for the non-reservation system:

$$\rho_d = \frac{1}{M} \sum_J \sum_I [\Pi]_I^{(I)} \sum_{l_d} \sum_{l_v} \sum_{n_d} \sum_{n_v} \sum_{c_v} P(l_d, l_v, n_d, n_v, c_v)(i_d + l_d - j_d), \quad (8)$$

$$\rho_v = \frac{1}{M} \sum_J \sum_I [\Pi]_I^{(I)} \sum_{l_d} \sum_{l_v} \sum_{n_d} \sum_{n_v} \sum_{c_v} P(l_d, l_v, n_d, n_v, c_v)c_v, \quad (9)$$

for the reservation system:

$$\rho_d = \frac{1}{M - R_v} \sum_J \sum_I [\Pi]_I^{(II)} \sum_{l_d} \sum_{l_v} \sum_{n_d} \sum_{n_v} \sum_{n} \sum_{c_{v_1}}$$
$$\cdot \sum_{c_{v_2}} P(l_d, l_v, n_d, n_v, n, c_{v_1}, c_{v_2})(i_d + l_d - j_d), \quad (10)$$

$$\rho_v = \frac{1}{M} \sum_J \sum_I [\Pi]_I^{(II)} \sum_{l_d} \sum_{l_v} \sum_{n_d} \sum_{n_v} \sum_{n} \sum_{c_{v_1}} \sum_{c_{v_2}} P(l_d, l_v, n_d, n_v, n, c_{v_1}, c_{v_2})$$
$$\cdot (c_{v_1} + c_{v_2}) \quad (11)$$

where for the non-reservation system $0 \le i_d \le N$, $0 \le i_v \le N - i_d$, $\max(0, i_d - M) \le j_d \le N - i_v$, $0 \le j_v \le \min(N - j_d, N - i_d)$, $0 \le l_d \le N - i_d - i_v$, $0 \le l_v \le N - i_d - i_v - l_d$, $l_d \le n_d \le i_d + l_d$, $l_v \le n_v \le i_v + l_v$, $0 \le c_v \le \min(n_v, M - i_d - l_d + j_d)$, and for the reservation system $0 \le i_d \le N$, $0 \le i_v \le N - i_d$, $\max(0, i_d - M + R_v) \le j_d \le N - i_v$, $0 \le j_v \le \min(N - j_d, N - i_d)$, $0 \le l_d \le N - i_d - i_v$, $0 \le l_v \le N - i_d - i_v - l_d$, $l_d \le n_d \le i_d + l_d$, $l_v \le n_v \le i_v + l_v$, $0 \le n \le n_v$, $0 \le c_{v_1} \le \min(n, R_v)$, $0 \le c_{v_2} \le \min(n_v - n, M - R_v - i_d - l_d + j_d)$. The following equations for both systems are in the same ranges.

Let $\bar{\rho}_v$ denote the average number of voice packets which are involved in collision on an arbitrary channel per slot. In the same manner as in the derivation of ρ_v, for the non-reservation system, it is given by

$$\bar{\rho}_v = \frac{1}{M} \sum_J \sum_I [\Pi]_I^{(I)} \sum_{l_d} \sum_{l_v} \sum_{n_d} \sum_{n_v} \sum_{c_v} P(l_d, l_v, n_d, n_v, c_v)$$
$$\cdot (n_v - c_v), \quad (12)$$

for the reservation system,

$$\bar{\rho}_v = \frac{1}{M} \sum_J \sum_I [\Pi]_I^{(II)} \sum_{l_d} \sum_{l_v} \sum_{n_d} \sum_{n_v} \sum_n \sum_{c_{v_1}} \sum_{c_{v_2}} P(l_d, l_v, n_d, n_v, n, c_{v_1}, c_{v_2})$$
$$\cdot (n_v - c_{v_1} - c_{v_2}) \tag{13}$$

where for the non-reservation system j_d is $\max(0, i_d - M) \leq j_d \leq N - i_v$, and for the reservation system j_d is $\max(0, i_d - M + R_v) \leq j_d \leq N - i_v$.

We define the loss probability of voice packets as the probability that a voice packet is involved in collision. Let L_v denote the loss probability of voice packets. By using ρ_v and $\bar{\rho}_v$, the loss probability of voice packets for both non-reservation and reservation systems is given by:

$$L_v = \frac{\bar{\rho}_v}{\rho_v + \bar{\rho}_v}. \tag{14}$$

We define the average packet delay as the average time elapsed from arrival of a data packet to the completion of its transmission and let $E[D]$ denote the average packet delay. As all N users are stochastically homogeneous, it is easy to see that the following balance equation holds true.
For the non-reservation system:

$$M\rho_d = \frac{\frac{\lambda_d}{\lambda_d + \lambda_v} N}{\frac{1}{\lambda_d + \lambda_v} - 1 + \frac{\lambda_d}{\lambda_d + \lambda_v} E[D] + \frac{\lambda_v}{\lambda_d + \lambda_v} \frac{1}{q_v}}. \tag{15}$$

For the reservation system:

$$(M - R_v)\rho_d = \frac{\frac{\lambda_d}{\lambda_d + \lambda_v} N}{\frac{1}{\lambda_d + \lambda_v} - 1 + \frac{\lambda_d}{\lambda_d + \lambda_v} E[D] + \frac{\lambda_v}{\lambda_d + \lambda_v} \frac{1}{q_v}}. \tag{16}$$

Solving Eqs.(15) and (16) for $E[D]$, we can obtain as
for the non-reservation system:

$$E[D] = \frac{N}{M\rho_d} + 1 - \frac{1}{\lambda_d}\left(1 - \lambda_v + \frac{\lambda_v}{q_v}\right), \tag{17}$$

for the reservation system:

$$E[D] = \frac{N}{(M - R_v)\rho_d} + 1 - \frac{1}{\lambda_d}\left(1 - \lambda_v + \frac{\lambda_v}{q_v}\right). \tag{18}$$

6   Numerical Results

Using the equations in Section 5, we can numerically compare the performance of the systems with and without multiple channels and reserved channels in several cases. For all cases, the number of users is $N = 30$ and all channels have the same bandwidth $v = V/M$, where V is the total bandwidth available to the system. A packet transmission time τ over a channel with bandwidth v, therefore, becomes $\tau = M\tau_0$, where τ_0 is the transmission time of a packet over a channel with full bandwidth V. The other parameters for the voice traffic are determined by referring to [6], where the mean talkspurt duration is 1.00 sec., the mean silent gap duration is 1.35 sec., and slot duration is 0.8 msec. In our model, the mean silent gap duration $1/\lambda_v$ corresponds to 1,688 slots, and $1/q_v$, the mean length of a user who is in speech mode, corresponds to 1,250 slots, so that we use $\lambda_v = 0.0006$ and $q_v = 0.001$ to provide performance curves.

In Figs. 1 and 2, we plot ρ_d and ρ_v for the non-reservation system as a function of the arrival rate of data packets λ_d ranging from 0.0 to 1.0. In general, numerical results may change with the retransmission rate of data packets μ_d and the generation rate of voice packets in the speech mode μ_v. As an example, we use $\mu_d = \mu_v = 0.1$ for all cases in Figs. 1 and 2. From Fig. 1, we observe that for smaller λ_d, the single-channel system ($M = 1$) offers higher channel utilization and has a peak in the neighborhood $\lambda_d=0.03$. But when λ_d is continuously increased over 0.03, the channel utilization of data traffic ρ_v decreases considerably. On the other hand, the multi-channel systems ($M > 1$) can maintain higher channel utilization even if λ_d increases up to 1.0. These are due to the fact that when λ_d is small, the access rate of users per channel in the case of $M = 1$ is actually greater than in the systems of $M > 1$, so that the utilization per channel becomes higher. However, as λ_d increases more, many more collisions of packets occur and the number of retransmission attempts increases in the single-channel system while the collisions of packets are actually lower per channel in the systems of $M > 1$, which results in the channel utilization of the single-channel system falling heavily. The behaviors of Fig. 1 are also observed in Fig. 2. The channel utilization of voice packets ρ_v in the single-channel ($M = 1$) system has the maximum utilization when λ_d is small and it decreases considerably with λ_d increasing. The reason is that in a non-reservation system all channels can be accessed by all voice and data packets. For the fixed arrival rate of voice packets λ_v, when λ_d is small, channel access rate of the users who have voice packets is in practice greater than that of the users who have data packets, so the systems have the higher channel utilizations. Packet collisions with data packets and therefore packet retransmissions become large when λ_d increases, the channel utilizations fall to low levels. Particularly the channel access rate of users for a channel in the single-channel system is the greatest, the channel utilizations fall to the lowest level when λ_d increases over 0.03. However, for the total same bandwidth, the multi-channel integrated systems perform better than the single-channel system and performance improvement of the multi-channel integrated systems can be obtained for larger λ_d.

In Figs. 3 and 4, we give ρ_d and $E[D]$ by setting $R_v = 0$ (non-reservation system) and $R_v = 2$ for $M = 5$ to compare the system's performance with and without reserved channels as a function of λ_d, where $\mu_d = \mu_v = 0.1$. In Fig. 3, we can see the case of $R_v = 2$ always offers higher channel utilization than the case of

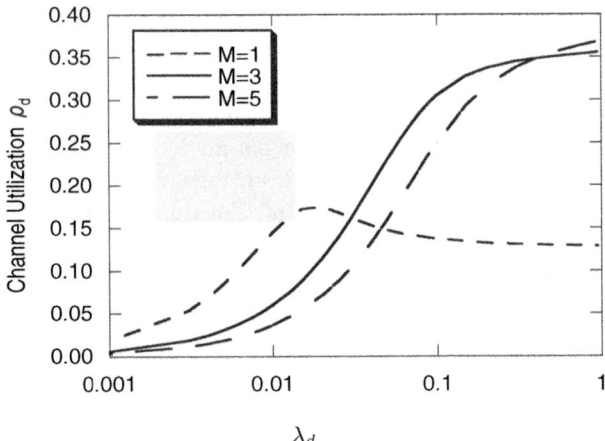

Fig. 1. Channel utilization ρ_d vs. λ_d for $M = 1, 3, 5$.

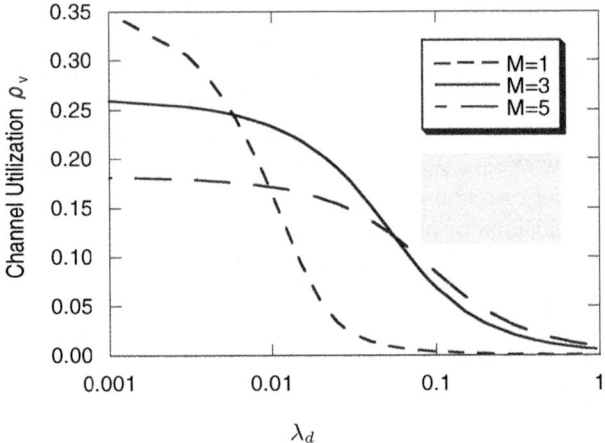

Fig. 2. Channel utilization ρ_v vs. λ_d for $M = 1, 3, 5$.

$R_v = 0$. The reason is because the access rate of data packets per channel for the reservation system is greater than that for the non-reservation system, so that the channel utilization of data packets per channel becomes higher. Fig. 4 shows the average delay of data packets $E[D]$ versus λ_d. The figure shows that if λ_d is small, the difference between the non-reservation system and the reservation system is negligible, but as λ_d increases, the difference becomes large. It should be noted here that in the integrated systems, by lending some reserved channels to voice traffic, we can obtain a significant impact on the system performance. Namely, (1) The effect of reserved channel in the multi-channel integrated system is more significant at higher data traffic; (2) The reservation system shows that

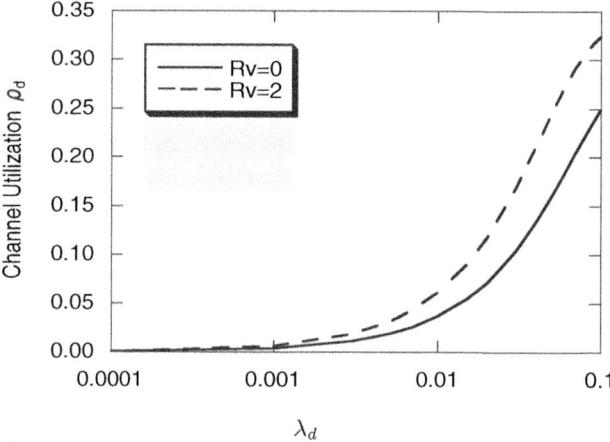

Fig. 3. Channel utilization ρ_d vs. λ_d for $M = 5, R_v = 0, 2$.

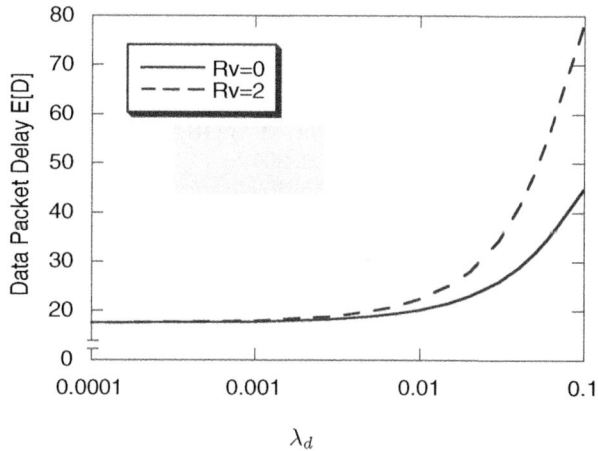

Fig. 4. Average data packet delay $E[D]$ vs. λ_d for $M = 5, R_v = 0, 2$.

it has high channel efficiency without sacrificing delay of data packets so much. In general, in an integrated multi-channel system of voice and data, the delay of data packets may be allowed to some extent, but if we increase the number of reserved channels for voice traffic too much, the loss of data packets will not be negligible.

7   Conclusions

This paper presented two multi-channel multiple access schemes for the slotted ALOHA operation with both data and voice traffic and their exact analysis to numerically evaluate the performance of the systems. In the non-reservation system, there was no limitation among data transmissions and voice transmissions, i.e., all channels can be accessed by all transmissions. In the reservation system, a channel reservation policy was applied, where a number of channels are used exclusively for voice packets, while the remaining channels are shared by both data packets and voice packets. Performance characteristics we obtained include channel utilization for both data and voice traffic, loss probability of voice packets and average delay of data packets. From the numerical results, we can conclude that the proposed multi-channel reservation scheme can improve the system's performance considerably. The analysis presented in this paper will be not only useful for the performance evaluation and the optimum design of multi-channel multi-traffic systems in wireless environments, but also applicable to evaluate other performance measures in priority networks, cellular mobile radio networks or multi-hop wireless networks.

References

1. Namislo, C. : Analysis of Mobile Radio Slotted ALOHA Networks. IEEE J. Selected Areas Communication, **SAC-2** (1984) 199–204
2. Szpankowski, W.: Packet Switching in Multiple Radio Channels: Analysis and Stability of a Random Access System. J. Computer Networks, **7** (1983) 17–26
3. Yue, W., Matsumoto, Y., Takahashi, Y., Hasegawa, T.: Analysis of a Multichannel System with Slotted ALOHA Protocol. J. The Transactions of the Institute of Electronics, Information and Communication Engineers, **J72-B-I**, No. 8 (1989) 632–641 (in Japanese)
4. Pountourakis I. E., Sykas, E. D.: Analysis, Stability and Optimization of ALOHA-type Protocols for Multichannel Networks. J. of Computer Communications, **15**, No.10 (1992) 619–629
5. Goodman,D.J., Valenzuela, R.A., Gayliard, K.T., Ramamurthi, B.: Packet Reservation Multiple Access for Local Wireless Communications. IEEE Transactions on Communications, **37**, No.8 (1989) 885–890
6. Feller, W.: An Introduction to Probability Theory and Its Applications Vol.I, Second Edition. John Wiley & Sons, Inc., New York (1957)

Distributed Quality of Service Routing

Donna Ghosh, Venkatesh Sarangan, and Raj Acharya

Department of Computer Science and Engineering,
State University of New York at Buffalo, NY 14260
{donnag, sarangan, acharya}@cse.buffalo.edu

Abstract. The goal of QoS routing algorithms is to find a loopless
path that satisfies constraints on QoS parameters such as bandwidth,
delay etc. In distributed QoS algorithms, the path computation is shared
among various routers in the network. These can be classified into two
categories depending on whether the routers maintain a global state or
not. Algorithms based on a global state information have less message
overhead in terms of finding a path. However they have inherent drawbacks like routing with imprecise global state information, frequent message exchanges to maintain the global state etc. Hence such algorithms
are not scalable. On the other hand, algorithms based on a local state
information rely on flooding techniques to compute the path. Hence they
have high overhead for finding a path. In this paper, we propose a distributed QoS routing algorithm, that maintains a partial global state
and finds a path based on this limited information. Experimental results
show that, the overhead of our algorithm is lesser than those that rely on
flooding. The results also show that the impreciseness introduced does
not affect the call admission ratio greatly.

1   Introduction

Currently, the networks provide routing mechanisms based on the shortest path
approach in terms of number of hops. However, future high speed networks
need to provide Quality of Service (QoS) guarantees to the connections. Hence,
they are likely to offer connection oriented services and routing mechanisms
based on the QoS requirements of the connection. These routing algorithms are
known as QoS Routing algorithms and can be categorized into the three main
classes of Source, Distributed and Hierarchical. Henceforth in this paper, the
term "routing algorithm" will refer to QoS routing algorithm unless specified.
Also, the terms "nodes" and "routers" have been used interchangeably in this
paper. In Source routing algorithms, the entire path computation is done at the
source. Hence, every router is required to maintain an upto date information
regarding the global network connectivity and the resource availability, known
as the global state information. However, global state information introduces
problems like routing with imprecise state information and need for frequent
message exchange in order to maintain the global state. Also, the computation
overhead at the source router increases.

M. Valero, V.K. Prasanna, and S. Vajapeyam (Eds.): HiPC 2000, LNCS 1970, pp. 439–448, 2000.

In Distributed routing algorithms (DRAs), the path computation is shared by all the intermediate routers. Hence, it is not computationally intensive for the source router. Various DRAs are proposed in [2], [5], [6], [7] and [8]. They can be broadly divided into two categories based on whether all the routers need to maintain global state information or not. However, DRAs which require global state information, when compared to Source routing algorithms, still do not overcome the problems of routing with impreciseness in the information and the frequent message exchange required for maintaining the information. On the other hand, DRAs which do not require global state information do away with these problems and rely on techniques like selective flooding [2] to establish a path. In this approach, an incoming request is flooded on all the outgoing links of the router (excluding the incoming link) which can satisfy its QoS requirements. However, the overhead in establishing a path for the connection could be very high due to selective flooding of the request at every intermediate router . In this paper, we propose a DRA in which every router stores a partial state information i,e. information about its immediate neighbors (reachable in one hop) and second degree neighbors (neighbors of a neighbor). The advantage of this approach is two-fold. Firstly, the message exchange required to maintain this partial state information and the impreciseness introduced are much lesser when compared to maintaining a global state information at every router. Secondly, using this partial state information, a router can now look two hops downstream. Hence it can now flood the incoming request more intelligently when compared to the selective flooding approach. Hence the overhead in connection establishment is reduced. This approach can also be extended so that a router stores information about its nth degree neighbor. However as we increase n, the impreciseness in the information stored by a router also increases proportionally.

The routing algorithm we propose has two separate tasks namely *Packet forwarding* and *Routing table maintenance*. The packet forwarding mechanism is responsible for forwarding the connection requests and uses a "two-level" routing table to do so. The table maintenance mechanism is responsible for constructing and maintaining the entries in this two-level routing table. This involves updating all the neighbors with a node's current state and processing the updates from the neighbors. This is explained in detail in Section 2. The packet forwarding mechanism is explained in Section 3. Experimental results aregiven in Section 4 and we conclude in section 5.

2   Routing Table Maintenance

We assume that (a)All nodes store their local metrics and (b)All nodes know when to send updates to their neighbors. Assumption (a) is valid since a node always knows the resources available in its outgoing links. Assumption (b) would become valid, if an update policy is prescribed. Various update policies are discussed in [1] and any of them could be used. In this paper, we have used an update policy based on *Thresholding* which will be discussed in detail in section 3.3.

Each node maintains a *Link-to-Node (LTN)* table. The *LTN* table basically gives which link to use to reach a given neighbor and the corresponding metric that the link can support. Links are assumed to be asymmetric i.e., the metric available in the forward direction need not be the same as that in the reverse direction. This table could easily be contructed by exchanging some kind of *Hello* packets. Each node on booting up constructs a Link-to-Node (LTN) table.

2.1 Building the Routing Table

Apart from maintaining a LTN table, each router also maintains a *Routing* (or) *Forwarding* table. On receiving a connection request probe, a router uses this forwarding table to decide on what outgoing links the probe must be forwarded. Let us consider a node v. Let,

- $N^1(v)$ denote those nodes that are adjacent to v in the network
- $E^1(v)$ denote the links that connect v to nodes in $N^1(v)$
- $N^2(v)$ denote those nodes that are adjacent to nodes in $N^1(v)$
- $E^2(v)$ denote the links that connect nodes in $N^1(v)$ to nodes in $N^2(v)$.

The forwarding table of v contains information about the metrics of all the links in $E^1(v)$ and $E^2(v)$. Entries corresponding to $E^1(v)$ are called the *first-level entries* R_v^1 of v; Entries corresponding to $E^2(v)$ constitute the *second-level entries* R_v^2 of v. The second level entries are represented as a tuple of the form $< l_i^1, l_j^2 >$ where $l_i^1 \in E^1(v)$ and $l_j^2 \in E^2(v)$. If a node say u is a member of both $N^1(v)$ and $N^2(v)$, it is represented only as a first level entry. In order to construct and maintain the routing table, node v must receive updates about the QoS metrics in all its second-level entries from the nodes in $N^1(v)$. This is done by exchanging special messages called *Hello2* packets at a frequency determined by the update policy used. These *Hello2* packets are constructed by copying the neighbor list and the available QoS metrics from the *LTN* table of the router. At a node v, the first-level entries in the routing table are made by copying the *LTN* table of v. The second-level entries in the routing table are made by inspecting the received *Hello2* packets. Node v on receiving a *Hello2* packet from node u, reads from it the neighbor list of u and the available QoS metrics on the corresponding links of u, and then discards the *Hello2* packet. Node v updates all its existing second-level entries in the routing table by using the information read from the *Hello2* packet. Also, any new entry is added to the existing second-level entries.

2.2 Update Policies

The update policy used decides when these *Hello2* packets are sent. A simple update policy could be based on timers such that an update is sent every T seconds. The disadvantage with this approach is that, the efficacy is dependent on the rate at which the connection requests arrive. Also it is very difficult to model the impreciseness in the table entries with such an update mechanism.

A detailed survey on various update policies can be found in [1]. The update policy used in this work is the one suggested in [4]. Each node remembers the last advertised metric on each link. If the ratio between the last advertised value and the current value is above (or below) a threshold T, an update is triggered. The node constructs *Hello2* packets and sends them to all its neighbors. The advantage of using this policy is that, the impreciseness can be easily modeled using probabilities. If bandwidth b is advertised on a link, then at any time, the actual metric available on that link could be modeled as a random variable with an uniform distribution in $[b/T, bT]$. Once the impreciseness is modeled, there are approaches given in [4] and others to do efficient routing with such imprecise information. However, our algorithm assumes that the information available in the tables is accurate and forwards the probes accordingly. As a result, the performance of the algorithm in terms of call establishment might be poorer. Experimental results in Section 4 show that, the performance degradation is not much.

3   Packet Forwarding Mechanism

The forwarding mechanism suggested could be used for any QoS metric. In this paper, bandwidth is taken as the QoS metric and all discussions and results are with respect to bandwidth. An outline of the packet forwarding mechanism is given in Section 3.1. and a more formal presentation is given in the appendix.

3.1   An Outline

Each node v maintains a routing/forwarding table in which two kinds of entries are present, $R^1(v)$ and $R^2(v)$. R^1_v is the set of entries corresponding to the immediate neighbors of v. R^2_v is the set of entries corresponding to the second degree neighbors, namely $N^2(v)$. A neighbor u, of node v, is said to be *eligible*, if the link (v, u) can support the requested bandwidth. The connection set-up process has three phases namely *Probing, Ack and Failure handling*. The probing phase is essentially the QoS routing and it establishes a *tentative path* between the source s and destination t such that the path satisfies the QoS requirements of the request. Each connection is identified by a unique connection id, *cid*. The probes are represented by a tuple $[k, QoS(Bandwidth = B), s, t, cid, \{l\}]$. This is interpreted as s is the source requesting a connection *cid* with metric requirements $QoS(Bandwidth = B)$ to destination t and k is the router that has forwarded the probe to the current node v. $\{l\}$ refers to the list of nodes to which v much forward the probe. All the nodes process only the first probe they receive for a connection and discard all the duplicates. On getting the first probe for a connection, node v determines whether the destination t is present in $R^1(v)$ or $R^2(v)$. If t is present in either of these, v forwards the probe along all the corresponding, eligible links. If t is not in the routing table, v examines the list $\{l\}$. It forwards the probe to all eligible neighbors in the list. If t is not in the routing table and if $\{l\}$ is empty, v constructs a list on its own. For an eligible neighbor u, a list $\{l_u\}$ of u's eligible neighbors is constructed. Node v then forwards a probe with $\{l_u\}$ to u. This is repeated for all eligible neighbors. This process continues in all the routers until the destination gets a probe. The path taken by the first probe to reach the destination is called as *tentative path*. During the probing phase only the resource availability is checked and no reservations are made. A router v stores the upstream router's id, $p_v(cid)$, that had forwarded the probe to it, in a table.

The destination t, sends an acknowledgement to the first probe it receives and discards all the duplicates. This ensures that only one path is established. The ack is propagated upstream along the *tentative path* (using the $p_v(cid)$ values) and resources are reserved along the way. A connection is established when the source gets an acknowledgement. All nodes store the downstream router's id referred to as $n_v(cid)$ (for node v) that had forwarded the ack to them. If any router in the *tentative path* is unable to reserve the required resources, it starts the failure handling phase by sending a *fail* message to the downstream router using the $n_v(cid)$ values. A router on receiving the *fail* message tears down the resources reserved for connection cid and forwards the *fail* message to its downstream router.

The destination sends an ack only to the first probe it receives and discards all the duplicates. This makes sure that the resources are reserved only along one path. Also, a router does not forward a probe more than once. This means that the tentative path found by the algorithm will be loop free. We also assume that these control messages are never dropped in the case of a congestion.

A more formal description of the algorithm is given in the appendix.

3.2 Iterative Two-Level Forwarding

Flooding based approach finds a tentative path through the competition among probes. If the network load is light, it is not a wise idea to blindly flood the probes on all eligible links. Often, it is only the shortest eligible path that is preferred. To direct the search along the shortest path, an approach was suggested in [2]. Each probe is assigned an *age*. The age of a probe p is defined as the number of hops the probe has traveled. Initially, $age(p) = 0$. Whenever p reaches a node, the age field is incremented by 1. In order to direct the search along the shortest path, the probe forwarding condition on link (i, j) at node i is modified as:-

$$forward\ condition\ on\ link(i,j) \rightarrow bandwidth(i,j) \geq B \wedge (age(p)+d_{j,t}+1 \leq L)$$

where $d_{j,t}$ is the shortest distance in terms of hops between the node j and destination t; L is a constant at least as large as $d_{s,t}$ and is the maximum age attainable by a probe. This forwarding condition would make the nodes flood the requests only along the shortest paths. Hence, this results in a much less overhead when the network load is light. When the network becomes heavily loaded, it is unlikely that this shortest path approach will succeed in establishing a path. Hence, if no path is established using $L = d_{s,t}$, the source makes a second attempt for the same connection with $L = \infty$. Flooding with $L = \infty$ is equivalent to flooding the probes blindly to all eligible nodes. In our simulation, all the sources make only one attempt with $L = d_{s,t}$.

Similar to the iterative flooding approach, we could also have an iterative approach for forwarding using the two-level table. The definition of an eligible node is modified similarly. If i is the current node of interest, a neighbor j is eligible if $bandwidth(i, j) \geq B$ and $d_{j,t} + age(p) + 1 \leq L$.

4   Experimental Results and Discussion

The motivation behind opting for a two-level routing table based forwarding mechanism (henceforth referred to in the paper as two-level forwarding) is to save in the message overhead due to flooding during the connection setup phase. However, in the process of maintaining this two-level routing table, a cost of message exchange is incurred. For an overall reduction in the message overhead when compared to selective flooding approaches [2], the number of messages exchanged for maintaining the two-level table should be much less than those saved due to intelligent flooding during the call setup phase. Our simulation results show that this is indeed the case.

Extensive simulations were done on various network topologies to compare the total message overhead in the two-level forwarding and the selective flooding approaches. Due to space constraints, results are reported only from the network topology of a standard ISP [3] given in figure 4.1b, and a triangulated mesh network topology given in figure 4.1c. We believe that these two collectively give a fair representation of the various real life topologoies that could be encountered. The simulations were done using OPNET. Each link is duplex and has a capacity of 155Mbps (OC-3). The background traffic has been set in the range [0,155Mbps]. All simulations were run for 2000 connection requests, which were generated as per a Poisson distribution and are for a constant time duration. The bandwidth requests are uniformly distributed in the range [64Kbps, 1.5Mbps]. Each node in the network generates a connection request for every other node with equal probability. The results could be divided into two sets. The first set is the comparison between the non-iterative versions of selective flooding and two-level forwarding. Second set is the comparison between the iterative versions.

4.1   Performance of the Non-iterative Versions

The graph given in figure 2a shows the performance of the non-iterative versions of the two approaches on MESH-I. In this graph and the subsequent, T is the threshold value used in the update policy. It is clear that, two-level forwarding has very low overhead (per call-admitted) when compared to selective flooding. When the available bandwidth in the network is less, the overhead increases significantly as the threshold T is reduced. However, when the network load is light (or the available bandwidth is high), the value of T does not affect the overhead. This behavior could be explained as follows: When the available bandwidth is less, the *Current Available bandwidth / Last advertised bandwidth* ratio will fluctuate significantly with each admitted call. As a result, if a low T value is used, the routers will send updates more frequently than at high T values. Hence, at high network loads, the overhead is high for low T values. If the load on the network is less, the available bandwidth in each link will be high and the *Current Available bandwidth / Last advertised bandwidth* ratio will not fluctuate much with each admitted call. Hence, the routers tend to send updates less frequently irrespective of the T value.

The graph in figure 2b shows the *bandwidth admission ratio* for the two non-iterative versions on MESH-I. Bandwidth admission ratio is defined as the ratio of bandwidth admitted into the network to the total bandwidth requested. The graph shows that, when the load is light, both the approaches perform almost equally well. However, when the traffic is heavy, forwarding based on the two-level table admits less bandwidth into the network than flooding. Also, the bandwidth admitted by the two-level approach reduces as T increases. The reason is that, the impreciseness in the table information increases with the value of T. This impreciseness makes the routers have a conservative estimate of bandwidth available in the second level links. Hence, a router discards probes even though the second level links are capable of supporting these requests. At low T values, the impreciseness reduces. So the routers tend to be less conservative and admit more bandwidth into the network. Figure 3 shows the performance of the non-iterative version of the two approaches on an ISP topology. On the ISP topology, the bandwidth admitted by the two-level forwarding is slightly lesser than the blind flooding. This could be attributed to reduced average node connectivity in the network.

4.2 Performance of the Iterative Versions

The motivation behind the iterative technique is to reduce unnecessary flooding in the network. In the iterative approach, the probes are given a maximum age of $L = d_{s,t}$, where $d_{s,t}$ is the smallest number of hops between the source s and destination t. Our algorithm's perfromance on MESH-I is given in figure 4. It is clear that the two-level forwarding has much less overhead when compared to

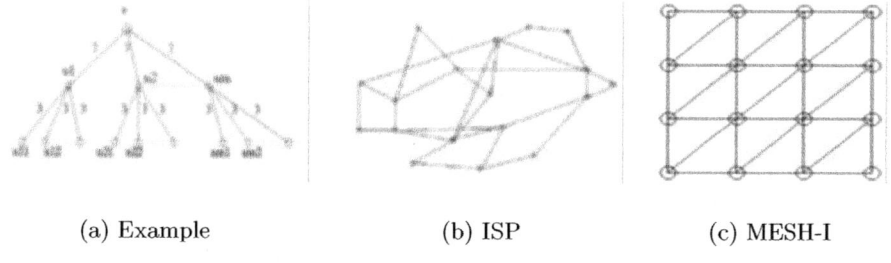

(a) Example   (b) ISP   (c) MESH-I

Fig. 1.

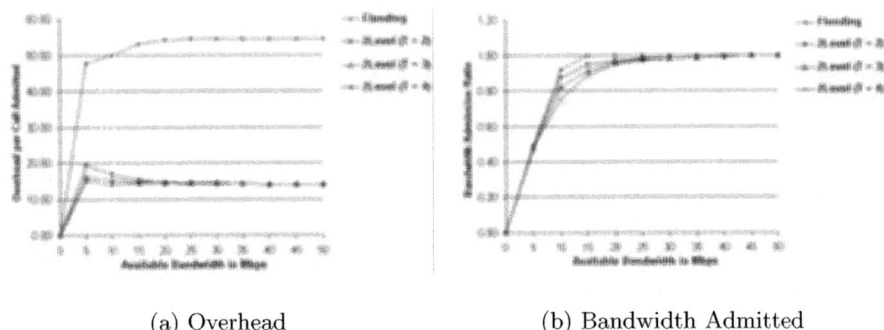

(a) Overhead   (b) Bandwidth Admitted

Fig. 2. Non-iterative : Performance on MESH - I

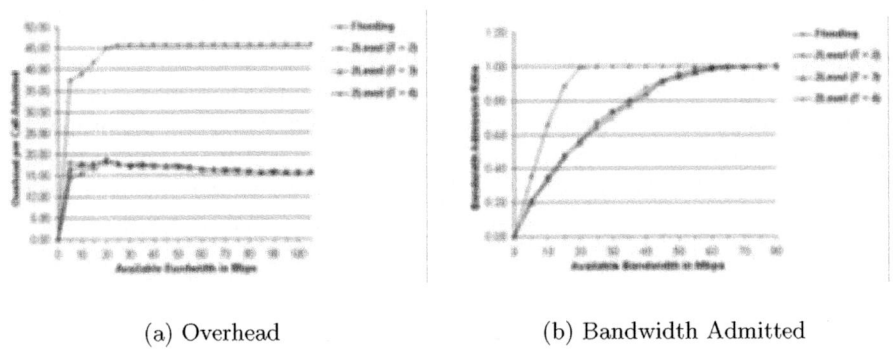

(a) Overhead   (b) Bandwidth Admitted

Fig. 3. Non-iterative : Performance on ISP

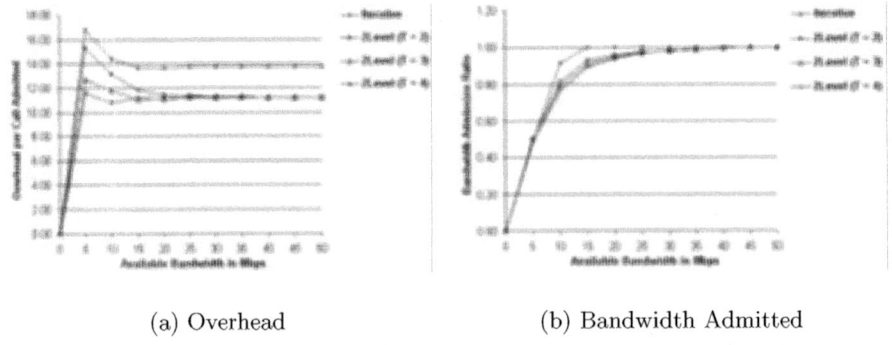

(a) Overhead   (b) Bandwidth Admitted

Fig. 4. Iterative (L) : Performance on MESH - I

flooding. Also, the iterative overhead is much lesser than the corresponding non-iterative overhead. This is consistent with the fact that, the probe forwarding is limited by the age. The behavior of the iterative versions with respect to bandwidth admitted in the network is very much like that of the non-iterative versions. However, at a given network load, the bandwidth admitted by the iterative versions is lesser than or equal to that of the non-iterative versions. This is also understandable, since in the iterative versions, the probes are directed only along the shortest path(s) . If the links along the shortest path(s) are unable to support the request, the call is dropped. On the other hand, in the non-iterative versions, probes are free to take any path that satisfies the bandwidth requirement. Hence, more bandwidth is admitted by the non-iterative versions. Figure 5 shows the performance of the iterative versions on an ISP topology.

5   Conclusion and Future Work

In this paper, we have proposed a new distributed QoS routing algorithm, in which a probe is routed based on the knowledge about resource availability in the second-level links. This two-level forwarding has less overhead when compared to a flooding based call set-up. Also, the impreciseness in the information stored, does not greatly affect the total bandwidth admitted in the network. Future work would involve extending the proposed algorithm to do forwarding, taking into consideration the impreciseness in the table entries.

References

1. Apostolopoulos G., Guerin R., Kamat S., and Tripathi S. K., Quality of Service based routing: A performance perspective in *Proceedings of SIGCOMM, Vancouver, Canada*, September 1998.
2. Chen S., Nahrstedt K., Distributed QoS Routing in High-Speed Networks based on Selective probing. *Technical Report, University of Illinois at Urbana-Champaign, Department of Computer Science*, 1998.
3. Comer, D. E., *Internetworking with TCP/IP Volume I, Principles, Protocols and Architecture*, Prentice Hall, 1995.
4. Guerin R., and Orda A., QoS Routing in Networks with Inaccurate Information: Theory and Algorithms, *INFOCOMM '97, Japan April 1997.*
5. Salama F., Reeves D.S., and Vinotis Y., A distributed algorithm for delay-constrained unicast routing. *Proceedings of the 4th International IFIP Workshop on QoS*, March 1996
6. Shin K.G., and Chou C.C., A Distributed Route Selection Scheme for Establishing Real-Time Channel. *Sixth IFIP Int'l Conference on High Performance Networking (HPN '95)*,pages 319-325, Sep. 1995.
7. Su Q., Langerdorfer H., A new distributed routing algorithm with end-to-end delay guarantee, *Proc. of 4th International IFIP Workshop on QoS*, March 1996.
8. Wang Z., and Crowcroft J., QoS Routing for Supporting Resource Reservation, *IEEE Journal on Selected areas in Communications*, September 1996.

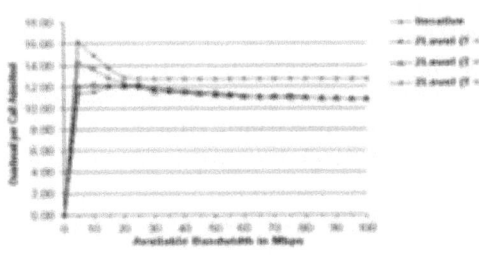

(a) Overhead

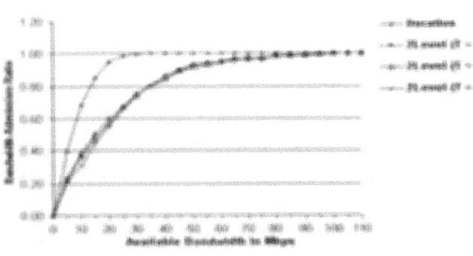

(b) Bandwidth Admitted

Fig. 5. Iterative (L) : Performance on ISP

Forwarding at node i
begin
while true **do**
block until receiving a message
switch (the received message at i)
case 1: $probe[k, Q, s, t, cid, \{l\}]$
if this is the first probe for
this cid
mark that i has forwarded a
probe for cid and $p_i(cid) = k$
if $i \neq t$
if t is present in R_i^1
if $bandwidth(i, t) \geq B$
send t a $probe$
$[i, Q, s, t, cid, \{ø\}]$
else
for every $j \in R_i^1$ & $j \neq t$
do
if $bandwidth(i, j) \geq B$
then
send j a $probe$
$[i, Q, s, t, cid, \{ø\}]$
end if
end for
end if
else if t is present in R_i^2,

for all $\{(i, l).(l, t)\} \in R_i^2$, **do**
if $bandwidth(i, l) \geq B$ &
$bandwidth(l, t) \geq B$
send l a $probe$
$[i, Q, s, t, cid, \{ø\}]$
end if
end for
else if probe is marked with
next hops
for every hop l **do**
if $bandwidth(i, l) \geq B$
send l a $probe$
$[i, Q, s, t, cid, \{ø\}]$
end if
end for
else if probe is not marked
with next hops,
for all $\{(i, l), (l, m_r)\} \in R_i^2$
do
if $bandwidth(i, l) \geq B$ &
$bandwidth(l, m) \geq B$
aggregate all paths through
the same next hop l;
construct a probe for
every such aggregation;
send l a $probe$
$[Q, s, t, cid, \{l\}]$ where
$l = \{m_1, .., m_r, .., m_n\}$
end if
end for
end if
else if $i = t$
send k an $ack[i, Q, s, t, cid]$
end if
else
the probe is a duplicate &
discard it
end if
case 2: $ack[k, Q, s, t, cid]$
if $bandwidth(i, k) \geq B$,
reserve a bandwidth of B on
link (i, k) for connection cid
$n_i(cid) = k$
if $i \neq s$
send $p_i(cid)$ an $ack[i, Q, s, t, cid]$
else
the connection is established
end if
else
send k a $failure[i, Q, s, t, cid]$
end if
case 3: $failure[k, Q, s, t, cid]$
if $i \neq t$
release the bandwidth reserved
on link $(i, n_i(cid))$ for connection
cid
send $n_i(cid)$ a $failure[i, Q, s, t, cid]$
end if
end switch
end while
end

Partitioning PCS Networks
for Distributed Simulation *

Azzedine Boukerche and Alessandro Fabbri

Parallel Simulations and Distributed Systems (PARADISE) Research Lab
Department of Computer Sciences, University of North Texas, Denton, TX. USA
{boukerche, fabbri}@cs.unt.edu

Abstract. In this paper, we study the load balancing problem for PCS simulation systems, and focus upon static strategies in order to reduce the synchronization overhead of SWiMNet, a parallel PCS simulation testbed devoloped at UNT. Performance results on a cluster of workstations indicate clearly the significant impact of load balancing on the efficient implementation of SWiMNet and its two stage design.

1 Introduction

Modeling and Simulation are frequently used as an approach to cope with the variety of details typical of complex wireless and mobile systems. Recently, in [3,4], we proposed a distributed simulation testbed (*SWiMNet*) for PCS wireless networks that uses a hybrid approach of conservative and time warp schemes. Our approach differs from previous work [5,8,10,12,14] in that it exploits event precomputation, and model independence within the PCS model. Experiments were conducted to validate the developed approach and promising results were reported.

As is well known, time warp is unstable due to many factors such as rollbacks and its devastating effect (e.g., serie of cascading rollbacks). In this paper, we focus upon upon the problem of partitioning the PCS model using SWiMNet, and mapping the model onto a cluster of workstations. The algorithm is based upon numerical estimate of cell-related loads computed as a function of PCS model parameters. Hence, our primary goal is to study the impact and the importance of partitioning in our PCS model while reducing significantly the number of rollbacks. Reducing the rollback; will help to stabilize Time warp. Future work will be directed at studying the other factors, such as memory management resulting from the necessity to save state saving. We must point out that the performance of Time warp doesn't result necessarily only in Speed. Stability, reducing rollbacks (and its devastating effect), memory managements, etc.. are very important issues that any time warp developers must deal with.

2 The Simulator's Architecture

In [3], we presented and studied SWiMNet. It is a parallel PCS network simulator which exploits event precomputation due to the following assumption: *mobility*

* This work was supported by the UNT Faculty Research Grant

M. Valero, V.K. Prasanna, and S. Vajapeyam (Eds.): HiPC 2000, LNCS 1970, pp. 449–458, 2000.
© Springer-Verlag Berlin Heidelberg 2000

*and call arrivals are independent of the state of the PCS network, and they
are independent of each other.* Processes composing SWiMNet are grouped into
two *stages*: a *precomputation stage* (Stage 1), and a *simulation stage* (Stage 2).
Processes in Stage 1 precompute all possible events for all mobile hosts (MH)
assuming all channel requests are satisfied. Possible events are: (1) *call_arrival*,
(2) *call_termination*, (3) *move_in*, and (4) *move_out*. Precomputed events are
sent to Stage 2, where their occurrence is checked according to the state of the
PCS network simulation. Stage 2 processes cancel events relative to calls which
turn out to be blocked or dropped due to channel unavailability.

The model partitioning in Stage 1 is mobile host-based, because each Stage 1
process produces all the precomputed events of a group of MHs. Events relative
to different MHs can be precomputed independently, therefore Stage 1 processes
do not interact. On the other hand, the partitioning of the model in Stage 2 is
cell-based, since each Stage 2 process elaborates precomputed events occurring
at a set of cells. Since events for the same call can span various cells, Stage 2
processes need notify other Stage 2 processes of blocked calls.

Process coordination in SWiMNet is based on a *hybrid* approach, using both
conservative and optimistic techniques. A conservative scheme is used for communication from Stage 1 to Stage 2: Stage 1 processes send precomputed events
in occurrence time order, and periodically broadcast their local simulation time
to Stage 2 processes through *null messages*. Stage 2 processes can consequently
establish which precomputed events are safe to be simulated. The low percentage of blocked calls in PCS networks is exploited by an optimistic scheme for
communication within Stage 2. A Stage 2 process optimistically assumes that
all handed-off calls to that cell are still allocated in the previous cell. If this
assumption holds, no message must be sent. Otherwise, a notification of the
contrary must be sent, which may trigger a *rollback* to correct effects of a wrong
assumption if the message is *straggler*.

In our mobility model, MHs are classified into groups of *workers, wanderers,
travelers* or *static users*. MHs of a specific group show similar behavior. The call
model is specified through intervals of different call ratios during the simulation,
to represent congested hours as well as under-loaded hours. The cell topology is
specified by a grid embedded in the PCS coverage area, with fine resolution of
50 meters which allows the representation of arbitrary cell sizes and shapes.

In the course of our experiments, we investigated several partitioning strategies. The primary goal of these partitioning schemes is to minimize the rollback
overhead. In the next section, we describe two possible partitioning strategies
for which we report results.

3   Load Balancing in SWiMNet

Partitioning and subsequent load balancing of the simulator described in Section
2 must be concerned with MH-based partitioning in Stage 1, and cell-based
partitioning in Stage 2. The general problem can be stated as follows: given
that the simulator must be composed of n_1 Stage 1 processes and of n_2 Stage
2 processes, find a partitioning of the set of n_{MH} MHs to Stage 1 processes,

and of the set of n_{cell} cells to Stage 2 processes. The partition should be such that if each process is executed on a different processor, and the communication between processes is homogeneous, the global performances of the simulator are optimized.

Finding a partion of the model for Stage 1 processes which is close to optimal from the point of view of load balancing is fairly easy. In fact, MHs of a group are modeled in such a way that they have similar behavior throughout the simulation. In addition, MHs do not interact with each other. Therefore, whatever load measure is used that assigns similar loads to similar MHs, the contribution to the total load of each process given by MHs of a given group can be balanced by simply partitioning MHs of that group evenly among all Stage 1 processes. Therefore, Stage 1 is partitioned by evenly distributing MHs of the same group to all Stage 1 processes.

On the other hand, loads for different cells which are at the basis of the load for processes in Stage 2, can vary a lot. In particular, cells can be very different relatively to their coverage, the channels they are assigned, and MH movements in that area. In a round robin scheme, cells are assigned to the n_2 Stage 2 processes through a round robin scheme, considering cells in the order supplied by the user in the PCS model parameters.

The Load-based Assignment scheme we prpose to investigate tries to improve the round robin scheme. It relies on an estimation of cell loads as we shall see later. The algorithm consists of computing an initial cell assignment, which is then improved in a sequence of steps. The purpose of this algorithm is to minimize the maximum load of all Stage 2 processes. Therefore, it tries to find an approximate solution to the NP-hard partitioning problem, where the objective function F to be minimized is: $F(Stage2) = \max_{P \in Stage2} Load(P)$, where $Stage2$ is the set of n_2 Stage 2 processes, and $Load(P)$ is the estimated load of process P.

The method we chose for computing an initial solution consists of assigning cells to processes in a round robin fashion, considering cells in order of estimated load. After estimating their loads, cell numbers are pushed onto a stack. Then, cell numbers are popped from the stack and assigned to Stage 2 processes. The pseudo code of this algorithm is described in Figure 1.

At each subsequent step of the algorithm, one cell is moved from the process with the largest total estimated load to the process with the smallest total estimated load. The algorithm chooses the cell to move by trying all the cells of the process with the largest load, one cell at a time. The cell movement which gives the smallest maximum load is moved at this step, and the algorithm proceeds to the next step. The algorithm terminates when no improvement is possible. The pseudo code of this algorithm is described in Figure 2.

The load contribution of each cell to Stage 2 processes of the simulator is due to the receipt and elaboration of precomputed events and and the receipt of simulation events, and subsequent rollbacks.

If the unit of measure of the load is assumed to be the time required for receiving and elaborating one precomputed event, a fair estimation of the load for cell k is given by the number of precomputed events, E_k, plus the cost of rollbacks

Compute estimated cell loads
Order cells in increasing estimated loads
Push cells numbers into a stack
Initialize the current processor to 1
While *(Stack S is not empty)* **Begin**
   Pop 1 cell nber from S and assign it to the current processor.
   Increment the current processor number
   If *current process number* $> n_2$ **Then**
     Set current process to 1
End While

Fig. 1. Initialization Algorithm

Find the initial partitioning $S_{Initial}$
Set $S_{Curr} = S_{Initial}$
Find $P_{MaxL} = MaxProc(S_{Curr})$
Find $P_{MinL} = MinProc(S_{Curr})$
Set $F_{Prev} = Load(P_{MaxL})$
Set $F_{Curr} = 0$
While $(F_{Curr} < F_{Prev})$ **Begin**
   Set $Trials = \emptyset$
   For Each $c \in P_{MaxL}$ **Begin**
     Set $S_{New} = S_{Curr}$
     Move c from P_{MaxL} to P_{MinL} in S_{New}
     Save S_{Trial} into $Trials$
   End
   Find $S_{Best} = BestPartition(Trials)$
   Set $S_{Curr} = S_{Best}$
   Set $P_{MaxL} = MaxProc(S_{Curr})$
   Set $P_{MinL} = MinProc(S_{Curr})$
   Set $F_{Prev} = F_{Curr}$
   Set $F_{Curr} = Load(P_{MaxL})$
End
S_{Curr} is the computed partition

Fig. 2. Partitioning Algorithm

as number of re-elaborated precomputed events, R_k. Thus, the total load for a process P managing a set of cells $Cells(P)$ is $Load(P) = \sum_{k \in Cells(P)}(E_k + R_k)$. The next subsections present the estimations of E_k and R_k for a cell, given the parameters dsiplayed in Table 1.

Precomputed events are of four types: (1) MH move-in during a call; (2) MH move-out during a call; (3) call-arrival; and (4) call-termination. The estimation of move-in and move-out events requires the estimation of the handoff probability at a cell. Given that the average density of MHs in the cell is known, together with their average speed and call duration, mobile hosts which contribute to handoffs are those residing close enough to the cell border that their movement brings them across the border during a call. Therefore, MHs prone to handoffs are only those inside the cell perimeter up to the distance traveled during a call.

Table 1. Notations

D	Average MH density of the cell (MH /Km^2).
S	Average speed of MHs in the cell (Km/hr).
L	Average call duration (hr).
C	Average call traffic per MH (calls/hr/MH).
T	Total simulation time.
A_k	Surface area of cell k.
P_k	Perimeter of cell k.

If α is the incidence angle of a MH path to the border, the MHs moving at an angle α that are prone to handoff are averagely those at a distance from the border equal to $S \times L \times \sin \alpha$. Supposing that all directions are equally likely for MHs, only half of MHs must be considered, i.e., those directed toward the border. Therefore, the average distance at which MHs are prone to handoffs is:

$$\frac{1}{2} \times \int_0^\pi S \times L \times \sin x \, dx = S \times L$$

Considering such a distance from the perimeter, a rough estimation of the average ratio of the MHs residing in cell k causing handoffs during a call is:

$$\frac{P_k \times S \times L}{A_k}$$

Such a ratio must be multiplied by the ratio of time that each MH spends calling, i.e., $L \times C$. Hence, the probability H_k that a call at cell k has an handoff is given by:

$$H_k = \frac{P_k \times S \times L^2 \times C}{A_k}$$

The estimated total number of calls occuring at cell k along simulation time T, C_k, is thus $C_k = A_k \times D \times C \times T$. The total expected number of calls causing handoff at cell k is given by $H_k \times C_k$. Therefore, the total precomputed event load E_k can be estimated as $E_k = C_k \times (1 + H_k) \times 2$.

The multiplication by two is due to the fact that there exist two events per call (call-arrival, call-termination), and per handoff (move-in, move-out).

Let us define by the number of rollbacks, the number of events causing a rollback. Recall that a rollback is caused by a call that was supposed to be on due to the pre-computation assumption, and turned out to be off (blocked) instead. Therefore, the number of rollbacks correspond to the number blocked calls. Let us also denote by rollback load the elaboration cost per rollback. Recall that for each rollback there is a re-elaboration of those events occurring during the call that caused the rollback. It is possible that such re-elaboration involves rollbacks for other local calls, thereby further re-elaborations. However, the number of additional re-elaborations is hard to estimate analytically. Thus, we estimate the rollback load as the average number of events occurring during a call at a cell.

Since E_k is the estimated total number of events occuring during the total simulation time, the number of events occuring during a call with average call duration L, r_k, is

$$r_k = \frac{E_k \times L}{T}$$

where L/T gives the fraction of total simulation time of an average call.

Given that B_k is the blocking probability at a cell (see next subsection), the total number of blocked calls for a cell k is given by $B_k \times C_k$. Hence, the total rollback load R_k for a cell k is given by: $R_k = B_k \times C_k \times r_k$

The blocking probability is computed from the Erlang-B model [13] as

$$Pr[\text{Blocking}] = \frac{Tr^{Ch}}{Ch!} \Big/ \sum_{i=0}^{Ch} \frac{Tr^i}{i!}$$

where Tr is the traffic unit in Erlangs offered to the system, and is given by $\frac{Q \times L}{60}$, Q is the number of calls per hour, estimated as $A_k \times D \times C$. and Ch is the number of channels in the cell.

4   Experimental Results

In our experiments, we selected a real suburban city area serviced by a TDMA based PCS network[1]. The PCS coverage map is of $11,000 \times 11,000$ meters, and is composed of a total of 54 cells, each with 30 channels. Call arrivals to each MH are a Poisson process with arrival rate $C = 4$ calls/hr/MH. The average call holding time is exponentially distributed with a mean $L = 120$ seconds. A granularity of 50 meters for direction and location change is used.

Concerning the mobility model, we used one group of workers, one group of wanderers, two groups of static users and three groups of travelers, located according to the presence of a town on the map. The downtown is placed at the center of the map, where base stations are more dense, and three heavy traffic highway branches are supposed to join to a loop around the town. The description of each mobile group is as follows:

- 2591 Workers reside in an annular area around downtown (the periphery), with a density of 100 users/km^2.
- 3430 Wanderers reside in a square area which includes the periphery of the town where they are uniformly distributed, with a density of 70 wanderers/km^2.
- Static users reside in the square area depicted at the bottom/top 3), uniformly distributed with a density of 20 users/km^2, where 880 static users were assigned to the bottom of the map and 1120 static users to the top left of the map.
- 1000 Travelers are distributed with a density of 40 users per km^2.

These parameters correspond a total of 9051 MHs, and a density of approximately 75 MH/km^2.

---

[1] The BS deployment model was acquired from a telecommunication company.

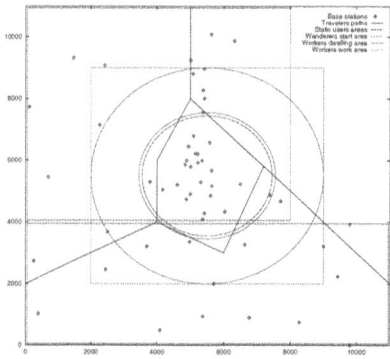

Fig. 3. Mobility and Cell Model

The goal of our experiments is to study the impact of our partitioning algorithm on the performance of SWiMNet. SWiMNet is implemented on a cluster of 16 Pentium II (200Mhz) workstations running Debian/Linux, connected via 100 Megabit Ethernet. The cluster has 8 nodes that are configurable for gigabit Ethernet connection. We made use of the LAM/MPI parallel processing system [11] to implement message passing in the simulator. A principal motivation for using MPI is to have a flexible and easily maintainable simulator.

In our experiments, we varied the call arrival from 1 to 4 calls. We consider a heavy and light load, i.e. 30 and 50 channels, and varied the MH density from 50, to 60, 75, and 90 to study the behavior of SWiMNet. In our experiments, we varied the number of processors from 4 to 16. The experimental data which follows was obtained by averaging several trail runs. We present our results below in the form of graphs of the rollback overhead as a function of the number of processors employed in the model for several arrival calls, and mobile density hosts.

First, some general comment about rollback messages. As Figures 4-7 show, by increasing the MH population (from 50 to 90 MH per square Km), SWiM-Net experiences in general higher rollback loads. This is evident in each figure, because curves for larger MH densities are generally above those for smaller densities. This is due to an indirect increase in the call traffic at the BSs due to the presence of more MHs. An increased call traffic results in a larger number of blocked calls, which in turn results in a larger number of precomputed events to be cancelled by the simulation, hence a larger number of messages which possibly cause rollbacks.

On the other hand, differences (if measured in percentage) among different MH densities tend to disappear as the call traffic per MH increases. In fact, by moving to figures relative to larger numbers of calls/hr/MH, the relative gap between curves for different MH densities seem to flatten out, although the number of rollback messages increases. This is mainly due to the model we used, which may cause a fixed amount of blocked calls, hence of rollback loads, almost independently of channel availability but dependent on MH density. As

the channel utilization is stressed by increasing call traffic, the induced rollback
load overwhelms the density-dependent one.

Our experiments (to our surprise) indicate that the rollback messages decrease as we increase the number of processors. This is evident in Figures 4, 5
and 7. Since each figure represents all results obtained by using one choice of
model parameters, by distributing the model over a larger number of processors
there should be more non-local messages, hence the relative drift of processes
and message timestamps should be larger. These causes should all increase the
chances of rollbacks, on the contrary rollback loads seem to decrease. We believe
that this behavior is due to the necessary rearrangement of cells to processes
due to the partitioning. We are planning to further investigate the causes of this
phenomenon.

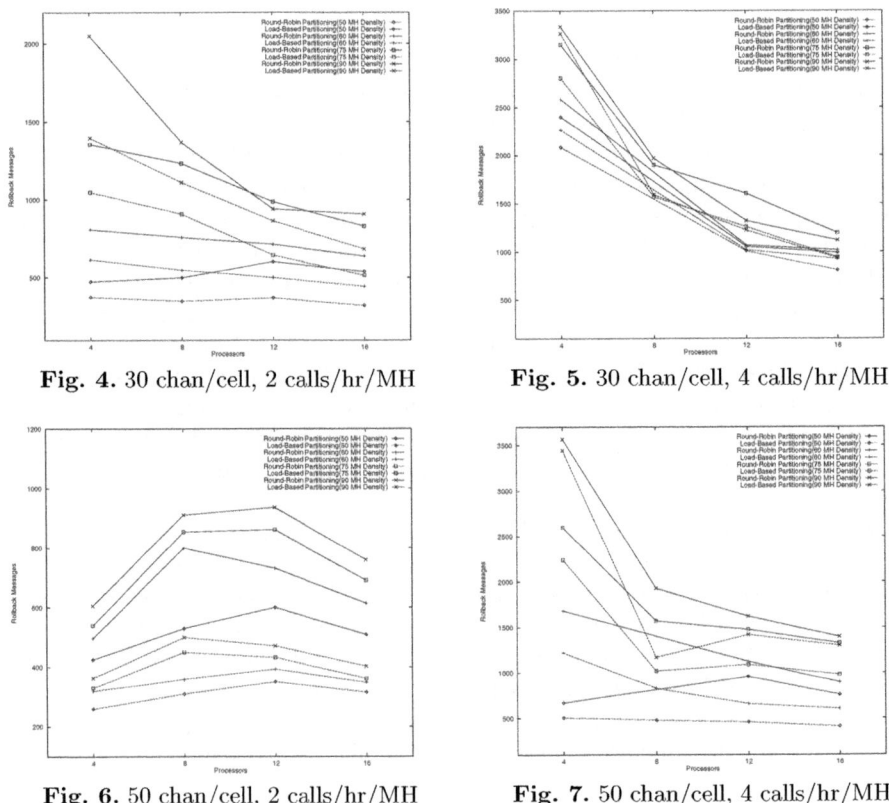

Fig. 4. 30 chan/cell, 2 calls/hr/MH   **Fig. 5.** 30 chan/cell, 4 calls/hr/MH

Fig. 6. 50 chan/cell, 2 calls/hr/MH   **Fig. 7.** 50 chan/cell, 4 calls/hr/MH

Figure: Rollback Messages Vs Number of Processors

Now, let us turn to the comparison of the two partitioning algorithms. It is
important to point out the improvement obtained on the rollback loads by using
the partitioning algorithm based on the load estimate, over the simple round

Table 2. Speedup Vs. Number of Processors (30 Channels/Cell, with $\lambda = 4$ calls/MH/hr)

Partitioning Strategies	50 MH Density			
Number of Processors	2	4	8	16
Round Robin	1.22	2.3	5.1	11.7
Load Based	1.61	3.1	6.9	13.2

robin-based partitioning. We observe a maximum rollback load reduction around 60% (see for instances Figures 6-7, curves for MH density of 90 MH/Km^2). The load reduction has a peak of 60% for MH density of 90, 1 call/hr/MH, both 30 and 50 channels per cell. For the same models, the reduction amounts to around 40-50% for smaller MH densities. The improvement tends to decrease as the call traffic load per MH increases. In fact, the factor of improvement between curves relative to the round robin-based partitioning and curves relative to load-based partitioning is smaller as we increase the number of calls/hr/MH. Models with 30 channels per cell, show reductions around 20-30% for 2 calls/hr/MH, around 15-25% for 3 calls/hr/MH, and around 10-15% for 4 calls/hr/MH. Models with 50 channels per cell still show reductions around 50-60% for 2 calls/hr/MH, around 30% for 3 calls/hr/MH, and around 20-30% for 4 calls/hr/MH. Again, we believe this is due to the same phenomenon we previously mentioned, i.e., the fact that ever growing additional load due to more blocked calls tends to mask the absolute improvement.

One last comment concerns a qualitative observation rather than a quantitative one. By increasing the number of channels per cell (from 30 to 50), there is an increase in the number of calls/hr/MH for which the percentage improvement seems to start decreasing.

Let us now examine the speedup obtained for all MH densities, for both heavy and light loads and for both partitioning schemes (Round Robin and Load-Based). The speedup $SP(n)$ achieved is calculated as the execution time T_1 required for a sequential simulator to perform the same type of simulation on one processor divide by the time T_n required for parallel simulation to perform the same simulation on n processors in addition to the time it takes to perform the partitioning task (T_{part}), i.e., $SP(n) = T_1/(T_n + T_{part})$. As indicated in Table 1, the results indicate clearly that our Load-Based partitioning exhibits a better performance and show that careful static partitioning is important in the success of our PCS simulation model.

5 Conclusion

Wireless communication systems grow very fast these years. These systems become very large and complicated. That makes the analysis and evaluation of capacity and performance of wireless systems very difficult, if it is not impossible. Simulation is a frequently used as an approach to evaluate wireless and mobile systems, while distributed simulation is used to reduce its (simulation) execution time. In this paper, we focused upon a load balancing study to evaluate SWiMNet, a parallel simulation testbed for PCS wireless networks that makes uses of both conservative and optimistic schemes.

Our results indicate clearly the significant impact of load balancing on the efficient implementation of SWiMNet and of its two-stage design. A significant reduction of the synchronization overhead of time warp was obtained. We would like to evaluate the relationship on inter-communication and computational load with the execution time of the simulation in the quest for dynamic load balancing algorithms.

References

1. Boukerche, A., "Time Management in Parallel Simulation", Chapter 18, pp. 375-395, in High Performance Cluster Computing, Ed. R. Buyya, Prentice Hall (1999).
2. Boukerche, A., and Das, S.K., Efficient Dynamic Load balancing Algrorithms for Conservative Parallel Simulations, *PADS'97*
3. Boukerche, A., Das, S.K., Fabbri, A., and Yildiz, O., Exploiting Model Independence for Parallel PCS Network Simulation, *Proc. of Parallel And Distributed Simulation*, PADS'99, pp. 166-173.
4. A. Boukerche, Das, S. K., Fabri, A., and Yildiz, O., "Design and Analysis of a Parallel PCS Network Simulation", IEEE HiPC'99, pp. 189-196, December 1999.
5. Carothers, C., Fujimoto, R.M., Lin, Y.-B., and England, P., Distributed Simulation of Large-Scale PCS Networks, *Proc. of the 2nd Workshop on Modeling, Analysis and Simulation of Computer and Telecommunication Syst.*, 1994 pp.2-6.
6. Fujimoto, R.M., Parallel Discrete Event Simulation, *Communications of the ACM*, Vol.33, No.10, October 1990, pp.30-53.
7. Jefferson, D.R., Virtual Time, *ACM Transactions on Programming Languages and Systems*, Vol.7, No.3, July 1985, pp.404-425.
8. Liljenstam, M., and Ayani, R., A Model for Parallel Simulation of Mobile Telecommunication Systems, *Proceedings of the 4th Workshop on Modeling, Analysis and Simulation of Computer and Telecommunication Systems* (MASCOTS'96).
9. Liljenstam, M., and Ayani, R., Partitioning PCS for Parallel Simulation *Proceedings of the 5th Symposium on Modeling, Analysis and Simulation of Computer and Telecommunication Systems*, pp.38-43, 1997.
10. Meyer, R.A., and Bagrodia, R.L., Improving Lookahead in Parallel Wireless Network Simulation, *Proceedings of the 6th Workshop on Modeling, Analysis and Simulation of Computer and Telecommunication Systems* (MASCOTS'98), pp.262-267.
11. MPI Primer / Developing with LAM, Ohio Supercomputer Center, *The Ohio State University*, 1996.
12. Panchal, J., Kelly, O., Lai, J., Mandayam, N., Ogielski, A., and Yates, R., WIP-PET, A Virtual Testbed for Parallel Simulations of Wireless Networks, *Proc. of the 12th Workshop on Parallel And Distributed Simulation*, 1998 , pp.162-169.
13. Rappaport, T.D., Wireless Communications: Principles and Practice, *Prentice Hall*, 1996.
14. Zeng, X., Bagrodia, R.L., GloMoSim: A Library for the Parallel Simulation of Large Wireless Networks, *Proc. the 12th Workshop on Parallel And Distributed Simulation* 1998, pp.154-161.

Providing Differentiated Reliable Connections for Real Time Communication in Multihop Networks

Madhavarapu Jnana Pradeep and C. Siva Ram Murthy

Department of Computer Science and Engineering
Indian Institute of Technology, Madras - 600 036, INDIA
mjpradeep@yahoo.com, murthy@iitm.ernet.in

Abstract. Several real-time applications require communication services with guaranteed timeliness and fault tolerance at an acceptable level of overhead. Different applications need different levels of fault tolerance and differ in how much they are willing to pay for the service they get. So, there is a need for a way of providing the requested level of reliability to different connections. We propose a new scheme based on the Primary-Backup approach for providing such service differentiation in a resource efficient manner. In our scheme, we provide *partial backups* for varying lengths of the primary path to enhance the reliability of the connection. We demonstrate the effectiveness of our scheme using simulation studies.

1 Introduction

Any communication network is prone to faults due to hardware failure or software bugs. It is essential to incorporate fault tolerance into QoS requirements for distributed real-time multimedia communications such as video conferencing, scientific visualization, virtual reality and distributed real-time control. Applications/users differ in their willingness to pay for such a service which gives fault tolerance. So, different applications/users paying differently should get different qualities of service. Typically, a network service provider would be providing such services. The service provider needs a way of providing different levels of fault tolerance to different applications/users. In this paper, we incorporate *reliability* of connections as a parameter of QoS and describe a scheme for establishing connections with such QoS guarantees.

Conventional applications which use multihop packet switching easily overcome a local fault but experience varying delays in the process. However, real-time applications with QoS guaranteed bounded message delays require a priori reservation of resources (link bandwidth, buffer space) along some path from source to destination. All the messages of a *real-time session* are routed through over this static path. In this way the QoS guarantee on timeliness is realized but it brings in the problem of fault tolerance for failure of components along its predetermined path. Two *pro-active* approaches are in vogue to overcome

M. Valero, V.K. Prasanna, and S. Vajapeyam (Eds.): HiPC 2000, LNCS 1970, pp. 459–468, 2000.
© Springer-Verlag Berlin Heidelberg 2000

this problem. The first approach is forward error recovery method [1], in which multiple copies of the same message are sent along disjoint paths. The second approach is to reserve resources along a path, called *backup path* [2], which is disjoint with the primary, in anticipation of a fault in the primary path. The second approach is far more inexpensive than the first if infrequent transient packet losses are tolerable. In reactive approaches, considerable amount of time is required for reestablishing the communication channel in case of a failure, which is avoided in the second pro-active approach. In [2] the authors consider *local detouring*, in which spare resources in the vicinity of the failed component are used to reroute the channel. In [3,4,5] *end to end detouring* is considered. It is more resource efficient, but has the additional requirement that the primary and backup paths be totally disjoint except for the source and destination. In our proposed scheme here, the span of the backup is varied: it can be either from source to destination, or for only a part of the primary path.

We now explain our scheme of providing Differentiated Reliable Connections. The parameter we use to denote the different levels of fault tolerance is, *reliability* of the connection. Reliability of a component is the probability that it functions correctly in a particular duration. Reliability of a connection is the probability that enough components reserved for this connection are functioning properly in order to communicate from source to destination in that duration. The application/user specifies the level of reliability required for a particular connection. In our scheme, we establish connections with a primary path and an optional backup. A backup is provided only when the reliability specified by the application requires that a backup be provided, and it can either be an end-to-end backup or a *partial backup* which covers only a part of the primary path. The length of the primary path covered by the partial backup can be chosen to enhance the reliability of the connection to the required level. For providing a backup we have to reserve resources along the backup path as well. This is an added cost and our scheme preserves resources by using only the required amount of backups. This improves resource utilization and thereby increases the ACAR (Average Call Acceptance Rate is the ratio of the number of calls accepted to the total number of calls requested).

Using our scheme, many connections will have only a partial backup rather than an end-to-end backup. This means that if there is a fault in the part of the primary path which is not covered by a backup, then the connection cannot be restored immediately: the whole path has to be reestablished. In conventional approaches to fault tolerance [3,4,5], end-to-end backups are provided, and they are able to handle any component failure under the single link failure model. In that model, only one link in the whole network can fail at any instant of time. Since the failure of components is probabilistic, such a model is not realistic, especially for large networks. Even connections with end-to-end backups will have to be reestablished in case more than one link fails simultaneously. In such a probabilistic environment, the network service provider cannot give any absolute guarantees but only probabilistic guarantees. Considering the requirements of different applications/users it is essential to provide services with different levels of reliabilities, and in this paper we give one such way of service differentiation.

In section 2, we describe partial backups. In section 3, we explain how differentiated reliable connections are provided. In section 4, we discuss route selection. Failure recovery procedure and delay are discussed in section 5. Results from simulation experiments are presented in section 6 and finally we summarize our work in section 7.

2   Partial Backups

A *primary segment* is any sequence of contiguous components along the primary path. A partial backup covers only a primary segment, i.e. the backup can be used when a component along the primary segment develops a fault. The partial backup and primary segment should not have any component in common other than the two end nodes.

Figure 1 shows a *partial backup*. A connection has to be established from the source to the destination. The primary path consists of 6 links. Here, links 2, 3 and 4 and their end nodes form a primary segment. The partial backup, consisting of links 7, 8 and 9 and their end nodes covers the above primary segment.

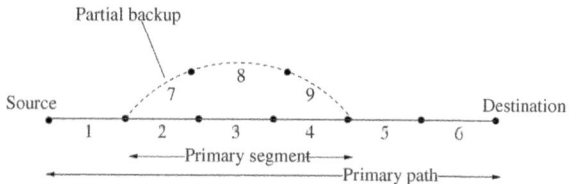

Fig. 1. Illustration of a Partial Backup

We now find the reliability of this connection. For simplicity, we take nodes to be fully reliable i.e. nodes cannot fail, only links are prone to faults. The subsequent discussion can be easily extended to include node failures also. The reliability of a segment consisting of links with reliabilities $r_1, r_2, ...r_n$ will be $\prod_{i=1}^{n} r_i$. Now, let r_p denote the reliability of the primary path, r_s denote that of the primary segment which is covered by a backup, r_b that of the backup and r_c that of the composite path comprising of the primary and the backup. Here, $r_p = \prod_{i=1}^{6} r_i$, $r_s = r_2.r_3.r_4$ and $r_b = r_7.r_8.r_9$. Now, $r_c = $ (*reliability of part of primary path not covered by backup*).(*reliability of primary segment and partial backup together*)

$$\Rightarrow r_c = \frac{r_p}{r_s}.(r_s + r_b.(1 - r_s)) \tag{1}$$

Let r_r denote the reliability required by the application. Let us see how the partial backups are useful. Suppose the reliabilities of the links is 0.9700 each, and the required reliability r_r is 0.900. Then, for the connection shown in Figure 1, $r_p = 0.8330$, $r_s = r_b = 0.9127$. Then, using Equation 1, we calculate r_c as 0.9057.

Thus, having a partial backup for any 3 links is just enough in this case as the required reliability is 0.900. In this example, we have taken all the links to have the same reliability, but in a practical network different links will have different reliabilities. So, partial backups can be used effectively by identifying primary segments which have low reliability and providing partial backups for them only. This results in reserving lesser amount of spare resources than those required for the traditional end-to-end backups. Further, it is easier to find partial backups for primary segments than to find an end-to-end backup, especially under heavy load conditions. This results in an increased call acceptance rate.

3   Providing Differentiated Reliable Connections

When an application requests a connection from a source to a destination, we try to provide a connection with the desired reliability using no backup at all, or a partial backup, or an end-to-end backup. In our scheme, we want to minimize the amount of spare resources while providing the required reliability. As the resources reserved for the backup are used only when there is a fault in the primary path, providing a backup would mean a large amount of extra resource reservation. Hence, we establish a backup only when it is not possible to find a primary path with the required reliability. In this paper, we take the delay along a path and network resources reserved to be proportional to the length of the path. Thus amount of resources and delay are synonymous with path length.

We propose to use the usual algorithms for finding the shortest path (to minimize resource reservation or delay) from source to destination. If the reliability of the shortest path so found is below the requirement, we try to find a path with the required reliability, using a *modified path-selection algorithm* (described in the next section). If this modified algorithm fails to find a suitable path, we try to establish a connection by using backups (partial or end-to-end) to achieve the required reliability. We outline our algorithm in detail below. r_p, r_r, r_b and r_c are as described in the earlier section.

1. Find a primary path from src to dest, using a shortest path algorithm.
2. **If** $r_p \geq r_r$ accept this path and **return** success.
   Else goto step 3.
3. Use the *modified path selection algorithm* (described in the next section) to find a primary path from the source to the destination again.
4. **If** new $r_p \geq r_r$, accept this path and **return** success.
   Else goto step 5.
5. Reconsider the primary path found in step 1.
6. *Identify* some primary segments for which we can take a backup to enhance their reliability. Find backups for the identified primary segments and *select* one which satisfies the reliability requirement (whether a backup satisfies the reliability requirement or not can be decided by evaluating r_c using eq. 1). If such a backup exists, accept that primary and backup and **return** success.
7. **return** failure.

In step 5, we reconsider the shortest path found in step 1 rather than that in step 3 to decrease the load on links with high reliability, which would be preferentially chosen by the modified path selection algorithm. The main issues involved here are given below and are discussed in the next section.

1. The modified path selection algorithm to find a path with higher reliability in step 3.
2. Identification of the segments of the primary path in step 6.
3. Selection of a suitable backup among all the eligible backups in step 6.

Although in our algorithm we establish only one backup, it can be easily adapted to establish multiple backups to further enhance the reliability of the connection. For example, in step 7, we can have

Establish one end-to-end backup and one partial backup. This primary with two backups might satisfy the reliability requirement. **If** it satisfies, accept this path with two backups and **return** success
Else return failure.

4   Route Selection

In this section we present simple solutions to the issues raised in the previous section. More complex and efficient solutions can exist and can be investigated.

4.1   Modified Path Selection Algorithm

Finding a path, subject to multiple constraints on routing metrics, is NP-complete [7]. Here, we are interested in minimizing delay and maximizing reliability. There is no provably efficient algorithm for doing this, and so we resort to heuristics. In this algorithm, we attempt to find paths with higher reliability at the expense of greater path length. To do this, we define a cost function for each link which is dependent both on its reliability and delay along it. We then use Dijkstra's minimum cost path algorithm to find a path from source to destination. Delay is an additive metric whereas reliability is a multiplicative one, i.e. the delay along a path is the sum of the delays along each link in it, whereas the reliability of the path is the product of the reliabilities of the links in it. Since Dijkstra's algorithm takes costs to be additive, we propose to take the logarithm of the reliability in the cost function. Thus, a suitable cost function would be,

$$cost = delay - relWeight * log(reliability) \tag{2}$$

where $relWeight$ is a parameter. By varying the value of $relWeight$, we can control the trade-off between reliability and delay along the path chosen.

4.2   Identification of Primary Segments

As described in the previous section, we identify some suitable primary segments and find backups for them to enhance their reliability to the desired level. So we identify primary segments whose reliability is less than $estRel$ which is calculated as given below. r_p, r_s, r_r and r_c are as described in section 2.

$r_c = \frac{r_p}{r_s}.(r_s + r_b.(1 - r_s)) \geq r_r \Rightarrow r_s \leq \frac{r_p}{r_r}.(r_s + r_b.(1 - r_s))$

Now, $r_b < 1$. Therefore, $r_s < \frac{r_p}{r_r}.(r_s + 1.(1 - r_s))$

$$\Rightarrow r_s < estRel = r_p/r_r \tag{3}$$

Among primary segments of a given length, it would be advantageous to provide backups for primary segments with low reliability because, as seen from equation 1, r_c increases as r_s decreases assuming $r_b \approx r_s$.

4.3   Selection of Suitable Backup

A number of segments, up to a maximum of $segmentTrials$ are found as de-scribed above and are remembered. We also add the whole primary path as an alternative in case an end-to-end backup is very convenient. We try to find back-ups for them which satisfy the reliability requirement. We use the modified path selection algorithm to find a backup path between the end nodes of the primary segment taking care to exclude all the components of the primary other than the end nodes of the primary segment. Among these backups (for the different primary segments), the backup reserving lesser amount of resources is preferable. However, in case of backups reserving slightly different amounts of resources, it might be better to choose one which gives higher composite reliability. So, we select a backup based on an $expense$ function given below.

$$expense = pathLength - compositeRelFactor * r_c \tag{4}$$

Here, $compositeRelFactor$ is a parameter which allows us to trade-off between composite reliability and extra resource reservation. We choose the backup with the least $expense$ value.

5   Failure Recovery and Delay

When a fault occurs in a component in the network, the connections passing through it have to be rerouted through their backup paths. This process is called failure recovery, and is required only when a component in the primary path of the connection fails. If the failed component is in the primary segment covered by a backup, the backup is activated, and if it is not covered by a backup, the whole connection is rerouted. This is done in three phases: $fault\ detection$, $failure$ $reporting$ and $backup\ activation$ or $connection\ rerouting$.

In our model, we assume that when a link fails, its end nodes can detect the failure. For failure detection techniques and their evaluation refer to [6]. Af-ter fault detection, the nodes which have detected the fault, report it to the concerned nodes. Failure reports are sent in both directions: towards the source and the destination. This failure reporting needs to use control messages. For this purpose, we assume a real-time control channel (RCC) [3] for sending control messages. In RCC, separate channels are established for sending control messages, and it guarantees a minimum rate of sending messages.

After failure reporting, if the failed component is covered by a backup, backup activation is done. In this case, the end nodes of the primary segment initiate the recovery process on receiving the failure report. They send the activation message along the backup path and the communication is resumed. The delay suffered here is low as required by most real-time applications. This process is illustrated in Figure 2.

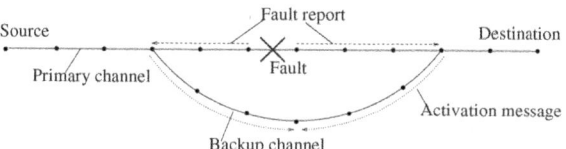

Fig. 2. Illustration of Failure Recovery

If the failed component is not covered by a backup, the source initiates the recovery process on receiving the failure report. The source again requests a reliable connection to be setup. This takes a much longer time.

In real-time communication, it is essential to have the delays along both the primary and backup to be as low as possible. Our routing algorithm attempts to minimize the delay from source to destination. In addition, depending on how delay-critical the application is, we can adjust the *relWeight* parameter to trade-off between delay and reliability. Even in selecting backups we try to minimize the *pathLength* or delay using the *expense* function.

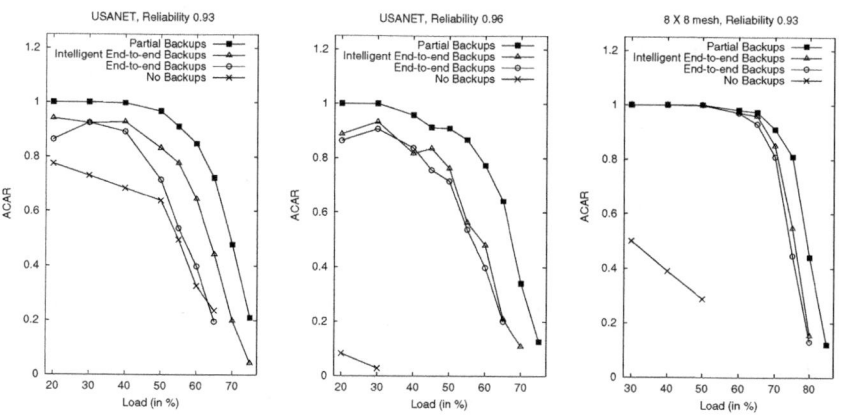

Fig. 3. ACAR vs. Load for Reliable Connections

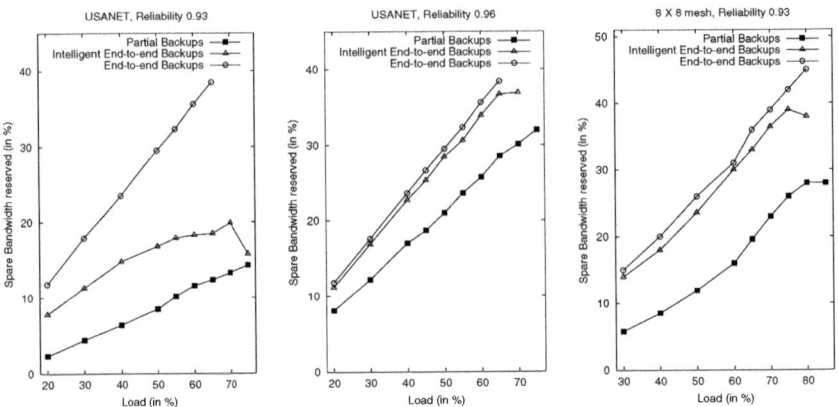

Fig. 4. Percentage of Network Bandwidth Reserved for Backups vs. Load for Reliable Connections

6   Performance Evaluation

We evaluated the proposed scheme by carrying out simulation experiments similar to those in [3], on the USANET and on a 8 X 8 mesh. We also implemented end-to-end backup and no backup schemes for comparative study.

In the simulated networks, neighbor nodes are connected by two simplex links, one in each direction, and all links have identical bandwidth. The delay of each link was set to 1. The reliability of the links was set as a uniformly distributed random value between 0.97 and 1.0. For simplicity, the bandwidth requirement of all connections was put equal to 1 unit. The route selection was done as described in section 4. Connections were requested randomly with a certain mean inter-arrival time for connections from each node as source. All connections were taken to be of identical duration and the simulation was run long enough to reach steady state. To avoid very short connections, source and destination of connections were chosen such that the length of the shortest path between them was at least $minPathLen$, which was chosen to be 3 for experiments using USANET and 5 for those using 8 X 8 mesh network. The number of segments identified for finding backups, $segmentTrials$ was taken as 15 in our experiments. The $relWeight$ parameter was taken as 500 and $compositeRelFactor$ was taken as 1.

The results are shown in Figures 3, 4, and 5. In Figure 3, we plot the ACAR against network load for our scheme, intelligent end-to-end backup scheme, end-to-end backup scheme and for no backups. The network load is taken as the percentage of total network bandwidth reserved for reliable connections. In intelligent end-to-end backup scheme, a backup is established only if the primary path fails to meet the requirement. In the end-to-end backup scheme, full backups are established for all the connections, irrespective of the reliability of the primary path. In Figure 4, we show the percentage of the network resources reserved as spare resources for backups. All connections were requested with a

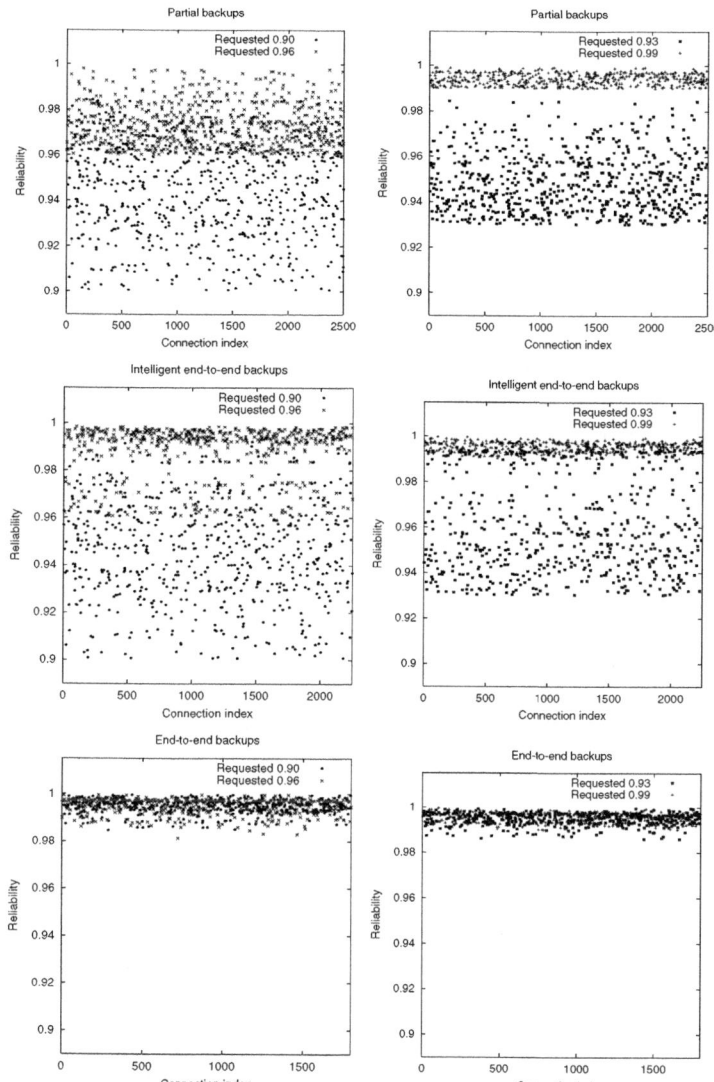

Fig. 5. Reliability Distribution of Connections against Connection Index for USANET

constant reliability requirement. Graphs are plotted for different values of the requested reliability. The readings were taken at steady state.

We note that the ACAR is highest for our method in all the cases. The ACAR is acceptably high even at high load levels, thus indicating high resource utilization. Our scheme reserves the least amount of spare resources. Our scheme is able to accept calls with a wide range of reliability requirements as it uses a modified routing algorithm to find higher reliability paths. It is easier to find partial backups rather than end-to-end backups, and this explains the high ACAR of our scheme even at high load levels.

Figure 5 shows the level of reliability got by connections requesting different reliabilities. The different rows of the graphs show the results for our partial backup scheme, intelligent end-to-end backup scheme, and for the end-to-end backup scheme. Figure 5 shows the results for the USANET topology. The experiment is started with zero connections and connections are established as well as released incrementally. Connections were requested with 4 different values of requested reliability: 0.90, 0.93, 0.96 and 0.99. We plot the reliability got by each connection against the connection index, for the different values of requested reliability. Each graph shows the distribution for 2 values of requested reliability, and the two graphs together showing for the 4 values of reliability. The band-like distribution of the reliabilities provided shows that a good level of service differentiation has been achieved using our scheme. In the case of end-to-end backups, most of the reliabilities tend to be at the higher end, since backups are established for all connections. In intelligent end-to-end backups, some of the reliabilities are again very high because of the full backup provided. The partial backup scheme provides connections with reliability close to the required amount. This is especially true for connections with high reliability requirements.

7   Summary

In this paper, we have proposed a resource efficient scheme for service differentiation in reliable connections. Our method of partial backups provides connections with different reliabilities as requested, and offers a high level of resource utilization. We evaluated the proposed scheme through simulations and demonstrated its usefulness.

References

1. P. Ramanathan and K. G. Shin, "Delivery of time-critical messages using a multiple copy approach," *ACM Trans. Computer Systems,* vol. 10, no. 2, pp. 144-166, May 1992.
2. Q. Zheng and K. G. Shin, "Fault-tolerant real-time communication in distributed computing systems," in *Proc. IEEE FTCS,* pp. 86-93, 1992.
3. S. Han and K. G. Shin, "A primary-backup channel approach to dependable real-time communication in multihop networks," *IEEE Trans. on Computers,* vol. 47, no. 1, pp. 46-61, January 1998.
4. S. Han and K. G. Shin, "Efficient spare-resource allocation for fast restoration of real-time channels from network component failures," in *Proc. IEEE RTSS,* pp. 99-108, 1997.
5. C. Dovrolis and P. Ramanathan, "Resource aggregation for fault tolerance in integrated services networks," *ACM SIGCOMM Computer Communication Review,* 1999.
6. S. Han and K. G. Shin, "Experimental evaluation of failure detection schemes in real-time communication networks," in *Proc. IEEE FTCS,* pp. 122-131, 1997.
7. Zheng Wang, "On the Complexity of Quality of Service Routing," *Information Processing Letters,* vol. 69, pp. 111-114, 1999.

A Multicast Synchronization Protocol for Multiple Distributed Multimedia Streams

Abderrahim Benslimane

Département d'Informatique- Sévenans
Université de Technologie de Belfort-Montbéliard
90010 Belfort cedex France
abder.benslimane@utbm.fr

Abstract. Recent advances in networking, storage, and computer technologies are stimulating the development of multimedia and distributed applications. In this paper, we propose a synchronization protocol for multiple stored multimedia streams in a multicast group. The proposed method consists of intra-stream, inter-stream and inter-receiver synchronization mechanisms. The scheme is very general because it only makes a single assumption, namely that the jitter is bounded. The protocol can be used in the absence of globally synchronized clocks without time stamped media units.
Verification and simulation results which we carried out show that streams synchronization is maintained. Moreover, group synchronization is respected.

1 Introduction

Since most multimedia applications (e.g., teleconferencing, TV broadcast, etc.) are inherently multicast in nature, support for point-to-point video communication is not sufficient. Unfortunately, multicast video transport is severely complicated by variation in the amount of bandwidth available and different delays throughout the network.

On the other hand, the development of storage devices having large capacity allows emergence of multimedia services on demand. In such services, data are stored in multimedia servers equipped with large storage devices, transmitted through integrated networks and played back by receivers. When data is a multimedia object constituted of audio and video, it must not only necessary to ensure continuity during the restitution of each of two streams, but also to preserve the temporal relations between them.

However, the major problem is the asynchronous communication which tends to interrupt these temporal relationships in packet switched network such as ATM. Four sources of asynchrony can disrupt multimedia synchronization : delay jitter, local clock drift, different initial collection times and different initial playback times. This paper is devoted to delay jitter problem. Controlling delay jitter helps when there is a requirement for media synchronization.

As shown in the next section, many researchers have addressed media synchronization issues and a number of media synchronization protocols have been proposed to satisfy diverse requirements [2].

M. Valero, V.K. Prasanna, and S. Vajapeyam (Eds.): HiPC 2000, LNCS 1970, pp. 469–476, 2000
© Springer-Verlag Berlin Heidelberg 2000

However, most of these protocols suppose the case where several streams are synchronized and mixed onto one stream and outputted to the participants. As an example, we find the well known RTP Protocol [10]. Effectively, this method allows to reduce bandwidth. But decreases the performance and the QoS of streams. Therefore, it is dependent of the mixer server. If a failure occurs for the mixer then the real-time application will be stopped.

To overcome this problem, we propose a multimedia synchronization protocol for several sources of stored multimedia streams. All streams are broadcasted directly to the receivers. The synchronization is accomplished by the destinations.

As the opposite of RTP, we use a distributed solution to multicast multiple streams. This allows to benefit from the proper performances of each server (power for coding, compression, etc) and to let streams independents.

The proposed method consists of intra-stream, inter-stream and inter-receiver synchronization mechanisms. The scheme is very general because it only makes a single assumption, namely that the jitter is bounded. The protocol can be used in the absence of globally synchronized clocks without time stamped media units.

The rest of paper is organized as follows. Section 2 gives related works carried out in the domain of multimedia synchronization. Section 3 proposes a new synchronization protocol when the system is composed of several sources of multimedia stream and several receivers. we show that the suggested protocol allows to maintain the synchronization and to respect multicast in a group of receivers. An example with its simulation results are given. Finally, section 4 presents the conclusion of paper.

2   Related Work

Several papers were presented in the multimedia synchronization domain [3, 7, 9]. These protocols differ mainly by the assumptions that they make on the communication manner, the topological structure of network, and the way of ensuring sending, delivery and playback of media units.

The approaches to stream synchronization proposed in literature differ in the stream configurations supported. Some of the proposals require all receivers of the synchronization group to reside on the same node (e.g., Multimedia Presentation Manager [4], ACME system [1]). Others assume the existence of centralized server, which stores and distributes data streams.

The protocol proposed in [3] allows to maintain synchronization of multiple sites distributed geographically. It is based on the clocks synchronization through all the network. The protocol timestamps temporally each media unit at the source and balance waiting time delay at destination to impose constant transit delay network among synchronized data streams. The equalization delay can be constant or adaptive. Rangan et al. propose in [7] a hierarchical algorithm for media mixing in multimedia conferences. This algorithm correct jitter and local drift clock with absence of globally synchronized clocks. But it does not give any accuracy on the way in which multimedia application is initiated synchronously.

In their paper [6], Rangan et al. propose a protocol allowing broadcast and playback of stored data streams. The receivers, which are mediaphones, must send periodically feedback messages to the server, which uses these messages to consider the temporal state of various streams. Since clocks are not supposed to be synchronous, the quality of these analyses depends on the delays of feedback messages. This protocol is well adapted for LAN networks.

The protocol proposed by K. Rothermel and T. Helbig in [8] allows a quick adaptation to network alterations and to change of user requirements for current session. It supposes the synchronization of network clocks, and supports two types of configurations : point-to-point and point-to-multipoint.

Authors in [5] propose a lip synchronization scheme for multimedia applications such as the video-on-demand. This protocol is adaptive and regulates the problem of non deterministic behavior in operating system or network. It allows intra-stream synchronization with minimum delay jitter in a continuous stream and inter-stream synchronization with the minimum drift between dependent data streams.

3   Synchronization Protocol

3.1 System Environment

As shown in figure 1, the communication network considered can be modeled as a graph connecting p sources to n receivers. A source arranges a multimedia flow that is decomposed in media units. The generation rate is equal to τs. Receivers restore periodically media units with the same rate as the generation. We assume that clock ticks at the sources and the destination have the same advancement, but that the current local times may be different (i.e., locally available clocks) [9].

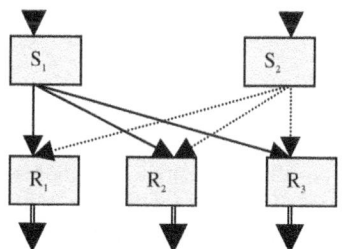

Fig. 1. System of 2 Sources of streams (S_1 and S_2) and 3 Receiver (R_1, R_2, R_3).

In this model, we are interesting in the case where several sources broadcast different streams to a group of receivers. These receivers must restore all streams at the same time while respecting intra-stream synchronization. This model is very interesting to study since in some multimedia applications, such as videoconference, several streams, resulting from various sources geographically distributed through the network, can be played back.

For each source S_i (i=1..p) and each receiver R_j (j=1..n), we adopt the following QoS parameters :

- d_{ij} : delay between the source S_i and the receiver R_j,
- D_{ij}^{min} : minimum delay experimented between S_i and R_j,

- D_{ij}^{max} : maximum delay experimented between S_i and R_j,
- D_i^{max} : greater value among maximum delays between S_i and all receivers (one stream from S_i),
- D^{max} : greater value among of maximum delays between all sources and all receivers (all streams).

3.2 Buffer Size

- To maintain intra-stream synchronization a receiver R_j needs to buffer a number of media units equal to : $\left| \delta_{ij} * \tau_s \right|$, where $\delta_{ij} = D_{ij}^{max} - D_{ij}^{min}$.
- To maintain, for the same stream, inter-receiver synchronization it is necessary to add the following quantity : $\left| \delta_i * \tau_s \right|$ media units where $\delta_i = D_i^{max} - D_{ij}^{max}$.
- Finally, to maintain the inter-stream synchronization, for the stream whose origin is source S_i, we must add : $\left| \delta * \tau_s \right|$ media units where $\delta = D^{max} - D_i^{max}$.

Then, the indispensable buffer size for receiver R_j to maintain synchronization of stream S_i is : $B_{ij} = \left\lceil (D^{max} - D_{ij}^{min}) * \tau_s \right\rceil$

The total buffer size for receiver R_j to maintain global synchronization for all streams is : $B_j = \sum_{j=1}^{p} \left\lceil (D^{max} - D_{ij}^{min}) * \tau_s \right\rceil$

3.3 Protocol Description

We use an evaluation phase which allows computation of network communication delays and receivers restitution times of the first media unit while tacking into account intra-stream and inter-stream synchronization mechanisms.

The evaluation phase is start-up by a source elected as the master. Other sources (i.e. slaves) communicate directly with the master only during the evaluation phase. There are no communications between slave sources. The synchronization concerns first, the receivers of the same stream, and second, the sources of various streams.

The first step allows intra-stream and inter-receiver synchronization for the same stream S_i (i=1..p) towards all receivers R_j (j=1..n). This stage relates to all streams separately. It is accomplished in parallel manner.

The second stage allows inter-stream and inter-receiver synchronization for all streams. This stage is accomplished in a coordinated way by the master source. This centralized computation can be distributed between sources. That is without great interest because it is used temporarily and only in the evaluation phase.

Evaluation Phase

In the evaluation phase, we suppose that communications between each two entities take the same path in both directions. This assumption can be easily removed since we can take roundtrip delay instead of end-to-end delay. This will be addressed in a next work. Also, a time between the receipt of a message and the corresponding acknowledgement is supposed to be zero. Moreover, the processing time is negligible.

Text of master source S_m

{This subprogram of the master concerns only the coordination between different slaves.}

At time $tstart_m$, S_m broadcasts REQUEST_M message to the other sources S_i (i=1..p) to initiate the evaluation phase.

At time $trecv_i$, S_m receives RESPONSE_M(D_i^{max}, $DphaseE_i$) from S_i.

At time $trecv^{max} = \max_{i=1}^{p}(trecv_i)$, S_m receives the last response from all sources.

It computes then :

- $D^{max} = \max_{i=1}^{p}(D_i^{max})$: as previously specified,

- $\psi_i = trecv^{max} - trecv_i$: difference between arrival time of the last response and that of S_i,

- $dm_i = (trecv_i - tstart_m - DphaseE_i)/2$: delay between sources S_m et S_i,

- $dm^{max} = \max_{i=1}^{p}(dm_i)$: greatest value among end-to-end delays between S_m and other sources S_i (i=1..p).

It is obvious that delay between S_m and S_m is nul. $dm_m = 0$.

At time $trecv^{max}$ S_m sends at each source S_i (i=1..p) BEGIN_M(D^{max}, dm^{max}, dm_i, ψ_i) message.

Text of each source S_i (i=1..p)

At time $tstart_i$, when receiving REQUEST_M message from S_m, S_i broadcasts REQUEST message to all receivers R_j (j=1..n).

At time a_{ij}, S_i receives the message RESPONSE(D_{ij}^{max}) from the receiver R_j.

At time $a_i^{max} = \max_{j=1}^{n}(a_{ij})$, S_i receives the last response, computes :

- $D_i^{max} = \max_{j=1}^{n}(D_{ij}^{max})$: greater value among maximum delays from S_i to all receivers,

- $DphaseE_i = (a_i^{max} - tstart_i)$, and

sends a RESPONSE_M($D_i^{max}, DphaseE_i$) message to the master source S_m.

Then, it computes :

- $\Psi_{ij} = a_i^{max} - a_{ij}$: difference between arrival time of the last response and that of R_j,

- $d_{ij} = (a_{ij} - tstart_i)/2$: computed delay between source the S_i and the receiver R_j, this delay must be bounded between D_{ij}^{max} and D_{ij}^{min}, and

- $d_i^{max} = \max_{j=1}^{n}(d_{ij})$: greater value among computed delays.

At the reception of the message BEGIN_M($D^{max}, dm^{max}, dm_i, \Psi_i$), S_i computes :

- $\delta_{m_i} = dm^{max} - dm_i$, and

- $\Delta_i = 2dm_i + \Psi_i + \delta_{m_i}$ {intermediate variable}

Then, it sends, message PLAY($D^{max}, d_{ij}, \Delta_i, \Psi_{ij}$) to each receiver R_j (j=1..n).

Finally, S_i computes the broadcast time of the first media unit :

$$Tbegin_i^1 = a_i^{max} + \Delta_i$$

Text of each receiver R_j (j=1..n) executed in reply to each source S_i (i=1..p)

At time s_{ij}, R_j receives REQUEST message from S_i.

At the same time, it replies to S_i by sending RESPONSE(D_{ij}^{max}) message.

At the reception of PLAY($D^{max}, d_{ij}, \Delta_i, \Psi_{ij}$) message from S_i, R_j computes start playback time of the stream from S_i :

$$Trest_{ij}^1 = s_{ij} + d_{ij} + \Psi_{ij} + \Delta_i + D^{max} \tag{4}$$

This time must be the same for each stream S_i (i=1..p).

We have showed that multimedia synchronization for multiple stored streams in a multicast group is respected. The proof was presented in the submitted version of the paper. A reader can find a complete version of the paper in [2].

3.4 Simulation Results

We use a network constituted of 2 sources S_1 and S_2 and 3 receivers R_1, R_2 and R_3. It is supposed that source S_2 is the master.
We retained the following network configurations :

For stream from S_1 :

$D_{11}^{max} = 24$ $D_{12}^{max} = 16$ $D_{13}^{max} = 38$

$D_{11}^{min} = 20$ $D_{12}^{min} = 10$ $D_{13}^{min} = 30$

For stream from S_2 :

$D_{21}^{max} = 30$ $D_{22}^{max} = 25$ $D_{23}^{max} = 45$

$D_{21}^{min} = 24$ $D_{22}^{min} = 18$ $D_{23}^{min} = 38$

In the same way, delays between the master source S_m $(= S_2)$ and the others are :

$dm_1 = 14$ $dm_2 = 0$

Generation rate of media units at sources : $T_s = 100$ media units/s.

The evaluation phase determines the following values :
- the time at which sources begin the output of media units is equal to 118,
- the time at which receivers start playback of media units composing each stream is equal to 163.
- buffer size for each stream, at R_1 : $B_{11}=3$, $B_{12}=4$, $B_{13}=2$, and at R_2 : $B_{21}=3$, $B_{22}=3$ and $B_{23}=1$.

These analytical results are confirmed by the diagram below corresponding to the arrival times of media units to the receivers. We notice that at restitution time of the first media unit of each stream, there is the same number of media units in the buffers as that computed.

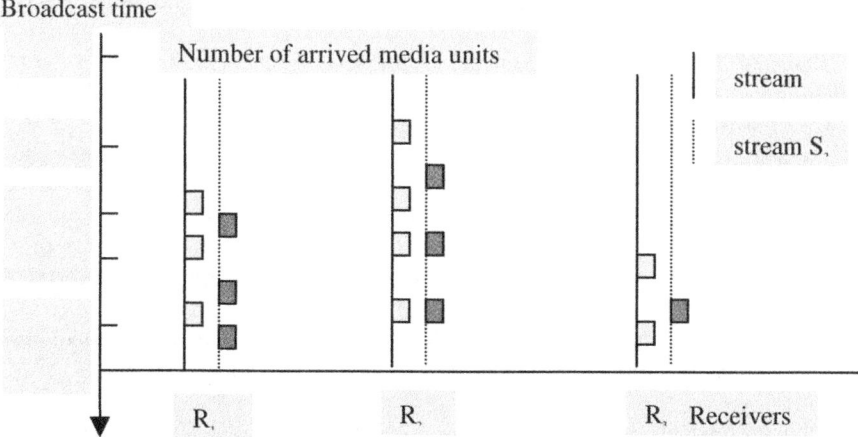

Fig. 2. Arrival diagram of media units before the restitution of the first one.

4   Conclusion

Mixing different types of stream is necessary when clients are located in a heterogeneous network with different priorities (i.e., bandwidth). However, by mixing different types of streams, the QoS of individual streams my be lost. Moreover, the multimedia application is dependent of the mixer. To overcome these problems, we proposed a multimedia synchronization protocol for several sources of stored multimedia streams. All streams are broadcasted directly to the receivers who accomplish the synchronization. The proposed method consists of intra-stream, inter-stream and inter-receiver synchronization mechanisms. The scheme is very general because it only makes a single assumption, namely that the jitter is bounded. The protocol can be used in the absence of globally synchronized clocks without time stamped media units.

Real implementation of the protocol and its integration in a tele-teaching application are in progress. However, technical issues such as scalability need to be addressed further.

References

1. D. P. Anderson, G. Homsy, "A Continuous Media I/O Server and its Synchronization Mechanism", IEEE Computer, 10 1991.
2. A. Benslimane. "A Multicast Synchronization Protocol for Multiple Distributed Multimedia streams", Technical Report, Computer Department, University of Technology of Belfort-Montbéliard, 06 2000.
3. J. Escobar, Debra Deutsch, and Craig Partridge, "Flow Synchronization Protocol", IEEE Global Communication Conferences, 12 1992, pp. 1381-1387.
4. IBM Corporation, "Multimedia Presentation Manager Programming Reference and Programming Guide", 1.0, IBM Form:S41G-2919-00 and S41G-2920-00, 3 1992.
5. Klara Nahrstedt et Lintian Qiao. "Lip Synchronization within an Adaptive VOD System", International Conference on Multimedia Computing and Networking, San Jose, California, 02 1997.
6. S. Ramanathan and P. V. Rangan, "Adaptive Feedback Techniques for Synchronized Multimedia Retrieval over Integrated Networks", IEEE/ACM Transactions one Networking, Vol. 1, N°. 2, p. 246-260, 04 1993.
7. P. V. Rangan, H. MR. Vin, and S. Ramanathan, "Communication Architectures and Algorithms for Media Mixing in Multimedia Conferences", IEEE/ACM Transactions one Networking, Vol. 1, No. 1, 02 1993, pp. 20-30.
8. Kurt Rothermel et Tobias Helbig. "An Adaptive Stream Synchronization Protocol", Stuttgart University, {rothermel, helbig}@informatik.uni-stuttgart.de, 1994.
9. H. Santoso, L. Dairaine, S. Fdida, and E. Horlait, "Preserving Temporal Signature : A Way to Convey Time Constrained Flows", IEEE Globcom, 12 1993, pp. 872-876.
10. H. Schulzrinne, RFC 1890: "RTP Profile for Audio and Video Conferences with Minimal Control", RFC 1890, 25 011996.

Session V-A

Wireless and Mobile Communication Systems
Chair: Azzedine Boukerche
University of North Texas, Denton

Improving Mobile Computing Performance by Using an Adaptive Distribution Framework

F. Le Mouël, M.T. Segarra, and F. André

IRISA Research Institute
Campus de Beaulieu
35042 Rennes Cedex, FRANCE
{flemouel,segarra,fandre}@irisa.fr

Abstract. Portable devices using wireless links have recently gained an increasing interest. These environments are extremely variable and resource-poor. Executing existing applications in such environments leads to a performance degradation that makes them unsuitable for this kind of environments. In this paper, we propose the utilization of AeDEn, a system that performs a global distribution of applications code and data in order to take benefit from the underutilized resources of the mobile environment. This is achieved by designing AeDEn as part of the MolèNE system, a more general framework that we have built to encapsulate a set of generic and adaptive services that manage wireless issues.

1 Introduction

Portable devices (laptops, PDAs, mobile phones, etc) using wireless links (GSM, satellites) have recently gained an increasing interest. Simultaneously, many applications are designed requiring more and more resources (memory, processor, etc) and assuming an environment that ensures the availability of all of them. Executing such applications on a wireless environment leads to a performance degradation for three main reasons: 1) portable devices are generally battery-limited, resource-poor entities compared to their counterparts in the fixed network, 2) wireless networks are subject to important variations and suffer from frequent disconnections due to the interaction with the environment, and 3) when moving, stations (mobile or not), devices (scanners, printers) and functionalities (proxy cache) that can be accessed by a portable computer change. Existing applications are not designed to tolerate such variations and their presence generally leads to applications finishing abruptly.

Approaches that deal with the performance degradation problem propose the transfer of resource-consuming tasks to a fixed station [4,5]. However, all these solutions lack from providing a systematic approach to exploit unused or under-utilized external resources. In this paper, we propose the AeDEn system (Adaptive Distribution Environment) [1] as providing the necessary means to effectively distribute applications code and data in a rapidly varying environment. Changes in the environment are taken into account at two very different levels: the application and the distribution system itself. The first one concerns

M. Valero, V.K. Prasanna, and S. Vajapeyam (Eds.): HiPC 2000, LNCS 1970, pp. 479–488, 2000.
© Springer-Verlag Berlin Heidelberg 2000

the dynamic layout of the application onto a set of stations and the second one allows to dynamically change the distribution algorithm used to decide about the layout.

In order to ensure the generic nature and the adaptability of AeDEn, we have designed it as part of a more general system called MolèNE (<u>M</u>obile <u>Net</u>working <u>E</u>nvironment) [10]. We have designed MolèNE as a system providing a set of generic functionalities that abstract concepts usually utilized by mobile applications. Moreover, MolèNE provides the necessary mechanisms to track environmental changes and to modify the behavior depending on these changes and applications needs. All these mechanisms are used by AeDEn making it a framework into which applications can insert their own distribution preferences.

The paper is organized as follows. Available mechanisms for the construction of adaptive services in MolèNE are presented in Section 2. Adaptive and dynamic distribution schemes of AeDEn are introduced in Section 3. Section 4 compares our systems to existing approaches. Finally, Section 5 concludes and gives future research directions.

2   Adaptation in MolèNE

Existing approaches to adapt applications behavior to a mobile environment are based on a set of common concepts which implementation is usually application-dependent. We have identified these concepts and their abstraction constitutes the foundation of MolèNE. Application-independent concepts, called *tools* are implemented as concrete object-oriented classes. Other functionalities are application-dependent and have to be customized to each application. Such functionalities, that we call *services*, are provided as an object-oriented framework. Some of its classes remain abstract allowing developers to customize services to different applications [10]. This is usually a simple task and allows the framework to perform other tasks much more complex.

Static and dynamic characteristics of mobile environments vary significantly over time making it necessary to adapt services behavior to the execution conditions. On one hand, many portable computers exist from laptops to palmtops that make different tradeoffs between resources availability and portability. On the other hand, different wireless networks can be used by a portable computer that provide a varying transmission quality.

Dynamically adapting services behavior to these characteristics allows to offer the user a better QoS (e.g. response time). Therefore, we have considered the adaptation as a necessary property and we have built the framework as a set of adaptive services managing issues related to the mobile environment on which applications execute. Three mechanisms are available in MolèNE to perform the dynamic adaptation of the services:

- a Detection & Notification Service that detects hardware and software *changes* and notify the interested applications;
- an adaptation interface allowing applications to describe their adaptation strategy: which are the significant changes and which are the *reactions* to be

performed when the changes occur. Therefore, application-dependent adaptations can be performed by MolèNE;

— a *reactive system* that takes into account the adaptation strategy provided by the application and uses the Detection & Notification Service to ensure the execution of the strategy.

2.1 Services Design

Services have been designed so that: 1) they can be customized to different applications, 2) they are extensible in the sense that they are able to consider application-provided strategies, and 3) they ensure the flexibility of the framework since different strategies can be considered for a service.

Functionalities of a service are implemented in MolèNE by *components*. A component is made up of a set of objects and constitutes the smallest software unit in MolèNE. For example, a cache management service provides two functionalities: the *data replacement* that is implemented by a MolèNE component deciding which data to delete from the cache and the *cache loading* that is implemented by a component that decides the data to be loaded into the mobile computer cache.

Figure 1 shows the architecture of a MolèNE component. The *Interaction* object receives functional requests from other components by means of events (synchronous or asynchronous) containing the necessary information to manage the request. The *Implementation* object provides the functionality of the component. Finally, applications can build an adaptive component by adding a *Controller* object to the component architecture. This object is the main element of the reactive system: it ensures the execution of the adaptation strategy for the component by using the Detection & Notification Service to be informed of changes on the environment and the *Adaptation Strategy* to know the reactions to be carried out.

Designing services in this way 1) facilitates their construction since a well-defined set of issues is identified that ease the choices of developers and limit redundant solutions, and 2) allows a fine-grained customization and dynamic adaptation of the services.

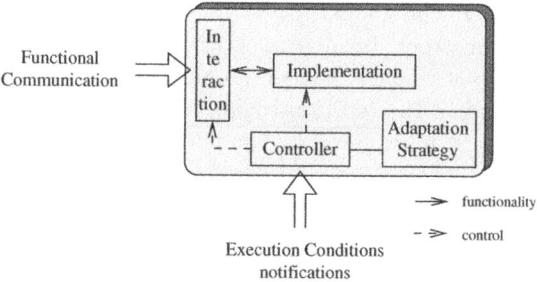

Fig. 1. MolèNE component architecture

2.2   Describing the Adaptation Strategy

The key element of the adaptation mechanisms of MolèNE is a set of automata that describes the reactions of the MolèNE components to changes on the execution conditions. These automata constitute the distributed adaptation service of MolèNE. Each automaton is described by application designers as a set of states that represent execution conditions and transitions which determine the reaction to be performed when changing from one state to another. The reactive system is then responsible for interpreting the automata and for executing the corresponding reactions.

The automaton for a component is described as an initial state and a set of reactions. The former specifies the initial behavior for the component and the execution conditions (the state of the automaton) on which it can be used. The latter describes the reaction to be performed for each transition in the automaton. Two types of reactions may be associated to a transition. The first one modifies some of the parameters of the component behavior. The second one changes the behavior itself.

Modifying the parameters of the component behavior.   When applied by the *Controller*, this reaction changes the value of some parameters of the *Implementation* object of the component. Applications must specify on the reaction, the parameters to be changed and the new values.

Modifying the component behavior itself.   This reaction asks the *Controller* to replace the *Implementation* object of the component. One important aspect of this replacement is the transfer of information between the old and the new object. As an example, consider the component responsible to decide which data to load into the mobile computer cache. This component requires a connection to the data management system in order to store the data when they arrive. Changing the behavior of this component consists on replacing the algorithm used to choose the data to be load. However, the connection to the data management system can be used by the new algorithm and, thus, must be transfered to it.

MolèNE provides the mechanisms allowing the transfer of information between *Implementation* objects. These mechanisms are performed by the objects to be replaced as well as the adaptation strategy. When an object is replaced by another, it must construct its *internal state* that is composed of the set of informations that it uses when it is functional. This state is modified by the *state adapter* associated to the reaction by applications in order to build a new internal state that can be interpreted by the new behavior.

2.3   The Reactive System

The *Controller* of a component (see Section 2.1) ensures the execution of the adaptation strategy defined by applications. The most difficult reaction to implement is the replacement of behavior since relations of the component with other elements must be considered. In order to perform this type of reactions, *Implementation* object collaborates with the *Controller* by providing its internal state that will be modified by the state adapter associated to the reaction.

Modifying the behavior of a component implies the collaboration of three elements in MolèNE: 1) the *Controller* that decides when to execute a reaction, 2) the *Implementation* object that asks for information to other components and builds its internal state, and 3) the state adapter associated to the reaction that implements the passage from one state to another. All of them are used by the Distribution Service available in MolèNE and presented in the next section.

3   The Distribution Service

Designing a distribution algorithm that satisfies all types of applications needs and takes all possible environment changes into account is not a realistic approach. Therefore, we aim at providing not a particular distribution algorithm but a framework, AeDEn, into which designers can insert their own distribution algorithm according to their applications needs and changes of the execution environment.

Traditionally, a distribution algorithm is composed of three policies [12]: *the information policy* which specifies the nature of informations needed for the election (CPU load, etc) and how they are collected, *the election policy* which elects processes which must be suspended, placed or migrated according to the collected informations, and *the placement policy* which chooses the station for each elected process.

Distribution functionalities are provided by AeDEn services which implement the different policies as illustrated in Fig. 2. The election and placement policies are part of the Distribution Service while the information policy is shared between the Data Manager Service and the Detection & Notification Service. The former stores the resources and services states. The latter detects environment changes and notifies the Data Manager Service accordingly.

AeDEn services benefit from two dimensions of adaptation. The first one allows applications to customize AeDEn to their needs. The second one refers to the capabilities of AeDEn to adapt to fluctuations in the environment.

3.1   Application-Adaptive Distribution

Applications interaction with AeDEn takes place at launching time. Figure 2 illustrates the steps performed by AeDEn at this time. Applications should provide the Distribution Service, and more concretely the election policy, with the set of components that can be distributed (1,2). The election policy interacts with the Data Manager Service in order to ask for resources and services states (3) so that load indices can be calculated (4). The components are then submitted to the placement policy (5) which determines the most suitable stations for them and performs their transfer (6). It registers the components in the Data Manager Service (7) and activates the monitoring of these components (8).

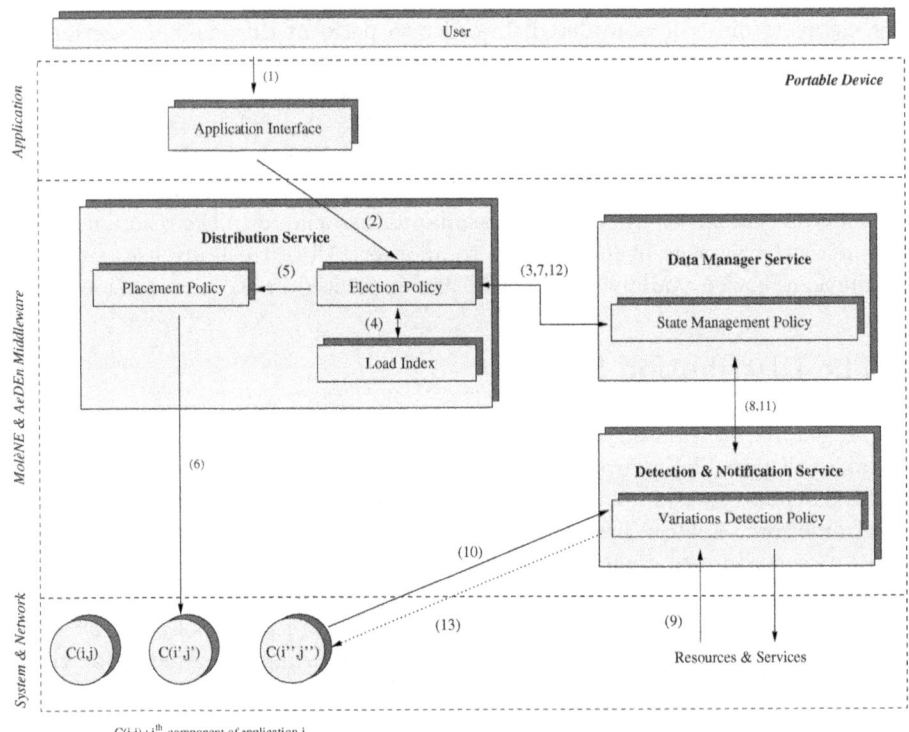

Fig. 2. AeDEn architecture

3.2   Environment-Aware Distribution

Changes in the environment are taken into account by AeDEn services at two
different levels: the application and the distribution system itself. The first one
concerns the mapping of the application onto a set of stations. According to environment changes and to policies decisions, AeDEn services dynamically change
the placement of applications components. The second level allows to dynamically change each policy of the distribution algorithm used to decide about the
mapping.

Dynamic Placement.   When the Detection & Notification Service detects
a change in resources or services of the environment, step (9) in Fig. 2, or in
components (for example a removal) (10), it notifies the Data Manager Service
which updates its information database (11) and notifies the Distribution Service if some of the distributed components are involved[1] (12). Then, the election

---

[1] Application components can also be designed to adapt themselves according to environment changes. In this case, they are directly notified by the Detection & Notification Service (13).

policy determines if these changes affect the placement. In this case, a new placement is performed (3,4,5,6) which may imply the placement of new components, the suspension, and/or the migration of existing ones.

Dynamic Policies.   As we said in the beginning of the section, a particular policy can be more or less relevant depending on the application needs and constraints. In order to ensure the suitability of AeDEn in all cases, we introduce *dynamic policies* which change their implementation as changes occur in the environment. Each dynamic policy has only one implementation at time t but can dynamically change and have a different implementation at time $t + 1$.

This opportunity to dynamically change the implementation of a policy requires well-adapted and efficient programming tools. We use the MolèNE object-oriented framework which provides these adaptation facilities. Each AeDEn policy is implemented as a MolèNE component. Interactions among themselves and with the application and the environment are therefore performed by sending/receiving events and the different possible implementations for a policy are provided as *Implementation* objects.

We illustrate the interest of dynamic policies by taking the example of the election policy (see Fig. 3). The events received by the *Interaction* object of this policy originate from the application level or the Detection & Notification Service while the placement policy receives those that are sent. These events correspond to those described in Sections 3.1 and 3.2. The notification of changes in the environment are also implemented as events that are sent by the Detection & Notification Service to the *Controller* of the election policy. They indicate that the current election *Implementation* object may be no longer adequate. The current *Implementation* is then changed by the *Controller* according to the *Adaptation Strategy* object provided by the application.

For example, let's consider a user working with a laptop. As long as the laptop is connected via a dock-station to a local network with a central server, it is reasonable to have a centralized and total election policy. When the laptop moves into the building, keeping the connection via a wireless link, a better policy would be centralized and partial in order to reduce the network traffic. If the laptop moves out-door and is connected to an ad-hoc network, the policy should be switched to a distributed and partial one.

This mechanism of dynamic policy is also used for the other policies of the distribution algorithm and for the load index. The load index can dynamically change to take into account mobility criteria such as the low bandwidth, the probability of disconnection, and the battery index.

4   Related Work

Adaptation.   Dynamic adaptation to execution conditions has become a hot topic in the last years and several systems propose their dynamic adaptation to changing conditions. Odyssey [8] allows applications to specify a data access strategy depending on the type of the accessed data and current resources

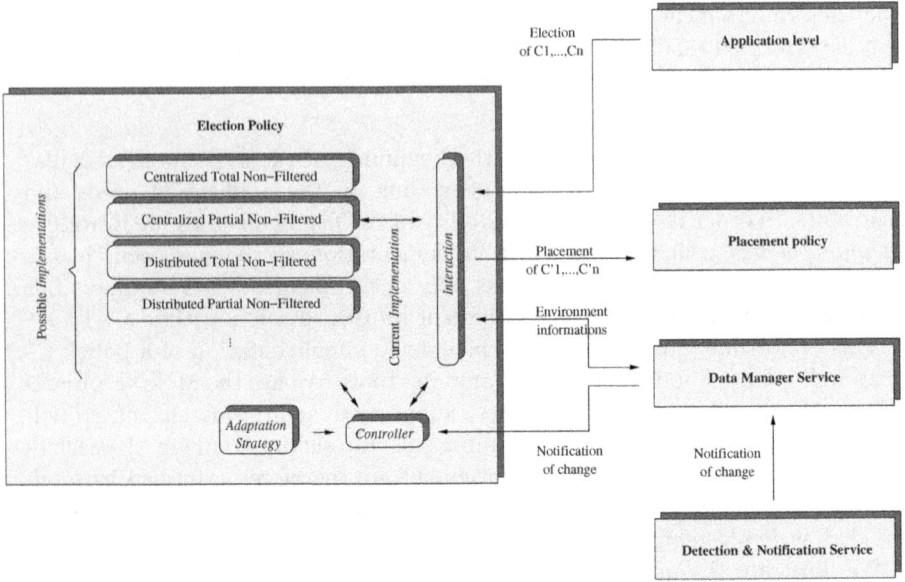

Fig. 3. Strategies for the election policy

availability. Mobiware [2] proposes a programmable mobile network on which mobile computers, access points and mobile-capable routers are represented by objects that can be programmed by applications to support adaptive quality of service assurances. Ensemble [9] is a network protocol architecture constructed from simple *micro-protocol* modules which can be stacked in a variety of ways to meet the communication demands of its applications. The adaptation mechanisms proposed by these approaches are specific to the problem they address: the bandwidth variability. MolèNE mechanisms such as the adaptation interface or the *Controller* element of a component are general and can provide so lution to different types of problems.

A more generic mechanism is proposed by the 2K system [6]. It performs the reconfiguration of components based on the reification of components dependences. This mechanism is complementary of the one provided by the *Controller* of a MolèNE component. In our system we have emphasized the adaptation of the components themselves which is not defined in the 2K system. Moreover, we are currently formalizing a way to manage relations among adaptive components.

Distribution.   Few approaches propose the utilization of remote resources to overcome their lack on portable devices. The mobile agent model is one of them [5]. This model allows to delegate actions such as information gathering and filtering to the mobile agent on the fixed network so that only the resulting data are transmitted to the mobile client. In the Client/Proxy/Server model, a proxy runs on a fixed station and uses its resources to provide facilities for the mobile user such as efficient new network protocols, storage of results while disconnections or filtering such as degradation or display media conversion [11].

Proxies can also be distributed to increase the number of available resources. For example, distribution techniques are used in Daedalus [4] to deploy adaptation-based proxy services on a cluster of PCs. The applied strategy consists in increasing the number of proxies according to the number of requests in waiting queues. This approach is nevertheless specific: distributed processes are exclusively proxies defined by the adaptation system designer for a particular task; the distribution strategy is also fixed and cannot take new criteria into account. To distribute an application in a more general way, distribution techniques proposed in the distributed computing area should rather be considered.

Many distribution algorithms exist which ensure good performance, load-balancing and stability [3,7]. However, these algorithms perform load sharing among stations based only on CPU criterion. Multi-criteria approaches also exist [13] but they do not take into account the new criteria introduced by mobile computing such as a low and variable bandwidth, the probability of disconnection (previous approaches assume to have a high and continuous connection) or a battery index.

5   Conclusion

In this paper, we have presented AeDEn, a system which provides adaptive and dynamic mechanisms to distribute applications in an extremely varying environment. This distribution allows to overcome the poorness of available resources on portable devices by using resources and services of the environment. It also allows to reduce variability effects by executing applications components on stations less sensitive to variations. Our mechanisms are application-adaptive in the sense that they take into account application needs. Moreover, environmental changes are considered not only by allowing the dynamic placement of the application but also by dynamically changing the policies used to decide about the placement.

AeDEn has been constructed as a part of a more general system, MolèNE, that has been designed in order to facilitate the construction of adaptive solutions to manage wireless issues. MolèNE is structured as a set of self-adaptive components thanks to the *Controller* object encapsulated into their architecture. MolèNE extensibility is ensured by the distinction between the interface and the implementation of a component and the utilization of introspection and reflexion techniques at the interface level.

We have implemented a preliminary AeDEn prototype using the MolèNE components developed in Java. We are currently integrating it in the MolèNE system and we plan to test the different distribution mechanisms with a real electronic press application. This application is a good candidate for our distribution system since it is interactive, manipulates an important volume of data and requires quite a lot of computation power.

References

1. F. André, A.-M. Kermarrec, and F. Le Mouël. Improving the QoS via an Adaptive and Dynamic Distribution of Applications in a Mobile Environment. In *Proc. of the 19th IEEE Symp. on Reliable Distributed Systems*, Nürnberg, Germany, October 2000.
2. O. Angin, A.T. Campbell, M.E. Kounavis, and R.R.-F. Liao. The Mobiware Toolkit: Programmable Support for Adaptive Mobile Networking. *IEEE Personal Communications Magazine*, August 1998.
3. B. Folliot and P. Sens. GATOSTAR: A Fault Tolerant Load Sharing Facility for Parallel Applications. In *Proc. of the 1st European Dependable Computing Conf.*, volume 852 of *Lecture Notes in Computer Science*. Springer-Verlag, October 1994.
4. A. Fox, S.D. Gribble, Y. Chawathe, and E.A. Brewer. Adapting to Network and Client Variation Using Infrastructure Proxies: Lessons and Perspectives. *IEEE Personal Communications*, 5(4), August 1998.
5. R. Gray, D. Kotz, S. Nog, D. Rus, and G. Cybenko. Mobile agents for mobile computing. Technical Report PCS-TR96-285, Department of Computer Science, Dartmouth College, May 1996.
6. F. Kon and R.H. Campbell. Supporting Automatic Configuration of Component-Based Distributed Systems. In *Proc. of the 5th USENIX Conf. on Object-Oriented Technologies and Systems*, California, USA, May 1999.
7. M.J. Litzkow and M. Livny. Experience With The Condor Distributed Batch System. In *Proc. of the IEEE Work. on Experimental Distributed Systems*, Alabama, USA, October 1990.
8. B.D. Noble, M. Satyanarayanan, D. Narayanan, J.E. Tilton, J. Flinn, and K.R. Walker. Agile Application-Aware Adaptation for Mobility. In *Proc. of the 16th Symp. on Operating Systems Principles*, St. Malo, France, October 1997.
9. R. Van Renesse, K. Birman, M. Hayden, A. Vaysburd, and D. Karr. Building Adaptive Systems Using Ensemble. *Software-Practice and Experience. Special Issue on Multiprocessor Operating Systems*, 28(9), July 1998.
10. M.T. Segarra and F. André. A Framework for Dynamic Adaptation in Wireless Environments. In *Proc. of the Technology of Object-Oriented Languages and Systems*, Saint Malo, France, June 2000.
11. Wireless Internet Today. *Wireless Application Protocol - White Paper*, October 1999.
12. S. Zhou. A Trace-driven Simulation Study of Dynamic Load Balancing. *IEEE Transactions on Software Engineering*, 14(9), September 1988.
13. S. Zhou, X. Zheng, J. Wang, and P. Delisle. Utopia: A Load Sharing Facility for Large, Heterogeneous Distributed Systems. *Software - Practice and Experience*, 23(12), December 1993.

Optimal Algorithms for Routing
in LEO Satellite Networks with ISL

M.M. Riad[1] and M.M. Elsokkary[2]

[1] Cairo University,Cairo,Egypt.
mmriad@alpha1-eng.cairo.eun.eg
[2] Cairo University,Cairo,Egypt.
mo_elsokkary@yahoo.com

Abstract. This paper deals with the performance evaluation for the routing techniques of the ISL's (intersatellite links) for the LEO. We took the iridium LEO network as a model, and studied the routing in two cases. The first is based on the minimum delay, and the second takes into consideration the delay and handover of the ISL's. A simulation program is developed in order to compare between these two techniques. Another algorithm was developed for finding the optimal set of short disjoint paths in a network to improve the network reliability.

1 Introduction

Because of technological advances and the growing interest in providing personal communications services on a global scale, satellite links will naturally be part of the concept whether as an overlay network or integrated in the terrestrial network. LEO satellites permit relaxation of the constraints on the link budget, allow the use of low-power handheld mobile terminals, and ensure the earth coverage with smaller cells, so achieving a higher traffic capacity. As an example for the LEO constellations is the Iridium network which uses the intersatellite links for the call routing. Two routing parts can be clearly identified with different characteristics; UDL (Uplink and Downlink) and ISL (Intersatellite links). In case of UDL routing [6],[7], Uplink must be directed towards one satellite seen by the calling party. Maybe more than one satellite is seen by the calling party and could takeover the transmission. So decision-making is required at this point. The same situation for the downlink, there must be a decision of which of the satellites overtakes the downlink. In case of ISL, the interplane connection between two satellites can't be maintained for all the time due to the movement of the satellites. So the visibility problems appear in both ISL and UDL. There are some techniques used to overcome this visibility problem such as Handover and Satellite Diversity [1]. Our purpose in this paper is to decrease the ISL handover as possible as long as the side effects that are represented by increasing the path delay, is acceptable.

M. Valero, V.K. Prasanna, and S. Vajapeyam (Eds.): HiPC 2000, LNCS 1970, pp. 489–500, 2000
© Springer-Verlag Berlin Heidelberg 2000

2 Network Model

Our model is as follows:
1. Near-polar circular orbits, constant inclination = $86.4°$ to the equator.
2. There are N satellites or nodes (N=66 in Iridium, distributed in 6 orbits, so each orbit has 11 satellites). Each satellite or node is capable of four active links; two intraplane forward and back, and two to satellites in each neighboring plane. Only twenty-four of the thirty-six satellites involved in ISLs will be fully four-connected at any time.
3. The network topology is subjected to changes due to the discrete time activation /deactivation of links which take place due to the motion of the satellites. The periodical topological changes within the ISL subnetwork are completely deterministic since they are only dependent on the respective satellite positions. The complete topology dynamics is periodic with period T. The dynamic topology is considered as a periodically repeating series of k topology snapshots separated by step width $\Delta t=T/k$. Each of these snapshots is modeled as a graph G(k) of constant set of nodes and each link of this graph is associated with its cost metric. For each snapshot we calculate a set of m (max_paths) distinct loopless, shortest paths for each two nodes (source, destination) using forward search algorithm. In other words the algorithm setup for each snapshot (k) an instantaneous virtual topology upon G(k), which is done by the (m-path) shortest path algorithm for every source-destination node. The algorithm is extended to calculate, for each m (max_paths) for all (source-destination) nodes, the path continuity optimization procedure to minimize the *path handover rate*, with *acceptable delay* and *delay jitter* [2].

3 Simulation Approach

The simulation approach is split into three steps:

3.1 ISL Topology Setup

Due to the physical time variant ISL topology of the system, the topology and the existing links between each pairs of satellites have to be known for each interval Δt. It is assumed that the time variance of the ISL topology is deterministic. Therefore all the ISL can be calculated in advance for whole system period, which is in the design process prior to system operation. In other words, its possible to set up off-line, i.e. prior to the operational phase of the system, dynamic virtual cross connect network incorporating all satellites as pure VP switches, this provides a set of VPC's between any pair of terminating satellites at any time. The satellite in adjacent planes cross each other making the antenna tracking process much more difficult and increasing the pointing angle requirement, like in iridium, so we have permanent ISL's and non-permanent ISL's.

3.2 VPC Path Search

From the first ISL setup step, the ISL are identified, and the next step deals with the search for the end-to-end routes. This search may be based on different path search algorithms, considering variable cost metrics in order to defined a subnet of the least cost paths for all satellite pairs and all time intervals. The following assumption will be followed:

1. The cost metric of the links is the propagation delay.
2. Permanently active ISL's should be given preference over those being temporarily switched off. The permanence of ISLs is introduced into the cost function as parameter *perm*, assigning lower costs to permanent ISLs and thus resulting in preferential routing over such links.
3. The unequal traffic demand imposed from earth suggests some sophisticated traffic adaptive routing during system operation to improve traffic distribution shaping in space for the sake of overall traffic routing performance. To handle this task, we add the parameter *geogr*, which assigns lower costs to ISLs of satellite covering regions with lower traffic demand. The weighting function is therefore given by the following:

$$LW(t_i) = \frac{dis\,\tan ce}{C} \frac{1}{perm} \frac{1}{geogr} \tag{1}$$

Where C = speed of light, and (*perm* , *geogr*) are dimensionless parameters from]0,1]. Once the calculation of all LW is completed, we can calculate the total link weight TLW for one VPC at t = t_i using the following relation,

$$TLW_j(t_i) = hops * d_{switch} + \sum_{k=1}^{hops} LW_k(t_i) \tag{2}$$

Where, hops: is the number of ISL's on the route, and d_{switch}: is the switching time at each satellite node [5]. So the TLW is the cost metric for the shortest path algorithm used. This algorithm search the whole set of the possible VPC's per time interval and per satellite pair, then selects a subnet of the max_paths best candidates, and stores them, sorted by their TLW value, for the next optimization [Delay + HO] step. We used the forward search algorithm to find the max_paths best candidates as shown in fig.1. The ISL topology setup provides us with the four connections for each node at any time; nodes (0), nodes (1), nodes (2), nodes (3) as shown in the flow chart in fig.1.

3.3 Optimisation Procedure w.r.t. (Delay and Handover)

The goal for this optimisation (taking into account the delay and handover) is to minimize the (VPC-HO) rate with reference to one system period, while regarding the side effects arising in the form of instantaneous higher costs which can be evaluated in terms of TLW jitter. The optimisation is performed over the complete system period to minimize the path handover rate with acceptable TLW and TLW jitter. The implementation of this approach is illustrated in the flow chart fig.2. The rate of path handover will be less if the algorithm for choosing the path takes into consideration not only the delay, but also the history of the path. From the available set of best paths found by the [forward search] algorithm for a specific time interval and start / end satellite pair, one path must be selected to provide the corresponding calls during this interval, following two rules:
1. If the path chosen in the last time interval is also available for the present one, it should be maintained as long as its costs are good enough.
2. If the path chosen in the last time interval is not an element of the present set, or its costs are not good enough, a VPC handover is initiated toward the best path of this set. The decision whether the costs of a path are good enough is in both cases not only based on the actual TLW values, but in addition on a complex non-linear function value TC (Total Cost), which in corporate the history of the respective paths in terms of costs. If the old path is in the present set but not the best candidate, a choice may be made between actually lowest cost but new path (increasing HO) and an old (continuing) but higher cost path (reducing HO and increasing the delay). In order to handle this case in a well-defined manner, the parameter [Change_Factor] is introduced which specifies the actual decision threshold.

So the cost metric will be the delay (TLW) plus the (TC). The TC function takes into accounts the presence and absence of the path in the time intervals. There are different approaches to calculate the TC [3]. Our goal here is to reward the path (VPC$_j$) if it is present in the time step, and to punish it if it is not present. So we include the continuity (Continuity_Factor) for the path in this function. Our rule is, the more the path is present, the less the TC is calculated which leads to small cost metric. After that we take the minimum path for each time step. This is shown in equation (4),(5). If the path is not present in the time step, we want to punish it i.e., enlarge its cost as possible as we can, this can be done by taking its cost as the maximum cost for the paths between this start and end satellite. This is expressed by the array MAX_COST[end sat] that gives the maximum cost appearing among all VPC's in all time intervals. The magnitude of the steps_missed is a parameter that is calculated for each path, which expresses the number of consecutive time steps, that the path (VPC$_j$) is absent. From this magnitude we can calculate the missed_part, which affect the TC function as shown in equation (5).

We have two cases for the missed path, either the path is absent for consecutive time intervals or having many absences spread all over the time intervals. So the relation between the steps_missed and the missed_part must not be linear. We take the relation between them [3] as in equation (7), by this simple non-linear function.

$$TC_j = \sum_{i=1}^{i=n} TC_{ji} \qquad\qquad (3)$$

Where, n is the number of simulation steps and TC_{ji} is given by,

$$TC_{ji} = \frac{TLW_j(t_i)}{Continuity\_Factor(t_i)} \qquad \text{if } VPC_j \text{ is present in } t_i \qquad (4)$$

$$TC_{ji} = MAX\_COST[end\_sat]*missed\_part \qquad \text{if } VPC_j \text{ is not present in } t_i \quad (5)$$

Where,

$$Continuity\_Factor(t_i) = Continuity\_Factor(t_i - 1) + 1 \qquad \text{if } VPC_j \text{ is present in } t_i \qquad (6)$$

$$Continuity\_Factor(t_i) = 1 \qquad\qquad \text{if } VPC_j \text{ is not present in } t_i$$

where,

$$Continuity\_Factor(0) = 1$$

$$missed\_part = steps\_missed - [\frac{1}{2*steps\_missed}] + .5 \qquad (7)$$

The simplified flow chart diagram [5] is illustrated in fig.2. We applied this flow chart in our simulation program.

3.4 Developed Simulation Programs

The simulation program developed through this paper was performed using *Visual C++* language to calculate:
1. ISL topology Setup.
2. The path search algorithms to find *joint* and *disjoint* paths.
3. Optimisation w.r.t. delay [Shortest path algorithm].
4. Optimisation w.r.t. delay and path handover.
The input are the ISL topology for the 66 sat at t=0, and the rules for the connections. The outputs of the simulation programs are files out.txt and outopti.txt.

3.4.1 Parameter Setting

1. The parameter Change_Factor is used in the optimization procedure to fine-tune the decision threshold for maintaining an earlier selected VPC over the first choice VPC of the considered step. The value for this parameter should be chosen to achieve a reasonable trade off between VPC-HO reduction and the increasing in the delay jitter between the old and the new path.
2. The switching time in the satellite nodes (d_{switch}) is 10 ms.
3. The parameters (perm) and (geogr) are set to one for all simulations.
4. The simulation time step (Δt) is equal to 2 min for iridium, thus providing sufficient time resolution within the simulated system period.
5. The LW=14 ms between any two neighbour nodes.
6. $\Delta t=2$ min, k=50 topology for the total period 100 min.

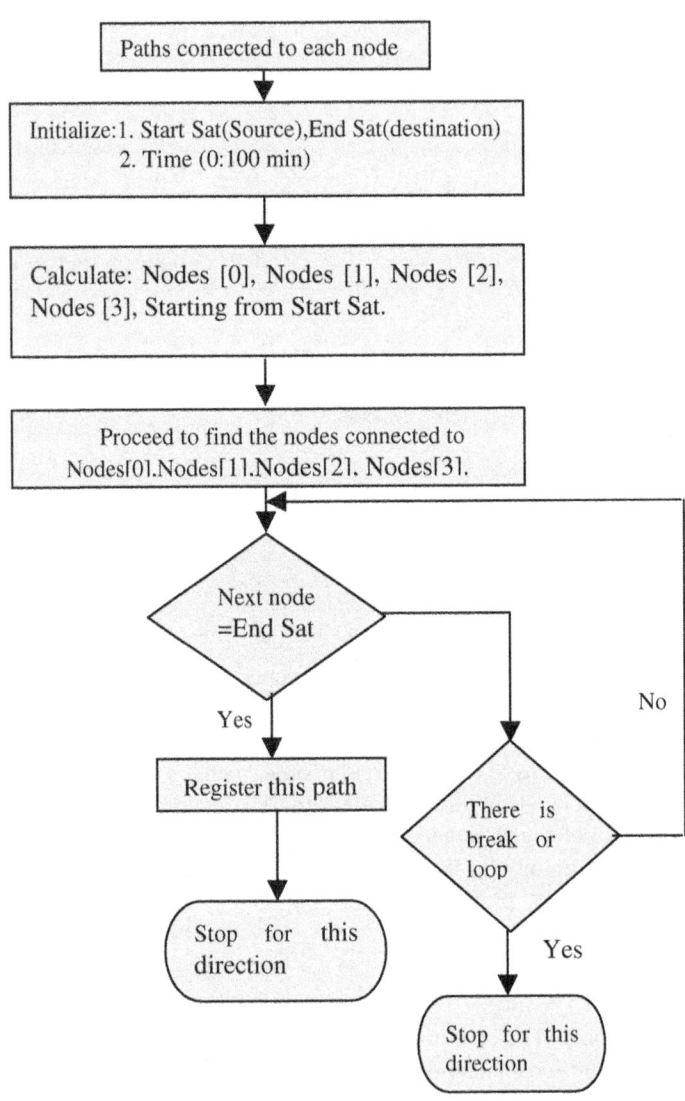

Fig. 1. The block diagram for the forward search algorithm

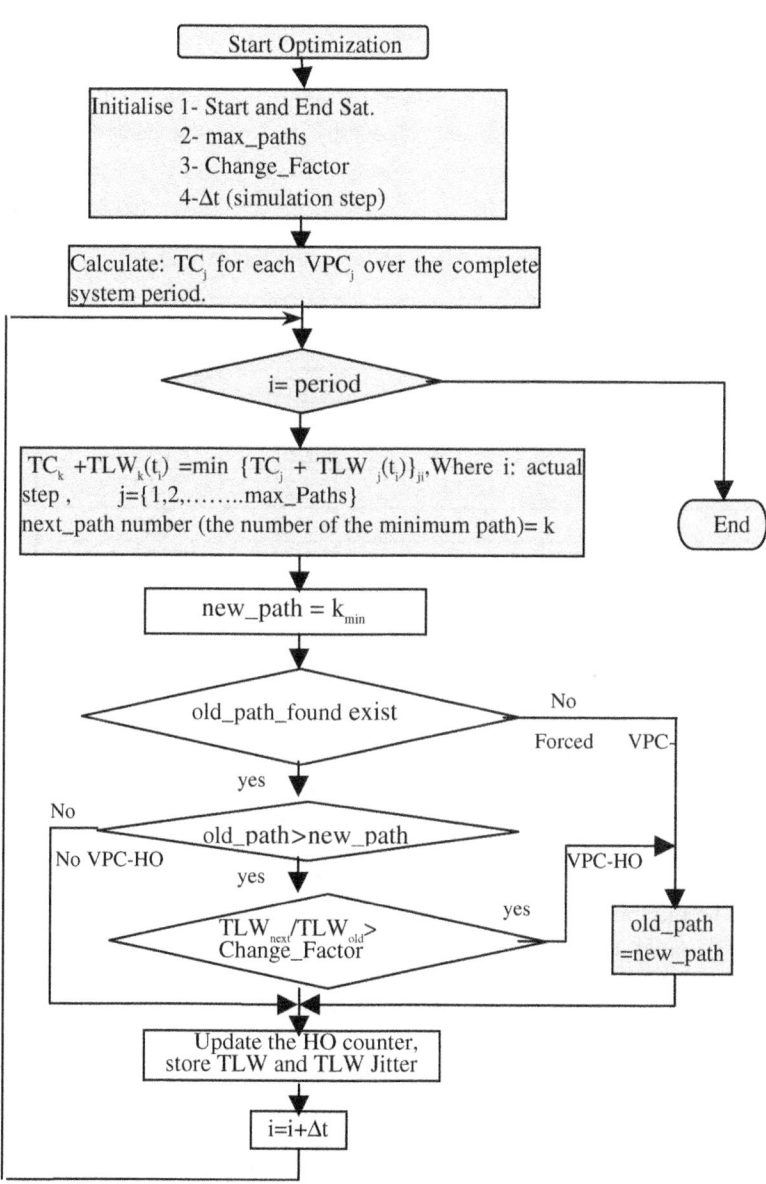

Fig. 2. Simplified Flow Chart for the VPC-HO optimisation procedure [5]

3.4.2 Results

It is clear that the more paths are available (max_paths) at a moment for the algorithm, the greater the chance that the path selected during the last time interval is also present in the actual interval therefore, theoretically, the more paths are available for the selection within a time interval, the better the optimization (HO and delay) algorithm will work. The real situation is then shown by the results presented in table (1) for our proposed system.

The values in table 1 are calculated with Change_Factor=0.2.

Table 1. Simulation results for different sets of best paths

Max_path	VPC-HO Number	HO decrease	Average TLW per step (ms)	TLW increase
1	577	-	108.78	-
2	523	9.36 %	111.32	2.33 %
3	509	11.79 %	112.63	3.54 %
4	501	13.17 %	113.26	4.12 %
5	500	13.34 %	113.38	4.23 %
6	496	14.04 %	113.45	4.29 %

From the results in this table, as the max_paths increases there is an improvement in the VPC-HO rate, and in the same time there is degradation in the quality (increase TLW).

For max_paths (3,4) there is a significant improvement in the VPC-HO as shown in the table, but for (5,6) the improvement in the VPC-HO rate is small, compared to the case of max_paths =4, and does not justify the computation of the additional paths. So, we will take the max_paths=4 as the best value. The results show that the delay jitter when applying the optimisation algorithm [HO+Delay] is very close to the optimisation using the forward search algorithm [Delay]. The maximum delay jitter was 24 ms. The delay jitter parameter is very important for the quality of service, the more the jitter the less the quality of service. The Change_Factor is the parameter that used to decide if the path can continue or VPC-HO must take place. The value for this parameter is]0,1[.

For our simulation, we take the case of max_paths =4, and investigate to study the influence of this parameter in the number of handover and the TLW. Chart (1) describes the increase of VPC-HO as the Chang_Factor parameter increases. Also chart (2) describes the decrease in the average TLW per simulation step, as the Change_Factor increases. From the two chart we can select some values for the Change_Factor which compromise the VPC-HO and TLW, i.e. achieve good reduction of VPC-HO with acceptable value of the side effects (increase of TLW). So we will choose Change_Factor =0.2.

Comparison between Joint and Disjoint Paths

The disjoint paths between two nodes are the paths share no nodes (vertices), and hence any links (edges), other than the source and terminal nodes. The advantages of finding two or more disjoint paths between source and terminal nodes are the enhancement the network reliability and survivability [4]. So by applying the forward search algorithm to find the paths for Change_Factor=0.2, and max_paths=4, we can reach to chart (1) in case of joint paths and chart (5) in case of disjoint paths. If we compare chart (5) by chart (1), it is obvious that in the disjoint paths the no. of VPC-HO reaches 576, but the maximum no. for the VPC-HO in the joint paths doesn't exceed 530. So we gain the reliability and pay the increasing no. of VPC-HO in case of using the algorithm with disjoint paths.

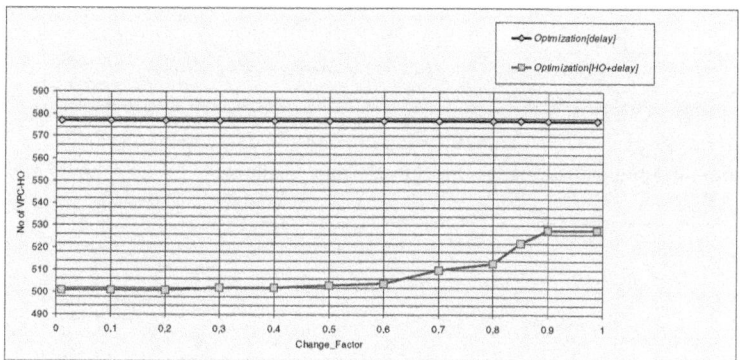

Chart 1. No. of VPC-HO vs. Change_Factor for one system period, Start satellite S11 and all satellite pairs for max_paths=4, Joint paths.

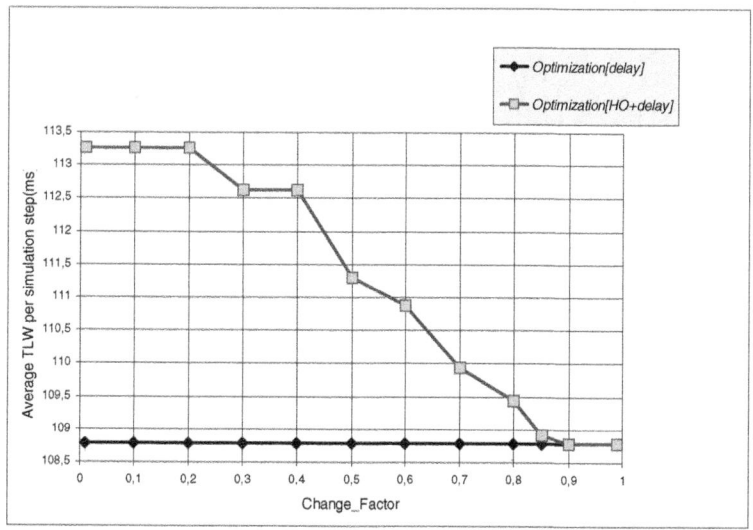

Chart 2. Average TLW per simulation step vs. Change_Factor for one system period, start satellite S11 and all satellite pair in Iridium for max_paths=4, Joint paths

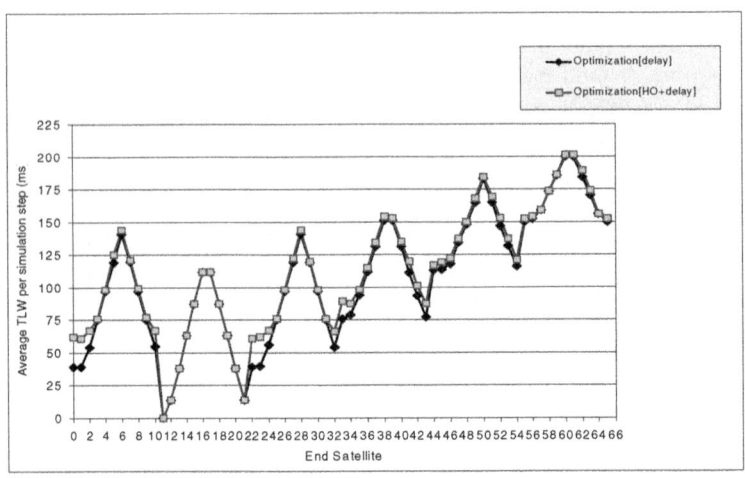

Chart 3. Average TLW over one system period for start satellite S11 in Iridium, Joint paths, Change_Factor=0.2, max_paths=4

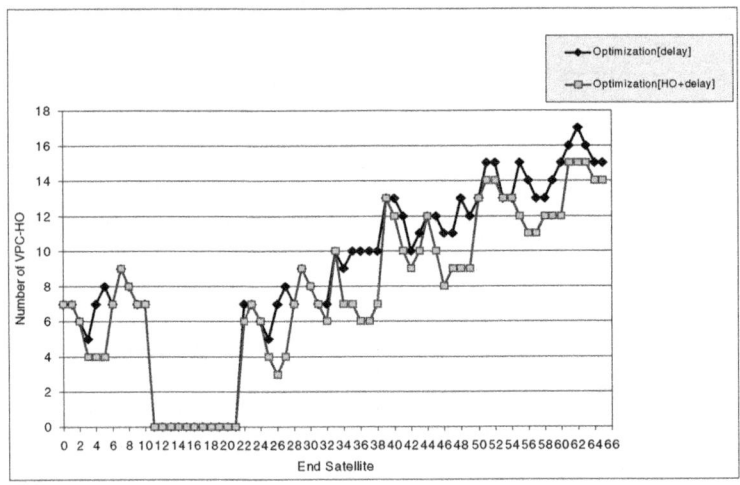

Chart 4. Number of VPC-HO over one system period for start satellite S11 in Iridium, Joint paths, Change_Factor=0.2, max_paths=4

4 Conclusion

The concept for routing of information over an ISL network of LEO satellite system has been proposed. DT-DVTR concept works completely off-line, i.e., prior to the operational phase of the system. In the first step, a virtual topology is set up for all successive time intervals, providing instantaneous sets of alternative paths between the source and destination nodes. In the second step, path sequence over a series of time intervals is chosen based on forward search shortest path algorithm. In the last step, an optimization strategy was developed to reduce the average number of route changes for an end-to-end connection, without significant increase of the path delay. This leads to improve the quality of service. Our future work will be focussed in the handover techniques in the UDL (up/down link) part of the proposed system. The used routing technique in the ISL and the handover in the UDL part affect the total performance and quality of service of the system.

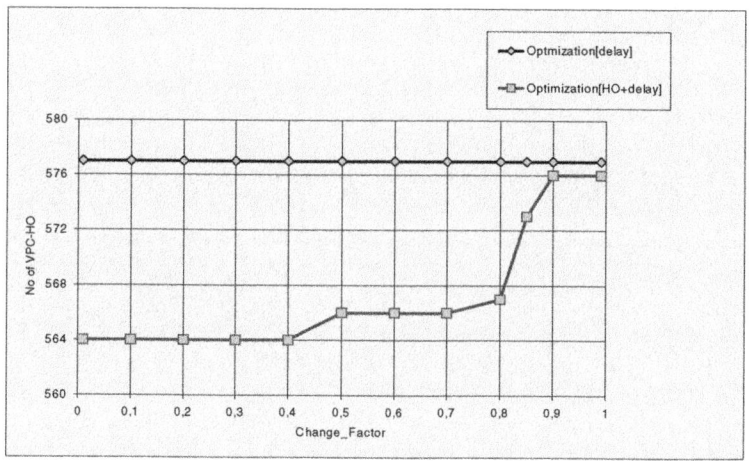

Chart 5. Number of VPC-HO vs. Change_Factor for one system period, start satellite S11 and all satellite pairs in Iridium for max_paths=4,disjoint paths.

References

[1]   "Satellite Diversity in Personal Satellite Communication Systems," Institute for Communications Technology of German Aerospace Research Establishment (DLR).

[2]   M. Werner "A dynamic Routing Concept for ATM-Based Satellite personal Communication Network," IEEE J. On Selected Areas in Communications, vol. 15,pp. 1636-1648, Oct 1997.

[3]   C.Delucchi, "Routing strategies in LEO/MEO satellite networks with intersatellite links," Master thesis, Technical University Munich, Institute of Communication Networks, Munich, Germany, Aug. 1995

[4]   D.Torrieri "Algorithms for finding an Optimal Set of Short Disjoint Paths in a Communication Network," IEEE Trans. on Communications, vol. 40, pp.1698-1702, Nov. 1992.

[5]   M. Werner, C.Delucchi, H.-J.Vogel, G.Maral, and   J.-J. De Ridder, "ATM-BASED Routing in LEO/MEO Satellite networks with intersatellite Links," IEEE J. Selected Areas Communications, vol. 15, pp. 69-82,Jan. 1997.

[6]   G. Maral, J. Restrepo, E. Del Re, R. Fantacci, and G. Giambene, "Performance Analysis for a Guaranteed Handover Service in an LEO Constellation with a Satellite-Fixed Cell System," IEEE Transactions on VT., vol.47, pp.1200-1214,Nov.1998.

[7]   E. Del Re, R. Fantacci, and G. Giambene "Handover Queuing Strategies with Dynamic and fixed Channel Allocation Techniques in Low Earth Orbit Mobile Satellite Systems" IEEE Transactions on Communications, vol.47, pp.89-102, Jan.1999.

Data Organization and Retrieval
on Parallel Air Channels

J. Juran, A.R. Hurson, N. Vijaykrishnan, and S. Boonsiriwattanakul

Computer Science and Engineering Department
The Pennsylvania State University
University Park, PA 16802, USA
{juran, hurson, vijay, boonsiri}@cse.psu.edu

Abstract. Broadcasting has been suggested as an effective mechanism to make public data available to the users while overcoming the technological limitations of the mobile access devices and wireless communication. In broadcasting, however, attempt should be made to reduce access latency at the mobile unit. Broadcasting over parallel channels and application of indices can be used to address these issues. This discusses how to order access to the data objects on parallel channels in order to improve access latencies. Several schemes have been proposed, simulated, and analyzed

1. Introduction

Traditional data management systems are based upon fixed clients and servers connected over a reliable network infrastructure. However, the concept of mobility, where a user accesses data through a remote connection with a portable, with a wide range of capability, has introduced several disadvantages for traditional database management systems (DBMS). These include: 1) a reduced network bandwidth, 2) frequent network disconnections, 3) limited processing and resources, and 4) limited power sources. In current practice, many of these portable devices are used to access public data and are characterized by: 1) the massive number of users, and 2) the similarities and simplicity of the requests generated by the users. In such an environment, broadcasting has been proposed as an effective method for distributing public data from central servers to many mobile users. In this work, we mainly concentrate on broadcasting as the underlying method of communication. Within the constraint of a limited bandwidth, the primary advantage of broadcasting is the fact that it scales well as the number of clients increases. Because the same data is sent to each client, there is no need to divide or multiplex the bandwidth amongst the clients. Furthermore, information on the air could be considered as "storage on the air", therefore, broadcasting is an elegant solution that compensates the limited resources (storage) at the mobile unit. Finally, the cost of communication is asymmetric, receiving information via this "push" method consumes about 10 times less energy than the traditional "pull" method of requesting information from a server and waiting for a response.

M. Valero, V.K. Prasanna, and S. Vajapeyam (Eds.): HiPC 2000, LNCS 1970, pp. 501–510, 2000
© Springer-Verlag Berlin Heidelberg 2000

Broadcasting has been used extensively in multiple disciplines. Figure 1 depicts an overview of the Mobile Data Access System (MDAS) environment. The data to be broadcast is available at the database server, and is transmitted to the mobile support stations (MSSs). The MSS broadcasts the data on air channels and is responsible for servicing the clients within its cell. The mobile hosts (MH) within the corresponding cell can retrieve the data from the air channels. In this paper, the term *broadcast* is referred to as the set of all broadcast data elements (the stream of data across all channels). A broadcast is performed in a cyclic manner. Since dependencies can exist among the data items within the broadcast, and since data items might be replicated within a broadcast, it is assumed that updates of data elements within a broadcast are only reflected at the following cycle. This constraint enforces the integrity among the data elements within the broadcast.

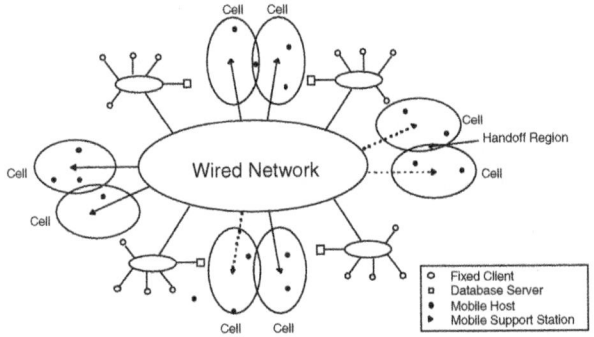

Fig. 1. Multidatabase System in the Mobile Computing Environment.

Several techniques can be used to reduce the response time for retrieving information from a broadcast. For example, information can be broadcast over parallel air channels and/or indexing technique can be used to determine the location and hence the time that the relevant information is available on air channel. One of the problems that arises from utilizing parallel air channels is the possibility for access conflicts between requested objects that are distributed among different channels. This results from the fact that the receiver at the mobile host can only tune into one channel at any given time and the length of time required for the receiver to switch from one channel to another. These conflicts require the receiver to wait until the next broadcast to retrieve the requested information, which will have a significant adverse impact on the response time and hence, power consumption.

This paper is aimed at analyzing the probability for conflicts between objects in parallel channels. It also shows that techniques for solving the traveling salesman problem can be utilized to organize object retrieval from a broadcast on parallel channels to minimize the number of broadcasts required, and thus to minimize the response time and power consumption at the mobile host. The next section models the conflicts on air parallel air channels and provides various schemes for efficient retrieval of objects. Section 3 concludes the paper and introduces some future research issues.

2.   Conflicts in Parallel Air-Channels

One of the problems associated with broadcasting information on parallel air channels is the possibility for conflicts between accessing objects on different channels. Because the mobile unit can tune into only one channel at a time, some objects may have to be retrieved on a subsequent broadcast. There is also a cost associated with switching channels. During the channel switch time, the mobile unit is unable to retrieve any data from the broadcast. Conflicts will directly influence the access latency and hence, the overall execution time. This section provides a mathematical foundation to calculate the expected number of broadcasts required to retrieve a set of objects requested by an application from a broadcast on parallel channels by formulating this problem as an Asymmetric Traveling Salesman Problem (TSP).

2.1   Model of Broadcasting System

Definition 1: A K-object request is an application request intended to retrieve K objects from a broadcast. In our model, we assume that each channel has the same number of pages of equal length and, without loss of generality; each object is residing on only a single page.

A single broadcast can be modeled as an N x M grid, where N is the number of pages per broadcast, and M is the number of channels. In this grid, K objects ($0 \leq K \leq MN$) are randomly distributed throughout the MN positions of the grid. The mobile unit can only tune into one channel at a time. The mobile host can switch channels, but it takes time to do this. Based on the common page size and the network speed, the time required to switch from one channel to another is equivalent to the time it takes for one page to pass in the broadcast. Thus, it is impossible for the mobile unit to retrieve both the ith page on channel A and the (i+1)th page on channel B (where A $\neq$ B). A diagram of the grid model described above is shown in Figure 2.

Definition 2: Two objects are defined to be in conflict if it is impossible to retrieve both objects on the same broadcast.

One method of calculating the number of broadcasts required is to analyze the conflicts between objects. For any particular object, all objects in the same or succeeding page (column) and in a different row (channel) will be in conflict. Thus, for any specific page (object) in the grid, there are $(2M - 2)$ conflicting pages (objects) in the broadcast. These $(2M - 2)$ positions are known as the conflict region. A diagram of the conflict region for a particular position is shown in Figure 3.

Although there are $(2M - 2)$ positions that may conflict with an object, there can be only $(M - 1)$ actual conflicts with a particular object. Two objects in the same row count as only one conflict because the mobile unit will be able to retrieve both objects in one broadcast. Because the number of broadcasts required is always one greater than the number of conflicts, the maximum number of broadcasts required is M.

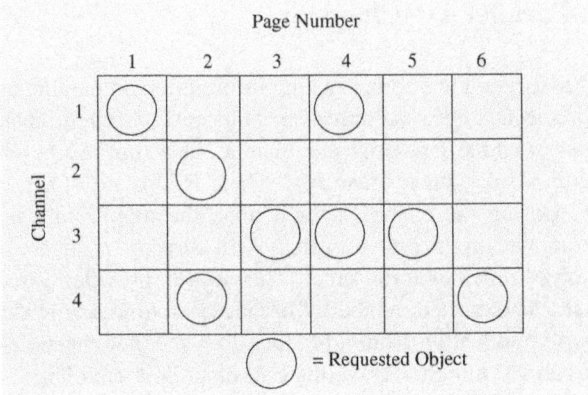

Fig. 2. Sample Broadcast with $M = 4$, $N = 6$, and $K = 8$.

For any particular object, it is possible to determine the probability of exactly i conflicts occurring (denoted as P(i)). Because the number of conflicts for any particular object is bounded by $(M - 1)$, the weighted average of these probabilities can be determined by summing a finite series. This weighted average is the number of broadcasts required to retrieve all K objects if all conflicts between objects are independent.

$$B = \sum_{i=0}^{M-1} (i + 1)^* P(i) \tag{1}$$

2.2   Enumerating Conflicts

In order to calculate P(i), it is necessary to count the number of ways the objects can be distributed while having exactly i conflicts divided by the total number of ways the K objects can be distributed over the parallel channels. Note that in our model, we consider only a conflict region of two pages (the current page and next page). In order to enumerate possible conflicting cases, we classify the conflicts as single or double conflicts as defined below.

Definition 3: A single-conflict is defined as an object in the conflict region that does not have another object in the conflict region in the same row. A double-conflict is an object that is in the conflict region and does have another object in the conflict region in the same row.

Examples of single- and double- conflicts are shown in Figure 3. The number of objects that cause a double-conflict, d, can range from 0 (all single-conflicts) up to the number of conflicts, i, or the number of remaining objects, (K $-$ i $-$ 1). When counting combinations, each possible value of d must be considered separately. The number of possible combinations for each value of d is summed to determine the total number of combinations for the specified value of i. When counting the number of ways to have i conflicts and d double-conflicts, four factors must be considered:

Whether each of the *(i – d)* objects representing a single-conflict is in the left or right column in the conflict region. Because each object has two possible positions, the number of variations due to this factor is $2^{(i-d)}$.

Which of the *(M – 1)* rows in the conflict region are occupied by the *(i – d)* single conflicts. The number of variations due to this factor is $\binom{M-1}{i-d}$.

Which of the *(M –1) – (i –d)* remaining rows in the conflict region are occupied by the *d* double- conflicts. *(i –d)* is subtracted because a double-conflict cannot occupy the same row as a single conflict. The number of variations due to this factor is $\binom{(M-1)-(i-d)}{d}$.

Which of the *(MN –2M +1)* positions not in the conflict region are occupied by the $(K - i - d - 1)$ remaining objects. The number of variations due to this factor is $\binom{MN-2M+1}{K-i-d-1}$.

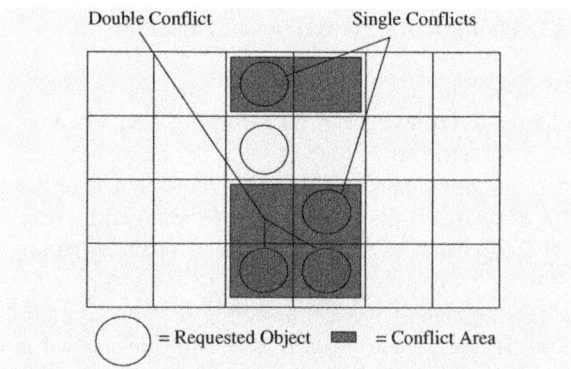

Fig. 3. Examples of Single and Double Conflicts

Note that these sources of variation are independent from each other and hence,

$$P(i) = \frac{\sum_{d=0}^{d \le MIN(i, K-i-1)} 2^{(i-d)} \binom{M-1}{i-d} \binom{(M-1)-(i-d)}{d} \binom{MN-2M+1}{K-i-d-1}}{\binom{MN-1}{K-1}} \qquad (2)$$

If the conflicts produced by one object are independent from the conflicts produced by all other objects, then Equation 2 will give the number of broadcasts required to retrieve all *K* requested objects. However, if the conflicts produced by one object are not independent of the conflicts produced by other objects, additional conflicts will occur which are not accounted for in previous equations. Equation 2 will thus underestimate the number of broadcasts required to retrieve all *K* objects.

2.3   Retrieving Objects from a Broadcast on Parallel Channels

The problem of determining the proper order to retrieve the requested objects from the parallel channels can be modeled as a TSP. Making the transformation from a broadcast to the TSP requires the creation of a complete directed graph G with K nodes, where each node represents a requested object. The weight w of each edge (i,j) indicates the number of broadcasts that must pass in order to retrieve object j immediately after retrieving object i. Since any particular object can be retrieved in either the current broadcast or the next broadcast, the weight of each edge will be either 0 or 1. A weight of 0 indicates that the object j is after object i in the broadcast and that objects i and j are not in conflict. A weight of 1 indicates that object j is either before or in conflict with objects i. An example of this conversion is shown in Figure 4.

The nature of the problem dictates that G is asymmetric; that is, the weight of edge (i,j) is not necessarily equal to the weight of edge (j,i). Thus, in solving this problem we can apply those techniques that are applicable to the Asymmetric TSP.

2.4   Experimental Evaluation of Data Retrieval Algorithms

A simulator was developed to randomly generate broadcasts and determine how many passes were required to retrieve a varying number of requested objects. Several algorithms for ordering the retrieval of objects from the broadcast, both TSP-related and non-TSP-related, are presented and analyzed. In addition to discussing algorithms for accessing pages in a broadcast, the efficiencies of the broadcast structure are analyzed. Issues such as the optimal number of broadcast channels to use are also discussed.

Simulation Model: The simulation models a mobile unit retrieving objects from a broadcast. A broadcast is represented as an N x M two-dimensional array, where N represents the number of objects in each channel of a broadcast and M represents the number of parallel channels. For each value of K, where K represents the number of requested objects and $1 \leq K \leq NM$, the simulation randomly generates 1000 patterns representing the uniform distribution of K objects among the broadcast channels. The K objects from each randomly generated pattern are retrieved using various retrieval algorithms. The number of passes is recorded and compared. To prevent the randomness of the broadcasts from affecting the comparison of the algorithms, the same broadcast is used for each algorithm in a particular trial. The mean value for each algorithm is reported for each value of K.

Algorithms for Retrieving Objects from Broadcast Channels: Several algorithms are used in our study to retrieve the objects from the broadcast. Because the TSP has been the subject of intensive research for decades, it is natural to expect TSP methods to give the best results. Two other heuristic methods developed for the purpose of this problem are also used.

TSP Methods: An exact TSP solution is simply too slow and too resource-intensive to use at a mobile unit. For example, some TSP problems with only 14 nodes took several minutes of CPU time in this experiment. While a better implementation of the algorithm may somewhat reduce the cost, it cannot change the fact that finding the exact solution will require exponential time for some inputs. Thus, a TSP heuristic based on the assignment problem relaxation is used. This heuristic requires far less CPU time and memory than the optimal tour finders, so it is suitable for use on a mobile unit. A publicly available TSP solving package named TspSolve [16] was used for the TSP algorithm implementation.

Next Object Access: Another heuristic, referred to as "Next Object", is included. This heuristic was developed independently of the TSP. The strategy used by this heuristic is simply to always retrieve the next available object in a broadcast (This was the scheme used in our analysis in simulations reported in prior sections). This can be considered as a greedy approach, because it assumes that retrieving as many objects as possible in each broadcast will result in the best solution. It is also similar to the Nearest Neighbor approach to solving TSP problems.

Row Scan: A simple row-scanning heuristic, referred to as "Row Scan", is also included. This algorithm simply reads all the objects from one channel in each pass. If a channel does not have any objects in it, it is skipped. This algorithm will always require as many passes as there are channels with requested objects in them. The benefit of this algorithm is that it does not require any time to decide on an ordering for objects. It can thus begin retrieving objects from a broadcast immediately. This is especially important when a large percentage of the objects in a broadcast have been requested.

2.5 Performance Analysis

As expected, the TSP heuristic provides much better results than both the Next Object and Row Scan heuristics. In Figure 5, for the TSP method, we find that the number of broadcasts required to retrieve all K requested objects from a broadcast is much greater than the number of broadcasts predicted by Equation 1. This is due to the fact that Equation 1 was based on the assumption that the conflicts among the requested objects are independent. Figure 5 used 5 parallel channels and 20 pages per channel. It is also interesting to note that the straightforward row scan nearly matches the performance of the TSP-based algorithm when more than about 45% of the total number of objects are requested. In this case, there are so many conflicts that it is virtually impossible to avoid having to make as many passes as there are parallel channels. When this occurs, it is better to do the straightforward row scan than to spend time and resources running a TSP heuristic.

2.6 Optimal Number of Broadcast Channels

One of the important questions that must be answered in designing a broadcasting system for the MDAS environment is how many parallel channels to use for broadcasting information. More channels mean that a given amount of information can be made available in a shorter period of time. However, there will be more conflicts when more channels are used.

Figure 6 shows the number of pages of data that must be broadcast, on average, to retrieve K objects from 4 different broadcasting schemes. The four broadcasting schemes have 1, 2, 4, and 8 parallel channels. This test was performed with 40 pages in each broadcast. The results show that it is always advantageous to use more broadcast channels. While there will be more conflicts between objects, this does not quite counteract the shorter broadcast length of the many-channel broadcasts. This is especially evident when only a few objects in a broadcast are being accessed.

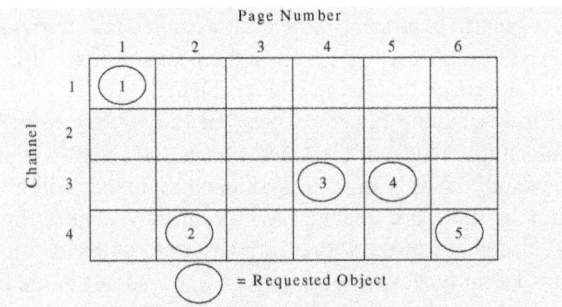

a) Broadcast representation

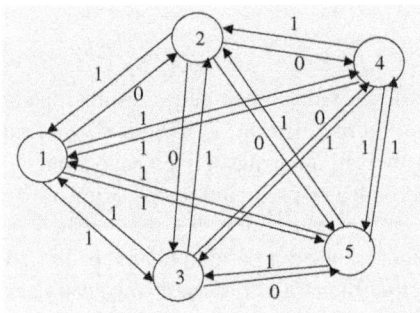

b) Graph representation

		Destination			
	1	**2**	**3**	**4**	**5**
1	X	0	0	0	0
2	1	X	0	0	0
3	1	1	X	0	0
4	1	1	1	X	1
5	1	1	1	1	X

(Source)

c) Matrix representation

Fig. 4. Representation of a broadcast as a Traveling Salesman Problem.

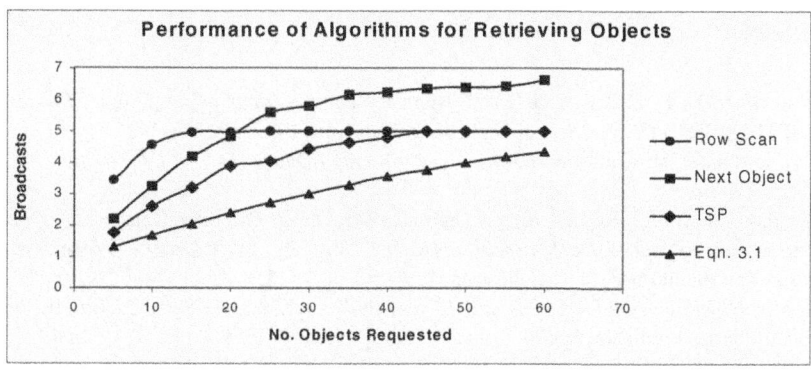

Fig. 5. Comparison of several algorithms for retrieving objects from parallel channels.

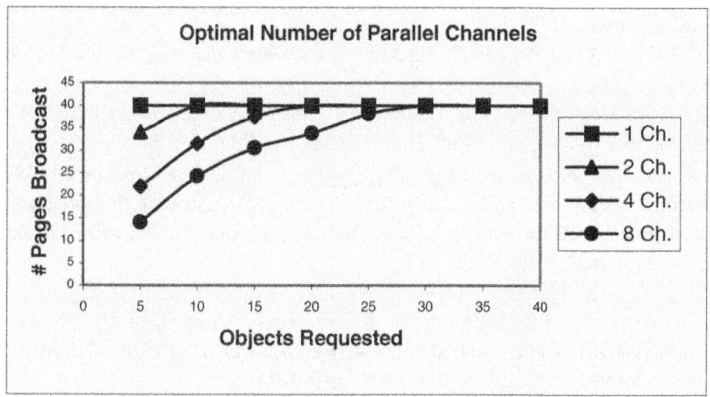

Fig. 6. Optimal Number of Broadcast Channels.

3. Conclusions

With the proliferation of mobile devices and the need for accessing data from these devices increasing rapidly, broadcasting has emerged has an essential information sharing scheme. In this paper, we developed and implemented different data retrieval algorithms for parallel air-channels. In particular, we found that existing solutions for TSP problem were applicable for solving this problem. The results from our study show that when a small (less than about 45%) percentage of the objects in a broadcast are requested, a TSP heuristic should be used to determine the order in which the broadcast objects should be retrieved. This technique reduced the number of broadcasts required to service all accesses in the parallel air channels. When a larger percentage of the objects is requested, the Row Scan method should be used to avoid the delay associated with switching between channels.

References

[1] R. Alonso and H. F. Korth, "Database System Issues in Nomadic Computing," *Proceedings ACM SIGMOD Conference on Management of Data,* 1993, pp. 388-392.

[2] E. Bertino, "Indexing Configuration in Object-Oriented Databases," *VLDB Journal,* No.3, 1994, pp.355-399.

[3] E. Bertino, "A Survey of Indexing Techniques for Object-Oriented Databases," *Query Processing for Advanced DataBase Systems,* J. C. Freytag, D. Maier, G. Vossen, eds., Maurgan Kaufmann Publishers, California, 1994, pp. 383-418.

[4] G. Booch, *Object-oriented Analysis and Design With Applications,* Second Edition, Benjamin/Cummings Publishing, 1994.

[5] Boonsiriwattanakul, S., Hurson, A.R., Vijaykrishnan,N., and Chehadeh, C., "Energy-Efficient Indexing on Parallel Air Channels in a Mobile Database Access System," 3rd World Multiconference on Systemics, Cybernetics and Informatics (SCI'99) And 5th International Conference on Information Systems Analysis and Synthesis (ISAS'99), Orlando, FL, pp. IV 30-38, August 1999.

[6] T. F. Bowen, "The DATACYCLE Architecture," *Communication of the ACM,* Vol. 35, No. 12, December 1992, pp. 71-81.

[7] M. W. Bright, A. R. Hurson, and S. H. Pakzad, "A Taxonomy and Current Issues in Multidatabase System," *IEEE Computer,* Vol. 25, No.3, March 1992, pp.50-60.

[8] NASDAQ World Wide Web Home Page, http://www.nasdaq.com, December 1997.

[9] Y. C. Chehadeh, A. R. Hurson, and L. L. Miller, "Energy Efficient Indexing on a broadcast Channel in a Mobile DataBase System," International conference on Information Technology: Coding and Computing, 2000, pp. 368-374

[10] Y. C. Chehadeh, A. R. Hurson, and M. Kavehrad, "Object Organization on a Single Broadcast Channel in the Mobile Computing Environment," *Journal of Multimedia Tools and Applications,* special issue on *Mobile Computing Environment for Multimedia Systems,* Kluwer Academic Publishers, Vol 9, 1999, pp. 69-94.

[11] G. H. Forman and J. Zahorjan, "The Challenges of Mobile Computing," *IEEE Computer,* Vol.27, No.4, April 1994, pp. 38-47.

[12] Hurson A.R., Chehadeh, Y.C., and Hannan J., "Object Organization on Parallel Broadcast Channels in a Global Information Sharing Environment", IEEE Conference on Performance, Computing, and Communications, pp. 347-353, 2000.

[13] A. R. Hurson, S. Pakzad, and J.-B. R. Cheng, "Object-Oriented Database Management Systems," *IEEE Computer,* Vol. 26, No. 2, 1993, pp. 48-60.

[14] C. Hurwitz, *TspSolve 1.3.6,* churritz@cts.com.

[15] T. Imielinski, S. Viswanathan, and B. R. Badrinath, "Data on Air: Organization and Access," *IEEE Transactions on Computers,* Vol.9, No.3, May/June 1997, pp. 353-372.

[16] Lim, J.B., Hurson, A.R., "Heterogeneous Data Access in a Mobile Environment — Issues and Solutions," *Advances in Computers,* Vol. 48, 1999, pp. 119-178.

A Weight Based Distributed Clustering Algorithm for Mobile ad hoc Networks*

Mainak Chatterjee, Sajal K. Das, and Damla Turgut

Center for Research in Wireless Mobility and Networking (CReWMaN)
Department of Computer Science and Engineering
University of Texas at Arlington
Arlington, TX 76019-0015
{chat,das,turgut}@cse.uta.edu

Abstract. In this paper, we propose a distributed clustering algorithm for a multi-hop packet radio network. These types of networks, also known as *ad hoc* networks, are dynamic in nature due to the mobility of the nodes. The association and dissociation of nodes to and from *clusters* perturb the stability of the network topology, and hence a reconfiguration of the system is often unavoidable. However, it is vital to keep the topology stable as long as possible. The *clusterheads*, which form a *dominant set* in the network, determine the topology and its stability. Our weight based distributed clustering algorithm takes into consideration the ideal degree, transmission power, mobility and battery power of a mobile node. We try to keep the number of nodes in a cluster around a pre-defined threshold to facilitate the optimal operation of the medium access control (MAC) protocol. The non-periodic procedure for clusterhead election gradually improves the load balance factor (LBF) which is a measure of the load distribution among the clusterheads. For lowering the computation and communication costs, the clustering algorithm is invoked on-demand which aims to maintain the connectivity of the network at the cost of load imbalance. Simulation experiments are conducted to evaluate the performance of our algorithm in terms of the number of clusterheads, *reaffiliation* frequency and dominant set updates. Results show that the our algorithm performs better than the existing algorithms and is also tunable to different types of ad hoc networks.

1 Introduction

Mobile multi-hop radio networks, also called *ad hoc* or *peer-to-peer* networks, play a critical role in places where a wired (central) backbone is neither available nor economical to build. Deployment of cellular networks takes time and cannot be set up in times of utmost emergency. Typical examples of ad hoc networks are law enforcement operations, battle field communications, disaster recovery

---

* This work is partially supported by Texas Advanced Research Program grant TARP-003594-013, Texas Telecommunications Engineering Consortium (TxTEC) and Nortel Networks, Richardson, Texas.

M. Valero, V.K. Prasanna, and S. Vajapeyam (Eds.): HiPC 2000, LNCS 1970, pp. 511–521, 2000.

situations, and so on. Such situations demand a network where all the nodes including the base stations are potentially mobile, and communication must be supported untethered between any two nodes.

A multi-cluster, multi-hop packet radio network for wireless systems should be able to dynamically adapt itself with the changing network configurations. Certain nodes, known as *clusterheads*, are responsible for the formation of *clusters* (analogous to *cells* in a cellular network) and maintenance of the topology of the network. The set of clusterheads is known as a *dominant set*. A clusterhead does the resource allocation to all the nodes belonging to its cluster. Due to the dynamic nature of the mobile nodes, their association and dissociation to and from clusters perturb the stability of the network and thus reconfiguration of clusterheads is often unavoidable. This is an important issue since frequent clusterhead changes adversely affect the performance of other protocols such as scheduling, routing and resource allocation that rely on it. Choosing clusterheads optimally is an NP-hard problem [2]. Thus, existing solutions to this problem are based on heuristic approaches and none attempts to retain the topology of the network [2,4]. We believe a good clustering scheme should preserve its structure as much as possible when the topology is dynamically changing. Otherwise, re-computation of clusterheads and frequent information exchange among the participating nodes will result in high computation cost overhead.

The concept of dividing a geographical region into smaller zones has been presented implicitly in the literature as *clustering* [9]. A natural way to map a "standard" cellular architecture into a multi-hop packet radio network is via the concept of a virtual cellular network (VCN) [4]. Any node can become a clusterhead if it has the necessary functionality, such as processing and transmission power. Nodes register with the nearest clusterhead and become member of that cluster. Clusters may change dynamically, reflecting the mobility of the nodes. The focus of the existing research has been on just partitioning the network into clusters [2,3,7,8,6], without taking into consideration the efficient functioning of all the system components. The lack of rigorous methodologies for the design and analysis of peer-to-peer mobile networks has motivated in-depth research in this area. There exist solutions for efficiently interconnecting the nodes in such a way that the latency of the system is minimized while throughput is maximized [6]. Most of the approaches [2,4,6] for finding the clusterheads do not produce an optimal solution with respect to battery usage, load balancing and MAC functionality.

This paper proposes a weight based distributed clustering algorithm which takes into consideration the number of nodes a clusterhead can handle ideally (without any severe degradation in the performance), transmission power, mobility and battery power of the nodes. Unlike existing schemes which are invoked periodically resulting in high communication overhead, our algorithm is adaptively invoked based on the mobility of the nodes. More precisely, the clusterhead election procedure is delayed as long as possible to reduce the computation cost. We show by simulation experiments that our method yields better results as compared to the existing heuristics in terms of the number of reaffiliations and dominant set updates.

2   Previous Work

To the best of our knowledge, three heuristics have been proposed to choose clusterheads in ad hoc networks. They include (i) Highest-Degree heuristic (ii) Lowest-ID heuristic and (iii) Node-Weight heuristic. In the assumed graph model of the network, the mobile terminals are represented as nodes and there exists an edge between two nodes if they can communicate with each other directly (i.e., one node lies within the transmission range of another). The performance of these heuristics were shown in [3,6] by simulation experiments where mobile nodes were randomly placed in a square grid and moved with different speeds in different directions. These heuristics are summarized below.

Highest-Degree Heuristic: This approach is a modified version of [10] which computes the degree of a node based on the distance between that node from others. A node x is considered to be a neighbor of another node y if x lies within the transmission range of y. The node with the maximum degree is chosen as a clusterhead and any tie is broken by the node ids which are unique. The neighbors of a clusterhead become members of that cluster and can no longer participate in the election process. This heuristic is also known as the highest-connectivity algorithm. Experiments demonstrate that the system has a low rate of clusterhead change but the throughput is low under this scheme. Typically, each cluster was assigned some resources which was shared among the members of that cluster on a round-robin basis [6,7,8]. As the number of nodes in a cluster is increased, the throughput of each user drops and hence a gradual degradation in the system performance is observed. This is the inherent drawback of this heuristic since the number of nodes in a cluster is not bounded.

Lowest-ID Heuristic: Gerla and Tsai [6] proposed a simple heuristic by assigning a unique id to each node and choosing the node with the minimum id as a clusterhead. However, the clusterhead can delegate its duties to the next node with the minimum id in its cluster. A node is called a *gateway* if it lies within the transmission range of two or more clusters. For this heuristic, the system performance is better compared with the Highest-Degree heuristic in terms of the throughput. Since the environment under consideration is mobile, it is unlikely that node degrees remain stable resulting in frequent clusterhead updates. The drawback of this heuristic is its bias towards nodes with smaller ids which leads to the battery drainage of certain nodes. Moreover, it does not attempt to balance the load uniformly across all the nodes.

Node-Weight Heuristic: Basagni et al. [2,3] assigned node-weights based on the suitability of a node being a clusterhead. A node is chosen to be a clusterhead if its weight is higher than any of its neighbor's node-weights. The smaller id is chosen in case of a tie. To verify the performance of the system [2], the nodes were assigned weights which varied linearly with their speeds but with negative slope. Results proved that the number of updates required is smaller than the

Highest-Degree and Lowest-ID heuristics. Since node weights were varied in each simulation cycle, computing the clusterheads becomes very expensive and there are no optimizations on the system parameters such as throughput and power control.

3   Our Approach

None of the above three heuristics leads to an optimal selection of clusterheads since each deals with only a subset of parameters which impose constraints on the system. For example, a clusterhead may not be able handle a large number of nodes due to resource limitations even if these nodes are its neighbors and lie well within its transmission range. Thus, the load handling capacity of the clusterhead puts an upper bound on the node-degree. In other words, simply covering the area with the minimum number of clusterheads will put more burden on each of the clusterheads. This will of course maximize the resource utilization. On the other hand, we could have all the nodes share the same responsibility and act as clusterheads. However, more clusterheads result in extra number of hops for a packet when it gets routed from the source to the destination, since the packet has to go via a larger number of clusterheads. Thus, this solution leads to higher latency, more power consumption and more information processing per node. The other alternative is to split the whole area into zones, the size of which can be determined by the transmission range of the nodes. This can put a lower bound on the number of clusterheads required. Ideally, to reach this lower bound, a uniform distribution of the nodes is necessary over the entire area. Also, the total number of nodes per unit area should be restricted so that the clusterhead in a zone can handle all the nodes therein. However, the zone based clustering is not a viable solution due to the following reasons.

The clusterheads would typically be centrally located in the zone, and if they move, new clusterheads have to be selected. It might so happen that none of the other nodes in that zone are centrally located. Therefore, to find a new node which can act as a clusterhead with the other nodes within its transmission range might be difficult. Another problem arises due to non-uniform distribution of the nodes over the whole area. If a certain zone becomes densely populated, the clusterhead might not be able to handle all the traffic generated by the nodes because there is an inherent limitation on the number of nodes a clusterhead can handle. We propose to select the minimum number of clusterheads which can support all the nodes in the system satisfying the above constraints.

In summary, choosing an optimal number of clusterheads which will yield high throughput but incur as low latency as possible, is still an important problem. As the search for better heuristics for this problem continues, we propose the use of a *combined weight* metric, that takes into account several system parameters like the ideal node-degree, transmission power, mobility and the battery power of the nodes.

3.1   Basis for Our Algorithm

The following features are considered in our clustering algorithm.

- The clusterhead election procedure is invoked as rarely as possible. This reduces system updates and hence computation and communication cost.
- Each clusterhead can ideally support M (a pre-defined threshold) nodes to ensure efficient MAC functioning. A high throughput of the system can be achieved by limiting or optimizing the number of nodes in each cluster.
- The battery power can be efficiently used within certain transmission range. Consumption of the battery power is more if a node acts as a clusterhead rather than an ordinary node.
- Mobility is an important factor in deciding the clusterheads. *Reaffiliation* occurs when one of the ordinary nodes moves out of a cluster and joins another existing cluster. In this case, the amount of information exchange between the node and the corresponding clusterhead, is local and relatively small. The information update in the event of a change in the dominant set is much more than a reaffiliation.
- A clusterhead is able to communicate better to its neighbors if they are closer to the clusterhead within the transmission range. This is due to signal attenuation with increasing distance.

3.2   Proposed Algorithm

Based on the preceding discussions, we propose an algorithm which effectively combines all the system parameters with certain weighing factors, the values of which can be chosen according to the system needs. For example, power control is very important in CDMA networks, thus the weight of that factor can be made larger. The flexibility of changing the weight factors helps us apply our algorithm to various networks. The procedure for clusterhead election is presented below. Its output is a set of nodes (dominant set) which forms the clusterheads for the network. According to our notation, the number of nodes that a clusterhead can handle ideally is M. The clusterhead election procedure is invoked at the time of system activation and also when the current dominant set is unable to cover all the nodes.

Clusterhead Election Procedure

Step 1: Find the neighbors of each node v (i.e., nodes within its transmission range). This determines the *degree*, d_v.

Step 2: Compute the *degree-difference*, $D_v = |d_v - M|$, for every node v.

Step 3: For all v, compute the *sum of the distances*, P_v, with all its neighbors.

Step 4: Compute the running average of the speed for every node. This gives a measure of its mobility and is denoted by M_v.

Step 5: Compute the total time, T_v, for which a node has been a clusterhead. T_v indicates how much battery power has been consumed which is more for a clusterhead than for an ordinary node.

Step 6: Calculate a *combined weight* $I_v = c_1 D_v + c_2 P_v + c_3 M_v + c_4 T_v$, for each node v. The coefficients c_1, c_2, c_3 and c_4 are the weighing factors for the corresponding system parameters.

Step 7: Choose v with the minimum I_v as the clusterhead. The neighbors of the chosen clusterhead can no longer participate in the election procedure.

Step 8: Repeat Steps 2 - 7 for the remaining nodes not yet assigned to any cluster.

3.3   System Activation and Update Policy

When a system is brought up, every node v broadcasts its id which is registered by all other nodes lying within v's transmission range, *tx_range*. It is assumed that a node receiving a broadcast from another node can estimate their mutual distance from the strength of the signal received. Thus, every node is made aware of its geographically neighboring nodes and their corresponding distances. Once the neighbors list is ready, our algorithm chooses the clusterhead for the first time. All the non-clusterhead nodes know the clusterhead they are attached to; similarly the clusterheads know their members. The topology of the system is constantly changing due to the movement of all the nodes, thus the system needs to be updated which may result in the formation of a new set of clusters. It may also result in nodes changing their point of attachment from one clusterhead to another within the existing dominant set, which is called *reaffiliation*. The frequency of update and hence reaffiliation is an important issue. If the system is updated periodically at a high frequency, then the *latest* topology of the system can be used to find the clusterheads which will yield a good dominant set. However, this will lead to high computational cost resulting in the loss of battery power or energy. If the frequency of update is low, there are chances that current topological information will be lost resulting in sessions terminated midway.

Every mobile node in any cellular system (GSM or CDMA) periodically exchanges control information with the base station. Similar idea is applied here, where all the nodes continuously monitor their signal strength as received from the clusterhead. When the mutual separation between the node and its clusterhead increases, the signal strength decreases. In that case, the mobile has to notify its current clusterhead that it is no longer able to attach itself to that clusterhead. The clusterhead tries to hand-over the node to a neighboring cluster and the member lists are updated. If the node goes into a region not covered by any clusterhead, then the clusterhead election procedure is invoked and the new dominant set is obtained.

The objective of our clusterhead election procedure is to minimize the number of changes in dominant set update. Once the neighbors list for all nodes are created, the degree-difference D_v is calculated for each node v. Also, P_v is computed for each node by summing up the distances of its neighbors. The mobility M_v is calculated by averaging the speed of the node. The total amount of time, T_v, a node remained as a clusterhead is also calculated. All these parameters are normalized, which means that their values are made to lie in a pre-defined range. The corresponding weights c_1, c_2, c_3 or c_4, which sum upto 1, are kept fixed for a given system. The weighing factors also give the flexibility of adjusting the effective contribution of each of the parameters in calculating the *combined weight* I_v. For example, in a system where battery power is more important, the weight c_4 associated with T_v can be made higher. The node with the minimum total weight, I_v, is elected as a clusterhead and its neighbors are no longer eligible to participate in the remaining part of the election process which continues until every node is found to be either a clusterhead or a neighbor of a clusterhead.

3.4   Balancing the Loads

It is not desirable to have any clusterheads to be overly loaded while some others are lightly loaded. At the same time, it is difficult to maintain a perfectly balanced system at all times due to frequent detachment and attachment of the nodes from and to the clusterheads. As a measure of how well balanced the clusterheads are, we define the *load balancing factor* (LBF) which is inversely proportional to the variance of the cardinality of the clusters. In other words,

$$LBF = \left(\frac{\sum_i x_i - \mu}{n_c} \right)^{-1/2}$$

where n_c is the number of clusterheads, x_i is the cardinality of cluster i, and μ is the average number of nodes a clusterhead has. Thus, $\mu = \frac{N - n_c}{n_c}$, N being the total number of nodes in the system.

3.5   Connecting the Clusters

As a logical extension to clustering, we investigate the connectivity of the nodes which is essential for any routing algorithm. *Connectivity* can be defined as the probability that any node is reachable from any other node. For a single component graph, any node is reachable from any other node and the connectivity becomes 1. For two clusters to communicate with each other, we assume that the clusterheads are capable of operating in *dual* power mode. It uses low power to communicate with its members within its transmission range and high power to communicate with the neighboring clusterheads because of higher range.

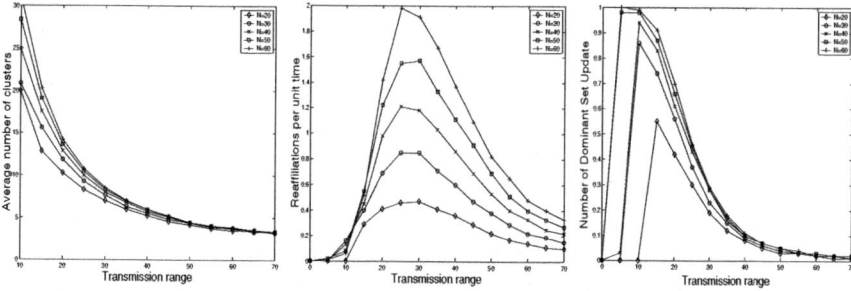

Fig. 1. $max\_disp=5$

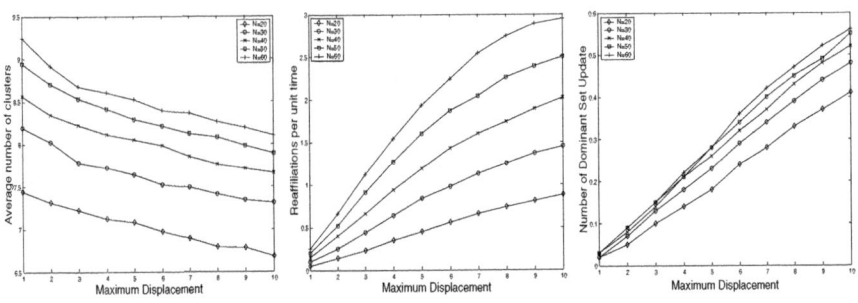

Fig. 2. $tx\_range=30$

4   Simulation Study

We simulate a system with N nodes on a $100{\times}100$ grid. The nodes could move in all possible directions with displacement varying uniformly between 0 to a maximum value (max_disp), per unit time. To measure the performance of our system, we identify three metrics: (i) the number of clusterheads, (ii) the number of reaffiliations, and (iii) the number of dominant set updates. Every time a dominant set is identified, its cardinality gives the number of clusterheads. The reaffiliation count is incremented when a node gets dissociated from its clusterhead and becomes a member of another cluster within the current dominant set. The dominant set update takes place when a node can no longer be a neighbor of any of the existing clusterheads. These three parameters are studied for varying number of nodes in the system, transmission range and maximum displacement. We also study how the load balance factor changes as the system evolves and how well connected the nodes are.

4.1   Summary of Experimental Results

In our simulation, N was varied between 20 and 60, and the transmission range was varied between 0 and 70. The nodes moved randomly in all possible directions with a maximum displacement of 10 along each of the coordinates. Thus, the maximum Euclidean displacement possible is $10\sqrt{2}$. We assume that each clusterhead can ideally handle 10 nodes in its cluster in terms of resource allocation. Therefore, the *ideal degree* was fixed at $M = 10$ for the entire experiment. Due to this, the weight associated with D_v was rather high. The next higher weight was given to P_v, which is the sum of the distances. Mobility and battery power were given low weights. The values used for simulation were $c_1 = 0.7$, $c_2 = 0.2$, $c_3 = 0.05$ and $c_4 = 0.05$. Note that these values are arbitrary at this time and should be adjusted according to the system requirements.

Figure 1 shows the variation of three parameters, namely average number of clusterheads, reaffiliation per unit time and the number of dominant set update with varying transmission range and a constant max_disp of 5. We observe that the average number of clusterheads decreases with the increase in the transmission range because a clusterhead with a large transmission range covers a larger area. For low transmission range, the nodes in a cluster are relatively close to the clusterhead, and a detachment is unlikely. The number of reaffiliations increases as the transmission range increases, and reaches a peak when transmission range is between 20 and 30. Further increase in the transmission range results in a decrease in the reaffiliations since the nodes, in spite of their random motion, tend to stay inside the large area covered by the clusterhead. The dominant set updates is high for a small transmission range because the cluster area is small and the probability of a node moving out of its cluster is high. As the transmission range increases, the number of dominant set updates decreases because the nodes stay within their cluster in spite of their movements. Figures 2 shows the variation of the same parameters but for varying *max_disp* and constant transmission range of 30. The average number of clusterheads is almost the same for different values of *max_disp*, particularly for larger values of N. This is because, no matter what the mobility is, it simply results in a different configuration, but the cluster size remains the same. We also see how the reaffiliations change per unit time with respect to the maximum displacement. As the displacement becomes larger, the nodes tend to move further from their clusterhead, detaching themselves from the clusterhead, causing more reaffiliation per unit time and more dominant set updates.

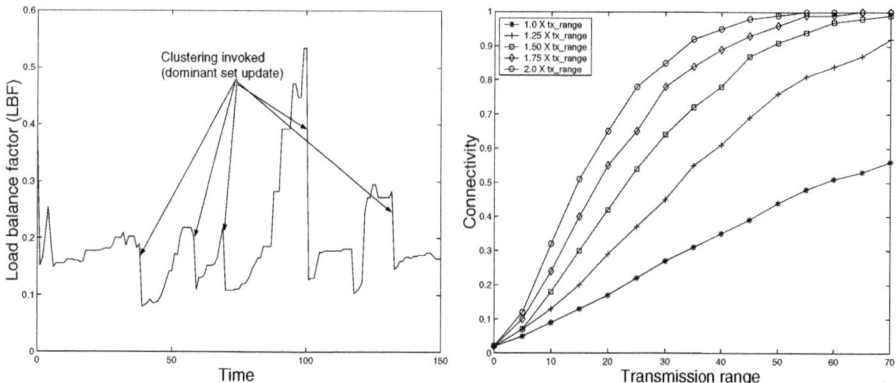

Fig. 3. Load distribution and connectivity

The non-periodic invocation of the clustering algorithm and the reachability of one node from another can be observed from Figure 3. There is a *gradual* increase in the load balance factor (LBF) due to the diffusion of the nodes among clusters. The improvement in LBF does not increase indefinitely because the nodes tend to move away from all possible clusterheads and the clustering algorithm has to be invoked to ensure connectivity. The clustering algorithm tries to connect all the nodes at the cost of load imbalance which is represented by the *sharp* decrease in LBF. As mentioned earlier, the lower power is used to communicate within the cluster whereas the higher power which effectively gives a higher transmission range is used to communicate with the neighboring clusterheads. To obtain the higher transmission range, we scaled the lower transmission

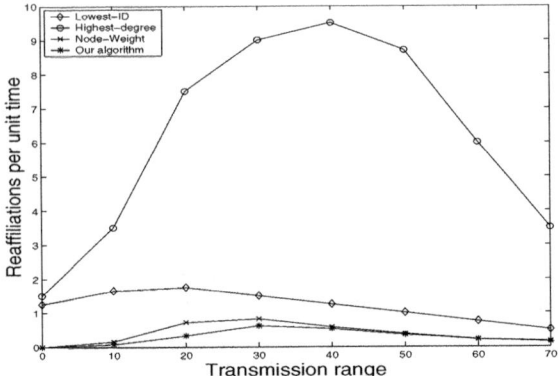

Fig. 4. Comparison of reaffiliations, N=30

range by a constant factor. Simulation was conducted for $N = 50$ and the constant factor was varied from 1.0 to 2.0 with increments of 0.25. It is observed that a well connected graph can be obtained at the cost of high power.

Figure 4 shows the relative performance of the Highest-Degree, Lowest-ID, Node-Weight heuristics and our algorithm in terms of reaffiliations per unit time. The number of reaffiliations for our algorithm is at most half the number obtained from the Lowest-ID. The main reason is that the frequency of invoking the clustering algorithm is lower in our case resulting in longer duration of stability of the network. Our algorithm performs marginally better than the Node-Weight heuristics which, however, does not give the basis of assigning the weights to the nodes. Our algorithm describes a linear model which takes into consideration the four important system parameters in deciding the suitability of the nodes acting as clusterheads. It also provides the flexibility of adjusting the weighing factors according to the system needs.

5   Conclusions

We propose a weight based distributed clustering algorithm which can dynamically adapt itself with the ever changing topology of ad hoc networks. Our algorithm has the flexibility of assigning different weights and takes into an account a combined effect of the ideal degree, transmission power, mobility and battery power of the nodes. The algorithm is executed only when there is a demand, i.e., when a node is no longer able to attach itself to any of the existing clusterheads. We see that there is a pattern of how the load balance factor changes to distribute the load and also ensure maximum connectivity. Our algorithm performs significantly better than both of the Highest-Degree and the Lowest-ID heuristics. In particular, the number of reaffiliations for our algorithm is about 50% of the number obtained from Lowest-ID heuristic. Though our approach performs marginally better than the Node-Weight heuristic, it considers more realistic system parameters and has the flexibility of adjusting the weighing factors.

References

1. D.J. Baker, A. Ephremides, and J.A. Flynn, "The design and simulation of a mobile radio network with distributed control", IEEE Journal on Selected Areas in Communications, vol. SAC-2, January 1984, pp. 226-237.
2. S. Basagni, I. Chlamtac, and A. Farago, "A Generalized Clustering Algorithm for Peer-to-Peer Networks", Workshop on Algorithmic Aspects of Communication (satellite workshop of ICALP), Bologna, Italy, July 1997.
3. S. Basagni, "Distributed Clustering for Ad Hoc Networks", International Symposium on Parallel Architectures, Algorithms and Networks", Perth, June 1999, pp. 310-315.
4. I. Chlamtac and A. Farago, "A New Approach to the Design and Analysis of Peer-to-Peer Mobile Networks", Wireless Networks, 5(3), Aug 1999, pp. 149-156.
5. A. Ephremides, J.E. Wieselthier, and D.J. Baker, "A design concept for reliable mobile radio networks with frequency hopping signaling", Proceedings of IEEE, vol. 75, Jan 1987, pp. 56-73.
6. M. Gerla and J. T.C. Tsai, "Multicluster, Mobile, Multimedia Radio Network", Wireless Networks, 1(3) 1995, pp. 255-265.
7. C.-H.R. Lin and M. Gerla, "A Distributed Control Scheme in Multi-hop Packet Radio Networks for Voice/Data Traffic Support", IEEE GLOBECOM, pp. 1238-1242, 1995.
8. C.-H. R. Lin and M. Gerla, "A Distributed Architecture for Multimedia in Dynamic Wireless Networks", IEEE GLOBECOM, pp. 1468-1472, 1995.
9. M. Joa-Ng and I.-T. Lu, "A Peer-to-Peer Zone-based Two-level Link State Routing for Mobile Ad Hoc Networks", *IEEE Journal on Selected Areas in Comm.*, Aug. 1999, pp. 1415-1425.
10. A.K. Parekh, "Selecting routers in ad-hoc wireless networks", ITS, 1994.

Session V-B

Large-Scale Data Mining
Chair: Gautam Das
Microsoft Research

A Scalable Approach to Balanced, High-Dimensional Clustering of Market-Baskets

Alexander Strehl and Joydeep Ghosh

The University of Texas at Austin,
Austin, TX 78712-1084, USA
{strehl,ghosh}@lans.ece.utexas.edu
http://www.lans.ece.utexas.edu

Abstract. This paper presents OPOSSUM, a novel similarity-based clustering approach based on constrained, weighted graph-partitioning. OPOSSUM is particularly attuned to real-life market baskets, characterized by very high-dimensional, highly sparse customer-product matrices with positive ordinal attribute values and significant amount of outliers. Since it is built on top of Metis, a well-known and highly efficient graph-partitioning algorithm, it inherits the scalable and easily parallelizeable attributes of the latter algorithm. Results are presented on a real retail industry data-set of several thousand customers and products, with the help of CLUSION, a cluster visualization tool.

1 Introduction

A key step in market-basket analysis is to cluster customers into relatively homogeneous groups according to their purchasing behavior. A large market-basket database may involve millions of customers and several thousand product-lines. For each product a customer could potentially buy, a feature (attribute) is recorded in the data-set. A feature usually corresponds to some property (e.g., quantity, price, profit) of the goods purchased. Most customers only buy a small subset of products. If we consider each customer as a multi-dimensional data point and then try to cluster customers based on their buying behavior, the problem differs from classic clustering scenarios in several ways [1]:

- *High dimensionality:* The number of features is very high and may even exceed the number of samples. So one has to face with the curse of dimensionality.
- *Sparsity:* Most features are zero for most samples. This strongly affects the behavior of similarity measures and the computational complexity.
- *Significant outliers:* Outliers such as a few, big corporate customers that appear in an otherwise small retail customer data, may totally offset results. Filtering these outliers may not be easy, nor desirable since they could be very important (e.g., major revenue contributors).

In addition, features are often neither nominal, nor continuous, but have discrete positive ordinal attribute values, with a strongly non-Gaussian distribution. Moreover, since the number of features is very high, normalization can

M. Valero, V.K. Prasanna, and S. Vajapeyam (Eds.): HiPC 2000, LNCS 1970, pp. 525–536, 2000.

become difficult. Due to these issues, traditional clustering techniques work poorly on real-life market-basket data. This paper describes OPOSSUM, a graph-partitioning approach using value-based features, that is well suited for market-basket clustering scenarios, exhibiting some or all of the characteristics described above. We also examine how this approach scales to large data sets and how it can be parallelized on distributed memory machines.

2   Related Work

Clustering has been widely studied in several disciplines, specially since the early 60's [2,3]. Some classic approaches include partitional methods such as k-means, hierarchical agglomerative clustering, unsupervised Bayes, and soft, statistical mechanics based techniques. Most classical techniques, and even fairly recent ones proposed in the data mining community (CLARANS, DBSCAN, BIRCH, CLIQUE, CURE, WAVECLUSTER etc. [4]), are based on distances between the samples in the original vector space. Thus they are faced with the "curse of dimensionality" and the associated sparsity issues, when dealing with very high dimensional data. Recently, some innovative approaches that directly address high-dimensional data mining have emerged. ROCK (Robust Clustering using linKs) [5] is an agglomerative hierarchical clustering technique for categorical attributes. It uses the binary Jaccard coefficient and a thresholding criterion to establish links between samples. Common neighbors are used to define inter-connectivity of clusters which is used to merge clusters. CHAMELEON [6] starts with partitioning the data into a large number of clusters by partitioning the v-nearest neighbor graph. In the subsequent stage clusters are merged based on inter-connectivity and their relative closeness. **Scalability Studies** on clustering have taken two directions:

1. Perform k-means or a variant thereof, on a single computer with limited main memory, with as few scans of the database as possible [7]. These algorithms implicitly assume hyperspherical clusters of about the same size, and thus the key idea is to update sufficient statistics (number of points, sum, sum-squared) about the potential clusters in main memory as one scans the database, and then do further refinement of cluster centers within main memory.
2. Parallel implementations. k-means is readily parallelizeable through data partitioning on distributed memory multicomputers, with little overhead [8]. At each iteration, the current locations of the k means is broadcast to all processors, who then independently perform the time consuming operation of finding the closest mean for each (local) data point, and finally send the (local) updates to the mean positions to a central processor that does a global update using a MPI_allreduce operation.

3   Domain Specific Features and Similarity Space

Notation. Let n be the number of objects (customers) in the data and d the number of features (products) for each sample $\mathbf{x}_j$ with $j \in \{1, \ldots, n\}$. The input data can be represented by a $d \times n$ product-customer matrix $\mathbf{X}$ with the j-th column representing the sample $\mathbf{x}_j$. Hard clustering assigns a label λ_j to each d–dimensional sample $\mathbf{x}_j$, such that similar samples tend to get the same label. The number of distinct labels is k, the desired number of clusters. In general the labels are treated as nominals with no inherent order, though in some cases, such as self-organizing feature maps (SOFMs) or top-down recursive graph-bisection, the labeling may contain extra ordering information. Let $\mathcal{C}_\ell$ denote the set of all customers in the ℓ-th cluster ($\ell \in \{1, \ldots, k\}$), with $\mathbf{x}_j \in \mathcal{C}_\ell \Leftrightarrow \lambda_j = \ell$ and $n_\ell = |\mathcal{C}_\ell|$.

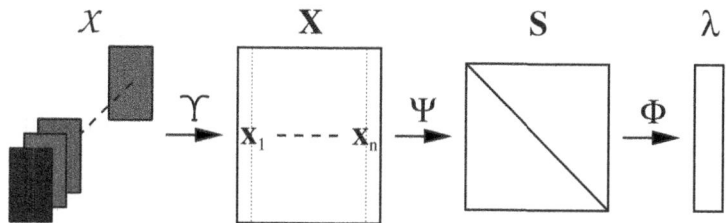

Fig. 1. Overview of the similarity based clustering framework OPOSSUM.

Fig. 1 gives an overview of our batch clustering process from a set of raw object descriptions $\mathcal{X}$ via the vector space description $\mathbf{X}$ and similarity space description $\mathbf{S}$ to the cluster labels λ: $(\mathcal{X} \in \mathcal{I}^n) \xrightarrow{\Upsilon} (\mathbf{X} \in \mathcal{F}^n \subset \mathbb{R}^{d \times n}) \xrightarrow{\Psi} (\mathbf{S} \in \mathcal{S}^{n \times n} = [0,1]^{n \times n} \subset \mathbb{R}^{n \times n}) \xrightarrow{\Phi} (\lambda \in \mathcal{O}^n = \{1, \ldots, k\}^n)$.

Feature Selection and Similarity Measures. While most of the well-known clustering techniques [3] have been for numerical features, certain recent approaches assume categorical data [5]. In general, non-binary features are more informative since they capture noisy behavior better for a small number of samples. For example, in market-basket data analysis, a feature typically represents the absence (0) or presence (1) of a particular product in the current basket. However, this treats a buyer of a single corn-flakes box the same as one who buys one hundred such boxes. In OPOSSUM, we extend the common Boolean notation to *non-negative, real-valued features*. The feature $x_{i,j}$ now represents the *volume* of product p_i in a given basket (or sample) $\mathbf{x}_j$. While "volume" could be measured by product quantity, we prefer to use *monetary value* (the product of price and quantity) to quantify feature volume. This yields an almost continuous distribution of feature values for large data sets. More importantly, monetary value represents a normalization across all feature dimensions. This normalization is highly desirable because it better aligns relevance for clustering with retail management objectives.

The key idea behind dealing with very high-dimensional data is to work in similarity space rather than the original vector space in which the feature vectors reside. A similarity measures $\in [0, 1]$ captures how related two data-points $\mathbf{x}_a$ and $\mathbf{x}_b$ are. It should be symmetric $(s(\mathbf{x}_a, \mathbf{x}_b) = s(\mathbf{x}_b, \mathbf{x}_a))$, with self-similarity $s(\mathbf{x}_a, \mathbf{x}_a) = 1$.

A brute force implementation does involve $O(n^2 \times d)$ operations, since similarity needs to be computed between each pair of data points, and involve all the dimensions. Also, unless similarity is computed on the fly, $O(n^2)$ storage is required for the similarity matrix. However, once this matrix is computed, the following clustering routine does not depend on d at all!

An obvious way to compute similarity is through a suitable monotonic and inverse function of a Minkowski distance, d. Candidates include $s = e^{-d^2}$, and $s(\mathbf{x}_a, \mathbf{x}_b) = 1/(1 + \|\mathbf{x}_a - \mathbf{x}_b\|_2)$. Similarity can also be defined by the angle or cosine of the angle between two vectors. The cosine measure is widely used in text clustering because two documents with equal classification because two documents with equal word composition but different lengths can be considered identical. In retail data this assumption loses important information about the life-time customer value by normalizing them all to 1.

OPOSSUM is based on **Jaccard Similarity**. For binary features, the Jaccard coefficient [2] measures the ratio of the intersection of the product sets to the union of the product sets corresponding to transactions $\mathbf{x}_a$ and $\mathbf{x}_b$:

$$s^{(J)}(\mathbf{x}_a, \mathbf{x}_b) = \frac{\mathbf{x}_a^\dagger \mathbf{x}_b}{\|\mathbf{x}_a\|_2^2 + \|\mathbf{x}_b\|_2^2 - \mathbf{x}_a^\dagger \mathbf{x}_b} \tag{1}$$

Since we want to analyze positive, real-valued features instead, an *extended Jaccard coefficient*, also given by equation 1, but using positive numbers for attribute values, is proposed. This coefficient captures a length-dependent measure of similarity. However, it is still invariant to scale (dilating $\mathbf{x}_a$ and $\mathbf{x}_b$ by the same factor does not change $s(\mathbf{x}_a, \mathbf{x}_b)$). A detailed discussion of the properties of various similarity measures can be found in [9], where it is shown that the extended Jaccard coefficient enables us to discriminate by the total value of market-baskets *as well as* to overcome the issues of Euclidean distances in high-dimensional sparse data.

4   CLUSION: Cluster Visualization

Since it is difficult to measure or visualize the quality of clustering in very high-dimensional spaces, we first built a CLUSter visualizatION toolkit, CLUSION, which is briefly described in this section. CLUSION first rearranges the columns and rows of the similarity matrix such that points with the same cluster label are contiguous. It then displays this permuted similarity matrix $\mathbf{S}$ with entries $s_{a,b} = s(\mathbf{x}_a, \mathbf{x}_b)$ as a gray-level image where a black (white) pixel corresponds to minimal (maximal) similarity of 0 (1). The intensity (gray level value) of the pixel at row a and column b corresponds to the similarity between the samples $\mathbf{x}_a$ and $\mathbf{x}_b$. The similarity *within* cluster ℓ is thus represented by the average intensity within a square region with side length n_ℓ, around the main diagonal of the matrix. The off-diagonal rectangular areas visualize the relationships *between* clusters. The brightness distribution in the rectangular areas yields insight towards the quality of the clustering and possible improvements. A bright off-diagonal region may suggest that the clusters in the corresponding rows and columns should be merged. In order to make these regions apparent, thin horizontal and vertical lines are used to show the divisions into the rectangular regions. Visualizing similarity space in this way can help to quickly get a feel for the clusters in the data. Even for a large number of points, a sense for the intrinsic number of clusters k in a data-set can be gained. Examples for CLUSION are given in Fig. 2. For further details, demos, and case studies that support certain design choices (using monetary value, extended Jaccard coefficient), see [1].

5   OPOSSUM

OPOSSUM (Optimal Partitioning of Sparse Similarities Using Metis) is based on partitioning a graph obtained from the similarity matrix, with certain constraints tailored to market-basket data. In particular, it is desirable here to have clusters of roughly equal importance for upper management analysis. Therefore OPOSSUM strives to deliver approximately equal sized (balanced) clusters using either of the following two criteria:

- *Sample balanced:* Each cluster should contain roughly the same number of samples, n/k. This allows retail marketers to obtain a customer segmentation with equally sized customer groups.
- *Value balanced:* Each cluster should contain roughly the same amount of feature values. In this case a cluster represents a k-th fraction of the total feature value $\sum_{j=1}^{n} \sum_{i=1}^{d} x_{i,j}$. If we use extended revenue per product (quantity × price) as value, then each cluster represents a roughly equal contribution to total revenue.

We formulate the desired balancing properties by assigning each sample (customer) a weight and then softly constrain the sum of weights in each cluster. For sample balanced clustering, we assign each sample $\mathbf{x}_j$ the same weight $w_j = 1$. To obtain value balancing properties, a sample $\mathbf{x}_j$'s weight is set to $w_j = \sum_{i=1}^{d} x_{i,j}$. The desired balancing properties have many application driven advantages. However, since natural clusters may not be equal sized, over-clustering (using a larger k) and subsequent merging may be helpful.

5.1   Single-Constraint Single-Objective Weighted Graph Partitioning

We map the problem of clustering to partitioning a weighted graph with a minimum number of edge cuts while maintaining a balancing constraint. Graphs are a well-understood abstraction and a large body of work exists on partitioning them. In [9] they have been shown to perform superior in high-dimensional document clustering. The objects to be clustered can be viewed as a set of vertices $\mathcal{V}$. Two web-pages $\mathbf{x}_a$ and $\mathbf{x}_b$ (or vertices v_a and v_b) are connected with an undirected edge $(a, b) \in \mathcal{E}$ of positive weight $s(\mathbf{x}_a, \mathbf{x}_b)$. The cardinality of the set of edges $|\mathcal{E}|$ equals the number of *non-zero* similarities between all pairs of samples. A set of edges whose removal partitions a graph $\mathcal{G} = (\mathcal{V}, \mathcal{E})$ into k pair-wise disjoint sub-graphs $\mathcal{G}_\ell = (\mathcal{V}_\ell, \mathcal{E}_\ell)$, is called an edge separator $\Delta\mathcal{E}$. Our objective is to find such a separator with a minimum sum of edge weights, as given by equation 2.

$$\min_{\Delta\mathcal{E}} \sum_{(a,b) \in \Delta\mathcal{E}} s(\mathbf{x}_a, \mathbf{x}_b) \, , \, \Delta\mathcal{E} = (\mathcal{E} \setminus (\mathcal{E}_1 \cup \mathcal{E}_2 \cup \ldots \cup \mathcal{E}_k)) \tag{2}$$

Without loss of generality, we can assume that the vertex weights w_j are normalized to sum up to 1: $\sum_{j=1}^n w_j = 1$ While striving for the minimum cut objective, the constraint $k \cdot \max_\ell \sum_{\lambda_j = \ell} w_j \leq t$ has to be fulfilled. The left hand side quantifies the load balance of the partitioning λ. The load balance is dominated by the worst cluster. A value of 1 indicates perfect balance. Of course, in many cases the constraint can not be fulfilled exactly (e.g., sample balanced partitioning with n odd and k even). However they can be fulfilled within a certain narrow tolerance. We chose the maximum tolerated load imbalance $t \geq 1$ to be 1.05, or 5%, for experiments in section 7.

Finding an optimal partitioning is an NP-hard problem. However, there are very fast, heuristic algorithms for this widely studied problem [10]. The basic approach to dealing with graph partitioning or minimum-cut problems is to construct an initial partition of the vertices either randomly or according to some problem-specific strategy. Then the algorithm sweeps through the vertices, deciding whether the size of the cut would increase or decrease if we moved this vertex v over to another partition. The decision to move v can be made in time proportional to its degree by simply counting whether more of v's neighbors are on the same partition as v or not. Of course, the desirable side for v will change if many of its neighbors switch, so multiple passes are likely to be needed before the process converges to a local optimum.

After experimentation with several techniques, we decided to use the Metis multi-level multi-constraint graph partitioning package because it is very fast and scales well. A detailed description of the algorithms and heuristics used in Metis can be found in Karypis et al. [11].

5.2   Optimal Clustering

This subsection describes how we find a desirable clustering, with *high overall cluster quality Γ* and a *small number of clusters k*. Our objective is to maximize intra-cluster similarity and minimize inter-cluster similarity, given by
$\text{intra}(\mathbf{X}, \lambda, i) = \frac{1}{(n_i - 1) \cdot n_i} \sum_{\lambda_a = \lambda_b = i, a > b} s(\mathbf{x}_a, \mathbf{x}_b)$ and
$\text{inter}(\mathbf{X}, \lambda, i, j) = \frac{1}{n_i \cdot n_j} \sum_{\lambda_a = i, \lambda_b = j} s(\mathbf{x}_a, \mathbf{x}_b)$,
respectively, where i and j are cluster indices. We define our *quality* measure

$\Gamma \in [0, 1]$ ($\Gamma < 0$ in case of pathological/inverse clustering) as follows:

$$\Gamma(\mathbf{X}, \lambda) = 1 - \frac{(n - k) \cdot \sum_{i=1}^{k} \sum_{j=i+1}^{k} n_i \cdot \text{inter}(\mathbf{X}, \lambda, i, j)}{n \cdot \sum_{i=1}^{k} (n_i - 1) \cdot \text{intra}(\mathbf{X}, \lambda, i)} \tag{3}$$

$\Gamma = 0$ indicates that samples within the same cluster are on average not more similar than samples from different clusters. On the contrary, $\Gamma = 1$ describes a clustering where every pair of samples from different clusters has the similarity of 0 and at least one sample pair from the same cluster has a non-zero similarity. Note that our definition of quality does not take the "amount of balance" into account, since balancing is already observed fairly strictly by the constraints in the graph-partitioning.

Finding the "right" number of clusters k for a data set is a difficult, and often ill-posed, problem. In probabilistic approaches to clustering, likelihood-ratios, Bayesian techniques and Monte Carlo cross-validation are popular. In non-probabilistic methods, a regularization approach, which penalizes for large k, is often adopted. To achieve a high quality Γ as well as a low k, the target function $\Lambda \in [0, 1]$ is the product of the quality Γ and a penalty term which works very well in practice. Let $n \geq 4$ and $2 \leq k \leq \lfloor n/2 \rfloor$, then there exists at least one clustering with no singleton clusters. The penalized quality gives the performance Λ and is defined as $\Lambda(k) = \left(1 - \frac{2k}{n}\right) \cdot \Gamma(k)$. A modest linear penalty was chosen, since our quality criterion does not necessarily improve with increasing k (as compared to the squared error criterion). For large n, we search for the optimal k in the entire window from $2 \leq k \leq 100$. In many cases, however, a forward search starting at $k = 2$ and stopping at the first down-tick of performance while increasing k is sufficient.

6   Scalability and Parallel Implementation Issues

The graph metaphor for clustering is not only powerful but can also be implemented in a highly scalable and parallel fashion. In the canonical implementation of OPOSSUM, the most expensive step (in terms of both time and space) is the computation of the similarity measure matrix, rather than the graph-based clustering or post-processing steps! In the straightforward implementation, every pair of samples need to be compared. Consequently, computational time complexity is on the order of $O(n^2 \cdot d)$. In practice, sparsity enables a better (but still $O(n^2)$) performance characteristic. Moreover, if space (memory) is a problem, the similarity matrix can be computed on the fly as the subsequent processing does not involve batch operations.

A given coarse clustering (with a smaller number of clusters than the final k) enables us to limit the similarity computation by only considering object pairs within the same coarse cluster. In retail data such clusterings are often already in place or can be induced fairly easily. Some examples include a pre-segmentation of the customers into geographical regions, a demographic segmentation or an a priori grouping by total revenue. Then, the complexity is reduced from $O(n^2)$ to

$O(\Sigma_i n_i^2)$; where n_i are the sizes of the coarse clusters. In particular, if k' coarse clusterings of comparable sizes are present, then computation is reduced by a factor of k'. If such a coarse clustering is not given a priori, a clustering could be computed on a small representative subset of the data. Further incoming data points are then pre-clustered by assigning it to the closest neighboring centroid using the extended Jaccard similarity semantic.

The graph partitioning algorithm for clustering is implemented using Metis. The complexity is essentially determined by the number of edges (customer-pairs with sufficient similarity), and thus it scales linearly with number of customers if the number of non-zero edges per customer is about constant. Note that while Euclidean based similarity induces a fully connected graph, non-Euclidean similarity measures induce several orders of magnitude fewer edges. Two approaches to reduce edges further have been prototyped successfully. On the one hand, edges that do not exceed a domain specific global minimum weight are removed. On the other hand, (if samples are approximately equally important) the v-nearest neighbor subgraph can be created by removing all but the strongest v edges for each vertex. Clearly, this reduces the number of vertices to the order of $O(kn)$. Extensive simulation results comparing Metis with a host of graph partitioning algorithms are given in [11].

Parallel Implementation Considerations. Parallel implementation of the all-pair similarity computation on SIMD or distributed memory processors is trivial. It can be done in a systolic (at step i, compare sample x with sample $x + i \bmod n$) or block systolic manner with essentially no overhead. Frameworks such as MPI also provide native primitives for such computations. Parallelization of Metis is also very efficient in [12], which reports partitioning of graphs with over 7 million vertices (customers) in 7 seconds into 128 clusters on a 128 processor Cray T3E. This shows clearly that our graph-based approach to clustering can be scaled to most real life customer segmentation applications.

7   Experimental Results

We experimented with real retail transactions of 21672 customers of a drugstore. For the illustrative purpose of this paper, we randomly selected 2500 customers. The total number of transactions (cash register scans) for these customers is 33814 over a time interval of three months. We rolled up the product hierarchy once to obtain 1236 different products purchased. 15% of the total revenue is contributed by the single item Financial-Depts which was removed because it was too common. 473 of these products accounted for less than $25 each in toto and were dropped. The remaining $d = 762$ features and $n = 2466$ customers (34 customers had empty baskets after removing the irrelevant products) were clustered using OPOSSUM.

OPOSSUM's results for this example are obtained with a 550 MHz Pentium II PC with 512 MB RAM in under 10 seconds when similarity is precomputed. Fig. 2 shows the similarity matrix (75% sparse) visualization before (a), after generalized k-means clustering using the standard Jaccard (b), after sample bal-

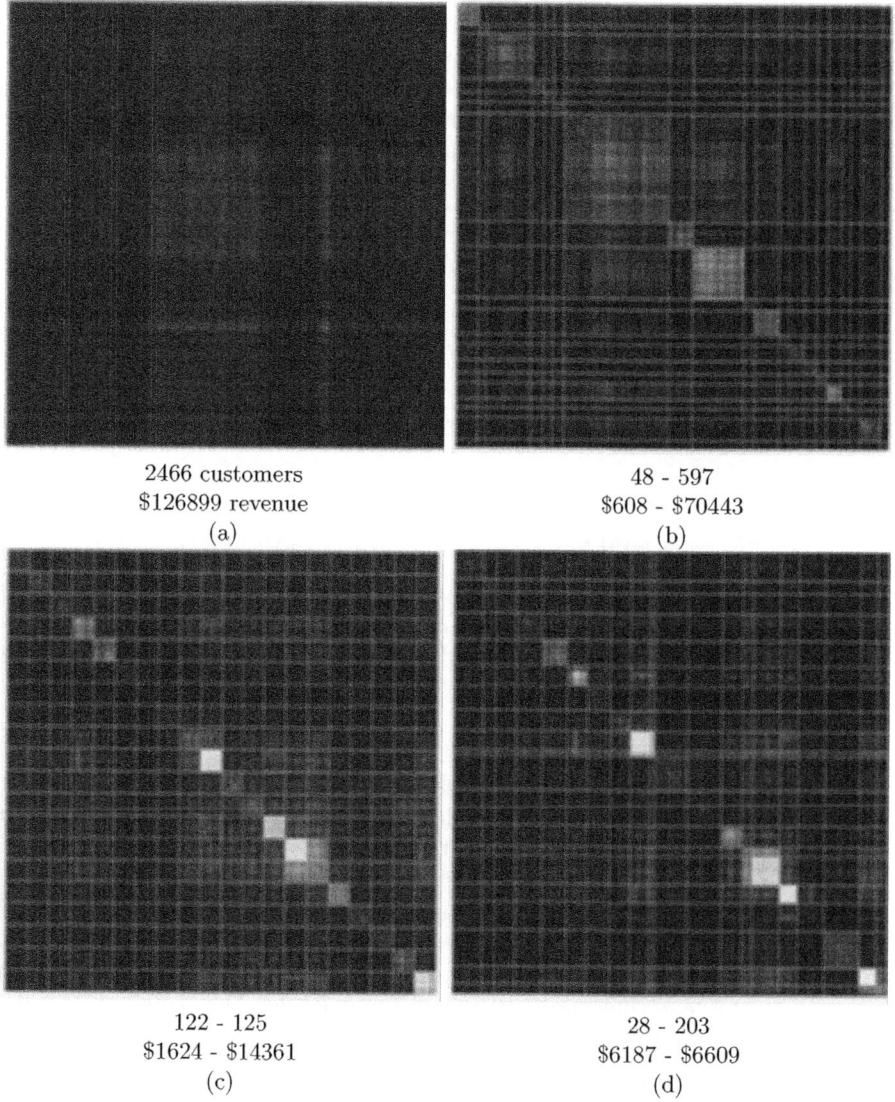

2466 customers
$126899 revenue
(a)

48 - 597
$608 - $70443
(b)

122 - 125
$1624 - $14361
(c)

28 - 203
$6187 - $6609
(d)

Fig. 2. Results of clustering drugstore data. Relationship visualizations using CLUSION (a) before, (b) after k-means binary Jaccard, (c) after sample balanced OPOSSUM (d) value balanced OPOSSUM clustering with $k = 20$. In (b) clusters are neither compact nor balanced. In (c) and (d) clusters are very compact as well as balanced.

anced (c), and after value balanced clustering (d). As the relationship-based CLUSION shows, OPOSSUM (c), (d) gives more compact (better separation of on- and off-diagonal regions) and perfectly balanced clusters as compared to, for example, k-means (b). In k-means, the standard clustering algorithm (which can be generalized by using $-log(s^{(J)})$ as distances), the clusters contain between 48 and 597 customers contributing between \$608 and \$70443 to revenue encumbering a good overview of the customer behavior by marketing. Moreover clusters are hardly compact: Brightness is only slightly better in the on-diagonal regions in (b). All visualizations have been histogram equalized for printing purposes. Fig. 3(a) shows how clustering performance behaves with increasing k. Optimal

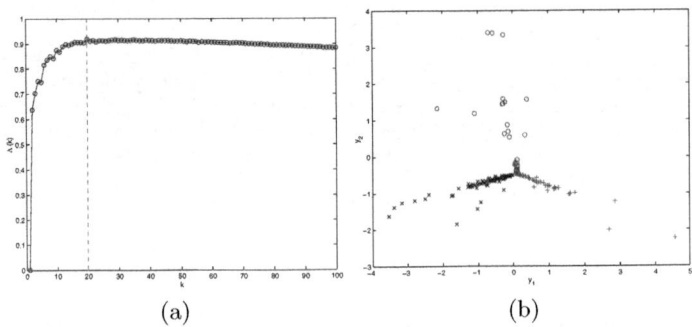

(a)   (b)

Fig. 3. Drugstore data. (a): Behavior of performance Λ for various k using value balanced clustering. The optimal k is found at 20 and is marked with a dashed vertical line. (b): 2–dimensional projection of 762–dimensional data-points on the plane defined by the centroids of the three value-balanced clusters 2 ($\circ$), 9 ($\times$) and 20 ($+$).

performance is found at $k = 20$ for value balanced clustering. In figure 3(b) the data points of the three clusters 2, 9 and 20 is projected onto a 2–dimensional plane defined by the centroids of these three clusters (a la CViz [13]). In this extremely low dimensional projection, the three selected clusters can be reasonably separated using just a linear discriminant function.

Table 1 gives profiles for two of the 20 value balanced customer clusters obtained. A very compact and useful way of profiling a cluster is to look at their most *descriptive* and their most *discriminative* features. This is done by looking at a cluster's highest revenue products and their most unusual revenue drivers. Revenue lift is the ratio of the average spending on a product in a particular cluster to the average spending in the entire data-set. In table 2 the top three descriptive and discriminative products for the customers in all clusters are shown. OPOSSUM identifies customer groups with very similar buying behavior. Marketing can use the customer groups to design and deploy personalized promotional strategies for each group. The clusters are balanced by revenue value and hence provide insight into the contributions and profiles of each customer group.

A detailed comparative study with other clustering methods has been omitted here for lack of space. However, an extensive comparison of several methods on high-dimensional data with similar characteristics, performed recently by us, can be found in [9]. It is very clear that L_p distance based or density estimation based clustering techniques that are representative of the vast majority of data mining approaches, do not work in the very high-dimensional spaces generated

Table 1. Most *descriptive* (left) and most *discriminative* (right) products purchased by the value balanced clusters C_2 and C_9. Customers in C_2 spent \$10 on average on smoking cessation gum and spent more than 34 times more money on peanuts than the average customer. Cluster C_9 seems to contain strong christmas shoppers probably families with kids.

<table>
<tr><td></td><td></td><td>value</td><td>lift</td><td></td><td>value</td><td>lift</td></tr>
<tr><td rowspan="10">C_2</td><td>SMOKING-CESSATION-GUM</td><td>10.153</td><td>34.732</td><td>SMOKING-CESSATION-GUM</td><td>10.153</td><td>34.732</td></tr>
<tr><td>TP-CANNING</td><td>2.036</td><td>18.738</td><td>BLOOD-PRESSURE-KITS</td><td>1.690</td><td>34.732</td></tr>
<tr><td>BLOOD-PRESSURE-KITS</td><td>1.690</td><td>34.732</td><td>SNACKS/PNTS-NUTS</td><td>0.443</td><td>34.732</td></tr>
<tr><td>TP-CASSETTE-RECORDER/PLAY</td><td>1.689</td><td>9.752</td><td>TP-RAZOR-ACCESSORIES</td><td>0.338</td><td>28.752</td></tr>
<tr><td>DIABETIC-TESTS</td><td>1.521</td><td>13.158</td><td>BABY-FORMULA-RTF</td><td>0.309</td><td>28.404</td></tr>
<tr><td>TP-TOASTER/OVEN/FRY/POP</td><td>1.169</td><td>7.016</td><td>TP-CANNING</td><td>2.036</td><td>18.738</td></tr>
<tr><td>BATT-ALKALINE</td><td>1.028</td><td>1.709</td><td>CONSTRUCTION-TOYS</td><td>0.855</td><td>18.230</td></tr>
<tr><td>TP-SEASONAL-BOOTS</td><td>0.991</td><td>1.424</td><td>PICNIC</td><td>0.379</td><td>18.208</td></tr>
<tr><td>SHELF-CHOCOLATES</td><td>0.927</td><td>7.924</td><td>CG-ETHNIC-FACE</td><td>0.350</td><td>17.335</td></tr>
<tr><td>CHRISTMAS-FOOD</td><td>0.926</td><td>1.988</td><td>TP-PLACEMT,NAPKN,CHR-PADS</td><td>0.844</td><td>16.743</td></tr>
</table>

<table>
<tr><td></td><td></td><td>value</td><td>lift</td><td></td><td>value</td><td>lift</td></tr>
<tr><td rowspan="10">C_9</td><td>CHRISTMAS-GIFTWARE</td><td>12.506</td><td>12.986</td><td>TP-FURNITURE</td><td>0.454</td><td>22.418</td></tr>
<tr><td>CHRISTMAS-HOME-DECORATION</td><td>1.243</td><td>3.923</td><td>TP-ART&CRAFT-ALL-STORES</td><td>0.191</td><td>13.772</td></tr>
<tr><td>CHRISTMAS-FOOD</td><td>0.965</td><td>2.071</td><td>TP-FAMILY-PLAN,CONTRACEPT</td><td>0.154</td><td>13.762</td></tr>
<tr><td>CHRISTMAS-LIGHT-SETS</td><td>0.889</td><td>1.340</td><td>TP-WOMENS-CANVAS/ATH-SHOE</td><td>0.154</td><td>13.622</td></tr>
<tr><td>BOY-TOYS</td><td>0.742</td><td>1.779</td><td>CHRISTMAS-GIFTWARE</td><td>12.506</td><td>12.986</td></tr>
<tr><td>AMERICAN-GREETINGS-CARDS</td><td>0.702</td><td>0.848</td><td>TP-CAMERAS</td><td>0.154</td><td>11.803</td></tr>
<tr><td>GAMES</td><td>0.694</td><td>1.639</td><td>COMEDY</td><td>0.154</td><td>10.455</td></tr>
<tr><td>CHRISTMAS-CANDY</td><td>0.680</td><td>2.585</td><td>CHRISTMAS-CANDOLIERS</td><td>0.192</td><td>9.475</td></tr>
<tr><td>TP-SEASONAL-BOOTS</td><td>0.617</td><td>0.887</td><td>TP-INFANT-FORMULA/FOOD</td><td>0.107</td><td>8.761</td></tr>
<tr><td>CHRISTMAS-CANDLES</td><td>0.601</td><td>4.425</td><td>CHRISTMAS-MUSIC</td><td>0.091</td><td>8.625</td></tr>
</table>

Table 2. Overview over *descriptive* (left) and *discriminative* products (right) dominant in each of the 20 value balanced clusters.

```
 1  bath gift packs  hair growth m   boutique island       1  action items      tp video comedy family items
 2  smoking cessati  tp canning item blood pressure         2  smoking cessati   blood pressure  snacks/pnts nut
 3  vitamins other   tp coffee maker underpads hea          3  underpads hea     miscellaneous k tp irons items
 4  games items      facial moisturi tp wine jug ite        4  acrylics/gels/w   tp exercise ite dental applianc
 5  batt alkaline i  appliances item appliances appl        5  appliances item   housewares peg  tp tarps items
 6  christmas light  appliances hair tp toaster/oven        6  multiples packs   christmas light tv's items
 7  christmas food   christmas cards cold bronchial          7  sleep aids item   kava kava items tp beer super p
 8  girl toys/dolls  boy toys items  everyday girls         8  batt rechargeab   tp razors items tp metal cookwa
 9  christmas giftw  christmas home  christmas food         9  tp furniture it   tp art&craft al tp family plan
10  christmas giftw  christmas light pers cd player        10  pers cd player    tp plumbing ite umbrellas adult
11  tp laundry soap  facial cleanser antacid h2 bloc       11  cat litter scoo   child acetamino pro treatment i
12  film cameras it  planners/calend hand&body thera       12  heaters items     laverdiere ca   ginseng items
13  tools/accessori  binders items   drawing supplie       13  mop/broom lint    halloween cards tools/accessori
14  american greeti  paperback items fragrances op         14  dental repair k   tp lawn seed it tp telephones/a
15  american greeti  christmas cards basket candy it       15  gift boxes item   hearing aid bat american greeti
16  tp seasonal boo  american greeti valentine box c       16  economy diapers   tp seasonal boo girls socks ite
17  vitamins e item  group stationer tp seasonal boo       17  tp wine 1.5l va   group stationer stereos items
18  halloween bag c  basket candy it cold cold items       18  tp med oint/liq   tp dinnerware i tp bath towels
19  hair clr perman  american greeti revlon cls face       19  hair clr perman   covergirl imple tp power tools
20  revlon cls face  hair clr perman headache ibupro       20  revlon cls face   telephones cord ardell lashes i
```

by real-life market-basket data, and that graph partitioning approaches have several advantages in this domain.

8   Concluding Remarks

OPOSSUM efficiently delivers clusters that are balanced in terms of either samples (customers) or value (revenue). Balancing clusters is very useful since each cluster represents a number of data points of similar importance to the user. The associated visualization toolkit CLUSION allows managers and marketers to get an intuitive visual impression of the group relationships and customer behavior extracted from the data. This is very important for the tool to be accepted and applied by a wider community. The OPOSSUM / CLUSION combine has been successfully applied to several real-life market-baskets. We are currently working on an on-line version of OPOSSUM that incrementally updates clusters as new data points become available. Moreover, modifications for improved scale-up to very large numbers of transactions, using parallel data-flow approaches are currently being investigated.

Acknowledgments. We want to express our gratitude to Mark Davis, Net Perceptions, (http://www.netperceptions.com) for providing the drugstore retail data set. This research was supported in part by the NSF under Grant ECS-9000353, and by KD1 and Dell.

References

1. Strehl, A., Ghosh, J.: Value-based customer grouping from large retail data-sets. Proc. SPIE Vol. 4057, (2000) 33–42.
2. Jain, A. K., Dubes, R. C.: Algorithms for Clustering Data. Prentice Hall, New Jersey (1988)
3. Hartigan, J.A.: Clustering Algorithms. Wiley, New York (1975)
4. Rastogi, R., Shim, K.: Scalable algorithms for mining large databases. In Jiawei Han, (ed), KDD-99 Tutorial Notes. ACM (1999)
5. Guha, S., Rastogi, R., Shim, K.: Rock: a robust clustering algorithm for categorical attributes. Proc.15th Int'l Conf. on Data Engineering (1999)
6. Karypis, G., Han, E., Kumar, V.: Chameleon: Hierarchical clustering using dynamic modeling. IEEE Computer, 32(8), (1999) 68–75
7. Bradley, P., Fayyad, U., Reina, C.: Scaling clustering to large databases. In Proc. KDD-98, AAAI Press (1998) 9–15 1998.
8. Dhillon, I., Modha, D.: A data clustering algorithm on distributed memory multiprocessors. KDD Workshop on Large-Scale Parallel Systems (1999)
9. Strehl, A., Ghosh, J., Mooney, R.: Impact of similarity measures on web-page clustering. In Proc. AAAI Workshop on AI for Web Search (2000) 58-64
10. Miller, G.L., Teng, S., Vavasis, S.A., A unified geometric approach to graph separators. In Proc. 31st Annual Symposium on Foundations of Computer Science (1991) 538–547
11. Karypis, G., Kumar, V.: A fast and high quality multilevel scheme for partitioning irregular graphs. SIAM Journal of Scientific Computing, 20 1 (1998), 359–392
12. Schloegel, K., Karypis, G., Kumar, V.: Parallel multilevel algorithms for multi-constraint graph partitioning. Technical Report 99-031, Dept of Computer Sc. and Eng, Univ. of Minnesota (1999)
13. Dhillon, I., Modha, D., Spangler, W.: Visualizing class structure of multidimensional data. In S. Weisberg, editor, Proc. 30th Symposium on the Interface: Computing Science and Statistics, (1998)

Dynamic Integration of Decision Committees

Alexey Tsymbal and Seppo Puuronen

Department of Computer Science and Information Systems,
University of Jyväskylä, P.O.Box 35, FIN-40351 Jyväskylä, Finland
{alexey, sepi}@jytko.jyu.fi

Abstract. Decision committee learning has demonstrated outstanding success in reducing classification error with an ensemble of classifiers. In a way a decision committee is a classifier formed upon an ensemble of subsidiary classifiers. Voting, which is commonly used to produce the final decision of committees has, however, a shortcoming. It is unable to take into account local expertise. When a new instance is difficult to classify, then it easily happens that only the minority of the classifiers will succeed, and the majority voting will quite probably result in a wrong classification. We suggest that dynamic integration of classifiers is used instead of majority voting in decision committees. Our method is based on the assumption that each classifier is best inside certain subareas of the whole domain. In this paper, the proposed dynamic integration is evaluated in combination with the well-known decision committee approaches AdaBoost and Bagging. The comparison results show that both boosting and bagging produce often significantly higher accuracy with the dynamic integration than with voting.

1   Introduction

Decision committee learning has demonstrated outstanding success in reducing classification error with an ensemble of classifiers [1,3,4,6,13-15,19]. This approach develops a classifier in the form of a committee of subsidiary base classifiers. Each of the base classifiers makes its own classification separately and the final classification of the committee is usually produced by majority voting, a way to combine individual classifications. Decision committee learning has especially been recommended for learning tasks when no prior opportunity to evaluate the relative effectiveness of alternative approaches exists, when no a priori knowledge is available about the domain area, and when the primary goal of learning is to develop a classifier with the lowest possible error [19].

Two decision committee learning techniques, boosting [13-15] and bagging [4], have received extensive attention lately. They both repeatedly build base classifiers using some learning algorithm by changing the distribution of the training set. Bagging changes the distribution using bootstrap samples drawn from the training set. Boosting changes the distribution sequentially, trying to pay more attention to the instances incorrectly classified by previous base classifiers.

M. Valero, V.K. Prasanna, and S. Vajapeyam (Eds.): HiPC 2000, LNCS 1970, pp. 537–546, 2000

Both bagging and boosting use a voting technique to combine the classifications of the committee members. Voting has a shortcoming that it is unable to take into account local expertise. When a new instance is difficult to classify, then it easily happens that only the minority of the classifiers will succeed, and the majority voting will quite probably result in a wrong classification. The problem may consist in discarding the base classifiers (by assigning small weights) that are highly accurate in a restricted region of the instance space because this accuracy is swamped by their inaccuracy outside the restricted region. It may also consist in the use of classifiers that are accurate in most of the instance space but still unnecessarily confuse the whole classification committee in some restricted areas of the space.

We suggest that the dynamic integration of classifiers [10] is used instead of voting to overcome this problem. Our method is based on the assumption that each committee member is best inside certain subareas of the whole domain space. In the training phase the information about the local errors of the base classifiers for each training instance is collected. This information is later used together with the initial training set as meta-level knowledge to estimate the local classification errors of the base classifiers for a new instance.

In this paper, the proposed dynamic integration method is evaluated in combination with the well-known decision committee algorithms AdaBoost and Bagging. In chapter 2 we review these algorithms. In chapter 3, the technique for dynamic selection of classifiers is considered. In chapter 4, results of experiments with the AdaBoost and Bagging algorithms are presented, and chapter 5 concludes with a brief summary and further research topics.

2   Bagging and Boosting for Decision Committee Learning

Two decision committee learning approaches, boosting [13-15] and bagging [4], have received extensive attention recently. In this chapter bagging and boosting algorithms to be used in comparisons are reviewed.

Given an integer T as the committee size, both bagging and boosting need approximately T times as long as their base learning algorithm does for learning a single classifier. There are two major differences between bagging and boosting. First, boosting changes adaptively the distribution of the training set based on the performance of previously created classifiers while bagging changes the distribution of the training set stochastically. Second, boosting uses a function of the performance of a classifier as a weight for voting, while bagging uses equal-weight voting. With both techniques the error decreases when the size of the committee increases, but the marginal error reduction of an additional member tends to decrease [19]. The error reduction has been a little mysterious for researchers, and several formal models to reveal the real background of the phenomenon have been suggested [3,6,13,15]. However many of the proposed models work only under strong assumptions, and there is still much work to do before the behavior of the methods is deeply understood.

The bagging algorithm (**B**ootstrap **agg**regat**ing**) [4] uses bootstrap samples to build the base classifiers. Each bootstrap sample of m instances is formed by uniformly sampling m instances from the training set with replacement. The amount of the different instances in the bootstrap sample is for large m about $1-1/e=63.2\%$. This results that dissimilar base classifiers are built with unstable learning algorithms (e.g., neural networks, decision trees) [1] and the performance of the committee can become better. The bagging algorithm generates T bootstrap samples B_1, B_2, ..., B_T and then the corresponding T base classifiers C_1, C_2, ..., C_T. The final classification produced by the committee is received using equal-weight voting where the ties are broken arbitrarily.

Boosting was developed as a method for boosting the performance of any weak learning algorithm, which needs only to be a little bit better than random guessing [15]. The AdaBoost algorithm (**Ada**ptive **Boost**ing) [6] was introduced as an improvement of the initial boosting algorithm and several further variants were presented. AdaBoost changes the weights of the training instances after each generation of a base classifier based on the misclassifications made by the base classifier trying to force the learning algorithm to minimize the expected error over different input distributions [1]. Thus AdaBoost generates during the T trials T training sets S_1, S_2, ..., S_T with weights assigned to each instance, and T base classifiers C_1, C_2, ..., C_T are built. A final classification produced by the committee is received using weighted voting where the weight of each classifier depends on its training set performance.

The AdaBoost algorithm requires a weak learning algorithm whose error is bounded by a constant strictly less than ½. In practice, the inducers we use provide no such guarantee [13]. In this paper, a bootstrap sample from the original data S is generated in such a case, and the boosting continues up to a limit of 25 such samples at a given trial, as proposed in [1]. The same is done when the error is equal to zero at some trial.

Breiman [3] has introduced the notion of the edge function to help in explanation of behaviour of boosting and, in general, arcing (*a*daptively *r*eweighting and *c*lassification) algorithms. This function is similar to the margin function by Schapire et.al. [13], and it is absolutely the same in the case of 2-concept classification. Breiman et. al. give a simple framework, based on the edge function, for understanding arcing algorithms. They then show that there is a large class of algorithms, including all of the current arcing ones, that work by minimizing the function of the edge. The edge function is defined as:

$$\text{edge}(\mathbf{c}, \mathbf{x}) = \sum_m c_m i(\mathbf{x}, \mathbf{E}_m), \tag{1}$$

where $\mathbf{c} = \{c_m\}$ is a collection of non-negative numbers (votes) summing to one, $\mathbf{x}$ is an instance of the instance space, the set $\mathbf{E}_m = \{\mathbf{x}_n ; C(\mathbf{x}_n) \neq y_n\}$ is the error set of the base classifier C_m, and $i(\mathbf{x}, \mathbf{E}_m) = I(\mathbf{x} \in \mathbf{E}_m)$ is the indicator function of $\mathbf{E}_m$ [3].

According to Breiman et.al. [3], all arcing and many other decision committee learning algorithms, given the set C of classifiers, and the training set $\mathbf{T}$ with the known class values $\{y_n\}$, work by finding appropriate $\mathbf{c}$ values to minimize the edge function (1). For example, some linear programming technique can be used to define the $\mathbf{c}$ values. After the $\mathbf{c}$ values are used as weights for the corresponding classifiers' votes. The implication of the Schapire et.al. work [13] is that values of **c** that give generally low values of edge(**c**,**x**) on the training set will also lead to low generalization error.

Both [13] and [3] as well as authors of most current decision committee learning algorithms consider the set **c** of weights to be constant over the instance space. However, as we show in this paper, in some situations it would be reasonable to discard this limitation. In the dynamic integration presented in this paper, we also try to minimize implicitly the edge function (1). However, we consider the votes' weights **c** to be varying over the instance space. We use heuristics to define those weights. In our approach, the **c** values depend on the evaluated local accuracy of the corresponding base classifiers. In the next section we consider those heuristics in more detail.

3   Dynamic Integration of Classifiers

The challenge of integration is to decide which classifier to rely on or how to combine classifications produced by several classifiers. The integration approaches can be divided into static and dynamic ones. Recently, two main approaches to the integration have been used: (1) combination of the classifications produced by the ensemble, and (2) selection of the best classifier. The most popular method of combining classifiers is voting [1]. More sophisticated selection approaches use estimations of the local accuracy of the base classifiers [9] or the meta-level classifiers ("referees"), which predict the correctness of the base classifiers for a new instance [8].

We have presented a dynamic integration technique that estimates the local accuracy of the base classifiers by analyzing the accuracy in near-by instances [10]. Knowledge is collected about the errors that the base classifiers make on the training instances and this knowledge is used during the classification of new instances. The goal is to use each base classifier just in that subarea for which it is most reliable.

The dynamic integration approach contains two phases. In the learning phase, information is collected about how each classifier succeeds with each training instance. With bagging and boosting this is done directly using the base classifiers produced. In the original version of our algorithm we have used cross-validation technique to estimate the errors of the base classifiers on the training set [10]. In the application phase, the final classification of the committee is formed so that for a new instance seven nearest neighbors are found out. The error information of each classifier with these instances is collected and then depending on the dynamic integration approach the final classification is made using this information. Two of these three different functions implementing the application phase were considered in [10]. The first function DS implements Dynamic Selection. In the DS application phase the classification error is predicted for each base classifier using the WNN [5] procedure and a classifier with the smallest error (with the least global error in the case of ties) is selected to make the final classification. The second function DV implements Dynamic Voting. In DV each base classifier receives a weight that depends on the local classifier's performance and the final classification is conducted by voting classifier predictions with their weights. The third function DVS was first presented in [21] and it applies a mixed approach. Using the collected error information, first about the better half of classifiers is selected and then the final classification is derived using weighted voting.

Thus, the learning phase of decision committee learning with dynamic integration takes almost the same time as ordinary decision committee learning with voting. We need just additional memory to store the performance matrix $n \times m$, where n is the number of instances in the training set, and m is the number of base classifiers being trained (the number of decision committee members). Each element of the matrix presents the classification error of the corresponding base classifier at the corresponding point of the instance space. The application phase in the case of dynamic integration besides lunching the base classifiers with the new instance includes also the calculation of the local accuracies of the base classifiers, which are used for definition of the classifiers' weights. The most time-consuming element at this stage is to find nearest neighbors among the training instances. The time taken at this stage depends on the number of instances considered, and on the number of their features. Some instance space indexing technique can be used to facilitate the process of nearest neighbor search.

A number of experiments comparing the dynamic integration with such widely used integration approaches as CVM (Cross-Validated Majority) and weighted voting were also conducted [10,11]. The comparison results show that the dynamic integration technique outperforms often weighted voting and CVM. In [17,18] the dynamic classifier integration was applied to decision committee learning, combining the generated classifiers in a more sophisticated manner than voting. This paper continues this research considering more experiments comparing bagging and boosting with and without the three dynamic integration functions.

4   Experiments

In this chapter we present experiments where the dynamic classifier integration is used with classifiers generated by AdaBoost and Bagging. First, the experimental setting is described, and then, results of the experiments are presented. The experiments are conducted on ten datasets taken from the UCI machine learning repository [2]. Previously the dynamic classifier integration was experimentally evaluated in [10] and preliminary comparison with AdaBoost and Bagging was made in [17, 18].

For each dataset 30 test runs are made. In each run the dataset is first split into the training set and the test set by random sampling. Each time 30 percent of the instances of the dataset are first randomly picked up to the test set. The rest 70 percent of the instances are then passed to the learning algorithm where the accuracy history of the committee classifiers on the initial training set is collected for later use in the dynamic integration. Cross-validation is not used here as it was in [10] for the case of classifiers built with different learning algorithms. Instead, the committee classifiers are simply applied to each instance of the initial training set, and their accuracy is collected into the performance matrix. The committee classifiers themselves are learnt using the C4.5 decision tree algorithm with pruning [12]. We make experiments with 5, 10, and 25 base classifiers in the committees.

We calculate average accuracy numbers for base classifiers, for ordinary bagging and boosting, and for bagging and boosting with the three dynamic integration methods described in chapter 3 separately. At the application phase of the dynamic integration the instances included in the test set are used. For each of them the classification errors of all the base classifiers of the committee are collected for seven nearest neighbors of the new instance. The selection of the number of nearest neighbors has been discussed in [10]. Based on the comparisons between different distance functions for dynamic integration presented in [11] we decided to use the Heterogeneous Euclidean-Overlap Metric, which produced good test results earlier. The test environment was implemented within the MLC++ framework (the machine learning library in C++) [7].

Table 1 presents accuracy values for Bagging with different voting methods. The first column includes the name of the corresponding dataset. Under the name, the accuracy of single plain C4.5 tree on this dataset is given, estimated over the 30 random train/test splits for the sake of comparison with the committee approaches. The second column tells how many base classifiers were included in the committee. The next three columns include the average of the minimum accuracies of the base classifiers (min), the average of average accuracies of the base classifiers (aver), and the average of the maximum accuracies of the base classifiers (max) over the 30 runs.

The next two columns include the average percentage of test instances where all the committee classifiers managed to produce the right classification (agree), and the average amount of test instances where at least one committee classifier during each run managed to produce the right classification (cover). The last four columns on the right-hand side of Table 1 include average accuracies for the base classifiers integrated with different voting methods. These are: equal-weighted voting (ordinary bagging, B), dynamic selection (DS), dynamic voting (DS), and dynamic voting with selection (DVS). DS, DV, and DVS were considered in chapter 3. Previous experiments with these two integration strategies [10] have shown that the accuracies of the strategies usually differ significantly; however, it depends on the dataset, what a strategy is preferable. In this paper we apply also a combination of the DS and DV strategies that was in [18] expected to be more stable than the other two.

From Table 1 one can see that in many cases at least some dynamic integration strategy works better than plain voting with Bagging. For example, on the MONK-1 dataset, Dynamic Selection (DS) from 25 Bagging classifiers gives 0.995 accuracy on average, while the usual Bagging with unweighted voting gives only 0.913. Dynamic integration strategies DV and DVS are also better on average than Bagging. DV and DVS are generally better than DS in these experiments. DS works surprisingly bad in these experiments, as was also in [17,18]. As it was supposed in [18], DVS works quite stably on the considered datasets. Sometimes, it overcomes even both DS and DV, as on the Glass and Lymphography datasets. The accuracies of decision committees are as was supposed in many cases better than the accuracies of the single C4.5 tree classifiers. The accuracies in Table 1 confirm also the known result that raising the number of committee members makes bagging more accurate, and the optimal committee size is 25 for many datasets. As a summary it can be said that on average over these datasets there is a benefit in the dynamic integration with bagging

(DV and DVS) but because there are many dataset-related differences it needs further systematic research before being able to make any firm conclusions.

Table 1. Accuracy values for Bagging decision committees with different voting approaches

	Decision committee: C4.5 trees with pruning									
DB	size	min	aver	max	agree	cover	B	DS	DV	DVS
Breast	5	0.649	0.691	0.739	0.464	0.875	0.710	0.697	0.710	0.706
0.691	10	0.625	0.686	0.746	0.384	0.906	0.717	0.693	0.712	0.702
	25	0.605	0.689	0.763	0.268	0.942	0.717	0.692	0.720	0.714
DNA	5	0.678	0.743	0.806	0.436	0.964	0.793	0.764	0.793	0.781
promoter	10	0.639	0.736	0.821	0.291	0.986	0.786	0.761	0.796	0.794
0.755	25	0.611	0.739	0.853	0.187	0.995	0.798	0.754	0.809	0.819
Glass	5	0.544	0.608	0.670	0.326	0.851	0.656	0.663	0.675	0.677
0.626	10	0.522	0.608	0.687	0.233	0.905	0.669	0.666	0.683	0.689
	25	0.500	0.610	0.711	0.150	0.930	0.682	0.681	0.696	0.705
LED7	5	0.568	0.625	0.669	0.432	0.795	0.660	0.688	0.688	0.690
0.665	10	0.548	0.627	0.689	0.386	0.845	0.674	0.698	0.701	0.701
	25	0.532	0.628	0.704	0.336	0.884	0.677	0.706	0.703	0.710
LED24	5	0.528	0.583	0.638	0.343	0.791	0.632	0.596	0.631	0.615
0.630	10	0.498	0.584	0.655	0.248	0.840	0.645	0.593	0.651	0.643
	25	0.483	0.585	0.667	0.160	0.889	0.660	0.592	0.660	0.654
Liver	5	0.557	0.610	0.657	0.243	0.911	0.645	0.621	0.653	0.637
0.614	10	0.541	0.606	0.669	0.117	0.958	0.649	0.622	0.663	0.657
	25	0.526	0.610	0.695	0.044	0.985	0.669	0.621	0.682	0.675
Lympho-	5	0.663	0.731	0.794	0.505	0.920	0.755	0.769	0.761	0.778
graphy	10	0.641	0.734	0.807	0.424	0.949	0.769	0.771	0.776	0.778
0.721	25	0.617	0.731	0.823	0.306	0.976	0.765	0.757	0.769	0.782
MONK-1	5	0.727	0.827	0.928	0.528	0.991	0.894	0.975	0.900	0.964
0.905	10	0.699	0.834	0.947	0.437	0.999	0.895	0.988	0.931	0.975
	25	0.666	0.826	0.976	0.320	1.000	0.913	0.995	0.942	0.976
MONK-2	5	0.510	0.550	0.588	0.203	0.870	0.567	0.534	0.565	0.547
0.525	10	0.499	0.551	0.600	0.127	0.941	0.600	0.536	0.566	0.540
	25	0.487	0.551	0.610	0.067	0.982	0.582	0.539	0.571	0.537
Zoo	5	0.798	0.857	0.908	0.735	0.944	0.883	0.910	0.905	0.912
0.908	10	0.778	0.858	0.925	0.693	0.966	0.886	0.920	0.914	0.927
	25	0.754	0.860	0.937	0.650	0.975	0.893	0.926	0.919	0.925
Average	5	0.681	0.732	0.782	0.493	0.919	0.766	0.759	0.770	0.769
	10	0.661	0.733	0.795	0.410	0.951	0.773	0.761	0.779	0.778
	25	0.643	0.733	0.811	0.331	0.971	0.779	0.762	0.786	0.786

Table 2 presents accuracies for AdaBoost with different voting methods. The columns are the same as in Table 1 except that B here means AdaBoost with weighted voting, and again the accuracies are averages over 30 runs.

Table 2. Accuracy values for AdaBoost decision committees with different voting approaches

DB	size	min	aver	max	agree	cover	B	DS	DV	DVS
					Decision committee: C4.5 trees with pruning					
Breast	5	0.577	0.637	0.703	0.253	0.933	0.684	0.664	0.686	0.689
0.691	10	0.548	0.628	0.708	0.127	0.978	0.686	0.658	0.693	0.689
	25	0.504	0.618	0.717	0.040	0.995	0.687	0.655	0.697	0.697
DNA	5	0.627	0.737	0.840	0.263	0.998	0.857	0.770	0.857	0.812
promoter	10	0.551	0.707	0.852	0.080	1.000	0.869	0.769	0.877	0.857
0.755	25	0.468	0.677	0.869	0.009	1.000	0.877	0.766	0.877	0.876
Glass	5	0.524	0.595	0.666	0.231	0.888	0.683	0.667	0.685	0.685
0.626	10	0.457	0.572	0.676	0.071	0.947	0.697	0.668	0.702	0.705
	25	0.371	0.533	0.683	0.003	0.976	0.703	0.661	0.704	0.708
LED7	5	0.487	0.573	0.656	0.310	0.802	0.648	0.680	0.680	0.683
0.665	10	0.453	0.551	0.658	0.224	0.831	0.631	0.682	0.670	0.688
	25	0.423	0.540	0.661	0.152	0.847	0.615	0.684	0.654	0.682
LED24	5	0.437	0.526	0.634	0.167	0.820	0.619	0.531	0.611	0.565
0.630	10	0.384	0.492	0.634	0.047	0.871	0.612	0.537	0.613	0.594
	25	0.338	0.465	0.634	0.010	0.915	0.615	0.523	0.609	0.601
Liver	5	0.550	0.601	0.650	0.205	0.928	0.642	0.609	0.642	0.624
0.614	10	0.526	0.593	0.656	0.070	0.982	0.639	0.606	0.640	0.634
	25	0.474	0.581	0.668	0.010	0.999	0.645	0.593	0.646	0.647
Lympho-	5	0.615	0.709	0.791	0.331	0.954	0.784	0.758	0.784	0.776
graphy	10	0.558	0.690	0.803	0.153	0.974	0.803	0.754	0.802	0.795
0.721	25	0.478	0.655	0.814	0.028	0.985	0.794	0.748	0.788	0.806
MONK-1	5	0.740	0.849	0.936	0.506	0.999	0.949	0.966	0.948	0.967
0.905	10	0.705	0.843	0.950	0.260	1.000	0.968	0.980	0.974	0.984
	25	0.570	0.803	0.965	0.091	1.000	0.989	0.983	0.990	0.993
MONK-2	5	0.473	0.535	0.594	0.170	0.928	0.509	0.572	0.509	0.559
0.525	10	0.458	0.535	0.607	0.115	0.978	0.503	0.575	0.517	0.538
	25	0.442	0.534	0.625	0.070	0.995	0.493	0.558	0.501	0.520
Zoo	5	0.830	0.893	0.938	0.769	0.962	0.934	0.936	0.937	0.935
0.908	10	0.810	0.893	0.945	0.716	0.969	0.938	0.936	0.941	0.942
	25	0.781	0.890	0.954	0.646	0.976	0.942	0.931	0.943	0.946
Average	5	0.586	0.666	0.741	0.321	0.921	0.731	0.715	0.734	0.730
	10	0.545	0.650	0.749	0.186	0.953	0.735	0.717	0.743	0.743
	25	0.485	0.630	0.759	0.106	0.969	0.736	0.710	0.741	0.748

From Table 2 one can see that in many cases at least some dynamic integration strategy works better than static weighted voting. For example, on MONK-2, DS from 10 AdaBoost classifiers gives 0.575 accuracy, while the usual AdaBoost gives only 0.503. DV and DVS are better on average also than AdaBoost and DS. The accuracies of decision committees are in general better than the accuracies of the single C4.5 tree. Table 2 confirms also the known result that raising the number of committee members makes boosting more accurate, and the optimal committee size is usually 25. However, for some datasets (as MONK-2) 10 classifiers are already enough. The LED24 dataset is the same as LED7 with 17 irrelevant features added. On those datasets one can see how irrelevant features influence the dynamic integration – while on LED7 the dynamic integration performs significantly better, on LED27 there is no significant difference between the dynamic integration and ordinary voting. This is true both for Bagging and AdaBoost. Some feature selection technique can be used to overcome this. As a summary it can be said that on average over these datasets there is also a benefit in the dynamic integration with AdaBoost (DV and DVS) but because there are many dataset-related differences it needs further systematic research to confirm this.

5   Conclusion

Decision committees have demonstrated spectacular success in reducing classification error from learned classifiers. These techniques develop a committee of base classifiers, which produce the final classification using usually voting. Ordinary voting has however an important shortcoming that it does not take into account local expertise.

In this paper a technique for dynamic integration of classifiers was proposed for the use instead of voting to integrate committee classifiers. The technique for dynamic integration of classifiers is based on the assumption that each committee member is best inside certain subareas of the whole instance space.

The dynamic integration technique was evaluated with AdaBoost and Bagging, the decision committee approaches which have received extensive attention recently, on ten datasets from the UCI machine learning repository. The results achieved are promising and show that boosting and bagging have often significantly better accuracy with dynamic integration of classifiers than with simple voting. Commonly this holds true on the datasets for which dynamic integration is preferable to static integration (voting or cross-validated selection) as it was shown by previous experiments. For many domains, there is a benefit in the use of dynamic integration with Bagging and AdaBoost (especially with DV and DVS) but because there are many dataset-related differences it needs further systematic research to find formal dependencies.

Further experiments can be conducted to make deeper analysis of combining the dynamic integration of classifiers with different approaches to decision committee learning. Another potentially interesting topic for further research is the bias-variance analysis of decision committees integrated with the dynamic approaches.

Acknowledgments

This research is supported by the COMAS Graduate School of the University of Jyväskylä. We would like to thank the UCI machine learning repository for the datasets, and the machine learning library in C++ for the source code used in this study. We want to thank also the anonymous reviewers for their constructive and helpful criticism.

References

1. Bauer, E., Kohavi, R.: An Empirical Comparison of Voting Classification Algorithms: Bagging, Boosting, and Variants. Machine Learning, Vol.36 (1999) 105-139.
2. Blake, C.L., Merz, C.J.: UCI Repository of Machine Learning Databases [http://www.ics.uci.edu/ ~mlearn/ MLRepository.html]. Dep-t of Information and CS, Un-ty of California, Irvine CA (1998).
3. Breiman, L.: Arcing the Edge. Tech. Rep. 486, Un-ty of California, Berkely CA (1997).
4. Breiman, L.: Bagging Predictors. Machine Learning, Vol. 24 (1996) 123-140.
5. Cost, S., Salzberg, S.: A Weighted Nearest Neighbor Algorithm for Learning with Symbolic Features. Machine Learning, Vol. 10, No. 1 (1993) 57-78.
6. Freund, Y., Schapire, R.E.: A Decision-Theoretic Generalization of On-Line Learning and an Application to Boosting. In: Proc. 2nd European Conf. on Computational Learning Theory, Springer-Verlag (1995) 23-37.
7. Kohavi, R., Sommerfield, D., Dougherty, J.: Data Mining Using MLC++: A Machine Learning Library in C++. Tools with Artificial Intelligence, IEEE CS Press (1996) 234-245.
8. Koppel, M., Engelson, S.P.: Integrating Multiple Classifiers by Finding their Areas of Expertise. In: AAAI-96 Workshop On Integrating Multiple Learning Models (1996) 53-58.
9. Merz, C.: Dynamical Selection of Learning Algorithms. In: D.Fisher, H.-J.Lenz (eds.), Learning from Data, Artificial Intelligence and Statistics, Springer-Verlag, NY (1996).
10. Puuronen, S., Terziyan, V., Tsymbal, A.: A Dynamic Integration Algorithm for an Ensemble of Classifiers. In: Z.W. Ras, A. Skowron (eds.), Foundations of Intelligent Systems: ISMIS'99, Lecture Notes in AI, Vol. 1609, Springer-Verlag, Warsaw (1999) 592-600.
11. Puuronen, S., Tsymbal, A., Terziyan, V.: Distance Functions in Dynamic Integration of Data Mining Techniques. In: B.V. Dasarathy (ed.), Data Mining and Knowledge Discovery: Theory, Tools, and Techniques, SPIE-The International Society for Optical Engineering, USA (2000) 22-32.
12. Quinlan, J.R.: C4.5 Programs for Machine Learning. Morgan Kaufmann, San Mateo, CA (1993).
13. Schapire, R.E., Freund, Y., Bartlett, P., Lee, W.S.: Boosting the Margin: A New Explanation for the Effectiveness of Voting Methods. The Annals of Statistics, Vol. 26, No 5 (1998) 1651-1686.
14. Schapire, R.E.: A Brief Introduction to Boosting. In: Proc. 16th Int. Joint Conf. on AI (1999).
15. Schapire, R.E.: The Strength of Weak Learnability. Machine Learning, Vol. 5, No. 2 (1990) 197-227.
16. Skalak, D.B.: Combining Nearest Neighbor Classifiers. Ph.D. Thesis, Dept. of Computer Science, University of Massachusetts, Amherst, MA (1997).
17. Tsymbal, A., Puuronen, S.: Bagging and Boosting with Dynamic Integration of Classifiers. In: Proc. PKDD'2000 4th European Conf. on Principles and Practice of Knowledge Discovery in Databases, Lyon, France, LNCS, Springer-Verlag (2000) to appear.
18. Tsymbal, A.: Decision Committee Learning with Dynamic Integration of Classifiers. In: Proc. 2000 ADBIS-DASFAA Symp. on Advances in Databases and Information Systems, Prague, Czech Republic, LNCS, Springer-Verlag (2000) to appear.
19. Webb, G.I.: MultiBoosting: A Technique for Combining Boosting and Wagging. Machine Learning (2000) in press.

Incremental Mining of Constrained Associations

Shiby Thomas[1] and Sharma Chakravarthy[2]

[1] Data Mining Technologies, Oracle Corporation, 200 Fifth Avenue, Waltham, MA
Shiby.Thomas@oracle.com
[2] Computer Science & Engineering Department, University of Texas, Arlington TX
sharma@cse.uta.edu

Abstract. The advent of data warehouses has shifted the focus of data mining from file-based systems to database systems in recent years. Architectures and techniques for optimizing mining algorithms for relational as well as Object-relational databases are being explored with a view to tightly integrate mining into data warehouses. Interactive mining and incremental mining are other useful techniques to enhance the utility of mining and to support goal oriented mining. In this paper, we show that by viewing the negative border concept as a constraint relaxation technique, incremental data mining can be readily generalized to efficiently mine association rules with various types of constraints. In the general approach, incremental mining can be viewed as a special case of relaxing the frequency constraint. We show how the generalized incremental mining approach including constraint handling can be implemented using SQL. We develop performance optimizations for the SQL-based incremental mining and present some promising performance results. Finally, we demonstrate the applicability of the proposed approach to several other data mining problems in the literature.

1 Introduction

Data mining or knowledge discovery is the process of identifying useful information such as previously unknown patterns embedded in a large data set. Efficient data mining algorithms are extremely important because making multiple passes over very large data sets is computationally expensive. Although data mining is used in its own right for many applications, it is likely to become a key component of decision support systems and as a result will complement the data analysis/retrieval done through traditional query processing. Of all the data mining techniques, association rule mining has received significant attention by the research community. A number of algorithms [1,3,12] have been developed for mining association rules assuming the data is stored in files.

The need for applying association rule mining to data stored in databases/ data warehouses has motivated researchers to [2,7,11]: i) study alternative architectures for mining data stored in databases, ii) translate association rule mining algorithms to work with relational and object/relational databases, and iii) optimize the mining algorithms beyond what the current relational query optimizers are capable of. Although it has been shown in [7] that SQL-based

M. Valero, V.K. Prasanna, and S. Vajapeyam (Eds.): HiPC 2000, LNCS 1970, pp. 547–558, 2000.

approaches *currently* cannot compete with *ad hoc* file processing algorithms and its variants, as pointed out rightly in [13], it does not negate the importance of this approach. For this reason, we will formulate some of the optimizations proposed in this paper using SQL and evaluate their performance.

The process of generating *all* association rules is computationally expensive and lacks user exploration and control. Interactive mining that includes the specification of expressive constraints is proposed in [6]. They also develop algorithms for the optimal use of these constraints for the apriori class of algorithms. Incremental mining, which is the subject of this paper, provides another vehicle for optimizing the generation of frequent itemsets when either the data or the constraints change. Incremental mining is predicated upon retaining item combinations selectively at each level for later use. The negative border approach identifies the itemsets to be retained so that when new data is added, complete recomputation of frequent itemsets is avoided. When mining relational databases, incremental view materialization can be used to implement incremental data mining algorithms.

1.1 Overview of the Paper

This paper generalizes the use of incremental mining techniques to association rule mining with constraints. The negative border concept [12] is used as the basis of incremental mining [9]. By viewing the negative border concept as a constraint relaxation technique, incremental data mining can be readily used to efficiently mine association rules with various types of constraints. We divide the constraints into four categories and show how they are used in the algorithm. We show how the generalized incremental mining approach including constraint handling can be implemented using SQL. We develop performance optimizations for the SQL-based incremental mining and present some promising performance results. Finally, we demonstrate the applicability of the proposed approach to several other data mining problems in the literature.

The contributions of this paper are: i) the generalization of the negative border concept to a larger class of constraints, ii) extending the incremental frequent itemset mining algorithm to handle constraints, iii) optimizations of the generalized incremental mining algorithm in the relational context and concomitant performance gains, and iv) wider applicability of the approach presented to problems such as mining closed sets, query flocks etc.

Paper Organization: The remainder of this paper is organized as follows. Section 1.2 briefly discusses related work on association rule mining in the presence of constraints. Section 2 describes the types of constraints handled and the generalization of the incremental negative border algorithm for them. In Section 3, we present SQL formulations of incremental mining along with the performance results. In Section 4 we show the applicability of the approach presented in the paper to closed sets and query flocks. Section 5 contains conclusions.

1.2   Related Work

Domain knowledge in the form of constraints can be used to restrict the number of association rules generated. This can be done either as a post-processing step after generating the frequent itemsets or can be integrated into the algorithm that generates the frequent itemsets. Use of constraints and the ability to relax them can also be used to provide a focused approach to mining. In [8], approaches to integrate a set of boolean constraints over a set of items into the frequent itemsets generation procedure are discussed. Constraints to facilitate interactive mining and user exploration are discussed in [6] and are further generalized in [5]. Query flocks introduced in [13] consists of a parameterized query that selects certain parameter value assignments and a filter condition to further qualify the results.

2   Constrained Associations

In this section, we introduce associations with different kinds of constraints on the itemsets or constraints that characterize the dataset from which the associations are derived. Let $\mathcal{I} = \{i_1, i_2, \ldots, i_m\}$ be a set of literals, called items that are attribute values of a set of relational tables. A constrained association is defined as a set of itemsets $\{X | X \subseteq \mathcal{I} \& C(X)\}$, where C denotes one or more boolean constraints.

2.1   Categories of Constraints

We divide the constraints into four different categories[1] We illustrate each of them with examples. The data model used in our examples is that of a point-of-sale (POS) model for a retail chain. When a customer buys a product or series of products at a register, that information is stored in a transactional system, which is likely to hold other information such as who made the purchase, what types of promotions were involved etc. The data is stored in three relational tables *sales, products_sold* and *product* with respective schemas as shown in Figure 1.

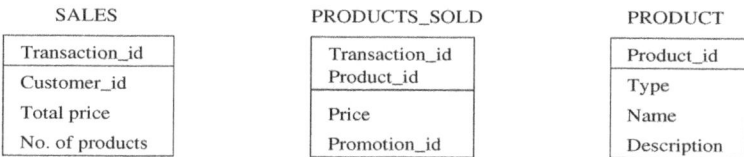

Figure 1. Point of sales data model

---

[1] We refer the reader to [8,6,5] for nice discussions of various kinds of constraints.

Frequency constraint. This is the same as the minimum support threshold in the "support-confidence" framework for association rule mining [3]. An itemset X is said to be frequent if it appears in at least s transactions, where s is the minimum support. In the point-of-sales data model a transaction correspond to a customer transaction. Most of the algorithms for frequent itemset discovery utilizes the downward closure property of itemsets with respect to the frequency constraint: that is, if an itemset is frequent, then so are all its subsets. Level-wise algorithms [1] find all itemsets with a given property among itemsets of size k (k-itemsets) and use this knowledge to explore $(k + 1)$-itemsets. Frequency is an example for a downward closed property. The downward closure property is similar to the anti-monotonicity property defined in [6]. In the context of the point-of-sale data model, the frequency constraint can be used to discover products bought together frequently.

Item constraint. Let $\mathcal{B}$ be a boolean expression over the set of items $\mathcal{I}$. The problem is to find itemsets that satisfy the constraint $\mathcal{B}$. Three different algorithms for mining associations with item constraints are presented in [8]. The item constraints enables us to pose mining queries such as "What are the products whose sales are affected by the sale of, say, barbecue sauce?" and "What products are bought together with sodas and snacks?".

Aggregation constraint. These are constraints involving aggregate functions on the items that form the itemset. For instance, in the POS example an aggregation constraint could be of the form $min(products\_sold.price) \geq p$. Here we consider a product as an item. The aggregate function could be *min, max, sum, count, avg* etc. or any other user-defined aggregate function. The above constraint can be used to find "expensive products that are bought together". Similarly $max(products\_sold.price) \leq q$ can be used to find "inexpensive products that are bought together". These aggregate functions can be combined in various ways to express a whole range of useful mining computations. For example, the constraint $(min(products\_sold.price) \leq p)\,\&\,(avg(products\_sold.price) \geq q)$ targets the mining process to inexpensive products that are bought together with the expensive ones.

External constraint. External constraints filter the data used in the mining process. These are constraints on attributes which do not appear in the final result (which we call external attributes). For example, if we want to find *products bought during big purchases where the total sale price of the transaction is larger than P*, we can impose a constraint of the form $sales.total\,price \geq P$.

2.2   Constrained Association Mining

Figure 2 shows the general framework for the kth level in a level-wise approach for mining associations with constraints. Note that we could also use SQL-based approaches for implementing the various operations in the different phases. The different constraints are applied at different steps in the mining process. The item

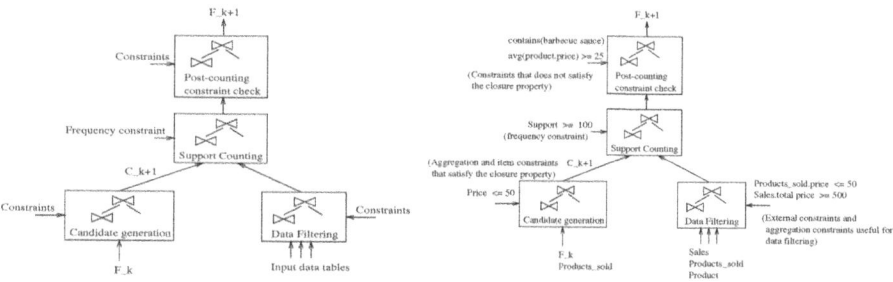

Figure 2. Framework for constrained association mining

Figure 3. Point-of-sales example for constrained association mining

constraints and the aggregation constraints that satisfy the closure property can be used in the candidate generation phase for pruning unwanted candidates. The aggregation constraints can be applied on the result of the candidate generation process. The external constraints and some of the aggregation and item constraints can be applied at the data filtering stage to reduce the size of the input data. The frequency constraint is applied during support counting to filter out the non-frequent itemsets. Finally, any unprocessed constraints are applied as a post-counting operation. The frequent itemsets before the post-counting constraint check are used for the next level candidate generation since these constraints do not satisfy the closure property.

Example: We illustrate the application of the various constraints here with a specific example using the point-of-sales data model in Section 2. The example shown in Figure 3 finds product combinations containing "barbecue sauce" where all the products cost less than $50 and the average price is more than $25 (some notion of similar priced products). The combinations should appear in at least 100 sales transactions with the total price of the transaction should be greater than $500. This gives an idea of "what other moderately priced products people buy with barbecue sauce in big purchases (perhaps for parties)". It could help the shop owner to decide on various promotions.

In the above example, the constraint on the total price of the transaction is an external constraint and is applied in the data filtering stage. Since the maximum price of a product in the desired combination is $50, we can also filter out records that does not satisfy this condition in the data filtering stage. $max(Price) \leq 50$ is an aggregation constraint which satisfies the closure property and can be applied in the candidate generation phase. The constraint to include barbecue sauce in the combination and the constraint on the average price are checked in the post-counting phase since they do not satisfy the closure property.

2.3   Incremental Constrained Association Mining

In this section, we outline how the algorithm for the incremental maintenance of frequent itemsets presented in [9] can be generalized to handle constrained associations.

The negative border based incremental mining algorithm is applicable for mining associations with constraints that are closed with respect to the set inclusion property, that is, if an itemset satisfies the constraint then so do all its subsets. For incremental constrained association mining we need to materialize and store all the itemsets in the negative border and their support counts (negative border is the set of minimal itemsets that did not satisfy the constraint). The reason why this is enough is that when new transaction data is added, only the support count of the itemsets could change. We assume that there is a frequency constraint, which is typically the case in association mining. We list below the various steps of the incremental algorithm.

1. Find the frequent itemsets in the increment database *db*. Simultaneously count the support of all itemsets $X \in FrequentSets \cup NegativeBorder$ in *db*. *For this we use the framework outlined in Section 2.2. The different constraints that are present can be handled at the various steps in the mining process as shown in Figure 2.*
2. Update the support count of all itemsets in $FrequentSets \cup NegativeBorder$ to include their support in *db*.
3. Find the itemsets in NegativeBorder that became frequent by the addition of *db* (call them Promoted-NegativeBorder, *PNb*).
4. Find the candidate closure with *PNb* as the seed (see section 3.2). During this computation, we use the candidate generation procedure with constraints as described in Section 2.2.
5. If there are no new itemsets in the candidate closure, skip this step. Otherwise, count the support of all new itemsets in the candidate closure against the whole dataset. *The dataset is subjected to the data filtering step before the support counting as shown in Figure 2.*

The negative border idea can be used to handle some cases of constraint relaxation also, especially relaxations to the external constraints and the frequency constraint.

3   SQL Formulations of Incremental Mining

In this section, we present SQL formulations for the incremental mining algorithm. This shows how the SQL-based mining framework in [7] can be extended for incremental mining. Note that incremental mining can be considered as a special case of relaxing the frequency constraint. It is possible to handle other kinds of constraints also within the same framework. The input transaction data is stored in a relational table T with the schema (tid, item). The increment transaction table δT also has the same schema. The frequent itemsets and negative border of size k are stored in tables with the schema $(item_1, \ldots, item_k, count)$. We present two approaches for support counting based on the Subquery and the Vertical approaches presented in [7]. We also outline the SQL-based candidate closure computation (step 4 of the incremental algorithm).

3.1   Subquery Approach

In this approach, support counting is done by a set of k nested subqueries where k is the size of the largest itemset. We present here the extensions to the subquery approach in [7] to count candidates of different sizes. Subquery Q_l finds all tids that support the distinct candidate l-itemsets. The output of Q_l is grouped on the l items to find the support of the candidate l-itemsets. Q_l is also joined with δT (T while counting the support in the whole database) and C_{l+1} to get Q_{l+1}. The SQL queries and the corresponding tree diagrams for the above computations are given in Figure 4. δB_l stores the support counts of all frequent and negative border itemsets in δT.

insert into δB_l select $item_1, \ldots, item_l$, count(*)
from   (Subquery Q_l) t
group by $item_1, \ldots, item_l$

Subquery Q_l (for any l between 1 and k):
    select $item_1, \ldots item_l$, tid
    from δT t_l, (Subquery Q_{l-1}) as r_{l-1}, C_l
    where $r_{l-1}.item_1 = C_l.item_1$ and ... and
        $r_{l-1}.item_{l-1} = C_l.item_{l-1}$ and
        $r_{l-1}.tid = t_l.tid$ and
        $t_l.item = C_l.item_l$

Subquery Q_0: No subquery Q_0.

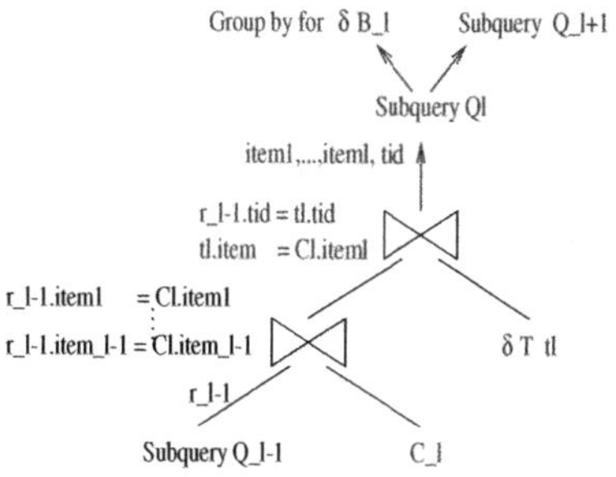

Figure 4. Support counting using subqueries

The output of subquery Q_l needs to be materialized since it is used for counting the support of l-itemsets and to generate Q_{l+1}. If the query processor is augmented to support multiple streams where the output of an operator can be piped into more than one subsequent operators, the materialization of Q_l's can be avoided. In the basic association rule mining, we do not have to count itemsets of different sizes in the same pass since C_{l+1} becomes available only after the frequent l-itemsets are computed.

For steps 1 and 5 of the incremental algorithm, we use queries outlined above. Tables B_l and δB_l store the frequent and negative border l-itemsets and their support count in T and δT respectively. Step 2 can be performed by joining B_l and δB_l and adding the corresponding support counts. We add another attribute to B_l to keep track of promoted negative borders and this can be done along with step 2. In step 3, we simply select the itemsets corresponding to the promoted negative border.

3.2   Computing Candidate Closure

In the apriori algorithm candidate itemsets are generated in two steps – the join step and the prune step. In the join step, two sets of $(k-1)$-itemsets called *generators* and *extenders* are joined to get k-itemsets. An itemset s_1 in generators joins with s_2 in extenders if the first $(k-2)$ items of s_1 and s_2 are the same and the last item of s_1 is lexicographically smaller than the last item of s_2. In the prune step, itemsets with non-frequent subsets are filtered out by joining with a set of itemsets termed *filters*.

In the incremental algorithm, we compute the candidate closure to avoid multiple passes while counting the support of the new candidates. It can be seen that all the new candidates will be supersets of promoted borders. Therefore, it is sufficient to use the itemsets that are promoted borders as generators. In order to generate C_k, the candidates of size k, we use $PB_{k-1} \cup C_{k-1}$ as generators and $PB_{k-1} \cup C_{k-1} \cup F_{k-1}$ as extenders and filters. PB_{k-1} and F_{k-1} denote promoted borders and frequent itemsets of size $(k-1)$ respectively. The candidate generation process starts with C_0 as the empty set and terminates when C_k becomes empty. It is straight-forward to derive SQL queries for this process and we do not present them here.

3.3   Vertical

In the Subquery approach, for every transaction that supports an itemset we generate (itemset, tid) tuples resulting in large intermediate tables. The Vertical approach avoids this by collecting all tids that support an itemset into a BLOB (binary large object) and generates (itemset, tid-list) tuples. Initially, tid-lists for individual items are created using a table function. The tid-list for an itemset is obtained by intersecting the tid-lists of its items using a user-defined function (UDF). The SQL queries for support counting are similar in structure to that of the Subquery approach except for the use of UDFs to intersect the tid-lists. We refer the reader to [7] for the details.

The increment transaction table δT is transformed into the vertical format by creating the delta tid-lists of the items. The delta tid-lists are used to count the support of the candidate itemsets in δT which are later merged with the original tid-lists. This can be accomplished by joining the original tid-list table with the delta tid-list table and merging the tid-lists with a UDF. If the incremental algorithm requires a pass over the complete data, the merged tid-lists are used for support counting.

3.4   Performance Results

In this section, we report performance results of the incremental algorithm. These experiments were performed on Version 5 of IBM DB2 Universal Server installed on a Sun Ultra 5 Model 270 with a 269 MHz UltraSPARC-IIi CPU, 128 MB main memory and a 4 GB disk. We report the results of the SQL formulations of the incremental algorithm based on the Subquery and Vertical approaches.

The experiments were performed on synthetic data generated using the same technique as in [3]. The dataset used for the experiment was T10.I4.D100K (Mean size of a transaction = 10, Mean size of maximal potentially large itemsets = 4, Number of transactions = 100 thousand). The increment database is created as follows: We generate 100 thousand transactions, of which $(100 - d)$ thousand is used for the initial computation and d thousand is used as the increment, where d is the fractional size (in percentage) of the increment.

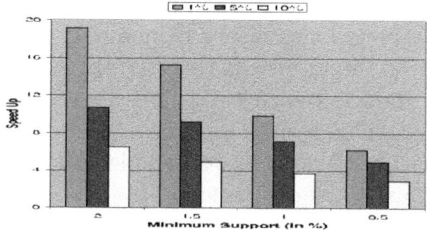
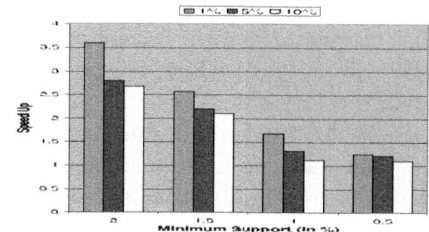

Figure 5. Speed up of the incremental algorithm based on the Subquery approach

Figure 6. Speed up of the incremental algorithm based on the Vertical approach

We compare the execution time of the incremental algorithm with respect to mining the whole dataset. Figures 5 and 6 shows the corresponding speed-ups of the incremental algorithm based on the Subquery and the Vertical approaches for different minimum support thresholds. We report the results for increment sizes of 1%, 5% and 10% (shown in the legend). We can make the following observations from the graphs. The incremental algorithm based on the Subquery approach achieves a speed-up of about 3 to 20 as compared to mining the whole dataset. However, the maximum speed-up of the Vertical approach is only about 4. For support counting the Vertical approach uses a user-defined function (UDF) to intersect the tid-lists. The incremental algorithm should also invoke the UDF at least the same number of times since the support of all the itemsets in the frequent set and the negative border needs to be found in the increment database. In cases where the support of new candidates needs to be counted the number of invocations will be even more. The time taken by the Vertical approach in the support counting phase is directly proportional to the number of times the UDF is called. However, the incremental algorithm saves in the tid-list creation phase since the size of the increment dataset is only a fraction of the whole dataset. This explains why the speed-up of the Vertical approach is low. In contrast the Subquery approach achieves higher speed-up since the time taken is proportional to the size of the dataset. The speed-up reduces as the minimum support threshold is lowered. At lower support values the chances of the negative border expanding is higher and as a result the incremental algorithm may have to compute the candidate closure and count the support of the new candidates in the whole dataset. The speed-up is higher for smaller increment sizes since the incremental algorithm needs to process less data. With respect to the absolute execution time, the Subquery and the Vertical approaches followed the same trend as reported in [7] – the Vertical approach was about 3 to 6 times faster than the Subquery approach.

3.5 New-Candidate Optimization

In the basic incremental algorithm, we find the frequent itemsets in the increment database db along with counting the support of all the itemsets in the frequent set and the negative border. However, the frequent itemsets in db is used only to prune the non frequent itemsets in db while computing the candidate closure. In the candidate closure computation we assume that the new candidate k-itemsets are frequent while generating the $(k+1)$-itemsets. At this step the new candidate k-itemsets that are infrequent in db are known to be infrequent in the whole dataset as well and can be pruned. This results in better speed-up as compared to the basic incremental algorithm. We have not included those performance results here due to space constraints and can be found in [10].

4   Applicability Beyond Association Mining

In this section, we discuss how the incremental approach presented in Section 2.3 can be extended to other data mining and decision support problems.

4.1 Mining Closed Sets

All the efficient algorithms for mining association rules exploits the closure property of frequent itemsets. Minimum support which characterizes frequent itemsets is downward closed: if an itemset has minimum support then all its subsets also have minimum support. The idea of negative border can be used for all incremental mining problems that possess closure properties. If the closure property is incrementally updatable also (for example, support), it is possible to limit the database access to at most one scan of the whole database as shown in the incremental frequent itemset mining example. A property is incrementally updatable if it is possible to derive its value from the corresponding values of different partitions of the input data. A few examples are COUNT, SUM, MIN, MAX etc. Mining sequential patterns, correlation rules [4] and maximal frequent itemsets are a few other data mining problems that have closure properties.

4.2 Query Flocks

A query flock is a parameterized query with a filter condition to eliminate the values of parameters that are "uninteresting". A query flock can be evaluated by dividing it into a set of safe subqueries [13]. The query flocks corresponding to the safe subqueries form a lattice with query containment as the partial order. During the execution of the subqueries of the query flock, all records with parameter values which satisfy the filter condition are propagated to the next higher subquery in the lattice for further evaluation. The parameter value combinations in this case are analogous to itemsets in boolean association rule mining.

The negative border in apriori-based query flock evaluation is the set of parameter value combinations that does not pass the filter condition test. These combinations, if materialized and stored along with their support counts can be effectively used to update the result of the query flock when the base tables over which it is defined are updated. We refer the reader to [10] for a representative example and further details.

4.3   View Maintenance

Incremental mining can be seen as materialized view maintenance. In boolean association rules, the frequent itemsets and the negative border are in fact aggregate views over the transaction table. In query flocks each element in the subquery lattice can be considered as a view defined on the base tables. Therefore, view maintenance techniques can be used for incremental mining. On the other hand, the negative border based change propagation can also be applied for the maintenance of views involving monotone aggregate functions that satisfy the apriori subset property. An itemset can also be treated as a point in the data cube defined by the items as dimensions and support as the measure. The data cube points can be arranged as a lattice according to the partial order on the itemsets.

5   Summary

In this paper, we have developed a general framework for incremental mining of association rules by applying the negative border concept to constraints having the downward closure property. Furthermore, we have shown how incremental mining can be implemented using SQL queries. The concept of negative border which is the key to the incremental algorithm has other applications also. It can be used for mining association rules with varying support and confidence values. For instance, the negative border can be used to determine the updated frequent itemsets if the support is changed. We have also outlined how the negative border idea can be applied to various other data mining problems. We believe that the ramifications of the incremental mining framework presented in this paper go far beyond the black-box mining and will be more useful in interactive mining.

References

1. R. Agrawal, T. Imielinski, and A. Swami. Mining association rules between sets of items in large databases. In *Proc. of the ACM SIGMOD Conference on Management of Data*, pages 207–216, Washington, D.C., May 1993.
2. R. Agrawal and K. Shim. Developing tightly-coupled data mining applications on a relational database system. In *Proc. of the KDD Conference*, Portland, Oregon, August 1996.
3. R. Agrawal and R. Srikant. Fast Algorithms for Mining Association Rules. In *Proc. of the 20th VLDB*, Santiago, Chile, September 1994.

4. S. Brin, R. Motwani, and C. Silverstein. Beyond market baskets: Generalizing association rules to correlations. In *Proc. of the ACM SIGMOD Conference on Management of Data*, May 1997.

5. L.V.S. Lakshmanan, R.T. Ng, J. Han, and A. Pang. Optimization of Constrained Frequent Set Queries with 2-variable Constraints. In *Proc. of the ACM SIGMOD Conference on Management of Data*, Philadelphia, Pennsylvania, June 1999.

6. R.T. Ng, L.V.S. Lakshmanan, J. Han, and A. Pang. Exploratory Mining and Pruning Optimizations of Constrained Association Rules. In *Proc. of the ACM SIGMOD Conference on Management of Data*, Seattle, Washington, June 1998.

7. S. Sarawagi, S. Thomas, and R. Agrawal. Integrating Association Rule Mining with Relational Database Systems: Alternatives and Implications. In *Proc. of the ACM SIGMOD Conference on Management of Data*, Seattle, Washington, June 1998.

8. R. Srikant, Q. Vu, and R. Agrawal. Mining Association Rules with Item Constraints. In *Proc. of the 3rd Int'l Conference on Knowledge Discovery in Databases and Data Mining*, Newport Beach, California, August 1997.

9. S. Thomas, S. Bodagala, K. Alsabti, and S. Ranka. An Efficient Algorithm for the Incremental Updation of Association Rules in Large Databases. In *Proc. of the 3rd Int'l Conference on Knowledge Discovery and Data Mining*, Newport Beach, California, August 1997.

10. S. Thomas and S. Chakravarthy. Incremental Mining of Constrained Associations. Technical Report TR 98-018, University of Florida, Gainesville, Florida, October 1998.

11. S. Thomas and S. Chakravarthy. Performance Evaluation and Optimization of Join Queries for Association Rule Mining. In *Proc. of the Int'l Conference on Data Warehousing and Knowledge Discovery*, Florence, Italy, August 1999.

12. H. Toivonen. Sampling large databases for association rules. In *Proc. of the 22nd Int'l Conference on Very Large Databases*, pages 134–145, Mumbai (Bombay), India, September 1996.

13. D. Tsur, J. Ullman, S. Abiteboul, C. Clifton, R. Motwani, S. Nestorov, and A. Rosenthal. Query Flocks: A Generalization of Association Rule Mining. In *Proc. of the ACM SIGMOD Conference on Management of Data*, Seattle, Washington, June 1998.

Scalable, Distributed and Dynamic Mining of Association Rules

V.S. Ananthanarayana, D.K. Subramanian, and M.N Murty

Dept. of Computer Science and Automation,
Indian Institute of Science, Bangalore 560 012, INDIA
{anvs, dks, mnm}csa.iisc.ernet.in

Abstract. We propose a *novel* pattern tree called Pattern Count tree (PC-tree) which is a *complete* and *compact* representation of the database. We show that construction of this tree and then generation of all large itemsets requires *a single database scan* where as the current algorithms need at least *two database scans*. The *completeness* property of the PC-tree with respect to the database makes it amenable for mining association rules in the context of changing data and knowledge, which we call *dynamic mining*. Algorithms based on PC-tree are *scalable* because PC-tree is *compact*. We propose a partitioned distributed architecture and an efficient distributed association rule mining algorithm based on the PC-tree structure.

1 Introduction

The process of *generating large itemsets* [1] is an important component of an Association Rule Mining (ARM) algorithm. This process involves intensive disk access activity necessitating the need for designing fast and efficient algorithms for generating large itemsets. Incremental mining algorithms [5] are proposed to handle additions/deletions of transactions to/from an existing database. In addition to the dynamic nature of data, knowledge and values of parameters like support also change dynamically. We call mining under such a dynamic environment, **dynamic mining**. In this paper, we propose a novel tree structure called Pattern Count tree (PC-tree) that is a complete and compact representation of the database; and is ideally suited for dynamic mining.

Existing parallel and distributed ARM algorithms [4] require (i) at least *two* database scans and (ii) they *generate and test* candidate itemsets to obtain large itemsets. However, the sequential ARM algorithm based on the frequent pattern tree (FP-tree) is shown to be superior [3] to the existing ones as it only *tests* candidate itemsets, for possible largeness, thus reducing the computational requirements significantly. However, the FP-tree based algorithm requires two database scans to generate frequent itemsets.

This motivated us to propose a scheme that employs an abstraction similar to FP-tree and more general and complete- that is Pattern Count tree (PC-tree). A PC-tree can be constructed based on *a single database scan* and can be *updated dynamically*. We show that we can construct a unique ordered FP-tree, called

M. Valero, V.K. Prasanna, and S. Vajapeyam (Eds.): HiPC 2000, LNCS 1970, pp. 559–566, 2000.

Lexicographically Ordered FP-tree (LOFP-tree) from a PC-tree without scanning the database. More specifically, we propose a distributed ARM algorithm based on this abstraction of the database, i.e., PC-tree, where the databases are distributed and partitioned across a loosely-coupled system. Here, we convert each partition of the database into a local PC-tree using a single database scan. Due to high data compression, it is possible to store local PC-trees in the main memory itself. This PC-tree, in turn, is converted into a local LOFP-tree using the knowledge of globally large 1-itemsets in the data. These local LOFP-trees are merged to form a single LOFP-tree, which can be used to generate all large itemsets.

2   Pattern Count Tree (PC-Tree)

PC-tree is a data structure which is used to store all the patterns occurring in the tuples of a transaction database, where a count field is associated with each item in every pattern which is responsible for a compact realization of the database.

A. Structure of a PC-tree :   Each node of the tree consists of the following structure: item-name, count and two pointers called child (c) and sibling (s). In the node, the item-name field specifies which item the node represents, the count field specifies the number of transactions represented by a portion of the path reaching this node, the c-pointer field represents the pointer to the following pattern; and the s-pointer field points to the node which indicates the subsequent other patterns from the node under consideration.

B. Construction of the PC-tree :   PC-tree construction requires *a single scan* of the database. We take each transaction from the database and put it as a new branch of the PC-tree if any sub pattern, which is a prefix of the transaction, does not exist in the PC-tree; else put it into an existing branch, e_b, by incrementing the corresponding count field value of the nodes in the PC-tree. We put the remaining sub patterns, if any, of the transaction by appending additional nodes with count field value equal to 1 to the path in e_b. We preprocess each transaction to put items in a lexicographical order. Refer to [6] for an algorithm for detailed construction.

We give below an example to explain the construction of a PC-tree. Consider the un-normalized, single attribute based relation (transaction database) shown in Figure 1A. The corresponding PC-tree is shown in Figure 1B.

C. Properties of the PC-tree :

Definition 2.1 Complete representation

A representation R is *complete* with respect to a transaction database DB, if every pattern in each transaction in DB, is present in R.

Lemma 2.1   PC-tree is a complete representation of the database.

Definition 2.2 Compact representation

Let D be a database of size D_s. Then, a complete representation R of DB is *compact* if $R_s < D_s$, where R_s is the size of R.

Lemma 2.2   PC-tree is a compact representation of the database.

Proof   Refer to [6] for proofs of Lemmas 2.1 and 2.2.

Transaction Id	Item numbers of items purchased (Transactions)
t_1	19, 40, 510, 527
t_2	19, 40, 179
t_3	19, 40, 125, 510
t_4	527, 740
t_5	527, 740, 795
t_6	19

FIGURE : A : Transaction Database, *D*

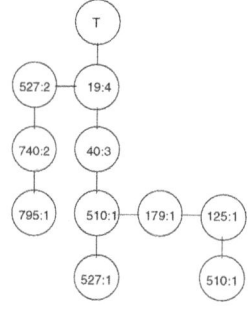

FIGURE : B   PC-TREE.

In Figure B, verticle link represents the c-pointer and the horizontal link represent the s-pointer. Each node has two parts : item-name and count value.

Fig. 1. Example transaction database and the PC-tree Structure for storing patterns

2.1   Construction of the LOFP-Tree from a Given PC-Tree

The PC-tree corresponding to a given database is unique, where as the FP-tree corresponding to a given database need not be unique. If all items in each transaction of the database are ordered lexicographically, and large 1-itemsets are also ordered lexicographically, when the items are having same frequency, we call the corresponding FP-tree the Lexicographically Ordered FP-tree (LOFP-tree) which is unique and the number of nodes in LOFP-tree is lesser then or equal to that of FP-tree. Since PC-tree is the complete representation of the database, LOFP-tree can be constructed directly from the PC-tree. Refer [6] for detailed construction process. Once the LOFP-tree is constructed from PC-tree, generation of large itemsets from such a structure is done by constructing a *conditional LOFP-tree*. Since the procedure for construction of conditional LOFP-tree is similar to that of conditional FP-tree [3], we do not discuss it here.

2.2   Experiments on Construction of PC-Tree and LOFP-Tree

We synthesize datasets using the synthetic data generator given in [1]. Figure 2A shows the parameter description and Figure 2B shows the data set used in our work with a fixed support value of 0.75%.

Experiment 1 - Memory requirements :   Figure 3 shows the memory requirements for database, PC-tree and LOFP-tree for the datasets Set_1 to Set_11. It is clear from the graphs that even though the size of PC-tree is greater than that of the LOFP-tree, the memory size requirement of PC-tree increases nominally with a significant increase in the database size. For example, the ratio of the storage space for the database to that of PC-tree keeps increasing from 7 to 10 as the database size increases from 8.5MB to 137MB, where $|T| = 10$. So, PC-tree requires a *lesser* memory space than the database.

|D|	Number of transactions.
|T|	Average size of a transaction.
|I|	Average size of maximal potentially large itemsets
|L|	Number of maximal potential large itemsets
NOI	Number of items.

FIGURE A : Parameters

NAME	|D|	|T|	|I|	|L|	NOI
Set_1	25K	10	4	1000	1000
Set_2	50K	10	4	1000	1000
Set_3	100K	10	4	1000	1000
Set_4	200K	10	4	1000	1000
Set_5	300K	10	4	1000	1000
Set_6	400K	10	4	1000	1000
Set_7	25K	15	4	1000	1000
Set_8	50K	15	4	1000	1000
Set_9	100K	15	4	1000	1000
Set_10	150K	15	4	1000	1000
Set_11	200K	15	4	1000	1000

FIGURE B : Data Set used

Fig. 2. Parameters and Data set used

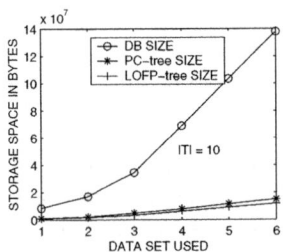
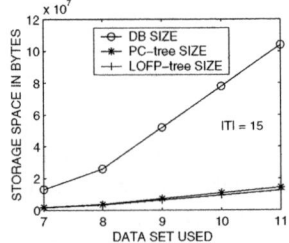

Fig. 3. Comparison based on the space requirements

Experiment 2 - Timing requirements : We conducted the experiment using the data sets Set_1, Set_2, Set_3 and Set_4 to compare the time required
to construct the LOFP-tree directly from the database which requires two
database scans and also from the PC-tree, which requires one database scan
and one scan of the PC-tree. Table 1 shows the values obtained. It can be
seen that time required to construct the LOFP-tree based on PC-tree is
lesser than that of its construction directly from the database. These results
indicate that the time required to construct LOFP-tree from the PC-tree is
smaller than that of direct construction of LOFP-tree by 2 to 22 sec.

Table 1. Comparison of both algorithms based on time requirement

Data Set	Set_1	Set_2	Set_3	Set_4
Time to construct LOFP-tree from PC-tree	22 sec.	45 sec.	88 sec.	173 sec.
Time to construct LOFP-tree directly from database	24 sec.	48 sec.	97 sec.	195 sec.

3   Dynamic Mining

The mining procedure for generating large itemsets which characterize *change of data*, *change of knowledge* and *change of values of parameters* like support is called **dynamic mining**. In the literature on ARM, mining under change of data scenario is called **incremental mining**. We explain how PC-tree handles the incremental mining, change of knowledge and change of support value below.

3.1 Incremental Mining of Large Itemsets : Incremental mining is a method for generating large itemsets over a changing transaction database in an incremental fashion without considering the part of the database which is already mined. Effectively there are two operations which lead to changes in a database; they are **ADD** and **DELETE** operations as explained below.

1. **ADD :-** A new transaction is added to the transaction database : this is handled in a PC-tree either by incrementing the count value of the nodes corresponding to the items in the transaction or by adding new nodes or by performing both the above activities.
2. **DELETE :-** An existing transaction is deleted from the transaction database : this is handled in a PC-tree by decrementing the count value of the nodes corresponding to the items in the transaction, if their count value is > 1; or by deleting the existing nodes corresponding to the items in the transaction.

Definition 3.1 Large-itemset equivalent : Let DB be a transaction database with L as the set of all large itemsets in DB. Let R_1 and R_2 be any two representations of DB. Let L_1 and L_2 be the sets of large itemsets generated from R_1 and R_2 respectively. R_1 and R_2 are said to be *large-itemset equivalent* if $L_1 = L_2 = L$.

Definition 3.2 Appropriate representation for incremental mining (ARFIM): Let R be a large-item equivalent representation of DB. Let DB be updated to $D\acute{B}$ such that $D\acute{B} = DB + db_1 - db_2$. R is an **ARFIM**, if there exists an algorithm that generates $\acute{R}$ using only R, db_1 and db_2 such that $\acute{R}$ is large-itemset equivalent to $D\acute{B}$. Note that the PC-tree can be incrementally constructed by scanning DB, db_1 and db_2 only once.

3.2 Change of Knowledge : Knowledge representation in the mining process is mainly by an 'is_a' hierarchy. Let K_1 be the knowledge represented using an 'is_a' hierarchy. Let PC be the PC-tree for the database D. We can construct generalized PC-tree (GPC-tree) by applying K_1 to the transactions obtained from PC without accessing D. If knowledge K_1 is changed to K_2, then by applying K_2 on PC, we get another generalized PC-tree without accessing D. This shows that PC-tree handles dynamic change of knowledge without accessing the database during change of knowledge. Change of support value can be handled in a similar manner.

Definition 3.3 A representation R is appropriate for dynamic mining of large itemsets if it satisfies the following properties :

1. R is a complete and compact representation of the database, D.

2. R can be used to handle changes in database, knowledge and support value without scanning D to generate frequent itemsets.

PC-tree is a representation that is appropriate for dynamic mining of large itemsets. For more details on dynamic mining, refer to [6].

4   Distributed Mining

Architecture

We propose an architecture based on Distributed Memory Machines (DMMs). Let there be m processors $P_1, P_2, \ldots, P_m$, which are connected by a high speed gigabit switch. Each processor has a local memory and a local disk. The processors can communicate only by passing messages. Let $DB^1, DB^2, \cdots, DB^n$ be n transaction databases at n sites and be partitioned in to m non-overlapping blocks $D^1, D^2, \cdots, D^m$, where m is the number of processors available $(m \geq n)$. Each block D^i has the same schema. Let t_j^i be the jth transaction in partition D^i.

Algorithm : Distributed PC to LOFP tree Generator (DPC-LOFPG)

Steps :

1. Each processor P^i makes a pass over its data partition, D^i and generates a local PC-tree, PC_1^i at level 1. At the end of this process, each processor generates a *local 1-itemset vector, LV*. It is an array of size s (where s is the number of items) whose each entry $LV^i[k]$ has a count value, corresponding to item k in the data partition D^i.

2. Process P^i exchanges local 1-itemset vector, LV^i with all other processors to develop a global 1-itemset vector, GV^i. Note that all $GV^i s (= GV)$ are identical and are large 1-itemsets. The processors are forced to synchronize at this step.

3. Each processor P^i constructs a local LOFP-tree using local PC_1^i and GV^i. The processors are forced to synchronize at this step.

4. Every two alternative LOFP-trees (viz. $LOFP_i^1$ and $LOFP_i^2$, $LOFP_i^3$ and $LOFP_i^4$ etc.) are given to a processor for merging and to generate a single merged LOFP-tree. This gives the LOFP-tree at the next level (viz. $i + 1^{th}$ level). After every iteration the processors are forced to synchronize. During each iteration, number of LOFP-trees generated and number of processors needed are reduced by a factor of 2 w.r.t the previous iteration. Iteration stops when one (global) LOFP-tree is generated.

5. Generation of conditional LOFP-tree to generate large itemsets. This step is the same as the one explained in [3].

Experiments

We compared **DPC-LOFPG** algorithm with its sequential counterpart. Figure 4 shows the block schematic diagram that depicts the stages with the corresponding timing requirements. From the block schematic shown in Figure 4, the time required to construct LOFP-tree from sequential mining, T

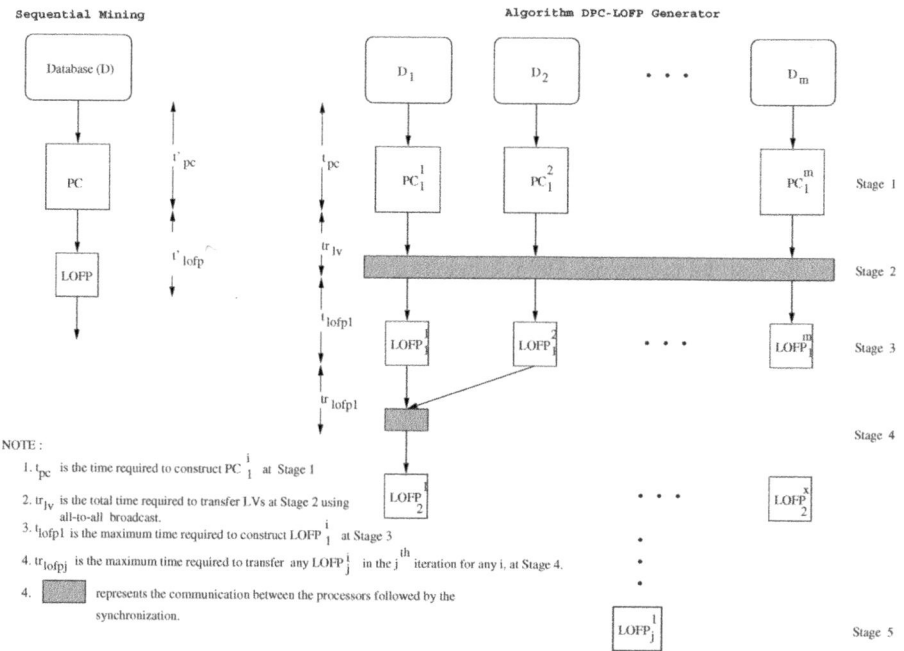

Fig. 4. Timing requirements for DPC-LOFPG and its sequential counterpart

and time for constructing global LOFP-tree from DPC-LOFPG algorithm, S are given by the following formulas : $S = t_{pc} + tr_{lv} + \{t_{lofp1} + tr_{lofp1} + t_{lofp2} + tr_{lofp2} + \cdots + t_{lofpj} + tr_{lofpj}\}$, if no. of processors $= 2^j$ at Stage 1, for $j > 0$. Corresponding to S, we have $T = \acute{t}_{pc} + \acute{t}_{lofp}$. To study the efficiency and response time of the parallel algorithm **DPC-LOFPG**, corresponding to its sequential counterpart, we conducted a simulation study by varying the number of processors from 1 to 8. Data sets used in our experiment are Set_3, Set_4 and Set_9, Set_11. Figure 5 shows the response time for different number of processors. *Efficiency of parallelism* is defined as $T/(S \times m)$, where

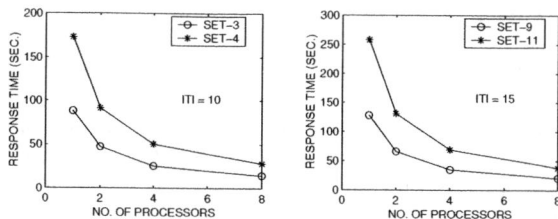

Fig. 5. Response time

S is the response time of the system when m (> 1) number processors are used; and T is the response time when $m = 1$. We show in Figure 6 the efficiency of parallelism for different number of processors. It may be observed from Figure 6 that the best value of efficiency is 98% for Set_11 and 95% for Set_4, both are exhibited by the 2-processor system.

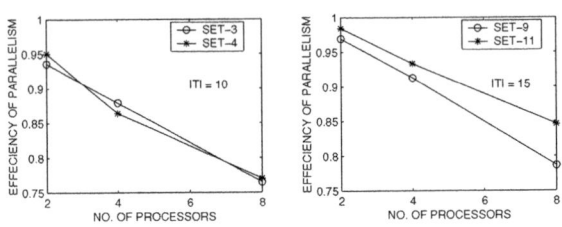

Fig. 6. Efficiency of Parallelism

5 Conclusions

In this paper, we proposed a novel data structure called PC-tree which can be used to represent the database in a *complete* and *compact* form. PC-tree can be constructed using *a single database scan*. We use it here for mining association rules. We have shown that the ARM algorithms based on PC-tree are *scalable*. We introduced the notion of *dynamic mining* that is a significant extension of *incremental mining* and we have shown that PC-tree is *ideally* suited for dynamic mining. The proposed distributed algorithm, **DPC-LOFPG** is found to be efficient because it scans the database **only once** and it does not generate any candidate itemsets.

References

1. Agrawal, R., Srikant, R. *Fast algorithms for mining association rules in large databases*, Proc. of 20th Int'l conf. on VLDB, (1994), 487 - 499.
2. Savasere, A., Omiecinsky, E., Navathe, S. *An efficient algorithm for mining association rules in large databases*, Proc. of Int'l conf. on VLDB, (1995), 432 - 444.
3. Han, J., Pei, J., Yin, Y. *Mining Frequent Patterns without Candidate Generation*, Proc. of ACM-SIGMOD, (2000).
4. Mohammed, J. Zaki. *Parallel and distributed association mining : A survey*, IEEE Concurrency, special issue on Parallel Mechanisms for Data Mining, Vol.7, No.4, (1999), 14 - 25.
5. Thomas, S., Sreenath, B., Khaled, A., Sanjay, R. *An efficient algorithm for the incremental updation of association rules in large databases*, AAAI, (1997).
6. Ananthanarayana, V.S., Subramanian, D.K., Narasimha Murty, M. *Scalable, distributed and dynamic mining of association rules using PC-trees*, IISc-CSA, Technical Report, (2000).

Author Index

Lecture Notes in Computer Science

For information about Vols. 1–1887
please contact your bookseller or Springer-Verlag

Vol. 1921: S.W. Liddle, H.C. Mayr, B. Thalheim (Eds.), Conceptual Modeling for E-Business and the Web. Proceedings, 2000. X, 179 pages. 2000.

Vol. 1922: J. Crowcroft, J. Roberts, M.I. Smirnov (Eds.), Quality of Future Internet Services. Proceedings, 2000. XI, 368 pages. 2000.

Vol. 1923: J. Borbinha, T. Baker (Eds.), Research and Advanced Technology for Digital Libraries. Proceedings, 2000. XVII, 513 pages. 2000.

Vol. 1924: W. Taha (Ed.), Semantics, Applications, and Implementation of Program Generation. Proceedings, 2000. VIII, 231 pages. 2000.

Vol. 1925: J. Cussens, S. Džeroski (Eds.), Learning Language in Logic. X, 301 pages 2000. (Subseries LNAI).

Vol. 1926: M. Joseph (Ed.), Formal Techniques in Real-Time and Fault-Tolerant Systems. Proceedings, 2000. X, 305 pages. 2000.

Vol. 1927: P. Thomas, H.W. Gellersen, (Eds.), Handheld and Ubiquitous Computing. Proceedings, 2000. X, 249 pages. 2000.

Vol. 1928: U. Brandes, D. Wagner (Eds.), Graph-Theoretic Concepts in Computer Science. Proceedings, 2000. X, 315 pages. 2000.

Vol. 1929: R. Laurini (Ed.), Advances in Visual Information Systems. Proceedings, 2000. XII, 542 pages. 2000.

Vol. 1931: E. Horlait (Ed.), Mobile Agents for Telecommunication Applications. Proceedings, 2000. IX, 271 pages. 2000.

Vol. 1658: J. Baumann, Mobile Agents: Control Algorithms. XIX, 161 pages. 2000.

Vol. 1766: M. Jazayeri, R.G.K. Loos, D.R. Musser (Eds.), Generic Programming. Proceedings, 1998. X, 269 pages. 2000.

Vol. 1791: D. Fensel, Problem-Solving Methods. XII, 153 pages. 2000. (Subseries LNAI).

Vol. 1799: K. Czarnecki, U.W. Eisenecker, Generative and Component-Based Software Engineering. Proceedings, 1999. VIII, 225 pages. 2000.

Vol. 1812: J. Wyatt, J. Demiris (Eds.), Advances in Robot Learning. Proceedings, 1999. VII, 165 pages. 2000. (Subseries LNAI).

Vol. 1932: Z.W. Raś, S. Ohsuga (Eds.), Foundations of Intelligent Systems. Proceedings, 2000. XII, 646 pages. (Subseries LNAI).

Vol. 1933: R.W. Brause, E. Hanisch (Eds.), Medical Data Analysis. Proceedings, 2000. XI, 316 pages. 2000.

Vol. 1934: J.S. White (Ed.), Envisioning Machine Translation in the Information Future. Proceedings, 2000. XV, 254 pages. 2000. (Subseries LNAI).

Vol. 1935: S.L. Delp, A.M. DiGioia, B. Jaramaz (Eds.), Medical Image Computing and Computer-Assisted Intervention – MICCAI 2000. Proceedings, 2000. XXV, 1250 pages. 2000.

Vol. 1937: R. Dieng, O. Corby (Eds.), Knowledge Engineering and Knowledge Management. Proceedings, 2000. XIII, 457 pages. 2000. (Subseries LNAI).

Vol. 1938: S. Rao, K.I. Sletta (Eds.), Next Generation Networks. Proceedings, 2000. XI, 392 pages. 2000.

Vol. 1939: A. Evans, S. Kent, B. Selic (Eds.), «UML» – The Unified Modeling Language. Proceedings, 2000. XIV, 572 pages. 2000.

Vol. 1940: M. Valero, K. Joe, M. Kitsuregawa, H. Tanaka (Eds.), High Performance Computing. Proceedings, 2000. XV, 595 pages. 2000.

Vol. 1941: A.K. Chhabra, D. Dori (Eds.), Graphics Recognition. Proceedings, 1999. XI, 346 pages. 2000.

Vol. 1942: H. Yasuda (Ed.), Active Networks. Proceedings, 2000. XI, 424 pages. 2000.

Vol. 1943: F. Koornneef, M. van der Meulen (Eds.), Computer Safety, Reliability and Security. Proceedings, 2000. X, 432 pages. 2000.

Vol. 1945: W. Grieskamp, T. Santen, B. Stoddart (Eds.), Integrated Formal Methods. Proceedings, 2000. X, 441 pages. 2000.

Vol. 1948: T. Tan, Y. Shi, W. Gao (Eds.), Advances in Multimodal Interfaces – ICMI 2000. Proceedings, 2000. XVI, 678 pages. 2000.

Vol. 1952: M.C. Monard, J. Simão Sichman (Eds.), Advances in Artificial Intelligence. Proceedings, 2000. XV, 498 pages. 2000. (Subseries LNAI).

Vol. 1954: W.A. Hunt, Jr., S.D. Johnson (Eds.), Formal Methods in Computer-Aided Design. Proceedings, 2000. XI, 539 pages. 2000.

Vol. 1955: M. Parigot, A. Voronkov (Eds.), Logic for Programming and Automated Reasoning. Proceedings, 2000. XIII, 487 pages. 2000. (Subseries LNAI).

Vol. 1960: A. Ambler, S.B. Calo, G. Kar (Eds.), Services Management in Intelligent Networks. Proceedings, 2000. X, 259 pages. 2000.

Vol. 1961: J. He, M. Sato (Eds.), Advances in Computing Science – ASIAN 2000. Proceedings, 2000. X, 267 pages. 2000.

Vol. 1963: V. Hlaváč, K.G. Jeffery, J. Wiedermann (Eds.), SOFSEM 2000: Theory and Practice of Informatics. Proceedings, 2000. XI, 460 pages. 2000.

Vol. 1966: S. Bhalla (Ed.), Databases in Networked Information Systems. Proceedings, 2000. VIII, 247 pages. 2000.

Vol. 1967: S. Arikawa, S. Morishita (Eds.), Discovery Science. Proceedings, 2000. XII, 332 pages. 2000. (Subseries LNAI).

Vol. 1968: H. Arimura, S. Jain, A. Sharma (Eds.), Algorithmic Learning Theory. Proceedings, 2000. XI, 335 pages. 2000. (Subseries LNAI).

Vol. 1969: D.T. Lee, S.-H. Teng (Eds.), Algorithms and Computation. Proceedings, 2000. XIV, 578 pages. 2000.

Vol. 1970: M. Valero, V.K. Prasanna, S. Vajapeyam (Eds.), High Performance Computing – HiPC 2000. Proceedings, 2000. XVIII, 568 pages. 2000.

Vol. 1971: R. Buyya, M. Baker (Eds.), Grid Computing – GRID 2000. Proceedings, 2000. XIV, 229 pages. 2000.

Vol. 1975: J. Pieprzyk, E. Okamoto, J. Seberry (Eds.), Information Security. Proceedings, 2000. X, 323 pages. 2000.

Vol. 1976: T. Okamoto (Ed.), Advances in Cryptology – ASIACRYPT 2000. Proceedings, 2000. XII, 630 pages. 2000.

GPSR Compliance

*The European Union's (EU) General Product Safety Regulation (GPSR)
is a set of rules that requires consumer products to be safe and our
obligations to ensure this.*

*If you have any concerns about our products, you can contact us on
ProductSafety@springernature.com*

In case Publisher is established outside the EU, the EU authorized
representative is:

Springer Nature Customer Service Center GmbH
Europaplatz 3
69115 Heidelberg, Germany

Batch number: 09624486

Printed by Printforce, the Netherlands